# QUANTUM WELL LASERS

# QUANTUM ELECTRONICS—PRINCIPLES AND APPLICATIONS

EDITED BY

*PAUL F. LIAO*
*Bell Communications Research, Inc.*
*Red Bank, New Jersey*

*PAUL L. KELLEY*
*Lincoln Laboratory*
*Massachusetts Institute of Technology*
*Lexington, Massachusetts*

A complete list of titles in this series appears at the end of this volume.

# QUANTUM WELL LASERS

Edited by

**Peter S. Zory, Jr.**

*University of Florida*
*Gainesville, Florida*

ACADEMIC PRESS, INC.
Harcourt Brace Jovanovich, Publishers

Boston San Diego New York
London Sydney Tokyo Toronto

This book is printed on acid-free paper. ∞



ACADEMIC PRESS, INC.
1250 Sixth Avenue, San Diego, CA 92101-4311

United Kingdom edition published by
ACADEMIC PRESS LIMITED
24-28 Oval Road, London NW1 7DX

Quantum well lasers / [edited by] Peter S. Zory, Jr.
p. cm. — (Quantum electronics—principles and applications)
Includes bibliographical references and index.
ISBN 0-12-781890-1 (alk. paper)
1. Semiconductor lasers. 2. Quantum electronics. I. Zory, Peter S., 1986- . II. Series.
TA1700.Q36 1993
621.36'6—dc20 92-43459
CIP

Printed in the United States of America
93 94 95 96 EB 9 8 7 6 5 4 3 2 1

# Contents

# Contributors

Numbers in parentheses indicate the pages on which the authors' contributions begin.

Masahiro Asada (97), *Department of Electrical and Electronic Engineering, Tokyo Institute of Technology, 2-12-1 O-okayama, Meguro-ku, Tokyo 152, Japan*

David P. Bour (415), *Xerox PARC, 3333 Coyote Hill Road, Palo Alto, CA 94304*

Larry A. Coldren (17), *Department of Electrical and Computer Engineering, University of California, Santa Barbara, Santa Barbara, CA 93106*

James J. Coleman (367) *Microelectronics Laboratory and Materials Research Laboratory, University of Illinois, Urbana, IL 61801*

Scott W. Corzine (17), *Department of Electrical and Computer Engineering, University of California, Santa Barbara, Santa Barbara, CA 93106*

Reinhart W. H. Engelmann (131), *Department of Electrical Engineering and Applied Physics, Oregon Graduate Institute of Science and Technology, Beaverton, OR 97006*

Dmitriy Z. Garbuzov (277), *Ioffe Physical-Technical Institute, Academy of Sciences of Russia, 194021, St. Petersburg, Russia*

Ali Ghiti (329), *Department of Physics, University of Surrey, Guildford, GU2 5XH, United Kingdom*

Charles H. Henry (1), *AT&T Bell Laboratories, Room 7B-202, 600 Mountain Avenue, Murray Hill, NJ 07974*

Eli Kapon (461), *Bellcore, NVC 3X-229, PO Box 7040, 331 Newman Springs Road, Red Bank, NJ 07701-7040*

Victor B. Khalfin (277), *Ioffe Physical-Technical Institute, Academy of Sciences of Russia, 194021, St. Petersburg, Russia*

Kam Y. Lau (189, 217), *Department of EECS, University of California at Berkeley, Berkeley, CA 94720*

Eoin P. O'Reilly (329), *Department of Physics, University of Surrey, Guildford, GU2 5XH, United Kingdom*

Chan-Long Shieh (131), *Phoenix Corporate Research Laboratories, Motorola Corporation, Tempe, AZ 85283*

Chester Shu (131), *Department of Electronic Engineering, The Chinese University of Hong Kong, Shatin, N.T., Hong Kong*

Ran-Hong Yan (17), *AT&T Bell Laboratories, 101 Crawfords Corner Road, PO Box 3030, Room 4D-429, Holmdel, NJ 07733-3030*

# Preface

Semiconductor diode lasers are the key components at the heart of many new high volume products such as compact disc players, laser printers, and fiber optic communication links. Most of the diode lasers used in these products are of the double heterojunction (DH) design in which the laser power is generated by electron–hole recombination in an active layer about 100 to 200 nm thick. This DH design, perfected in the late 1970s and early 1980s, is now giving way to a more complex design in which the active layer is about an order of magnitude thinner than in the DH design. As the active layer thins down to the 10 nm regime, the distribution of low energy, wave-like states available for electrons and holes confined to the active layer (potential well) changes from quasi-continuous to discrete. Since laser action is derived by stimulating electron–hole recombination between these discrete (quantum well) states, devices with this active layer design are called quantum well (QW) lasers.

The first ideas that quantum effects might be utilized to make better diode lasers were generated in the early 1970s. As recalled in the Foreword of this book by Charles B. Henry of AT&T Bell Laboratories, the inspiration for these ideas came from the new field of integrated optics, which was just starting to gather momentum at the time. While considering GaAs/AlGaAs waveguides, where refractive index differences are utilized to confine photons on discrete modes, Henry realized that the bandgap differences beween these materials might also be used to confine electrons on discrete modes or quantum states. He then went on to predict that the spectral absorption profile of thin GaAs layers sandwiched between layers of AlGaAs should show a series of bumps characteristic of the quantum well state distribution. The successful demonstration of this effect by R. Dingle using material grown by W. Wiegmann led to further predictions suggesting that diode lasers made with quantum wells should have performance characteristics superior to those of the standard DH lasers being made at that time. As discussed in the Foreword, it wasn't until the late 1970s and early 1980s that reports began to appear claiming that high performance diode lasers incorporating GaAs/AlGaAs quantum wells could in fact be realized. A review of the state of the QW laser art up to about 1985, written by W. T. Tsang, can be found in Volume 24 of the Semiconductor and Semimetals Series published by Academic Press in 1987 (edited by Raymond Dingle).

Since 1985, the published literature on QW lasers has increased enormously. Perhaps the most important development that has come out of all of this new work is the realization that QW lasers with built-in strain can have performance characteristics, including reliability, that are substantially superior to unstrained devices. This improved reliability finding was quite unexpected, since the experience of the 1970s and early 1980s taught that "strained lasers" were doomed to an early death. The fact that the infamous (100) dark line defect growth mechanism can be totally suppressed using strained QW designs has opened up the possibility that large diode arrays having long life expectancy can now be fabricated. Since individual QW diode lasers are now capable of producing cw output powers on the order of watts, the idea that one can build reliable arrays of such devices with no power limit has excited even the laser fusion experts. The reason for the excitement is that solid state lasers pumped by diode laser arrays can now be built with efficiencies high enough to make commercial power production by laser fusion feasible.

Whether or not QW lasers are ever utilized in fusion-powered electricity generators, it seems likely that they will eventually replace most laser types now in existence. Consequently, laser enthusiasts, from the advanced student to the experienced laser technologist, will want to study QW lasers. The aim of this book is to provide the information necessary to achieve a thorough understanding of QW lasers from the basic mechanism of optical gain, through the state of the art, to the future where quantum wire and quantum dot technologies may eventually replace the quantum well technology of today.

In the first chapter of this book the authors, Corzine, Yan, and Coldren

(University of California at Santa Barbara) teach the reader all the physics necessary to predict the spectral dependence of optical gain on carrier density and radiative current density in strained quantum wells. Since the strained QW design is likely to be the one used in most future diode lasers, this chapter should retain its value indefinitely. As mentioned at the end of Chapter 1, the main uncertainty in the model used is the proper choice of broadening function required by carrier scattering processes. These spectral broadening processes, collectively called the intraband relaxation effect, are discussed in considerable detail in Chapter 2 by Asada of the Tokyo Institute of Technology. Although the exact broadening function involves numerical integrations, Professor Asada provides an approximate closed-form expression, which is relatively easy to use. In Chapter 3, by Engelmann (Oregon Graduate Institute of Science and Technology), Shieh (Motorola Corporate Research Laboratories), and Shu (Center for Telecommunication Research Columbia University), the importance of nonradiative recombination and carrier leakage is discussed, and threshold current relations in multi-quantum well configurations are considered. Also introduced is a simplified model useful in making rough estimates of threshold current. Chapters 4 and 5 by Lau (University of California, Berkeley) deal respectively with very low threshold current and high speed devices. In discussing low threshold devices, Professor Lau makes the point that the observed reduction in threshold current is due in large part to "physical scaling" rather than quantum effects. In discussing high speed dynamics, he points out that while quantum well lasers have not yet fulfilled their promise as high speed devices, the reasons why are beginning to be understood. As of this writing, there appears to be a consensus forming that the mechanism limiting high speed operation in QW lasers involves an unexpected partitioning of carriers between the quantum well(s) and confinement (barrier) layers. This carrier partitioning effect as well as other unusual phenomena, such as lasing on higher lying subbands of the QW structure, are discussed in Chapter 6 by Garbuzov and Khalfin (Ioffe Physical-Technical Institute). Also discussed in Chapter 6 are InGaAsP/GaAs 808-nm QW lasers, of considerable interest as competitors to the standard AlGaAs/GaAs 808-nm QW lasers used as pumps for Nd:YAG solid state lasers.

In Chapter 7 by O'Reilly and Ghiti (University of Surrey), strained QW designs are discussed in which Auger recombination and intervalence band absorption, major loss mechanisms in longer wavelength diode lasers, can be virtually eliminated. These exciting ideas are just now being realized in 1.3/1.55-$\mu$m strained QW lasers, of use in fiber optic communication applications. Chapter 8 by Coleman (University of Illinois) also deals with strained

QW lasers, specifically InGaAs/GaAs lasers operating in the 0.9 to 1.1 $\mu$m range. Considerable interest exists in using these lasers as pumps for rare earth-doped fiber amplifiers, of use in long distance fiber optic communication systems. Chapter 9 by Bour (Xerox, Palo Alto Reseach) deals with visible AlGaInP/GaAs QW lasers (600–700 nm) and discusses recent results with strained QW devices. The improved performance obtained with the strained layer designs is another indicator that "strain" is here to stay. Additional support for this position comes from the fact that a strained QW design was utilized in fabricating the world's first blue diode lasers (reported by 3M Company personnel in June 1991). It remains to be seen if the operating characteristics of these blue diode lasers can be understood using the basic $k$-selection formalism presented in Chapter 1 of this book.

In concluding the book, Kapon (Bellcore) describes how it may be possible to utilize arrays of quantum wires or quantum dots to make semiconductor lasers with performance characteristics superior to even quantum well (QW) lasers. While quantum wire lasers are still in the laboratory, performance characteristics comparable with QW lasers have been demonstrated, and improvement seems imminent.

As one may gather, the field of QW lasers is already quite vast and considerable study is required to master it. I believe the chapter authors have done an admirable job in referencing the published literature and in capturing the essence of their respective topics. They have made it possible for the reader to find, in this one book, the answer to almost any question about QW lasers. My thanks to the authors for their fine efforts.

Peter S. Zory Jr.

# *Foreword*

# THE ORIGIN OF QUANTUM WELLS AND THE QUANTUM WELL LASER

**Charles H. Henry**

*AT&T Bell Laboratories*
*Murray Hill, New Jersey*



## 1. INTRODUCTION

Semiconductor lasers are attractive for research because they are both physically very interesting and technologically important. This is especially true of quantum well lasers. Quantum well technology allows the crystal grower for the first time to control the range, depth, and arrangement of quantum mechanical potential wells. This can be used not only to demonstrate examples of elementary quantum mechanics, but to make very good lasers. In the last decade, the importance of the quantum well laser has steadily grown until today it is preferred for most semiconductor laser applications. At the 12th International Semiconductor Laser Conference, held in Davos, Switzerland, more than half the conference papers were



ISBN 0-12-781890-1

concerned with quantum well lasers [1]. While the first quantum well lasers operated at a wavelength near 0.8 $\mu$m, they have now been demonstrated from the visible through the infrared (.49–10 $\mu$m).

This growing popularity is because, in almost every respect, the quantum well laser is somewhat better than conventional lasers with bulk active layers. One obvious advantage is the ability to vary the lasing wavelength merely by changing the width of the quantum well. A more fundamental advantage is that the quantum well laser delivers more gain per injected carrier than conventional lasers, which results in lower thresholds currents. Because the injected carriers are in large measure responsible for internal losses, quantum well lasers, which require fewer injected carriers, are more efficient and can generate more power than conventional lasers. Another advantage is that quantum well lasers deliver gain with less change in refractive index than bulk lasers, resulting in lower chirp. Both lower internal loss and lower refractive index change result in quantum well lasers having narrower linewidth than conventional lasers. The splitting of the light- and heavy-hole valence bands by spatial quantization as well as the ability to grow quantum wells with compressive and tensile strain result in greater control over the optical polarization than in bulk lasers. The differential gain, gain per injected electron, is greater in properly designed multi–quantum well lasers and should lead to higher speed than for bulk lasers. Whether quantum well lasers actually excel at high speed is still being debated.

In view of the great interest in this laser, it seems appropriate to look back and review its origin. In this article I will trace the history of the quantum well laser from the first calculations and experiments on quantum wells to the first quantum well lasers of high performance. In doing this, I will relate recollections of my own work as well as what I have been able to learn from colleagues and the published literature.

## 2. EARLY PROPOSALS

Modern semiconductor lasers incorporate a heterostructure in which the active layer is surrounded by higher bandgap material. The heterostructure laser concept was suggested in 1963 by H. Kroemer [2] in the U.S. and by R. F. Kazarinov and Zh. I. Alferov [3] in the Soviet Union. In 1967, J. M. Woodall *et al.* [4] of IBM succeeded in growing heterostructures of GaAs and AlGaAs by liquid phase epitaxy (LPE). Soon thereafter, semiconductor lasers made using this approach were reported by Alferov's group, H. Kressel and H. Nelson of RCA and I. Hayashi and M. B. Panish at Bell Laboratories [5]. By 1970, Alferov's group [6] and Hayashi and Panish [7] had demonstrated double heterostructure lasers continuously operating at room temperature.

The role of the heterostructure in the early lasers was described as providing a barrier that blocked the diffusion of electrons. However, in the absence of grading of interfaces and depletion effects, the heterostructure acts as a potential well for electrons in the conduction band and holes in the valence band. It was natural for those concerned with heterostructures to try to model what would happen when these layers became extremely thin and spatial quantization occurred. The first work in this direction was that of L. Esaki and R. Tsu of IBM in 1970 [8], who considered carrier transport in a superlattice, an additional periodic potential formed in a semiconductor by doping or alloy composition and having a period of order 100 Å. They concluded that a parabolic band would break into mini-bands separated by small forbidden gaps and having Brillouin zones associated with this period. A similar conclusion was reached earlier by L. V. Keldysh [9], who modeled a semiconductor in a periodic potential produced by an intense ultrasonic wave. Esaki and Tsu [8] predicted that, under a high applied field, a consequence of the mini-bands is negative differential resistance, and at still higher fields high frequency electron oscillations will occur. These oscillations have not yet been observed.

In 1971, Kazarinov and R. A. Suris [10] of the Ioffe Institute analyzed the effect of current transport through a superlattice, and they refined their analysis in subsequent papers [11, 12]. They considered the mini-bands from the tight binding viewpoint, where they can be thought of as coupled states of quantum wells. They showed that the transport between wells would consist of tunneling through the barrier layers separating the wells. They predicted two phenomena. The first was tunneling at field values where the ground state of one well coincides with an excited state of a neighboring well. The second was stimulated emission resulting from photon-assisted tunneling between the ground state of one well and an excited state of a neighboring well, which is lower in energy because of the applied electric field. Tsu and Esaki also modeled resonant tunneling through a number of quantum wells in 1973 [13]. Such a far infared laser has not been demonstrated, but resonant tunneling was observed in 1974, as I will mention.

## 3. THE FIRST OBSERVATIONS OF QUANTUM WELLS

At this point, I would like to trace my own route to quantum wells. In 1972, I was a newly appointed department head of the Semiconductor Electronics

Research Department at Bell Telephone Laboratories. Under the leadership of D. G. Thomas, this department had successfully developed GaP light emitting diodes. After transferring this technology to others, we went on to work on other wider bandgap semiconductors with the hope of making blue emitters, but this work did not look promising and was about to be abandoned. Consequently, we were looking for new areas of research. My director, J. K. Galt, head of the Solid State Electronics Research Laboratory in which the work of Hayashi and Panish had been carried out, suggested that we try integrated optics in the AlGaAs heterostructure material. The new field of integrated optics looked very interesting, and R. A. Logan and later J. L. Merz of my department began working in this area.

At that time, I was working experimentally on the study of deep levels in semiconductors by photocapacitance, but I took a keen interest in integrated optics. Logan and I attended the first integrated optics conference in 1972. Later that year, I studied a collection of papers on this subject by my coworkers at the Crawford Hill location of Bell Laboratories. Under the direction of S. E. Miller, they had been working on integrated optics for a number of years and summarized their work in the September 1969 issue of the *Bell System Technical Journal.* I am the type of physicist who gains more from making calculations than from reading. For that reason, I did a few exercises to get a feeling for optical waveguides. I made simple mathematical models of a directional coupler and a symmetric slab waveguide, and a calculation of waveguide bend loss.

In the course of these calculations, it suddenly occurred to me that "a heterostructure is a waveguide for electrons." On reflection, it was clear that there is a complete analogy between the confinement of light by a slab waveguide and the confinement of electrons by the potential well that is formed from the difference in bandgaps in a heterostructure. This then led me to think that there should be discrete modes (levels) in the quantum well, and a simple estimate showed that when the heterostructure was as thin as several hundred angstrom units, the electron levels would be split apart by tens of milli-electron volts. I then calculated how this quantization would alter the optical absorption, considering only that noninteracting electrons are bound in the well. My conclusion was that instead of the optical absorption increasing smoothly as the square root of energy, the absorption edge of a thin heterostructure would appear as a series of steps.

The early lasers were all grown by LPE. At that time, this method was suitable for forming bulk active layers about 2000 Å thick, but not layers an order of magnitude thinner than this. Fortunately, there was another method of growing heterostructures that was capable of forming much thinner layers, molecular beam epitaxy (MBE). This method had also been developed in Galt's laboratory. The method was originated by J. R. Arthur in 1968 [14] and was turned into a practical method of crystal growth by A. Y. Cho. By 1971, Cho had demonstrated the growth of heterostructures with optical absorption edge features that were just as well defined as in samples grown by LPE. Furthermore, by monitoring the crystal surface in situ with high energy electron diffraction, he was able to show that as the crystal grows the top surface becomes much smoother than the polished surface of the substrate [15]. The growth rates at that time were about 5 monolayers/min, making MBE capable of abrupt interfaces [16].

I attempted to organize an observation of the predicted steps in the optical absorption edge in late 1972. At that time, D. D. Sell had made an apparatus for making extremely sensitive measurements of reflectivity and optical absorption. I asked him to look for the expected structure in the absorption edge using thin heterostructures grown by Cho with MBE. The experiment was tried, but without success. Sell found no evidence of structure due to quantum effects.

In early 1973, R. Dingle returned to my department after a one year leave of absence. I immediately suggested to him that he look for steps associated with quantized levels in the absorption edge of thin GaAs-AlGaAs heterostructures. Dingle went to W. Wiegmann, who at that time was operating the second MBE machine in our laboratory, which machine he built under the direction of J. C. Tracy. Tracy had subsequently left, and Wiegmann was temporarily working on his own. In order to have adequate signal to noise in optical absorption, Wiegmann grew as many as 50 identical heterostructures using only a hand operated shutter and a stop watch. The GaAs substrate was then selectively etched away in a small area, and the optical absorption was measured at liquid helium temperature.

The results of these experiments were the remarkable spectra shown in Fig. 1. The steps in the absorption edge were observed. Associated with each step was a strong exciton peak. The thinner samples showed a splitting of the lowest exciton peak. Dingle identified this as the splitting of the levels associated with the confined light and heavy holes. We were able to fit all of the data by assuming that the difference in bandgaps produced two attractive wells, one in the conduction band and one in the valence band (Fig. 2). The fits showed that the walls of the well were abrupt to within 5 Å. Our paper reporting the optical absorption properties of quantum wells was submitted in June 1974 [17].

In retrospect, these fits were disappointing in one respect. We deduced that about 88% of the bandgap difference was associated with the conduction

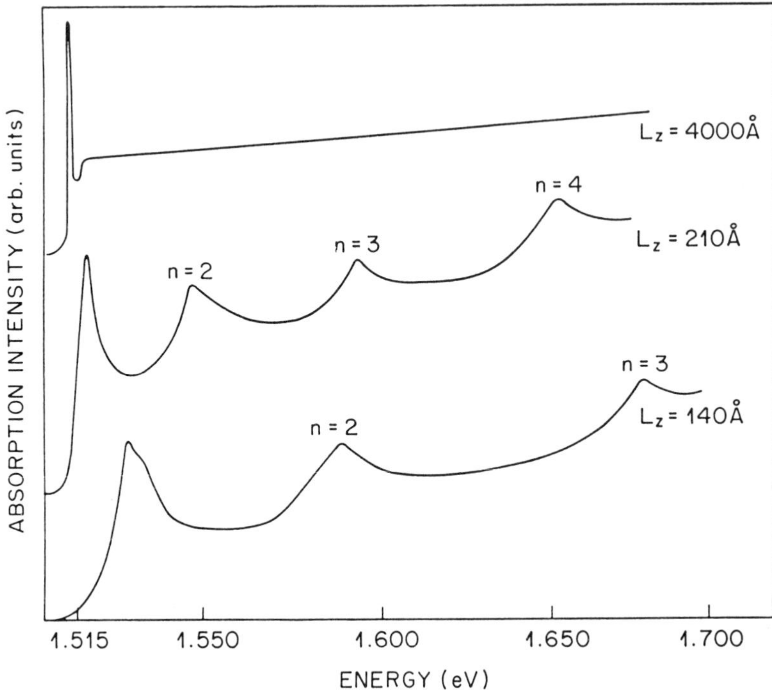


**Fig. 1.** The optical absorption spectra of bulk and quantum well heterostructures [after Dingle *et al.*, Ref. 17].

band. A value of 85% was used for nearly 10 years by subsequent workers until R. C. Miller *et al.* [18] showed that a much better fit is obtained with a value of 57% and altered valence band masses. This initial experiment was followed by a series of experiments by Dingle and A. C. Gossard in which parabolic wells, triangular wells, coupled wells, and coupled wells in a superlattice were investigated [19].

While this work was going on at Bell Laboratories, L. L. Chang and coworkers at the IBM Watson Research Center were also progressing toward an investigation of quantum wells by electrical conductivity measurements. In 1973, they completed the building of an MBE apparatus similar to Cho's, but with computer control for reproducible growth of complex arrays of thin layers [20]. In March 1974, three months before our submission on the optical absorption in quantum wells, Chang, Esaki, and Tsu submitted a paper on the resonant tunneling through levels of a quantum well [21]. This publication marks the first experimental demonstration of quantum well physics in heterostructures. They formed a quantum well separated from

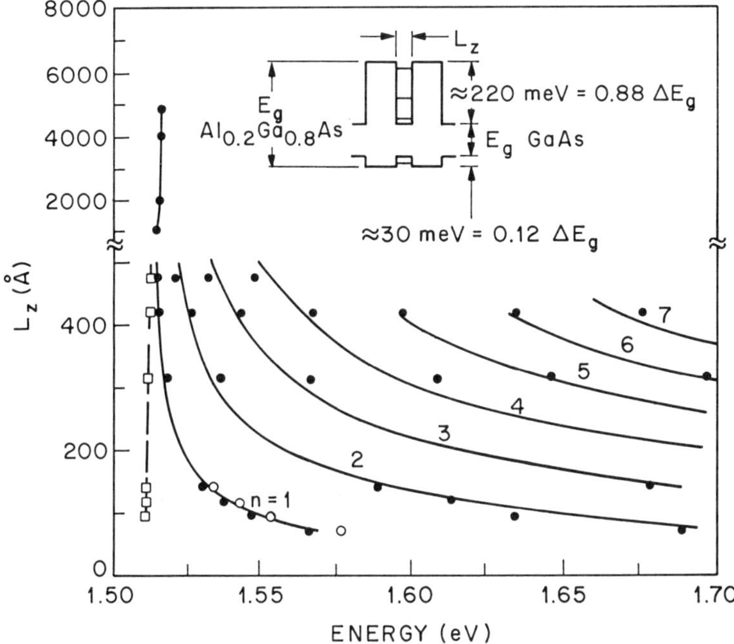


**Fig. 2.** Theoretical fit of the quantum well absorption spectra for different well thicknesses [after Dingle *et al.*, Ref. 17].

bulk GaAs by AlGaAs barrier layers. The two bound states of the quantum well could be probed by tunneling between the emitting layer and the quantum well levels. They found current maxima associated with this resonant tunneling (Fig. 3). This data is in accordance with the model of Tsu and Esaki [13], as well as earlier works on tunneling through double barriers [21]. Later, in July 1974, Esaki and Chang submitted an observation of resonant tunneling in a superlattice [22]. Current maxima were found when the applied field is such that the ground state of one quantum well is aligned with the excited state of a neighboring well, as predicted by Kazarinov and Suris [10].

The restriction of the motion of electrons to two dimensions by spatial quantization also occurs in field effect transistors, where electrons are trapped in an inversion layer by an applied electric field. This was first recognized by R. Schrieffer in 1957 [23]. The restriction of the trapped electrons in inversion layers to two dimensional motion was first verified by the magneto-conductance experiment of A. B. Fowler *et al.* in 1966 [24]. Spectral effects due to spatial quantization were observed in thin bismuth films in 1968 by V. N. Lutskii and L. A. Kulik [25].

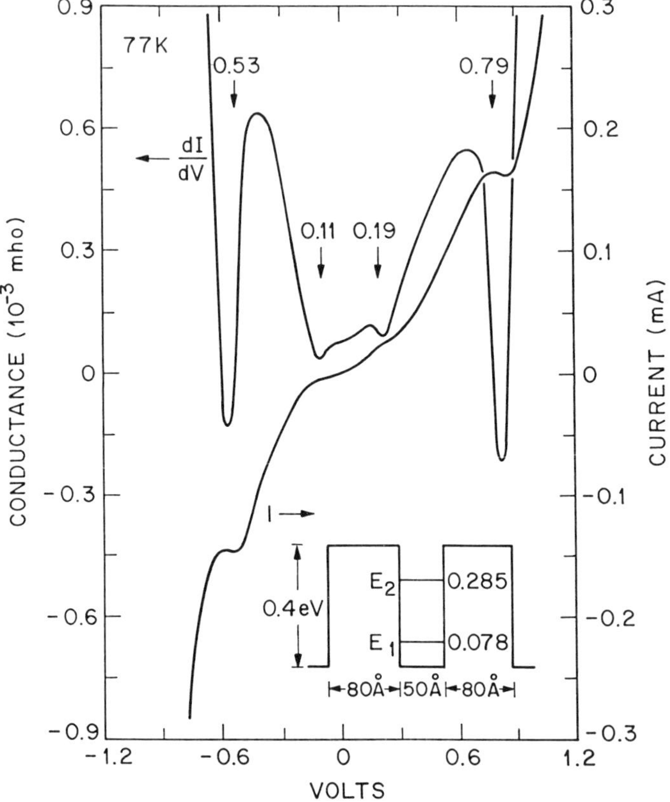


**Fig. 3.** Current and conductance characteristics of a double barrier structure of GaAs between two $Ga_{.3}Al_{.7}As$, as shown in the energy diagram. Both the thickness and the calculated quasistationary states of the structure are indicated in the diagram. The arrows indicate the observed singularities corresponding to these resonant states [after Chang *et al.*, Ref. 21].

## 4. THE QUANTUM WELL LASER

After hearing of Dingle's first measurements, vividly demonstrating the reality of quantum wells, I tried to think of some use for them. It seemed to me that quantum well confinement should result in lower threshold lasers as a result of the altered density of states that results from confinement in a plane. In a bulk semiconductor, the density of states increases as the square root of energy, while in quantum wells, the increase is step-like; see Figs. 4a and 4b. As a consequence of momentum conservation in optical transitions,

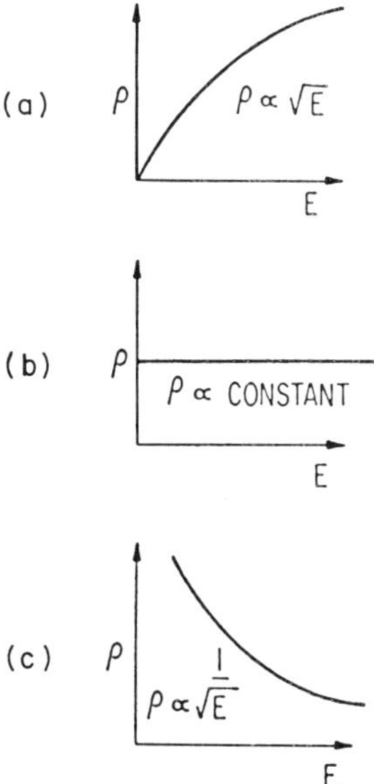


**Fig. 4.** Densities of states for (a) bulk layers, (b) quantum well layers, and (c) quantum wires [after Dingle and Henry, Ref. 26].

this is true both for the densities of states of the individual bands and of the joint density of states determining the spectra of optical absorption and emission. A high gain requires population inversion of levels with a high density of states. In a bulk active layer laser, this can only be achieved after first filling the lower lying energy levels. In a quantum well laser, this is unnecessary; the peak gain is associated with levels at the bottom of the bands. My estimates indicated that, for the same peak gain, there would be a carrier density reduction in a quantum well laser compared with a bulk laser. The calculated improvement was 1.4 (Fig. 5). On the basis of our work, including among other things this improvement and the tunability of the laser wavelength by changing the layer thickness, a patent application by Dingle and myself was filed by attorney M. J. Urbano in March 1975 [26]. We also pointed out in the patent application that further improvements in the density of states will occur with confinement to a wire (Fig. 4c).

Laser operation in quantum well structures was demonstrated less than one year after our observation of the quantum well absorption spectrum. However, it took six more years before quantum well lasers were made with properties superior to bulk active layer lasers. A major obstacle to this demonstration was the low quality of the early MBE layers containing aluminum. These layers appeared to have very short nonradiative lifetimes.

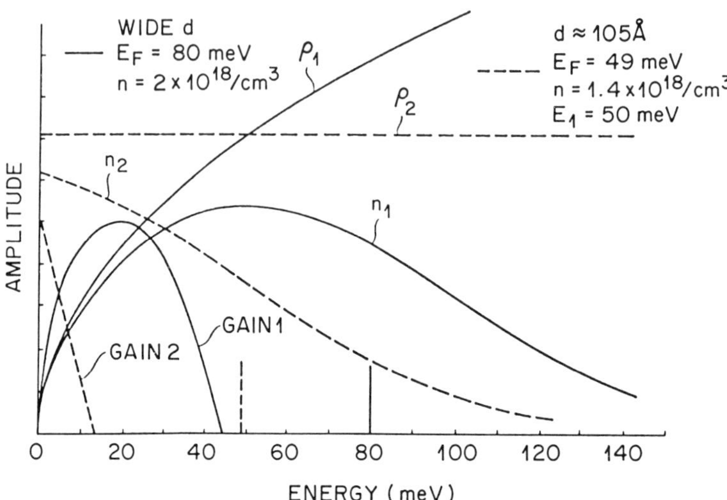


**Fig. 5.** Comparison of the carrier distributions and gains in bulk and quantum well lasers. The carrier densities are adjusted to give the same gain. The vertical lines mark the computed Fermi levels [after Dingle and Henry, Ref. 26].

While MBE heterostructures at that time were quite uniform with sharp interface transitions, they had very low radiative efficiency. Nevertheless, the first observation of quantum well laser operation was made by J. P. Van der Ziel *et al.* [27], using samples similar to those grown earlier for our investigation of optical absorption. Their paper was submitted in January 1975. They optically pumped the material at a temperature of 15 K and achieved a threshold of 36 kW $cm^{-2}$, a value many times greater than needed for LPE heterostructure lasers but still low enough to encourage further experimentation using MBE and other growth systems. These experiments were followed by additional optically pumped laser experiments at room temperature by Miller *et al.* [28], also with high thresholds.

A new method of crystal growth capable of forming very thin layers, metalorganic chemical vapor deposition (MOCVD), was first used to make AlGaAs heterostructure lasers by R. D. Dupuis and P. D. Dapkus of Rockwell International in 1977 [29]. This method was first published by H. M. Manesevit of Rockwell International in 1968 [30]. In a paper submitted in January 1978 [31], Dupuis and Dapkus showed that MOCVD grown lasers were as good as the best lasers grown by LPE. They achieved threshold currents as low as 590 $A/cm^2$ for lasers with high Al contents in the cladding layer. This was comparable with the lowest thresholds achieved by LPE by G. H. B. Thompson *et al.* [32] and H. Kressel and M. Ettenberg [33]. At the same time, Dupuis and Dapkus in collaboration with N. Holonyak and coworkers at the University of Illinois used the MOCVD technology to make improved quantum well lasers. The name "quantum well" became part of semiconductor laser vocabulary at that time by its usage in the papers of these authors. In November of 1977, they sumitted a paper first reporting the demonstration of a quantum well injection laser [34]. It had a single 200 Å quantum well and a threshold current of about 3 kA/cm$^2$ at 300 K when operated pulsed. The authors pointed out that the laser threshold could be improved by additional layers for separately confining the optical field. A similar single quantum well (SQW) laser photopumped to continuous operation was reported in a paper submitted in February 1978 [35]. In July 1978, they demonstrated cw operation of a photopumped multi–quantum well (MQW) laser having a threshold equivalent to 500 A/cm$^2$, showing the high quality of their layer interfaces [36]. In June and October 1978, they demonstrated the first report of cw operation of SQW and MQW injection lasers [37, 38]. The lowest threshold achieved was 1660 A/cm$^2$. External differential quantum efficiencies as high as 85% were observed.

In this time period, the growth of lasers by MBE gradually improved, and the first cw injection laser made by MBE was achieved by Cho *et al.* in December 1975 [39]. A great improvement in MBE grown lasers resulted from the research of W. T. Tsang. In a paper submitted in September 1978, he reported the first successful preparation of low threshold lasers by MBE [40]. A major improvement that he made was to increase the temperature of growth. He found that as the temperature increases above 620°C, the luminescence efficiency of the AlGaAs layers increased by more than an order of magnitude, and the lowest laser thresholds were achieved [41]. Another important factor was the use of a load lock sample exchange mechanism to reduce contamination of the vacuum chamber by water vapor. With these improvements, Tsang found that the bulk active layer thresholds of MBE grown lasers were about 1 kA/cm$^2$ [42], which was the same as achieved by Dupuis and Dapkus [31] and by Thrush *et al.* by MOCVD [43] for lasers of similar composition. In a paper submitted for publication in July 1979, Tsang *et al.* [44] reported MBE-grown AlGaAs injection lasers containing 14 quantum wells, 139 Å wide, with a room temperature threshold of 2 kA/cm$^2$. They concluded that the AlGaAs interfaces were essentially free of the nonradiative defects that had plagued earlier efforts to make low threshold quantum well lasers by MBE.

Tsang followed this paper with one in which he optimized the quantum well parameters and obtained enormously reduced threshold currents. Simply varying the well widths and the number of wells brought about only modest reductions in threshold. Tsang discovered that the high thresholds of MQW lasers were mainly due to the inefficiency in injecting carriers over the barriers existing between wells. In a paper submitted in July 1981 [45], he optimized barrier heights (Fig. 6), and thicknesses and used greater Al content in the outer layers to obtain separate confinement of the optical energy. This approach is adopted in all present day MQW lasers. He achieved MQW laser thresholds as low as 250 A/cm$^2$ with four 120-Å quantum wells and a 380 $\mu$m long cavity.

At the same time, Tsang worked at optimizing the design of single quantum well lasers, as reported in a paper submitted in August 1981 [46]. The modal gain of a single quantum well saturates at a relatively low value. To optimize the performance, it was necessary to reduce facet losses by lengthening the cavity and to reduce carrier absorption losses by reducing the doping of the layers surrounding the active layer. In addition to this, Tsang introduced the graded index waveguide for separate optical confinement (GRIN SCH) design. With these improvements, he found threshold currents as low as 160 A/cm$^2$ for 1125 $\mu$m long lasers. The internal loss of these lasers was as low as 3 cm$^{-1}$ and the internal quantum efficiency nearly 0.95.

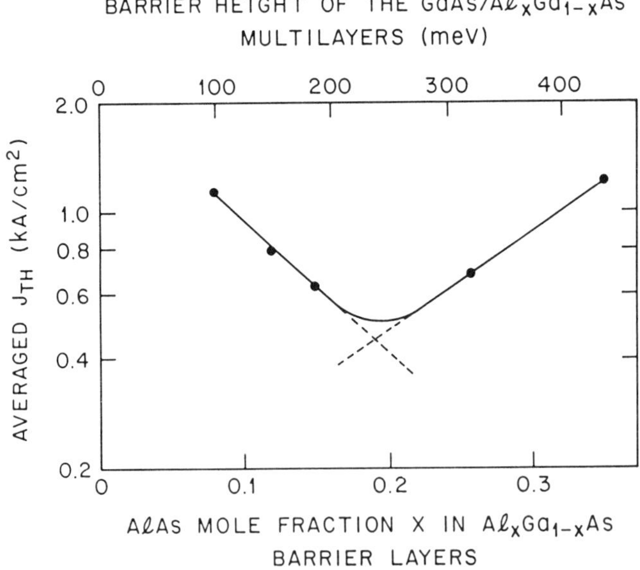


**Fig. 6.** Optimization of heights of the barriers between wells in a MQW laser [after Tsang, Ref. 45].

These two landmark papers by Tsang demonstrated the advantage of quantum well lasers over conventional bulk active layer lasers in achieving low laser threshold. This started a steady growth in the popularity of quantum well lasers, which is still continuing.

## ACKNOWLEDGMENTS

I wish to thank F. Capasso, L. L. Chang, A. Y. Cho, P. D. Dapkus, D. Gershoni, R. F. Kazarinov, B. F. Levine, M. B. Panish, and W. T. Tsang for their comments, which greatly improved the accuracy of this review.

## REFERENCES

1. *Proceedings of the* 12th *IEEE international semiconductor laser conference.* Davos, Switzerland, September 1990.
2. Kroemer, H. (1963). A proposed class of heterojunction lasers, *Proc. IEEE* **51**, 1782–1783.
3. Alferov, Zh. I., and Kazarinov, R. F. (1963). Semiconductor laser with electrical pumping, U.S.S.R. Patent 181737.
4. Woodall, J. M., Rupprecht, H., and Pettit, G. D. (1967). *Solid state device conference*, June 19, 1967, Santa Barbara, California. [Abstract reported in *IEEE Trans. Electron. Devices* **ED-14**, 630 (1967).]
5. Casey, H. C., and Panish, M. B. (1968). *Heterostructure lasers*, Chapter 1. Academic Press, New York.
6. Alferov, Zh., Andreev, V. M., Garbuzov, D. Z., Zhilyaev, Yu. V., Morozov, E. P., Portnoi, E. L., and Trofim, V. G. (1971). *Sov. Phys. Semicond.* **4**, 1573.
7. Hayashi, I., Panish, M. B., Foy, P. W., and Sumski, S. (1970). Junction lasers which operate continuously at room temperature, *Appl. Phys. Lett.* **17**, 109.
8. Esaki, L., and Tsu, R. (1970). Superlattice and negative differential conductivity, *IBM J. Res. Dev.* **14**, 61–65.
9. Keldysh, L. V. (1963). Effect of ultrasonics on the electron spectrum of crystals, *Sov. Phys. Solid State* **4**, 1658–1659.
10. Kazarinov, R. F., and Suris, R. A. (1971). Possibility of amplification of electromagnetic waves in a semiconductor superlattice, *Sov. Phys. Semicond.* **5**, 707–709.
11. Kazarinov, R. F., and Suris, R. A. (1972). Electric and electromagnetic properties of a superlattice, *Sov. Phys. Semicond.* **6**, 120–131.
12. Kazarinov, R. F., and Suris, R. A. (1973). Theory of electrical properties of semiconductors with superlattices, *Sov. Phys. Semicond.* **7**, 347–352.

13. Tsu, R., and Esaki, L. (1973). Tunneling in a finite superlattice, *Appl. Phys. Lett.*, **22**, 562–564.
14. Arthur, J. R. (1968). Interaction of Ga and $As_2$ molecular beams with GaAs surfaces, *J. Appl. Phys.* **39**, 4032–4034.
15. Cho, A. Y. (1971). Film deposition by molecular beam techniques, *J. Vac. Sci. Technol.* **8**, S31–S38.
16. Cho, A. Y. (1971). Growth of periodic structures by the molecular-beam method, *Appl. Phys. Lett.* **19**, 467.
17. Dingle, R., Wiegmann, W., and Henry, C. H. (1974). Quantized states of confined carriers in very thin $Al_xGa_{1-x}As$–GaAs–$Al_xGa_{1-x}As$ heterostructures, *Phys. Rev. Lett.* **33**, 827–830.
18. Miller, R. C., Klienman, D. A., and Gossard, A. C. (1984). Energy-gap discontinuities and effective masses for GaAs–$AlAsGa_{1-x}As$ quantum wells, *Phys. Rev. B* **29**, 7085–7087.
19. Gossard, A. C. (1982). Molecular beam epitaxy of superlattices in thin films, *Treatise Mater. Sci. Technol.* **24**, 13–65.
20. Chang, L. L., Esaki, L., Howard, W. E., and Ludeke, R. (1973). Growth of GaAs–GaAlAs superlattices, *J. Vac. Soc.* **10**, 11.
21. Chang, L. L., Esaki, L., and Tsu, R. (1974). Resonant tunneling in semiconductor double barriers, *Appl. Phys. Lett.* **24**, 593–595.
22. Esaki, L., and Chang, L. L. (1974). New transport phenomenon in a semiconductor "superlattice," *Phys. Rev. Lett.* **33**, 495–498.
23. Schrieffer, J. R. (1957). In *Semiconductor surface physics* (Kingston, R. H., ed.), p. 68. University of Pennsylvania Press, Philadelphia.
24. Fowler, A. B., Fang, F. F., Howard, W. E., and Stiles, P. J. (1966). Magneto-oscillatory conductance in silicon surfaces, *Phys. Rev. Lett.* **16**, 901–903.
25. Lutskii, V. N. (1970). Quantum size effect—present state and perspective on experimental investigations, *Phys. Stat. Sol. (a)* **1**, 199–220.
26. Dingle, R., and Henry, C. H., (1976). Quantum effects in heterostructure lasers, U.S. Patent 3982207, September 21.
27. van der Ziel, J. P., Dingle, R., Miller, R. C., Wiegmann, W., and Nordland, W. A., Jr. (1975). Laser oscillations from quantum states in very thin GaAs–$Al_{0.2}Ga_{0.8}As$ multilayer structures, *Appl. Phys. Lett.* **26**, 463–465.
28. Miller, R. C., Dingle, R., Gossard, A. C., Logan, R. A., and Nordland, W. A., Jr. (1976). Laser oscillation with optically pumped very thin GaAs–$Al_xGa_{1-x}As$ multilayer structures and conventional double heterostructures, *J. Appl. Phys.* **47**, 4509–4517.
29. Dupuis, R. D., and Dapkus, P. D. (1977). Room temperature operation of $Ga_{1-x}Al_xAs$/GaAs double-heterostructure lasers grown by metalorganic chemical vapor deposition, *Appl. Phys. Lett.* **31**, 466–468.
30. Manesevit, H. M. (1968). Single crystal GaAs on insulating substrates, *Appl. Phys. Lett.* **12**, 156–159.

31. Dupuis, R. D., and Dapkus, P. D. (1978). Very low threshold $Ga_xAl_xAs$–GaAs double-heterostructure lasers grown by metalorganic chemical vapor deposition, *Appl. Phys. Lett.* **32**, 473–475.
32. Thompson, G. H. B., Henshall, G. D., Whiteaway, J. E. A., and Kirkby, P. A. (1976). Narrow beam five-layer (GaAl)As/GaAs heterostructure lasers with low threshold and high peak power, *J. Appl. Phys.* **47**, 1501–1514.
33. Kressel, H., and Ettenberg, M. (1976). Low threshold double heterostructure AlGaAs/GaAs laser diodes, *J. Appl. Phys.* **47**, 3533–3537.
34. Dupuis, R. D., Dapkus, P. D., Holonyak, N., Jr., Rezek, E. A., and Chin, R. (1978). Room temperature operation of quantum-well $Ga_{1-x}Al_xAs$–GaAs laser diodes grown by metalorganic chemical vapor deposition, *Appl. Phys. Lett.* **32**, 295–297.
35. Holonyak, N., Jr., Kolbas, R. M., Dupuis, R. D., and Dapkus, P. D. (1978). Room temperature continuous operation of photopumped MOCVD $Al_xGa_{1-x}As$–GaAs–$Al_xGa_{1-x}$ laser quantum-well lasers, *Appl. Phys. Lett.* **33**, 73–75.
36. Holonyak, N., Jr., Kolbas, R. M., Laidig, W. D., Vojak, B. A., Dupuis, R. D., and Dapkus, P. D. (1978). Low-threshold continuous laser operation (300–337 K) of multilayer MOCVD $Al_xGa_{1-x}As$–GaAs quantum-well heterostructures, *Appl. Phys. Lett.* **33**, 737–739.
37. Dupuis, R. D., Dapkus, P. D., Chin, R., Holonyak, N., Jr., and Kirchoefer, S. W. (1979). Continuous 300 K operation of single-quantum-well $Al_xGa_xAs$–GaAs heterostructure diodes grown by metalorganic chemical vapor deposition, *Appl. Phys. Lett.* **34**, 265–268.
38. Dupuis, R. D., Dapkus, P. D., Holonyak, N., Jr., and Kolbas, R. M. (1979). Continuous room-temperature multiple-quantum-well $Al_xGa_{1-x}As$–GaAs injection lasers grown by metalorganic chemical vapor deposition, *Appl. Phys. Lett.* **35**, 487–490.
39. Cho, A. Y., Dixon, R. W., Casey, H. C., Jr., and Hartman, R. L. (1976). Continuous room-temperature operation of GaAs–$Al_xGa_{1-x}As$ double-heterostructure lasers prepared by molecular-beam epitaxy, *Appl. Phys. Lett.* **28**, 501–503.
40. Tsang, W. T. (1978). Low-current-threshold and high-lasing uniformity GaAs–$Al_xGa_{1-x}As$ double heterostructure lasers grown by molecular beam epitaxy, *Appl. Phys. Lett.* **34**, 473–475.
41. Tsang, W. T., Reinhart, F. K., and Ditzenberger, J. A. (1980). The effect of substrate temperature on the current threshold of GaAs–$Al_xGa_{1-x}As$ double-heterostructure lasers grown by molecular beam epitaxy, *Appl. Phys. Lett.* **36**, 118–120.
42. Tsang, W. T. (1980). Very low current threshold GaAs–$AlGa_{1-x}As$ double-heterostructure lasers grown by molecular beam epitaxy, *Appl. Phys. Lett.* **36**, 11–14.

43. Thrush, E. J., Selway, P. R., and Henshall, G. D. (1979). Metalorganic CVD growth of GaAs–GaAlAs double heterostructure lasers having low interfacial recombination and low threshold current, *Electron. Lett.* **15**, 156–158.
44. Tsang, W. T., Weisbuch, C., Miller, R. C., and Dingle, R. (1979). Current injection GaAs–$Al_xGa_{1-x}As$ multi-quantum-well heterostructure lasers prepared by molecular beam epitaxy, *Appl. Phys. Lett.* **35**, 673–675.
45. Tsang, W. T. (1981). Extremely low threshold (AlGa)As modified multiquantum well heterostructure lasers grown by molecular-beam epitaxy, *Appl. Phys. Lett.* **39**, 786–788.
46. Tsang, W. T. (1982). Extremely low threshold (AlGa)As graded-index waveguide separate-confinement heterostructure lasers grown by molecular-beam epitaxy, *Appl. Phys. Lett.* **40**, 217–219.

# *Chapter 1*

# OPTICAL GAIN IN III–V BULK AND QUANTUM WELL SEMICONDUCTORS

**Scott W. Corzine**
**Ran-Hong Yan**
**Larry A. Coldren**

*Department of Electrical and Computer Engineering*
*University of California, Santa Barbara*
*Santa Barbara, California*






ISBN 0-12-781890-1

## 1. INTRODUCTORY COMMENTS

The classic tale of the "one dimensional electron in a box" has finally become a problem with more than mere academic interest. With the ability to grow different semiconductor layers epitaxially with atomic scale precision in thickness, the material bandgap can be designed to confine electrons in much the same way as described in standard quantum mechanics textbooks [1]. This quantum confinement of the electron along the growth direction significantly alters the band structure of the semiconductor, altering almost every property of the material to one degree or another. With regard to quantum well lasers, we will be particularly interested in how the quantum confinement alters the optical gain of the material, both as a function of injection current and as a function of wavelength.

In this chapter, we examine the fundamental theory necessary to provide an adequate description of optical gain in bulk and quantum well semiconductor material. We start off in Sections 2–4 by deriving the standard general relationships for gain, applicable to both bulk and quantum well material. We then go into more detail regarding the band structure of III–V semiconductors. For example, in Section 5, we describe the bulk conduction and valence bands in the framework of the $k \cdot p$ theory [2, 3], enabling us to understand and predict polarization-dependent gain and absorption phenomena [4–15] in quantum-confined structures. In Sections 6 and 7, we show how quantum confinement alters the bulk band structure in a quantum well in the framework of the envelope function approximation [16–28]. In Section 8, we consider the beneficial effects of introducing strain into the quantum well [28–35]. We will see there that the formalism presented in Section 7 for predicting valence band-mixing effects is easily extendible to the study of strained material systems [26–29]. In Section 9, example calculations of gain in bulk and quantum well material will be presented by plugging the band structure calculated in Sections 5–8 into the gain expressions presented in Sections 2–4. Comparisons between different approximate methods will also be given. We will then summarize in Section 10.

## 2. ELECTRONIC STATES IN THE CRYSTAL

### 2.1. General Description

We begin our study of gain in III–V semiconductors with the simple description of electronic states in the conduction and valence bands of the crystal. The electron wavefunctions in the conduction and valence bands are found by solving the Schroedinger equation, which relates the system Hamiltonian $H_0$ of the crystal lattice to the energy $E$ of the electron. It can be written as

$$H_0\psi = \left[\frac{\mathbf{p}^2}{2m_0} + V(\mathbf{r})\right]\psi = E\psi, \tag{1}$$

where $\mathbf{p}$ is the momentum operator, $\mathbf{r}$ is the position vector, $m_0$ is the free electron mass, $\psi$ is the wavefunction of the electron, and $V(\mathbf{r})$ is the potential created by the crystal lattice. Due to the periodicity of $V(\mathbf{r})$, the solutions of this equation are given by Bloch waves [1, 36] of the form

$$\psi = e^{i\mathbf{k}\cdot\mathbf{r}}u(\mathbf{k}, \mathbf{r}), \tag{2}$$

where $\mathbf{k}$ is the wavevector of the electron, and $u$ is a Bloch function with the special property that it is periodic with the crystal lattice and hence repeats itself in each unit cell of the crystal. In defining "localized" solutions of (1) such as those in quantum wells, it is useful to consider linear combinations of the Bloch wave solutions in (2). Using an arbitrary set of expansion coefficients, $A(k)$, we can express a spatially localized wavefunction as

$$\psi = \int A(k)e^{i\mathbf{k}\cdot\mathbf{r}}u(\mathbf{k}, \mathbf{r})\, d^3\mathbf{k} \approx u(0, \mathbf{r}) \int A(k)e^{i\mathbf{k}\cdot\mathbf{r}}\, d^3\mathbf{k} \equiv F(\mathbf{r})u(\mathbf{r}), \tag{3}$$

The preceding description of localized states is known as the envelope function approximation [16]. The key assumption here is that, within a given energy band, the Bloch function is not a strong function of $\mathbf{k}$ (at least in the proximity of the band edge) and can thus be approximately represented by the band edge ($\mathbf{k} = 0$) Bloch function, $u(0, \mathbf{r}) \equiv u(\mathbf{r})$. This allows us to pull it out of the expansion and define an envelope function $F(\mathbf{r})$, whose Fourier spectrum is made up of the plane wave components of the solutions in (2). Thus, our generalized *approximate* solutions in a given energy band consist of two components: the band edge Bloch function, multiplied by a slowly varying envelope function. We choose the two components to be normalized such that, in Dirac notation [1], we have

$$\langle F|F\rangle \equiv \int_V F^*F\, d^3\mathbf{r} = 1, \qquad \langle u|u\rangle \equiv \frac{1}{V_{\text{uc}}}\int_{\text{unit cell}} u^*u\, d^3\mathbf{r} = 1. \tag{4}$$

For the envelope functions, $V$ is the volume of the crystal. For the Bloch functions, we only need to consider the volume of a single unit cell of the crystal, $V_{\text{uc}}$. Note that with our chosen definitions, the envelope functions have dimensions of $V^{-1/2}$, whereas the Bloch functions are dimensionless. [*Note*: Our convention for Dirac notation applied to envelope and Bloch functions throughout this chapter will be consistent with the operational definitions contained in (4).] In addition, the conduction band and valence band Bloch functions are orthogonal to each other such that

$$\langle u_c | u_v \rangle = \langle u_v | u_c \rangle = 0. \tag{5}$$

In bulk material, the normalized envelope functions are simply given by

$$F = \frac{1}{\sqrt{V}} e^{i\mathbf{k}\cdot\mathbf{r}}. \tag{6}$$

The envelope functions in quantum well material will be considered in Sections 6 and 7.

We are particularly interested in the relationship between the energy $E$ of the electron or hole, given in (1), and the electron's wavevector $\mathbf{k}$, given in (6). It is quite common to assume the bands to be parabolic in both the conduction and valence bands [37], allowing us to write

$$E_e(\mathbf{k}) = E_c + \frac{\hbar^2 k_e^2}{2m_c}, \qquad E_h(\mathbf{k}) = E_v - \frac{\hbar^2 k_h^2}{2m_v}, \tag{7}$$

where $E_{c,v}$ are the band edge energies, $m_{c,v}$ are the effective masses in the two bands, and $k_{e,h}$ are the magnitudes of the wavevectors of a given electron or hole. However, these expressions are oversimplifications to reality. First of all, $E(\mathbf{k})$ is not the same along all directions of the crystal, and we cannot simply say that $\mathbf{k} = k$. For example, the heavy-hole band in GaAs is known to be highly anisotropic as a function of the $\mathbf{k}$ vector direction [38]. (In comparison, the conduction and light-hole bands are much more isotropic at energies near the band edge.) Second, at energies away from the band edge, the band curvature does not remain perfectly parabolic, especially in the light-hole band [38]. Thus, in defining the relations for gain, we must keep in mind that (7) is not always a good approximation to reality.

## 2.2. Density of States

Another aspect of the wavefunctions in (2) that we must consider involves the concept of density of electronic states. To determine the density of states, we must first determine the existing electronic states in the crystal. Assume we have a bulk crystal of dimensions $L_x$, $L_y$, and $L_z$. One thing we can say is that we know the electron wavefunction must go to zero at the edges of the crystal. However, the plane wave solutions in (6) do not like to go to zero. Therefore, we must create standing wave solutions by superposing two or more counterpropagating plane waves. Applying the boundary conditions to the new sine wave solutions, we find that the electron's **k** vector components must be such that $\sin(k_x \cdot 0) = \sin(k_x \cdot L_x) = 0$, and likewise for the other two directions. Thus, we quickly conclude that the discrete set of **k** vector states an electron can have is given by

$$\mathbf{k} = n_x \mathbf{K}_x + n_y \mathbf{K}_y + n_z \mathbf{K}_z, \qquad \text{where } |\mathbf{K}_i| = \pi/L_i. \tag{8}$$

The $n_i$ are quantum numbers of the system, which can take on both positive and negative nonzero integer values. It should be noted, however, that physically distinct states depend only on $(|n_x|, |n_y|, |n_z|)$ when considering standing wave solutions. The $\mathbf{K}_i$ represent vectors along the three orthogonal directions in **k** space, whose magnitudes are inversely proportional to the length of the crystal in that direction. Thus, we should expect the distribution of **k** vector states in $k$ space to be dependent on the dimensions of the crystal. For example, in Fig. 1, we have illustrated three typical crystal geometries in the first column, along with the corresponding arrangement of **k** vector states in $k$ space in the second column. The top row illustrates a bulk crystal with sides of equal length and large scale dimensions ($>0.1$ $\mu$m), resulting in equal length $\mathbf{K}_i$ vectors in all three directions. The result is a uniform distribution of states throughout $k$ space. If we scale one of these dimensions (for example, $L_z$) down to the quantum regime ($<0.1$ $\mu$m), we obtain a quantum well, illustrated in the second row. The small $L_z$ in this case produces an elongated $\mathbf{K}_z$ vector, which transforms the uniform distribution of states in the bulk case into separated horizontal *planes* of uniformly distributed states, each plane corresponding to a different quantum number $n_z$. We can take this a step further by scaling $L_y$ down to the quantum regime to obtain a quantum wire [6, 13, 14, 18], as shown in the last row of Fig. 1. In this case, $\mathbf{K}_y$ also becomes elongated, transforming the planes of uniformly distributed states into separated *lines* of equally spaced states.

In bulk material, we will be interested in determining the density of states in all three dimensions. However, in a quantum well, the separation between planes in $k$ space is actually much greater than shown in Fig. 1 (typically, $\mathbf{K}_z$ is *orders of magnitude* larger than either $\mathbf{K}_x$ or $\mathbf{K}_y$). It is therefore convenient to treat each horizontal plane individually and concentrate on deriving a "two dimensional" density of states function within each plane. Similar reasoning applies to a quantum wire, where the spacing between each line in $k$ space is so large that it becomes convenient to consider a "one dimensional" density of states function along each line. Because the derivation for the density of states function follows the same line of reasoning in each of these cases, it is useful to discuss things in general terms by introducing the abstract concept of an $m$ dimensional "volume" of $k$ space,

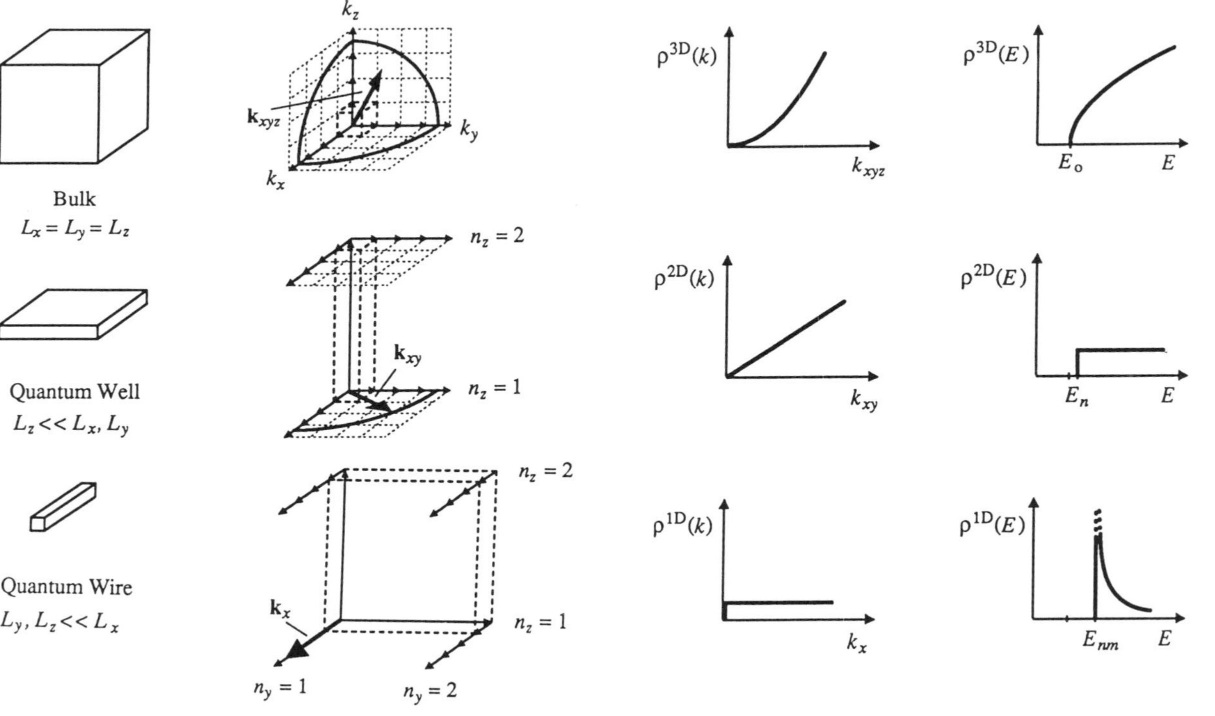


**Fig. 1.** First column: Illustration of three typical crystal geometries. Second column: Distribution of **k** vector states in $k$ space. For each case, a single unit cell is illustrated by the use of heavy dashed lines. Third column: The density of states function is plotted versus the magnitude of the $m$ dimensional **k** vector. Fourth column: The density of states function is plotted versus energy. The reference energy (i.e., the energy of the electron when the $m$ dimensional **k** vector is equal to zero) is indicated for each case.

which is actually a volume for $m = 3$, an area for $m = 2$, and a length for $m = 1$.

To derive the density of states function, we must first develop a method for counting the states. Let us begin by finding the total number $N_s$ of $\mathbf{k}$ vector states having $\mathbf{k}$ vector magnitudes less than some value $k$ (indicated in all three cases by the large arrow in column 2 of Fig. 1). For the bulk case, we see that a three dimensional (3D) vector $\mathbf{k}_{xyz}$ sweeps out a sphere of radius $k$ in $k$ space. For a quantum well, within one plane, a 2D vector $\mathbf{k}_{xy}$ sweeps out a *circle* of radius $k$ in $k$ space. In general terms, an $m$D vector sweeps out an $m$D volume $V_k$ of radius $k$ in $k$ space. To find the total number $N_s$ of states contained within this "volume," we simply divide the total volume $V_k$ by the volume of $k$ space occupied by one single $\mathbf{k}$ vector state. From Eq. (8), increments of $2\mathbf{K}_i$ along each direction define one state (the factor of two is necessary because positive and negative values, $\pm n_i$, define the same state). In bulk material, then, one state occupies a cube of volume $(2\pi/L)^3$ (where $L^3$ is shorthand for $L_xL_yL_z$). In a quantum well, one state occupies a square of area, $(2\pi/L)^2$. In general, a single state occupies an $m$D volume $(2\pi/L)^m$. Therefore, the expression for the total number of states, $N_s$, (neglecting spin degeneracy) can be written as

$$N_s = \frac{V_k}{(2\pi/L)^m}, \qquad \text{where } m = 1, 2, 3. \tag{9}$$

The specific form of $V_k$ for each value of $m$ is summarized in Table I. If the *density* of $\mathbf{k}$ vector states in $k$ space were known throughout all of $k$ space,

**Table I.**

Density of States for Bulk (3D), Quantum Well (2D), and Quantum Wire (1D) material

| $m$ Dimension | $V_k$ | $\frac{dV_k}{dk}$ | $\rho(k)^a$ | $\rho(E)^{a,b}$ |
|---|---|---|---|---|
| 3D | $\frac{4}{3}\pi k^3$ | $4\pi k^2$ | $\frac{k^2}{2\pi^2}$ | $\frac{\sqrt{E}}{4\pi^2}\left(\frac{2m}{\hbar^2}\right)^{3/2}$ |
| 2D | $\pi k^2$ | $2\pi k$ | $\frac{k}{2\pi}\left(\frac{1}{L_z}\right)$ | $\frac{m}{2\pi\hbar^2}\left(\frac{1}{L_z}\right)$ |
| 1D | $2k$ | $2$ | $\frac{1}{\pi}\left(\frac{1}{L_xL_y}\right)$ | $\frac{\rho(k)}{2\sqrt{E}}\left(\frac{2m}{\hbar^2}\right)^{1/2}$ |

[a] Spin not included.
[b] Assuming parabolic bands.

then we should also be able to obtain the total number $N_s$ of states contained within the $k$ space volume $V_k$ simply by integrating the density of states over that volume. Mathematically, then, we can state that

$$\int_{V_k} \rho^{m\mathrm{D}}(k)\, dk = \frac{N_s}{V}, \tag{10}$$

where $\rho^{m\mathrm{D}}(k)$ is defined as the $m$ dimensional density of states function. The volume of the crystal, $V$, appears such that integration of $\rho^{m\mathrm{D}}(k)$ yields the total number of **k** vector states *per unit volume* of the crystal. Equation (10) serves as an implicit definition for the density of states. Solving for $\rho^{m\mathrm{D}}(k)$ and inserting (9), we obtain

$$\rho^{m\mathrm{D}}(k) \equiv \frac{1}{V}\frac{dN_s}{dk} = \frac{(L)^m}{V}\frac{1}{(2\pi)^m}\frac{dV_k}{dk}. \tag{11}$$

Table I displays the specific form of $\rho^{m\mathrm{D}}(k)$ for bulk (3D), quantum well (2D), and quantum wire (1D) material. These expressions are also graphically illustrated in the third column of Fig. 1. It is interesting to note that $\rho^{m\mathrm{D}}(k)$ increases with $k$ as the surface area of the sphere, the perimeter of the circle, and the endpoints of the line in the three cases, respectively. [*Note*: While it is common practice to account for spin degeneracy in the definition of the density of states function (by adding a factor of two), we have chosen not to do it this way in the present chapter. To help avoid confusion in comparisons with other sources (including other chapters in this book), we have included an explicit reminder in Table I. We will account for spin degeneracy later when considering the transition matrix element in Section 5.2, where a discussion of spin degeneracy is more appropriate.]

Often times it is the energy of the electron that is important, not its **k** vector. We can transform the above derived density of states function into energy space through the relation

$$\rho(E)\, dE = \rho(k)\, dk \rightarrow \rho(E) = \frac{\rho(k)}{dE/dk}. \tag{12}$$

Expressions for $\rho(E)$ are also included in Table I assuming parabolic bands, as in (7). The electron's energy when $E = 0$ in Table I corresponds to when $k = 0$ in the three cases. In bulk material, when $k_{xyz} = 0$ the electron is resting at the edge of the conduction or valence band. Thus, in Table I, the expression for $\rho^{3D}(E)$ is relative to the material band edge, denoted in the top right of Fig. 1 by $E_0$ (which could represent either $E_c$ or $E_v$). In a quantum well, when $k_{xy} = 0$ in a given plane, $k_z$ remains finite, and the electron's energy can still be greater than the band edge. In fact, the quantum number $n_z$

determines the length of $k_z$ for each horizontal plane of **k** vector states, from Eq. (8). Thus, when $k_{xy} = 0$ the set of nonzero $k_z$'s is responsible for creating a quantized set of energy levels $E(n_z)$, or $E_n$ for short, within the quantum well. From (7) and (8), we can determine that $E_n \propto (n_z\mathbf{K}_z)^2$ in a quantum well with infinite barriers (such that the envelope function goes to zero at the boundaries). In Table I, then, the expression for $\rho^{2D}(E)$ for the $n$th horizontal plane is relative to the quantized energy $E_n$. Similar reasoning applies to the quantum wire case, where the nonzero $k_y$ and $k_z$ lead to quantized energy levels $E(n_y, n_z)$, or $E_{nm}$, from which $\rho^{1D}(E)$ is referenced.

We have mentioned that when $k_{xy} = 0$ a discrete set of energies $E_n$ within the quantum well are allowed. It is interesting to consider what happens as $k_{xy}$ increases from zero. Plotted on the left of Fig. 2 is the energy of the electron as a function of $k_{xy}$ for the three lowest energy levels of a quantum well. The parabolic energy bands, or *subbands*, as they will be referred to, are a result of the fact that within one plane, the energy of the electron still increases parabolically as a function of the in-plane **k** vector, in accordance with Eq. (7). To help picture how the set of energy subbands reduces to the

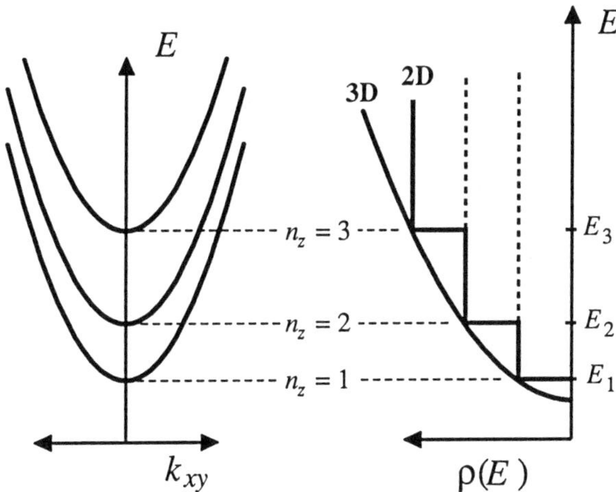


**Fig. 2.** On the left side, the lowest three energy subbands of an infinite barrier quantum well are plotted as a function of the in-plane **k** vector magnitude, $k_{xy}$. On the right side, the corresponding two dimensional (2D) density of states function for each subband is shown. The summation over all three $\rho^{2D}(E)$ is indicated by the heavy solid line labeled "2D." The three dimensional density of states function (indicated by the curve labeled "3D") is also shown, to indicate the correlation between the overall 2D and the 3D density of states functions.

single bulk energy band in the limit of large $L_z$, we could imagine the $k_z$ axis coming out of the page, in which case the three energy subbands would be successively displaced from each other by an amount $\mathbf{K}_z$ out of the page. As $L_z$ increases, $\mathbf{K}_z$ decreases, and the energy subbands become closer together along the $k_z$ axis. In addition, the quantized energy levels will become smaller since $E_n \propto (n_z\mathbf{K}_z)^2$. Eventually, in the limit of very large $L_z$, the subbands will form a quasi-continuous smooth surface along $k_z$, and since the energy levels $E_n$ increase as the square of $n_z$, they will form a *parabolic* curve along $k_z$, similar to the one in the $k_{xy}$ direction. Thus, the single bulk energy band is obtained. Another interesting relationship holds between $\rho^{3D}(E)$ in bulk material and $\rho^{2D}(E)$ in quantum wells *with infinite barriers*. Both are plotted versus energy on the right side of Fig. 2. As can be seen, if we stack the $\rho^{2D}(E)$ arising from each subband within the quantum well, the tips of the overall "staircase" function align perfectly with $\rho^{3D}(E)$. We leave it as an exercise for the reader to verify this relationship and show that as the width of the quantum well, $L_z$, increases to infinity, $\rho^{2D}(E)$ summed over all subbands does in fact reduce to $\rho^{3D}(E)$.

In Section 6, a more rigorous derivation of parabolic energy subbands in the conduction band of finite barrier quantum wells will be given. In Section 7, we will find out that much more complex subbands arise in the valence band, due to the interaction of the heavy and light hole bands.

There is one last point regarding the density of states functions we should address before moving on. For the quantum well (and quantum wire) structure previously discussed, we have assumed infinite barriers, so that the electrons are completely confined within the well width $L_z$. Thus, when we discuss a density of states *per unit volume*, it certainly seems valid in Eq. (10) to consider the physical volume occupied by the electrons as $L_z$ multiplied by the two transverse dimensions of the crystal. However, in more realistic situations where the barriers of the quantum well are finite, the electron actually penetrates into the barrier regions, and the volume they occupy is greater than we have accounted for in our expressions in Table I. The question is, should there be cause for concern here? The answer is no, as long as we are consistent with our interpretation of $L_z$ whenever we discuss physically measurable quantities.

For example, in a moment we will be deriving an expression for the material gain of a quantum well. The final result will contain $\rho^{2D}(E)$, which is inversely proportional to $L_z$. At first glance, then, it might appear that to calculate material gain correctly we should definitely use the appropriately increased $L_z$, which accounts for the penetration of the electron into the barrier regions. Before jumping to conclusions, however, we must realize that material gain in a quantum well is not a physically measurable quantity. We can only measure *modal* gain, which is the material gain adjusted to take into account the poor overlap that always exists between the optical mode and the electron envelope function in the quantum well. One interpretation would be to say that, in a finite barrier quantum well, the electron penetration into the barriers reduces the density of states *per unit volume*, which in turn reduces the material gain. At the same time, however, the overlap between the new "wider" electron envelope function and the optical mode (which could be traveling either along the plane of the well or perpendicular to it) is increased, exactly cancelling out the reduction in the material gain. Thus, the physically measurable modal gain is independent of the effect of electron penetration into the barriers. Another more common (and more convenient) intepretation is to assume that $\rho^{2D}(E)$ is unaffected by the electron penetration, so that $L_z$, defined as the physical well width, is always used in Table I. In turn, we must then make certain that whenever we calculate the overlap between the electron envelope function and the optical mode, we use the physical well width to define the spatial extent of the electron. This latter interpretation is by far the most common, and we will adhere to it thoughout the chapter.

Other examples where the issue of how to interpret $L_z$ might arise include derivations that yield expressions for the radiative recombination rate *per unit volume*, as well as the *volume* carrier density (as we will do in later sections). In both of these cases, when we want to compare our theoretical expressions with experiment, we need to convert the theoretical volume densities to a physically measurable current *per unit area* and a charge *per unit area*. In the conversion process, somewhere we will need to multiply by the width of the quantum well, which will inevitably cancel out the factor of $L_z$ appearing in the denominator of $\rho^{2D}(E)$ in Table I. Thus, again our interpretation of $L_z$ is purely academic (as long as we are consistent with our interpretation!). The bottom line is that a quantum well is inherently a two dimensional structure, and any concepts involving *per unit volume* are only temporary conceptual aids in obtaining final results. (For example, we could carry out the entire following analysis for a quantum well by using a density of states *per unit area*, which would not include $L_z$ in its definition, but this would lead to results that were specific to quantum wells. By defining the density of states function to be *per unit volume* independent of the dimensionality of the semiconductor, we can obtain expressions that apply equally to all cases given in Table I.)

## 3. BAND-TO-BAND TRANSITIONS

### 3.1. Fermi's Golden Rule

Optical gain in semiconductors is caused by photon-induced transitions of electrons from the conduction band to the valence band. Thus, in order to understand optical gain, we need to characterize electron–photon interactions in the crystal. To examine the interaction, we represent the photon classically by an electromagnetic wave. The wave's interaction with the electron enters into the Schroedinger equation through the vector potential [39], which can be expressed as

$$\mathbf{A}(\mathbf{r}, t) = \hat{\mathbf{e}} \operatorname{Re}[A(\mathbf{r})e^{-i\omega t}] = \hat{\mathbf{e}}\,\tfrac{1}{2}[A(\mathbf{r})e^{-i\omega t} + A^*(\mathbf{r})e^{i\omega t}], \tag{13}$$

where $\hat{\mathbf{e}}$ is the unit polarization vector in the direction of $\mathbf{A}$, and $\omega(\hbar\omega)$ is the angular frequency (energy) of the photon. The Schroedinger equation in (1) is now modified by the substitution [39]

$$\mathbf{p}^2 \to (\mathbf{p} + e\mathbf{A})^2 \approx \mathbf{p}^2 + 2e\mathbf{A} \cdot \mathbf{p}, \tag{14}$$

where $e$ is the magnitude of the electron charge. In expanding the square, we can neglect the squared vector potential term, since it does not affect our final results (orthogonality of the wavefunctions ensures that the operator $\mathbf{A}^2$ does not perturb the system, assuming we can neglect the spatial variation of $\mathbf{A}$ within one unit cell). Substituting (14) into (1), we can write the new Hamiltonian as

$$H = H_0 + [H'(\mathbf{r})e^{-i\omega t} + \text{h.c.}], \qquad H'(\mathbf{r}) \equiv \frac{e}{2m_0} A(\mathbf{r})\hat{\mathbf{e}} \cdot \mathbf{p}; \tag{15}$$

the h.c. stands for Hermitian conjugate [1], and it simply means that we take the complex conjugate of all terms except the Hermitian operator $\mathbf{p}$. The term in brackets can be viewed as a time-dependent perturbation to the original Hamiltonian. The effect of this perturbation is to induce electronic transitions between the conduction and valence bands.

In quantifying the gain, we need to know the number of transitions that will occur per second in the crystal in response to a given flux of photons in a given optical mode (determined by the magnitude of the vector potential for that mode). This is accomplished by studying the time evolution of a given electron wavefunction $\Psi$, initially in the conduction band state, $\psi_e$, as it makes a transition to the valence band. We expand the wavefunction into a linear combination of the initial state plus an array of possible final states in the valence band, where the expansion coefficients are time-dependent.

Inserting the wavefunction into the time-dependent Schroedinger equation [1], it is possible to obtain an approximate expression for the probability of finding the electron in a particular state in the valence band as a function of increasing time. The time derivative of this time-dependent probability then gives an approximate expression for the transition rate from the conduction band state $\psi_e$ to a particular valence band state $\psi_h$. The transition rate derived in most quantum mechanics textbooks [1, 36, 39] can be expressed (in units of $s^{-1}$) as

$$W_{e\to h} = \frac{2\pi}{\hbar} |H'_{eh}|^2 \, \delta(E_e - E_h - \hbar\omega), \tag{16}$$

$$H'_{eh} \equiv \langle \psi_h | H'(\mathbf{r}) | \psi_e \rangle = \int_V \psi_h^* H'(\mathbf{r}) \psi_e \, d^3\mathbf{r}. \tag{17}$$

Equation (16) is known as Fermi's golden rule [1] and is applicable to many systems where interaction with photons is of concern. [*Note*: In some derivations, Eq. (16) appears with an extra factor of four in the denominator [37]. This is because the factor of 1/2 in the definition of $H'(\mathbf{r})$ is sometimes pulled out and written explicitly into the expression for Fermi's golden rule.] The delta function (which has units of energy$^{-1}$) indicates that the difference between the initial and final energy ($E_e - E_h$) of the electron must be equal to the energy $\hbar\omega$ of the photon that induced the transition. [*Note*: We must be somewhat careful with our interpretation of the delta function, for it is not meant to imply that the transition rate is infinite when $E_e - E_h = \hbar\omega$. The use of the delta function here implicitly assumes that $W_{e\to h}$ refers to a single transition pair within a continuum of states. To correctly evaluate the total transition rate, we must *necessarily* sum over all of these continuum states. As we shall see in Section 3.2, when the sum is evaluated, the singularity introduced by the delta function disappears.]

Equation (17) can be further reduced by using the electron and hole wavefunctions defined in (3) and the perturbation term defined in (15) to obtain

$$\begin{aligned} H'_{eh} &= \frac{e}{2m_0} \int_V F_h^* u_v^* (A(\mathbf{r})\hat{\mathbf{e}} \cdot \mathbf{p}) F_e u_c \, d^3\mathbf{r} \\ &= \frac{e}{2m_0} \left\{ \int_V u_v^* u_c F_h^* (A(\mathbf{r})\hat{\mathbf{e}} \cdot \mathbf{p}) F_e \, d^3\mathbf{r} + \int_V [F_h^* A(\mathbf{r}) F_e] u_v^* \hat{\mathbf{e}} \cdot \mathbf{p} u_c \, d^3\mathbf{r} \right\} \end{aligned} \tag{18}$$

where we have used the fact that $\mathbf{p}AB = B\mathbf{p}A + A\mathbf{p}B$. In transitions from the conduction band to the valence band, the first integral within the braces vanishes due to the orthogonality condition expressed in Eq. (5) and due to the fact that the other terms in the integrand are, to a good approximation, constant in any one unit cell. [*Note*: For transitions within the *same* energy band, however, the Bloch function overlap is equal to unity, and the first integral may or may not be zero, depending on the envelope functions [40].] To evaluate the second integral, we break the integration over the crystal volume up into a sum of integrations over each unit cell. The terms collected in brackets in the second integral can again be taken as constant over the dimensions of a unit cell, and we can write

$$H'_{\mathrm{eh}} = \frac{e}{2m_0} \sum_j [\cdots]_{\mathbf{r}=\mathbf{r}_j} \int_{\text{unit cell}} u_{\mathrm{v}}^* \hat{\mathbf{e}} \cdot \mathbf{p} u_{\mathrm{c}} \, d^3\mathbf{r}, \tag{19}$$

where $j$ sums over all unit cells in the crystal, and $\mathbf{r}_j$ is a position vector to the $j$th cell. Because the Bloch functions $u$ repeat themselves in each unit cell, the integral can be pulled out of the summation to obtain

$$\begin{aligned} H'_{\mathrm{eh}} &= \frac{e}{2m_0} \left( \frac{1}{V_{\mathrm{uc}}} \int_{\text{unit cell}} u_{\mathrm{v}}^* \hat{\mathbf{e}} \cdot \mathbf{p} u_{\mathrm{c}} \, d^3\mathbf{r} \right) \sum_j [\cdots]_{\mathbf{r}=\mathbf{r}_j} V_{\mathrm{uc}} \\ &= \frac{e}{2m_o} \langle u_{\mathrm{v}} | \hat{\mathbf{e}} \cdot \mathbf{p} | u_{\mathrm{c}} \rangle \int_V F_{\mathrm{h}}^* A(\mathbf{r}) F_{\mathrm{e}} \, d^3\mathbf{r}, \end{aligned} \tag{20}$$

where, by assuming the volume of a unit cell to be very small, we have converted the summation back into an integral.

The envelope function overlap integral in (20) is quite general and can be applied to many types of problems. For example, we can treat bulk band-to-band transitions by using the plane waves in (6) in the overlap integral. In this case, if we also assume the vector potential to be a plane wave with photon wavevector $\kappa$ the exponents of the three plane waves must add up to zero or else the integral vanishes, implying that $\mathbf{k}_{\mathrm{h}} = \mathbf{k}_{\mathrm{e}} + \kappa$, which is just another way of saying that momentum must be conserved in the transition. However, typical photon wavelengths are much longer than electron wavelengths, and we can state to a very good approximation that we must have $\mathbf{k}_{\mathrm{h}} = \mathbf{k}_{\mathrm{e}}$, a condition known as **k**-conservation, or the **k**-selection rule [37]. It is important to realize that this rule only applies to transitions between two plane wave states (band-to-band transition).

In other types of transitions, **k**-selection does not have to apply. For example, in transitions to or from a localized state such as an electron bound to a donor impurity [37], **k**-selection does not have much meaning since the Fourier spectrum of the localized state contains a spread of **k** vectors in general. A case of particular interest to us involves transitions between *two* localized states, both bound to a quantum well along the $z$ direction. In this case, the dimensions of the well are much smaller than the wavelength of light, and the vector potential can be taken as a constant, $A_0$, across the region of the well, allowing us to write

$$\int_V F_h^* A(\mathbf{r}) F_e \, d^3\mathbf{r} \approx A_0 \int F_h^* F_e \, dz \equiv A_0 \langle F_h | F_e \rangle. \quad (21)$$

The quantum well envelope functions along $z$ will be found later in Sections 6 and 7. The envelope functions within the plane of the well are still given by plane waves, and hence **k**-conservation must still be maintained in those directions. Assuming this condition is met, the in-plane integral is equal to unity, justifying our conversion from a volume integral to an integration along $z$ in the preceding equation. Substituting (21) into (20), we finally obtain

$$|H'_{eh}|^2 = \left(\frac{eA_0}{2m_0}\right)^2 |M_T|^2, \qquad \text{where } |M_T|^2 \equiv |\langle u_v | \hat{\mathbf{e}} \cdot \mathbf{p} | u_c \rangle|^2 |\langle F_h | F_e \rangle|^2. \quad (22)$$

The term $|M_T|^2$ defined here will be referred to as the transition matrix element and will be studied in more detail in Section 5.

The overlap integral between the electron and hole envelope functions in (22) leads one to the interesting concept of forbidden transitions. For example, let us examine the transition probability in an infinite quantum well from a conduction band state with quantum number $n_c$ to a valence band state with quantum number $n_v$. To analyze the transition, we must remember two important facts about the envelope functions or *eigenfunctions* of an infinite quantum well: (a) the eigenfunctions are independent of the effective mass of the electron or hole; and (b) the eigenfunctions form an orthogonal set of functions [1]. When $n_c = n_v$, the envelope functions in both bands are identical from (a), and $|\langle F_h | F_e \rangle|^2 = 1$. Thus, these transitions are allowed. When $n_c \neq n_v$, the envelope functions in the two bands are orthogonal to each other from (a) and (b), and $|\langle F_h | F_e \rangle|^2 = 0$. Thus, these transitions are forbidden. The various allowed and forbidden transitions are graphically illustrated in Fig. 3.

For quantum wells with finite barriers, penetration of the wavefunction into the barriers depends on the mass of the particle, and statement (a) above becomes invalid. However, the eigenfunctions are only a weak function of the electron or hole effective mass, and, in most cases, when $n_c = n_v$ we still have $|\langle F_h | F_e \rangle|^2 \approx 0.95$–$1$. When $n_c \neq n_v$, orthogonality between electron and hole envelope functions no longer holds rigorously but does remain approximately true. Thus, forbidden transitions can occur in a finite barrier quantum well; however, the transition probability is typically very weak, and

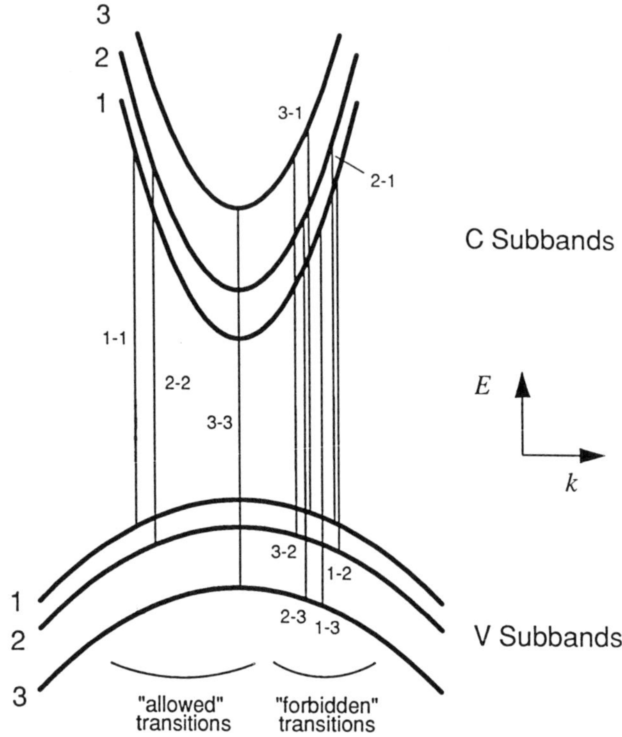


**Fig. 3.** Illustration of transitions between the conduction and valence subbands of a quantum well. All transitions are drawn with equal transition energy and equal in-plane **k** vector. The allowed transitions have very strong transition probabilities, whereas the forbidden transitions have zero transition probability in an infinite barrier quantum well and weak probability at best in a finite barrier quantum well.

$|\langle F_{\mathrm{h}}|F_{\mathrm{e}}\rangle|^2 \approx 0$–$0.1$. It should be noted, however, that even in the finite barrier quantum well, transitions between symmetric (cosine-like) functions, and antisymmetric (sine-like) functions are still strictly forbidden, since there is always othogonality between symmetric and antisymmetric functions. [*Note*: We will learn later that band mixing away from the band edge produces valence band states that are neither purely symmetric nor antisymmetric but a mixture of both. The result is that most transitions become allowed (to a small extent) whether strictly forbidden or not, as we will discuss in more detail in section 9.]

## 3.2. Reduced Density of States

Fermi's golden rule gives the transition rate for a single pair of conduction and valence band states. From the discussion of Section 2.2, we know that many $k$ states exist in both conduction and valence bands. To find the total transition rate, we must sum (16) over all transition pairs that are allowed. In this chapter, we will only consider undoped material such that band-to-band transitions dominate, allowing us to assume strict **k**-selection rules. Thus, only vertical transitions in $k$ space are allowed, and we can simply sum over all $N_s$ electronic states in *either* band, as illustrated in Fig. 4. The total transition rate *per unit volume* is then given (in units of $s^{-1}\,cm^{-3}$) by

$$
\begin{aligned}
W_{c\to v} &= \frac{1}{V}\int W_{e\to h}\,dN_s = \int W_{e\to h}\frac{1}{V}\frac{dN_s}{dk}\,dk \\
&= \frac{2\pi}{\hbar}\int |H'_{eh}|^2\,\delta(E_{eh}-\hbar\omega)\,\rho(k)\,dk,
\end{aligned}
\tag{23}
$$

where we have introduced the *transition* energy, $E_{eh} \equiv E_e - E_h$, into the delta function. We have also converted to an integration in $k$ space using the density of states function defined in (11). To handle the delta function, we

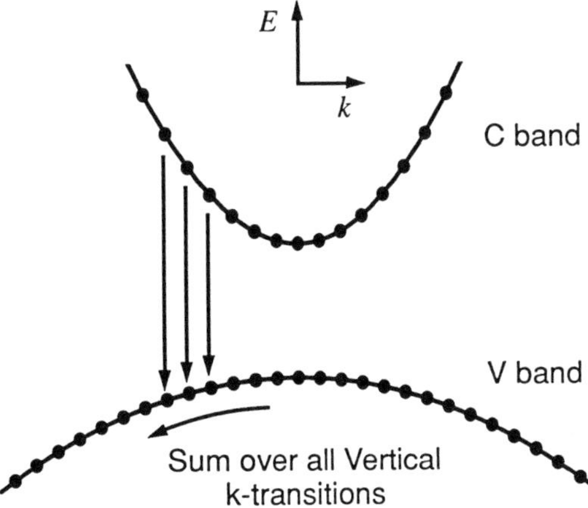


**Fig. 4.** Illustration of "vertical" transitions in $k$ space from the conduction (C) band to the valence (V) band in bulk material (or from a single C subband to a single V subband in a quantum well). Note that there is always a one-to-one correspondence in $k$ space between electronic states in the two bands.

make a variable substitution, letting $x = E_{eh} - \hbar\omega$. The integral then becomes

$$\begin{aligned} W_{c\to v} &= \frac{2\pi}{\hbar}\int |H'_{eh}|^2\,\delta(x)\,\rho(k)\,\frac{dx}{(dE_{eh}/dk)} \\ &= \frac{2\pi}{\hbar}|H'_{eh}|^2\left[\frac{\rho(k)}{dE_{eh}/dk}\right]_{E_{eh}=\hbar\omega}. \end{aligned} \tag{24}$$

The term in brackets is defined as the combined or *reduced* density of states [36]. When we are able to assume parabolic energy bands, we can further reduce the term in brackets. Using (7) and the fact that $k_e = k_h$, we can write

$$E_{eh} \equiv E_e - E_h = E'_g + \frac{\hbar^2 k^2}{2m_c} + \frac{\hbar^2 k^2}{2m_v}, \tag{25}$$

$$\frac{dE_{eh}}{dk} = \frac{\hbar^2 k}{m_r}, \qquad \text{where } \frac{1}{m_r} \equiv \frac{1}{m_c} + \frac{1}{m_v}. \tag{26}$$

The primed bandgap energy, $E'_g \equiv E_c - E_v$, is defined as the bandgap between two given subbands in a quantum well and is just the material band edge in bulk material. With the use of a reduced mass, $m_r$, we can conveniently adapt the definition for the density of states per unit *energy* given in (12) and write

$$W_{c\to v} = \frac{2\pi}{\hbar}|H'_{eh}|^2 \rho_{red}(E_{eh} - E'_g), \qquad \text{where } E_{eh} = \hbar\omega. \tag{27}$$

The form for the reduced density of states, $\rho_{red}(E)$, is given in Table I for bulk material and reduced dimensional structures. We interpret $m$ in the table as $m_r$, and $E$ as $E_{eh} - E'_g$.

In our summation over states we did not include the electron occupation probability, which is detemined from the Fermi–Dirac distribution [36] for electrons in the conduction and valence bands. It is defined here as

$$f_{c,v} = \frac{1}{1 + \exp[(E_{e,h} - E_{fc,fv})/k_B T]}, \tag{28}$$

where $E_{fc,fv}$ are the nonequilibrium quasi-Fermi levels [37] in the conduction and valence bands, $k_B$ is Boltzmann's constant, and $T$ is the temperature of the crystal. For parabolic bands, the individual energies $E_{e,h}$ are related to the transition energy $E_{eh}$ by

$$E_e = E_c + (E_{eh} - E'_g)\frac{m_r}{m_c}, \qquad E_h = E_v - (E_{eh} - E'_g)\frac{m_r}{m_v}. \tag{29}$$

Equation (27) should now be modified to include the probability of the conduction band state being occupied, $f_c$, multiplied by the probability of the valence band state being unoccupied, $1 - f_v$. With this modification, we can now write the total downward transition rate occurring in response to a flux of photons in a given optical mode (in units of $s^{-1}\ cm^{-3}$) as

$$W_{c\to v} = \frac{2\pi}{\hbar}|H'_{eh}|^2 \rho_{red} f_c(1 - f_v), \qquad |H'_{eh}|^2 = \left(\frac{eA_0}{2m_0}\right)^2 |M_T|^2. \tag{30}$$

The upward transition rate $W_{v\to c}$ (photon absorption) is obtained from (30) by interchanging the c and v subscripts for the Fermi functions. [*Note*: When dealing with nonparabolic bands (as we will in the valence band of quantum wells), Eq. (24) is actually more appropriate than Eq. (30) because $\rho_{red}(E)$ cannot be given in closed form in that case.]

## 4. GAIN AND SPONTANEOUS EMISSION RATE EXPRESSIONS

### 4.1. Gain

Equation (30) allows us to quantify the number of downward transitions occurring per second per unit volume in response to a flux of incoming photons in a given optical mode. Upward transitions or absorption of photons will also occur in response to the photon flux. Each downward transition generates a new photon, while each upward transition absorbs one. If the number of downward transitions per second exceeds the number of upward transitions, there will be a *net* generation of photons, and optical gain can be achieved. If we assume an electromagnetic wave propagating in the $z$ direction, then the net *increase* in photon flux per unit length along the $z$ direction can simply be written as

$$\frac{d\Phi}{dz} = W_{c\to v} - W_{v\to c}, \tag{31}$$

where $\Phi$ is the photon flux (in units of $s^{-1}\ cm^{-2}$). The optical gain $g$ of the material is usually defined as the *fractional* increase in photons per unit length, or simply

$$g \equiv \frac{1}{\Phi}\frac{d\Phi}{dz} = \frac{W_{c\to v} - W_{v\to c}}{\Phi}. \tag{32}$$

Let's assume that the vector potential generated by the photon flux is a plane wave of amplitude $A_0$. Recalling that the vector potential is related to the electric field through a time derivative [39] so that $E_0^2 = \omega^2 A_0^2$, we can write the photon flux as the stored energy per unit volume of the wave, multiplied by the group velocity of the wave [41], and divided by the energy per photon, or

$$\Phi(\omega) = \frac{1}{\hbar\omega}\left(\frac{c}{\bar{n}_g}\right)\left(\frac{1}{2}\bar{n}^2\varepsilon_0\omega^2 A_0^2\right), \tag{33}$$

where $\bar{n}$ is the index of refraction in the crystal, $c$ is the speed of light in free space, and $\varepsilon_0$ is the free space permittivity. The group velocity of the wave is characterized by the group index of refraction $\bar{n}_g$, which can be defined by [41]

$$\bar{n}_g = \bar{n}_{\text{eff}} + \omega(d\bar{n}_{\text{eff}}/d\omega), \tag{34}$$

where $\bar{n}_{\text{eff}}$ is the "effective" index (phase velocity index) of the guided mode [41]. For unguided optical modes, $\bar{n}_{\text{eff}}$ is just equal to $\bar{n}$. Hence, the second term in (34) includes both waveguide dispersion and material dispersion. Substituting the transition rate given in (30) along with the photon flux given before into (32), we can write the expression for gain as

$$g(\hbar\omega) = \left(\frac{1}{\hbar\omega}\right)\frac{\pi e^2\hbar}{\varepsilon_0 c m_0^2}\frac{\bar{n}_g}{\bar{n}^2}|M_T|^2\rho_{\text{red}}(E_{\text{eh}} - E_g')(f_c - f_v). \tag{35}$$

The gain is written as a function of photon energy to emphasize that the optical gain experienced by an incoming photon is very much dependent on the photon's energy. In fact, optical gain can typically only be achieved over a fairly narrow band of photon energies, as we shall see later in Section 9.

For bulk material, Eq. (35) tells us all we need to know in order to estimate the spectral gain (assuming undoped material such that only band-to-band transitions are important). In quantum well material, the situation is somewhat more complicated due to the many quantized levels from which transitions can occur, as illustrated previously in Fig. 3 for a particular incoming photon energy. In this case, the preceding gain relation corresponds to the gain produced within each subband transition in the well. If we index the subbands in the conduction and valence bands by the quantum numbers $n_c$ and $n_v$, then the spectral gain created by each subband transition pair can be denoted $g_{\text{sub}}(\hbar\omega, n_c, n_v)$. The total gain at a particular photon energy is found by summing over all subband transition pairs, or

$$g(\hbar\omega) = \sum_{n_c}\sum_{n_v} g_{\text{sub}}(\hbar\omega, n_c, n_v). \tag{36}$$

In calculating $g_{\text{sub}}(\hbar\omega, n_c, n_v)$, it is important to keep in mind that each subband transition will have its own set of envelope functions $F_i$ and subband gap $E_g'$.

From (35), we see that when $f_c(E_e) > f_v(E_h)$, $g(\hbar\omega)$ is positive, and an incoming light wave with photon energy $\hbar\omega$ will be amplified by the material. With a little thought, one can show [37] that this inequality is equivalent to saying that

$$E_g < \hbar\omega < E_{fc} - E_{fv} \tag{37}$$

is the requirement for gain at a photon energy $\hbar\omega$. In other words, the quasi-Fermi level separation must be greater than the bandgap to achieve optical gain in the material. Under equilibrium conditions, $E_{fc} = E_{fv}$, and optical gain is impossible to achieve. In addition, if the electron carrier

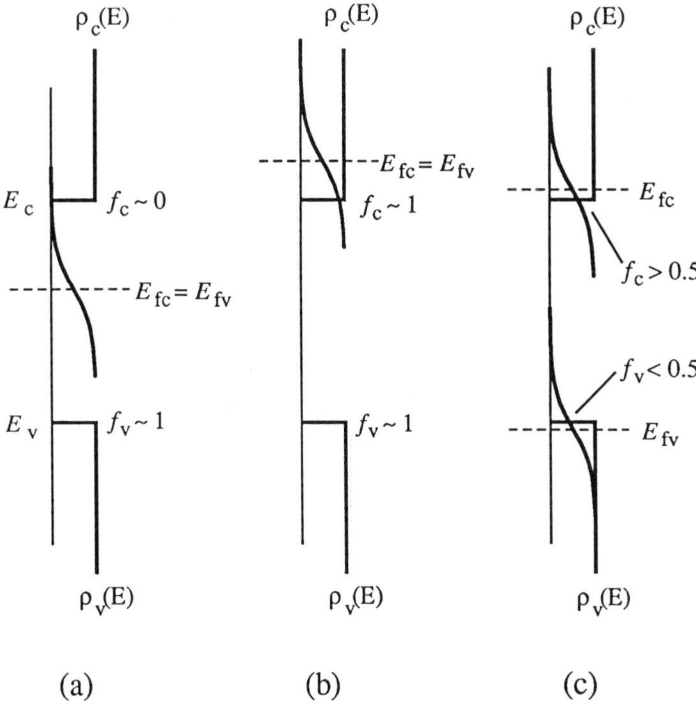


**Fig. 5.** Illustration of the quasi-Fermi functions superimposed on the two dimensional density of states functions of the conduction and valence subbands in a quantum well for three different cases: (a) lightly doped n-type material under equilibrium conditions; (b) degenerately doped n-type material under equilibrium conditions; and (c) undoped material under strong nonequilibrium conditions such that a large concentration of both electrons and holes exists.

density is small, $f_c$ will be close to zero and $f_c - f_v \approx -1$, as illustrated in Fig. 5a. Hence, the gain is negative, and the material is highly absorbing for photon energies greater than the bandgap. For extremely high electron carrier densities, we can at most set $f_c - f_v \approx 0$ under equilibrium conditions, making the material transparent at best, as shown in Fig. 5b. Therefore, in order to generate optical gain, we must create *nonequilibrium* conditions such that a very high electron *and* hole density can be *simultaneously* maintained, as illustrated in Fig. 5c. This can be achieved either within the depletion region of a p–n junction under strong forward bias or by optically pumping the material to generate a high density of electron–hole pairs. When the quasi-Fermi level separation is large enough so that $E_{fc} - E_{fv} = E_g$, the material will become transparent for photon energies equal to the bandgap, according to (37). The electron and hole carrier density that is required to provide this separation is known as the transparency carrier density, $N_{tr}$, and its magnitude is fundamentally related to the densities of states $\rho_c$ and $\rho_v$ of a given material. To illustrate this, we can use the fact that $E_{fc} - E_{fv} = E_g$ at transparency to draw some simple diagrams that show how $N_{tr}$ changes as we go from bulk to the quantum well regime when $m_v = m_c$ and how the quantum regime changes when $m_v = 5m_c$. The conclusions reached will be of more than academic interest since, as we shall see in Section 8, strain effectively allows us to change the relationship between $m_c$ and $m_v$.

The transparency condition is illustrated in Figs. 6a–c for the three simplistic situations just mentioned. Each case is drawn such that the *area* of the shaded regions is equal to the total carrier density in the band (see Section 4.3). From Figs. 6a–b, we see that the "soft" band edge of bulk material yields $N_{tr}(\text{bulk}) < N_{tr}(\text{QW})$. Note also that a smaller $\rho(E)$ in either case would produce a smaller $N_{tr}$ (a point we will return to in discussions of strained quantum wells). The last case shows the effect of increasing $\rho(E)$ in one band. The asymmetry shifts both Fermi functions toward the band with lighter effective mass in order to maintain equal numbers of carriers in both bands. The result is a larger $N_{tr}$ than either of the other two cases. (Note that this could be compensated by heavily doping the material p-type).

Optical gain in the material is attained when we inject a carrier density beyond $N_{tr}$ such that the quasi-Fermi levels are separated by an energy greater than the band gap (see (37)). The *rate* at which gain increases as we inject more carriers, known as the *differential* gain, $dg/dN$, is of great practical interest. In looking at (35), we find that, aside from some material constants, the rate at which we can change the difference $[\rho_{red} f_c + \rho_{red}(1 - f_v)] - \rho_{red}$ will determine how quickly the gain responds to changes in $E_{fc} - E_{fv}$ (or the total carrier density). For symmetric bands, as in Figs. 6a–b, $\rho_{red}$ is really no different from the density of states in either band (aside from a factor of two) and we can directly associate the products $\rho_{red} f_c$ and $\rho_{red}(1 - f_v)$ with the electron and hole carrier densities that exist *at the band edge* of the material. Thus, the differential gain is directly related to how quickly the "band edge carrier density" can be increased.

From Figs. 6a–b, we see that an increase in $E_{fc} - E_{fv}$ has a much more

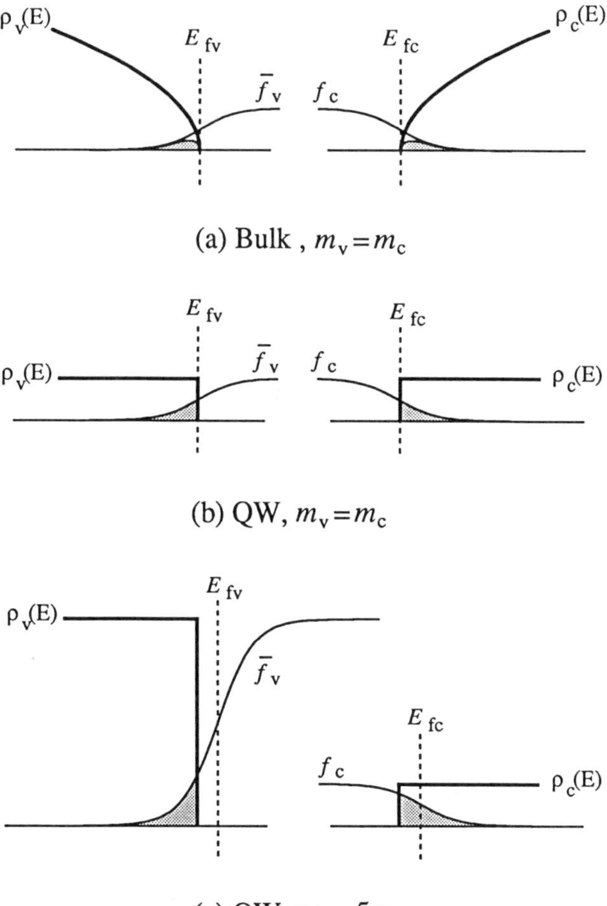


**Fig. 6.** Band edge transparency condition ($E_{fc} - E_{fv} = E_g$) illustrated for three different idealized cases. The densities of states $\rho_c$ and $\rho_v$ for each case are taken from Table I (or Fig. 1), assuming parabolic bands. The carrier filling of each band is illustrated by the shaded overlap region between the Fermi function and the density of states. The "hole" Fermi function $\bar{f}_v \equiv 1 - f_v$ is used in the figure for clarity.

profound effect on the band edge carrier density when $\rho(E)$ has a steep profile. Thus, the step-like density of states in a quantum well is expected to result in a much higher differential gain than can be obtained in bulk material. For symmetric bands, note that the maximum slope of the Fermi function is always coincident with the band edge at transparency. In Fig. 6c, we see that this is no longer true when the bands are asymmetric. The band edge electron density in the conduction band is now much less sensitive to the movement of $E_{fc}$, leading to a much lower differential gain. [*Note*: It might appear that the large $\rho(E)$ in the valence band should increase the differential gain. However, $E_{fc}$ changes much more than $E_{fv}$ in response to an increase in $E_{fc} - E_{fv}$, in an effort to keep the electron and hole densities equal. Thus, changes in the band edge *electron* density are more important when considering differential gain in the last case (this fact is reflected in the reduced density of states, where for large $m_v$, $\rho_{red} \to \rho_c$, from (26)).]

Conclusions we can draw from the preceding discussions are that to obtain low transparency carrier densities, we would like (1) a symmetric band structure ($m_v = m_c$), (2) a "soft" band edge, and (3) an overall small density of states. To obtain a high differential gain, we would like (1) a symmetric band structure ($m_v = m_c$), and (2) a "steep" band edge. [*Note*: There is no corresponding third condition for $dg/dN$, because both $g$ and $N$ increase with increasing $\rho(E)$, roughly cancelling out the effects of the magnitudes of $\rho_c$ and $\rho_v$ on the differential gain, $dg/dN$. The relative shapes of $\rho_c$ and $\rho_v$ are more important in this case.] From the standpoint of reducing transparency levels *and* increasing differential gain, then, we find that *matching* $\rho_c$ and $\rho_v$ and reducing their overall magnitudes in a semiconductor material is highly desirable. We will find later in Section 8 that strained quantum well material can provide us with both of these desirable features.

In deriving the preceding expression for gain, we have assumed that an electron in a conduction band state would remain in that state forever if it weren't for interactions with photons (i.e., the energy of the state is sharp). In reality, interactions with *phonons* and other electrons from time to time scatter the electron into another conduction band state [42–50]. Therefore, the lifetime of a given state is not infinite. In fact, it is presently believed that, on average, approximately every 0.1 ps [5, 44, 45] an electron (or hole) is "bumped" into a new state. If we assume the state decays exponentially with time, then the Fourier energy spectrum necessary to construct the time-dependent state has a Lorentzian lineshape, and hence the energy of each state (and each transition) is no longer sharp but has an energy spread over a range of $\Delta E \approx \hbar/0.1$ ps $\approx 7$ meV on each side of the "expected" energy of the state (or transition). This means that an incoming photon with energy $\hbar\omega$ will not only interact with transitions given by $E_{eh} = \hbar\omega$, but also with transitions within an energy spread $E_{eh} \approx \hbar\omega \pm \Delta E$. [*Note*: The previous argument is somewhat of a simplification, since in reality the conduction and valence band states are *independently* broadened. However, for the purposes of the present chapter, we shall make an approximation by assuming that it is the transition energy $E_{eh}$ that is broadened. See Chapter 2 for more details.]

To include the spectral broadening of each transition, we convolve the expression for gain with some spectral lineshape function over all transition energies $E_{eh}$ to obtain [42–50]

$$G(\hbar\omega) \equiv \int g(\hbar\omega) L(E_{eh})\, dE_{eh}, \tag{38}$$

$$L(E_{eh}) \equiv \frac{1}{\pi} \frac{\hbar/\tau_{in}}{(E_{eh} - \hbar\omega)^2 + (\hbar/\tau_{in})^2}, \tag{39}$$

where $g(\hbar\omega)$ is taken directly from (35) or (36), $L(E_{eh})$ is a normalized Lorentzian lineshape function, and $\tau_{in}$ is the *intraband relaxation time*, or simply the lifetime of each state, and is about 0.1 ps [5, 44, 45] in bulk material. $G(\hbar\omega)$ is now the spectrally "smoothed" gain (for comparisons between $g(\hbar\omega)$ and $G(\hbar\omega)$, see Section 9). [*Note*: The photon energy terms $\hbar\omega$ in (35) are not to be included in the convolution, since they are not related to the energy of the state but are the actual photon energy. In practice, however, it makes little difference whether they are included or not.]

The use of a Lorentzian lineshape function is common; more recently [46–50], however, more sophisticated lineshape functions have been derived. Yamanishi and Lee [47] have suggested that the state decays initially as a Gaussian but then takes on exponential behavior for larger times. The corresponding lineshape function was derived by Yamanishi and Lee and later approximated by Chinn *et al.* [48]. Asada [49] has also performed a detailed analysis of intraband scattering in quantum wells (see Chapter 2), arriving at an asymmetrical lineshape function that falls off much faster than a Lorentzian on the low energy side of the transition, similar to what Yamanishi and Lee had found. Kucharska and Robbins [50] have kept with a Lorentzian lineshape, but have theoretically derived an energy-dependent lifetime, arguing that the scattering rate out of a state is dependent on where the state is in the band and on how full the band is. At present there appears to be no consensus as to which lineshape most closely resembles reality.

## 4.2. Spontaneous Emission Rate

The expressions we have derived reveal that the downward and upward transition rates are proportional to the photon density in a given optical mode through $A_0^2$. A true quantum mechanical description of the optical field would reveal that this conclusion is not exactly correct [39]. The upward transition rate is still proportional to $\langle n_{ph} \rangle \hbar\omega$, where $\langle n_{ph} \rangle$ is the expected number of photons in a given optical mode; however, the downward transition rate is actually proportional to $(\langle n_{ph} \rangle + 1)\hbar\omega$ [39]. Surprisingly, with no photons in the mode, a finite downward transition rate still exists, implying that an electron in the conduction band will interact with an optical mode even if it does not contain any photons! This phenomenon is the origin of radiative spontaneous emission. And because it is not required that a given optical mode be occupied, the electron will interact with *all* optical modes in the energy range near $\hbar\omega \approx E_{eh}$.

The density of optical modes in a given energy range, $\rho_{opt}(\hbar\omega)$, can be found using the 3D $\rho(k)$ in Table I, and the general definition for $\rho(E)$ in Eq. (12). Using the $E$–$k$ relation for photons, we obtain

$$k_{opt} = \bar{n}\frac{\omega}{c} = \frac{\bar{n}}{\hbar c}\hbar\omega, \qquad \frac{dk_{opt}}{d(\hbar\omega)} = \frac{1}{\hbar c}\left[\bar{n} + \omega\frac{d\bar{n}}{d\omega}\right] \equiv \frac{\bar{n}'_g}{\hbar c}. \tag{40}$$

Using these in (12), we obtain the optical mode density (in units of energy$^{-1}$ cm$^{-3}$) as

$$\rho_{opt}(\hbar\omega) = 2\left[\frac{1}{2\pi^2}\left(\frac{\bar{n}(\hbar\omega)}{\hbar c}\right)^2 \frac{\bar{n}'_g}{\hbar c}\right] = \frac{1}{\pi^2}\frac{\bar{n}^2\bar{n}'_g}{(\hbar c)^3}(\hbar\omega)^2. \tag{41}$$

The preceding factor of two accounts for the fact that, for each plane wave, two polarization states exist. Note that the group index $\bar{n}'_g$ used here is the same as the definition given for $\bar{n}_g$ earlier if we set the effective index equal to the material index of refraction (compare (40) with (34)).

The total spontaneous emission rate per unit volume in a given energy range is found by multiplying the downward transition rate per unit volume given in (30) by the *number* of optical modes in the energy range, $d(\hbar\omega)$, or

$$R_{sp}(\hbar\omega)\, d(\hbar\omega) \equiv W_{c\to v}[V\rho_{opt}(\hbar\omega)\, d(\hbar\omega)]. \tag{42}$$

The vector potential $A_0$ in (22) is replaced by relating the energy stored in the wave to the energy of a single photon, or

$$\hbar\omega = \frac{1}{2}\bar{n}^2\varepsilon_0\omega^2 A_0^2 V \to A_0^2 = \frac{2\hbar\omega}{\bar{n}^2\varepsilon_0\omega^2 V}. \tag{43}$$

Substituting (43) into the expression for the transition rate given in (30), and using that in (42), we obtain (in units of $s^{-1}$ $cm^{-3}$ $energy^{-1}$)

$$R_{sp}(\hbar\omega) = \left(\frac{1}{\hbar\omega}\right)\frac{\pi e^2\hbar}{\bar{n}^2\varepsilon_0 m_0^2}|M_{ave}|^2\rho_{red}(E_{eh} - E_g')\rho_{opt}(\hbar\omega)\cdot f_c(1 - f_v), \quad (44)$$

$$|M_{ave}|^2 \equiv \frac{1}{3}\sum_{\substack{\text{all three}\\\text{polarizations}}}|M_T|^2. \quad (45)$$

In bulk material, the matrix element, $|M_T|^2$, has the same value for spontaneously emitted light polarized along $x$, $y$, or $z$, so that $|M_{ave}|^2 = |M_T|^2$. In quantum well structures, the matrix element is enhanced for certain electric field polarizations and reduced for others [4–15]. For total spontaneous emission, we are interested in the total output of light (or total radiative recombination of carriers), not which polarizations of light are being emitted spontaneously. Thus, we take the average strength of the transition matrix element over all three polarizations. We will discuss these issues more thoroughly in Section 5.

To obtain the spontaneous emission at a given photon energy, we again need to convolve $R_{sp}$ with the lineshape function over $E_{eh}$, as was done for the gain. However, we are usually interested in the *total* spontaneous emission output integrated over all photon energies. In this case, the convolution does not affect our final result. Thus, we can set $E_{eh} = \hbar\omega$ and integrate just once over the photon energy. The integrated spontaneous emission output is just equal to the total radiative recombination of carriers and is therefore equal to the radiative component of the volume current density required to achieve a given quasi-Fermi level separation. The relation can be written as

$$j_{rad} = e\int R_{sp}(\hbar\omega)\, d(\hbar\omega). \quad (46)$$

The current density is usually assigned units of $kA/(\mu m \cdot cm^2)$. In a laser, we would multiply by the active layer thickness (or quantum well width) to obtain the current density per unit area. Non-radiative current, such as Auger recombination current and leakage current, is also important in estimating threshold currents of semiconductor lasers, and these would have to be added to obtain the total current density. We will not consider these processes in the present chapter; however, discussion of these processes can be found elsewhere [51].

### 4.3. Carrier Density

Another important parameter is the carrier density required to achieve a given quasi-Fermi level separation. The carrier density in a given band can be found for a given quasi-Fermi level by integrating the density of states multiplied by the occupation probability over the entire band [36]. For parabolic bands, it is convenient to integrate over energy using $\rho(E)$ in Table I, so that we can write

$$N = 2\int_{E_c}^{\infty} \rho^{mD}(E - E_c) f_c \, dE, \qquad P = 2\int_{-\infty}^{E_v} \rho^{mD}(E_v - E)[1 - f_v] dE, \quad (47)$$

where $N$ and $P$ are the electron and hole carrier densities. The factor of two takes into account the spin degeneracy of each band, which we include here explicitly because of our definition of the density of states in Table I. For quantum-confined structures we must sum the above integrals over all subbands and, in the valence band, we must consider contributions from both the heavy- and light-hole bands.

For bulk material, the form of $\rho^{3D}(E)$ prevents us from obtaining a closed-form expression to the above integrals, and a numerical evaluation of Eq. (47) is required (the carrier densities usually dealt with are too high to be treated by analytical approximations). However, in a quantum well, $\rho^{2D}(E)$ is independent of energy, and the integral *can* be evaluated analytically. For parabolic subbands, such as in the conduction band of a quantum well, we can then write the electron concentration as

$$N = \frac{m_c k_B T}{\pi \hbar^2 L_z} \sum_n \ln\{1 + \exp[-(E_{cn} - E_{fc})/k_B T]\}. \quad (48)$$

The sum is over all quantized subbands within the conduction band of the quantum well, and the $E_{cn}$ are the quantized energy levels. The effective mass $m_c$ refers to the in-plane effective mass, to be defined later in Section 6.

For nonparabolic bands, for example, in the valence band of a quantum well, where the subband structure is far from parabolic (see Section 7), Eq. (48) is no longer valid because $\rho^{2D}(E)$ becomes a complicated function, including local maxima that can extend toward infinity. In addition, the band minimum is not always located at $k = 0$ (see Fig. 16), creating additional complications. Thus, for this case, it is more appropriate to find the carrier density by numerically integrating over $k$ space, because $k$ states are always uniformly distributed in $k$ space (see Section 2.2), independent of how complicated the energy bands become. For this case, we have

$$P = 2\sum_n \int_0^{k_{max}} \rho^{2D}(k_{xy})\{1 - f_v[E_{vn}(k_{xy})]\} \, dk_{xy}. \quad (49)$$

The sum is again over all quantized subbands $E_{vn}(k_{xy})$ within the valence band of the quantum well, including both heavy- and light-hole bands. Also, $k_{xy}$ is the **k** vector component in the plane of the well as discussed in Section 2.2, and $k_{max}$ represents some numerical limit beyond which the contribution to the integral can be neglected. A method for calculating the $E_{vn}(k_{xy})$ will be presented in Section 7.

With the relation between carrier density and the quasi-Fermi level in each band known, we can find the quasi-Fermi level separation, $E_{fc} - E_{fv}$, by requiring that our choice of $E_{fc}$ and $E_{fv}$ maintain charge neutrality in the material [37] or that $N + N_A^- = P + N_D^+$ (where $N_D^+$ and $N_A^-$ are the ionized donor and acceptor concentrations in the material).

To be exact in our quantum well analysis, we should also include the unbound states that have energies higher than the well barriers. This point has been discussed in detail by Nagarajan *et al.* [52]. Contributions from the $X$ and $L$ indirect valleys in the conduction band have also been considered by Chinn *et al.* [48]. For relatively low carrier injection, these contributions are small. However, at a high carrier injection, for example, when generating high gain in the second energy level of a thin quantum well [93], these contributions should be considered.

## 5. BAND STRUCTURE BASICS

### 5.1. Valence Band Bloch Functions

We now have relations for adequately describing gain and spontaneous emission in bulk and quantum well material. However, the transition matrix element $|M_T|^2$ is still an unknown parameter. To properly evaluate $|M_T|^2$, it will be important to have a good understanding of the Bloch functions, especially those in the valence band.

The conduction and valence bands are illustrated in Fig. 7. The three valence bands are commonly known as the heavy-hole (HH), light-hole (LH), and split-off hole (SO) bands [16]. We can view each energy band as originating from the discrete atomic energy levels of the isolated atoms that compose the crystal. In this sense, the conduction band can be thought of as a remnant of an $s$ atomic orbital, while the three valence bands are remnants of the three $p$ atomic orbitals: $p_x$, $p_y$, and $p_z$. The corresponding Bloch functions for these orbitals are denoted here as $u_s$ and $u_x$, $u_y$, and $u_z$ [2, 3, 53]. It is very useful to make this correspondence, because the Bloch functions retain many of the symmetries that the atomic orbitals possess [53].

For instance, the conduction band Bloch function $u_s$ has even symmetry in all three directions within each unit cell, similar to the spherically

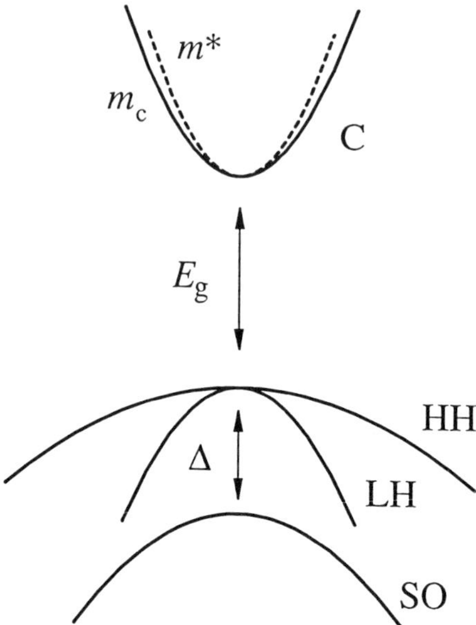


**Fig. 7.** Typical III–V semiconductor band structure schematic illustrating the conduction (C), heavy hole (HH), light hole (LH), and split-off (SO) bands and their relative energy separations. The true (solid) and approximate (dashed) C bands are shown with their corresponding effective masses, $m_c$ and $m^*$. (The relative scale of the four bands corresponds roughly to GaAs.)

symmetric $s$ atomic orbital [54]. In a similar manner, $u_z$ has odd symmetry along $z$ but even symmetry in the other two directions, within each unit cell, similar to the $p_z$ atomic orbital [54]. From these two facts alone, we can state that the net odd symmetry along $z$ must give $\langle u_s|u_z\rangle = 0$ (where the brackets indicate integration over a unit cell). However, operating on $u_z$ with the momentum operator $p_z$ inverts the symmetry along $z$ and the momentum matrix element $\langle u_s|p_z|u_z\rangle$ is, therefore, in general nonzero. From these and similar arguments, we can immediately write down some useful symmetry relations and definitions [53]:

$$\langle u_s|p_i|u_j\rangle = 0, \qquad \text{for } i \neq j; \tag{50a}$$

$$\langle u_s|\mathbf{p}|u_i\rangle = \langle u_s|p_i|u_i\rangle \equiv M; \tag{50b}$$

$$\langle u_s|\mathbf{p}|\bar{u}_i\rangle = 0; \tag{50c}$$

where $i = x, y, z$, and $u_i$, $\bar{u}_i$ indicate spin-up and spin-down functions. The third relation comes from the fact that Bloch functions of opposite spin do not interact [3, 53]. The constant $M$ is defined here as the basis function momentum matrix element. Thus, through simple symmetry arguments, the various momentum matrix elements between $u_s$ and the three $u_i$ can now all be related simply to one constant, $M$.

The valence band Bloch functions $u_{\mathrm{hh}}$, $u_{\mathrm{lh}}$, and $u_{\mathrm{so}}$ corresponding to the bands of Fig. 7 are usually written as linear combinations of the "basis" functions $u_x$, $u_y$, and $u_z$ [2, 3, 16] (in a manner analogous to the construction of hybrid orbitals in the study of molecular bonds). Spin degeneracy exists in all three bands, so we actually need to define six Bloch functions. Defining the electron's **k** vector to be directed along $z$, the valence band Bloch functions can be written as

$$\begin{aligned} u_{\mathrm{hh}} &= -\frac{1}{\sqrt{2}}(u_x + \mathrm{i}u_y), & \bar{u}_{\mathrm{hh}} &= \frac{1}{\sqrt{2}}(\bar{u}_x - \mathrm{i}\bar{u}_y), \\ u_{\mathrm{lh}} &= -\frac{1}{\sqrt{6}}(\bar{u}_x + \mathrm{i}\bar{u}_y - 2u_z), & \bar{u}_{\mathrm{lh}} &= \frac{1}{\sqrt{6}}(u_x - \mathrm{i}u_y + 2\bar{u}_z), \\ u_{\mathrm{so}} &= -\frac{1}{\sqrt{3}}(\bar{u}_x + \mathrm{i}\bar{u}_y + u_z), & \bar{u}_{\mathrm{so}} &= \frac{1}{\sqrt{3}}(u_x - \mathrm{i}u_y - \bar{u}_z). \end{aligned} \tag{51}$$

The prefactors are normalization constants that can have arbitrary phase (the phases chosen here are those of Broido and Sham [23]). For **k** directed along another direction, we would have to redefine these relations [3] (however, cyclic permutation of $x$, $y$, and $z$ do yield equivalent relations if we also include the direction of **k** in the permutation).

The preceding linear combinations of basis functions are known as the angular momentum representation [2, 3, 16]. They are useful when we consider the spin–orbit interaction between the angular momentum of the $p$ orbitals and the spin angular momentum of the electron (the spin–orbit interaction term is "diagonalized" in the above representation [2]). [*Note*: The SO band would be degenerate with the HH and LH bands if the spin–orbit interaction did not exist. As it is, the spin–orbit interaction partially removes the degeneracy [2, 3, 16], suppressing the SO band from the other two by a spin–orbit splitting energy $\Delta$, which in GaAs is $\sim$0.34 eV [38].]

With the above description of the valence bands, we can obtain a more complete description of the transition matrix element. However, the magnitude of the basis function momentum matrix element is still an unknown. Therefore, we will close this subsection with an example use of the above relations to obtain an approximate expression for the magnitude of $|M|^2$.

Evaluation of $|M|^2$ in bulk material was first obtained by theoretically relating it to the *curvature* of the conduction band [3, 37, 55, 56]. Using a second order perturbation technique known as the $k \cdot p$ method [2, 3, 36, 53], we can express the conduction band effective mass along the electron's **k** vector direction (which, for the definitions of Eqs. (51), is the $z$ direction) as

$$\frac{1}{m_{cz}} = \frac{1}{m_0}\left[1 + \sum_{n \neq c} \frac{2}{m_0} \frac{|\langle u_c|p_z|u_n\rangle|^2}{E_c - E_n}\right], \tag{52}$$

where the summation sums over all $n$ energy bands (not to be mistaken with energy subbands in quantum wells) of the crystal, and the $E_n$ and $E_c$ are the band edge energies of each band. From (52), we see that the deviation of the conduction band effective mass from the free electron mass arises from the interaction between the conduction band and all other energy bands in the crystal.

It is interesting to note that, due to the sign of the denominator, contributions from higher energy bands make the effective mass *heavier* and tend to flatten out the conduction band, while contributions from lower lying energy bands tend to decrease the effective mass, increasing the curvature of the conduction band. In either case, we find that the effect of a given band is to *repel* the conduction band away from it.

Also because of the denominator, only energy bands close in energy to the conduction band will contribute significantly to the summation. By neglecting all but the three valence bands in the summation, we can obtain an approximate closed form expression for the conduction band effective mass using the relations given in (50) and (51). Note that the HH Bloch function does not contain $u_z$, and hence its contribution to the sum is zero. Thus, summing over the LH and SO bands and using the energy separations defined in Fig. 7, we obtain

$$\frac{1}{m^*} = \frac{1}{m_0}\left\{1 + \frac{2|M|^2}{m_0}\left[\frac{2}{3}\frac{1}{E_g} + \frac{1}{3}\frac{1}{(E_g + \Delta)}\right]\right\}. \tag{53}$$

The approximate conduction band effective mass $m^*$ is expected to be lighter than the true effective mass, since the effects of any higher energy bands have been neglected in our approximation. The approximate conduction band (with curvature related to $m^*$) and the true conduction band are both illustrated in Fig. 7.

The true conduction band effective mass can be measured experimentally to a good degree of precision using cyclotron resonance techniques [38]. Thus, by assuming that $m^*$ is close to the true effective mass $m_c$, we can rearrange (53) to obtain

$$|M|^2 = \left(\frac{m_0}{m^*} - 1\right)\frac{(E_g + \Delta)}{2(E_g + \frac{2}{3}\Delta)} m_0 E_g. \tag{54}$$

So we see that the simple description of the valence bands given in (50) and (51) has led directly to a formula that can yield a rough estimate of $|M|^2$ [37]. And while the above relation is not exact, it does reveal that $|M|^2$ is roughly proportional to the *ratio* of the energy gap to the conduction band effective mass of the semiconductor [5].

We have derived Eq. (54) as an exercise in using the valence band Bloch functions. It is a useful formula for materials that have not been fully characterized. In more common materials such as GaAs, however, much more accurate methods of determining $|M|^2$ exist [57–61]. The inaccuracy of (54) stems from the fact that the contribution from higher lying energy bands can be significant, and thus $m^*$ is not always a very good approximation to the true effective mass $m_c$. For example, in GaAs, $m^*$ is approximately equal to $0.053m_0$, compared with the true effective mass $0.067m_0$ (listed in Table III). Thus, using the true effective mass in (54) leads to an *underestimation* of the matrix element by about 26%. Many previous calculations have failed to recognize this correction (see Yan *et al.* [62] for a discussion of this), implying that both calculated spontaneous emission (and hence, calculated radiative current density) and optical gain will be underestimated by 26%—a significant factor. The most accurate estimates of $|M|^2$ have actually been obtained using electron spin resonance techniques [57–61]. In Table II, we have tabulated the most accurately reported values of $|M|^2$ in several material systems commonly used in semiconductor laser applications. (Note that $2|M|^2/m_0$ has units of energy.)

**Table II.**

Magnitude of $|M|^2$ for Various Material Systems

| Material system | $\frac{2\|M\|^2}{m_0}$ (in eV) | Reference |
|---|---|---|
| GaAs | $28.8 \pm 0.15$ | 58, 59 |
| $Al_xGa_{1-x}As$ ($x < 0.3$) | $29.83 + 2.85x$ | 60 |
| $In_xGa_{1-x}As$ | $28.8 - 6.6x$ | 58, 59 |
| InP | $19.7 \pm 0.6$ | 58, 59 |
| $In_{1-x}Ga_xAs_yP_{1-y}$ ($x = 0.47y$) | $19.7 + 5.6y$ | 59, 61 |

## 5.2. Transition Matrix Element

With the magnitude of $|M|^2$ known from experiment, we are now in a position to quantify the transition matrix element $|M_{\mathrm{T}}|^2$ defined in Eq. (22). The difference between the two matrix elements is that $|M|^2$ determines the transition probability between $u_{\mathrm{s}}$ and the basis functions ($u_x$, $u_y$, $u_z$, or collectively $u_i$), whereas $|M_{\mathrm{T}}|^2$ determines the transition probability between $u_{\mathrm{c}}$ $(=u_s)$ and the valence band Bloch functions ($u_{\mathrm{hh}}$, $u_{\mathrm{lh}}$, $u_{\mathrm{so}}$, or collectively $u_{\mathrm{v}}$). By expanding the $u_{\mathrm{v}}$ in terms of the $u_i$ using Eq. (51), we can express $|M_{\mathrm{T}}|^2$ in terms of $|M|^2$. Before we do this, however, we need to discuss spin degeneracy and how to include it here.

Earlier, in Section 3.2, we did not include the spin degeneracy (a factor of two) in the definition of our reduced density of states function. We intentionally delayed the consideration of spin degeneracy until now to stress that there are subtleties involved, which are often overlooked when we simply include a factor of two in our equations. For example, to include the spin degeneracy in our evaluation of $|M_{\mathrm{T}}|^2$, we must obviously sum over both $u_{\mathrm{c}} \to u_{\mathrm{v}}$ and $\bar{u}_{\mathrm{c}} \to \bar{u}_{\mathrm{v}}$ transitions. However, what is not so obvious is that in our sum we must also include $\bar{u}_{\mathrm{c}} \to u_{\mathrm{v}}$ and $u_{\mathrm{c}} \to \bar{u}_{\mathrm{v}}$ transitions! This is necessary because the LH and SO valence band Bloch functions are made up of both spin-up *and* spin-down basis functions, as seen from their definitions in Eq. (51). Therefore, a total of *four* transitions must be considered, as shown in Fig. 8. Summing over these, Eq. (22) becomes

$$|M_{\mathrm{T}}|_{\mathrm{v}}^2 = \sum_{u_{\mathrm{c}}, \bar{u}_{\mathrm{c}}} \sum_{u_{\mathrm{v}}, \bar{u}_{\mathrm{v}}} |\langle u_{\mathrm{v}} | \hat{\mathbf{e}} \cdot \mathbf{p} | u_{\mathrm{c}} \rangle|^2. \tag{55}$$

We have set the envelope function overlap integral $|\langle F_{\mathrm{h}} | F_{\mathrm{e}} \rangle|^2$ equal to unity, because for the moment we will be interested in transitions between two bulk plane wave electron states. In Section 7 we will return to a more general form of Eq. (55), which does include the envelope function overlap integrals.

To simplify Eq. (55), we can first of all replace the dot product between the unit polarization vector and the electron momentum operator, $\hat{\mathbf{e}} \cdot \mathbf{p}$, with the expansion, $e_x p_x + e_y p_y + e_z p_z$. Then, by using the selection rules given in Eqs. (50a–c), in combination with the expansions of the valence band Bloch functions given in Eq. (51), we can reduce the expression for $|M_{\mathrm{T}}|^2$ to a very simple form. To aid the reader in following the derivation, we give here the intermediate step in simplifying $|M_{\mathrm{T}}|$ for the three valence band transitions:

$$\begin{aligned}
|M_{\mathrm{T}}|_{\mathrm{hh}}^2 &= \tfrac{1}{2}|M|^2[|-e_x - ie_y|^2 + 0 + 0 + |e_x - ie_y|^2], \\
|M_{\mathrm{T}}|_{\mathrm{lh}}^2 &= \tfrac{1}{6}|M|^2[|2e_z|^2 + |e_x - ie_y|^2 + |-e_x - ie_y|^2 + |2e_z|^2], \\
|M_{\mathrm{T}}|_{\mathrm{so}}^2 &= \tfrac{1}{3}|M|^2[|-e_z|^2 + |e_x - ie_y|^2 + |-e_x - ie_y|^2 + |-e_z|^2].
\end{aligned} \tag{56}$$

Each of the terms within brackets corresponds to one of the four spin-degenerate transitions (the ordering of terms from left to right corresponds to the numbering shown in Fig. 8). Note that in every case the first term is equal to the fourth, and the second term is equal to the third (leading to the standard factor of two for spin degeneracy).

To make the final expression as general as possible, we first replace every occurrence of $e_x^2 + e_y^2$ with the equivalent expression $1 - e_z^2$ (since $\hat{\mathbf{e}}$ is a unit vector). This substitution places everything in terms of $e_z$. We can then interpret $e_z$ as the component of $\hat{\mathbf{e}}$ that is parallel to the electron $\mathbf{k}$ vector, since $\mathbf{k}$ is directed along $z$ (an assumption made in defining (51)). In other words, we can set $e_z = \hat{\mathbf{k}} \cdot \hat{\mathbf{e}}$, where $\hat{\mathbf{k}}$ is a unit vector directed along $\mathbf{k}$. Using these substitutions, we find that

$$|M_{\mathrm{T}}|_{\mathrm{v}}^2/|M|^2 = \begin{cases} e_x^2 + e_y^2 = 1 - |\hat{\mathbf{k}} \cdot \hat{\mathbf{e}}|^2 & \text{for HH band,} \quad (57\text{a}) \\ \frac{1}{3}(e_x^2 + e_y^2 + 4e_z^2) = \frac{1}{3} + |\hat{\mathbf{k}} \cdot \hat{\mathbf{e}}|^2 & \text{for LH band,} \quad (57\text{b}) \\ \frac{2}{3}(e_x^2 + e_y^2 + e_z^2) = \frac{2}{3} & \text{for SO band.} \quad (57\text{c}) \end{cases}$$

The *relative* transition strengths given in Eqs. (57a–c) allow us to relate the transition matrix element $|M_{\mathrm{T}}|^2$ needed in our gain calculations to the experimentally measurable matrix element $|M|^2$. Note that the use of a dot product has allowed us to drop any reference to a coordinate system, and hence drop the constraint that the electron $\mathbf{k}$ vector be directed along $z$ (in other words, the physics does not lie in the coordinate system we choose, but in the relative orientation between the field polarization $\hat{\mathbf{e}}$ and the electron $\mathbf{k}$ vector).

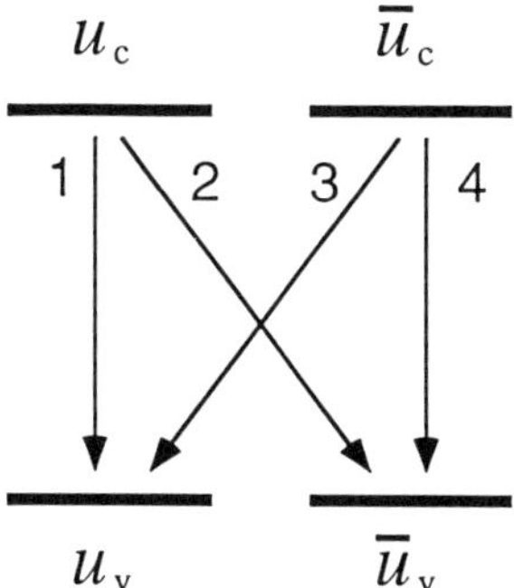


**Fig. 8.** Diagram illustrating the four possible transitions between the spin-degenerate C and V bands, which must be considered when estimating the transition matrix element.

To examine the dependence of Eq. (57) on the field polarization in a more visual fashion, we have plotted the relative transition strengths for C–HH and C–LH transitions in Fig. 9 as a function of the angle between the electron **k** vector and the electric field polarization, $\hat{\mathbf{e}}$. These three dimensional renderings reveal that the strength of interaction between each electron plane wave state and photon is highly polarization-dependent. However, the striking features of Fig. 9 do not reveal themselves in bulk material because photons of a given polarization interact with a great number of electrons, all with **k** vectors pointing in different directions. The average over all these interactions transforms the interesting shapes in Fig. 9 into uniform spheres [15]. In fact, the average of the dot product $|\hat{\mathbf{k}} \cdot \hat{\mathbf{e}}|^2$ for $\hat{\mathbf{k}}$ sweeping over all three dimensions is equal to 1/3. Thus, for all three valence band transitions, the bulk material transition matrix element is just equal to $2/3 \times |M|^2$ (spin included) for any electric field polarization.

### 5.3. Polarization-Dependent Effects

The derivation presented in Section 5.2 assumed plane wave states for the envelope functions, which then led to the polarization dependence illustrated in Fig. 9. In quantum-confined structures, the envelope functions are typically constructed from two (or more) plane wave states. The magnitude squared of the transition matrix element in (55) will, in general, then contain cross-terms between the various plane waves that make up the confined state. In the following discussion we will ignore these cross-terms, making the analysis simpler. In Section 9, we will see that, in quantum wells, the conclusions derived in the present section are consistent with the band-mixing model.

Neglect of the cross-terms in (55) implies that we can treat the plane waves that make up the confined states as independent from each other. In this approximation, each plane wave's **k** vector direction will then have a corresponding polarization dependence similar to that derived in the previous subsection and shown in Fig. 9. Near the band edge in quantum-confined structures, the **k** vectors are quantized along certain directions, and the situation will be as shown in Fig. 10 for a typical quantum well and quantum wire. In the quantum well, all **k** vectors point along the same axis, and the polarization dependence is simply proportional to Fig. 9. However, in the quantum wire case, we must average over the polarization dependence

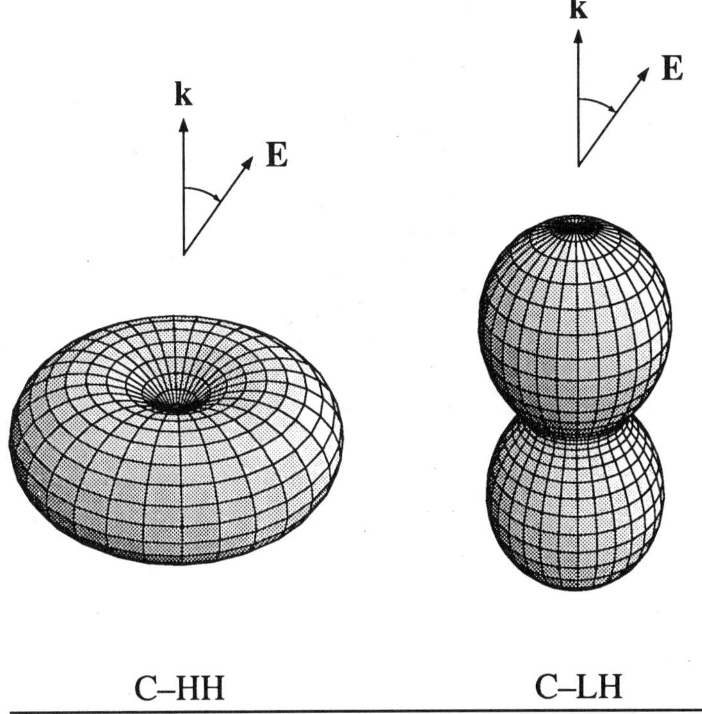


**Fig. 9.** Dependence of the transition strength $|M_T|^2$ on angle between the electron's **k** vector and the incident electric field vector **E** for C–HH and C–LH transitions (C–SO) transitions are independent of angle). For C–HH transitions, $|M_T|^2$ is zero when $\mathbf{E} \parallel \mathbf{k}$ and becomes a maximum of $1 \times |M|^2$ when $\mathbf{E} \perp \mathbf{k}$. For C–LH transitions, when $\mathbf{E} \parallel \mathbf{k}$, $|M_T|^2$ has a peak value of $(4/3) \times |M|^2$ and is reduced to $(1/3) \times |M|^2$ when $\mathbf{E} \perp \mathbf{k}$.

of each plane wave (as indicated by the dashed curve in the lower right side of Fig. 10). To quantify the *average* polarization dependence, we choose our coordinate system along the confinement axis (axes) of the structure. It is then possible to evaluate the average transition strength along the three orthogonal field polarizations $\hat{\mathbf{e}}_x$, $\hat{\mathbf{e}}_y$, and $\hat{\mathbf{e}}_z$, simply by replacing $\hat{\mathbf{k}}$ in Eqs. (57) with some appropriate *average* **k** vector direction $\hat{\mathbf{k}}_{\text{ave}}$.

We leave it to the reader to justify that $\hat{\mathbf{k}}_{\text{ave}}$ is obtained simply by finding the average direction of all allowed **k** vectors within the *first octant* of our coordinate system (if we included all octants in our average, $\hat{\mathbf{k}}_{\text{ave}}$ would always be zero!). In the following, we list $\hat{\mathbf{k}}_{\text{ave}}$ for bulk and various quantum-confined structures for band edge states (where *band edge* implies that the

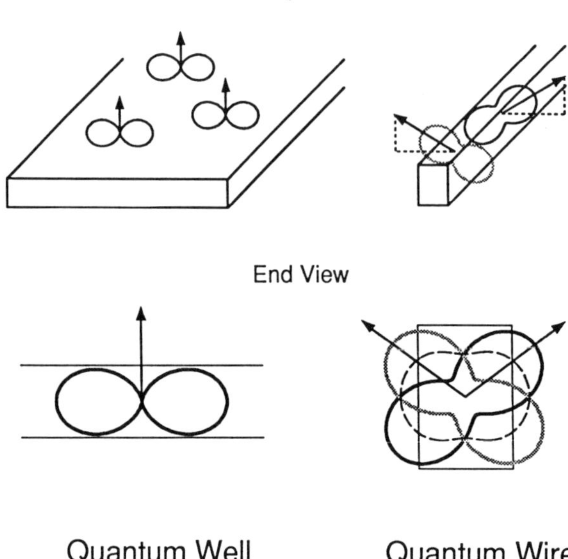


**Fig. 10.** Illustration of how quantum confinement in a quantum well and a quantum wire serves to "polarize" the momentum of band edge electrons along certain directions. The C–HH transition strength is superimposed on each electron's **k** vector in the quantum well case. The end view of the quantum well suggests that C–HH interactions with light are strongest when the light is polarized in the plane of the well. In the quantum wire case, the C–LH transition strength is used. The end view shows two possible **k** vector directions for electrons. The average C–LH transition strength (indicated by the dashed curve) reveals that the polarization dependence in the plane perpendicular to the wire axis is fairly weak (if the well widths in both directions were equal, the average C–LH transition strength would be a circle, and no polarization dependence would be observable).

total **k** vectors are simply equal to the quantized **k** vectors):

$$\hat{\mathbf{k}}_{\text{ave}} = \frac{1}{\sqrt{3}}(\hat{\mathbf{k}}_x + \hat{\mathbf{k}}_y + \hat{\mathbf{k}}_z), \qquad \text{Bulk;} \tag{58a}$$

$$\hat{\mathbf{k}}_{\text{ave}} = \hat{\mathbf{k}}_z, \qquad \text{Quantum well } (L_z); \tag{58b}$$

$$\hat{\mathbf{k}}_{\text{ave}} = \frac{1}{\sqrt{2}}(\hat{\mathbf{k}}_x + \hat{\mathbf{k}}_z), \qquad \text{Quantum wire } (L_x = L_z); \tag{58c}$$

$$\hat{\mathbf{k}}_{\text{ave}} = \frac{1}{\sqrt{5}}(\hat{\mathbf{k}}_x + 2\hat{\mathbf{k}}_z), \qquad \text{Quantum wire } (L_x = 2L_z). \tag{58d}$$

The prefactors are normalization constants since $\hat{\mathbf{k}}_{\text{ave}}$ is a unit vector. Equation (58d) was obtained assuming $k \propto 1/L$ (which is exactly true only for an infinitely deep well).

Figure 11 illustrates the band edge transition strengths for the three orthogonal polarizations in the four structures listed here, obtained by substituting the $\hat{\mathbf{k}}_{\text{ave}}$ defined for each structure into (57) [15]. When multiplied by $|M|^2$, the numbers in Fig. 11 give the magnitude of the transition matrix element $|M_{\text{T}}|^2$ for each particular case (spin included). Note that the sum of the transition strengths over the three polarizations for each type of transition is always equal to $2 \times |M|^2$, as is the case in bulk material (the factor of two arising from the spin degeneracy). Thus, the "total" transition

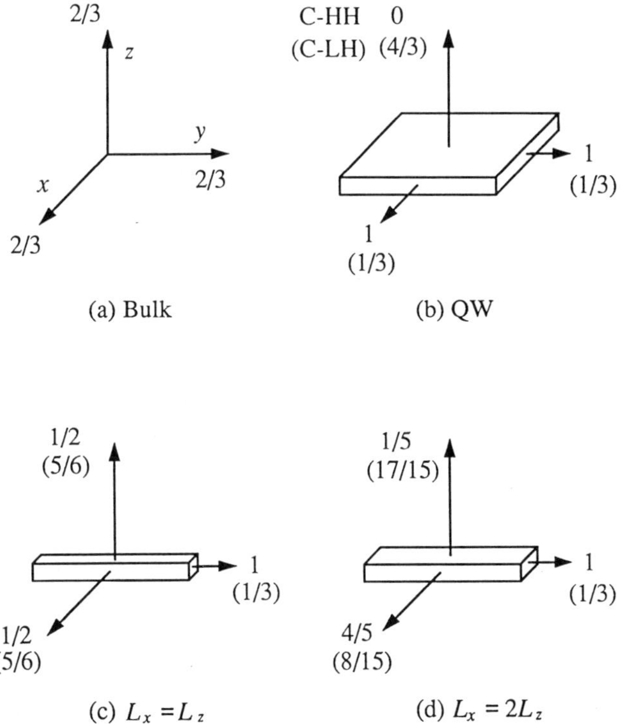


**Fig. 11.** Relative band edge transition strengths for various quantum confinement structures. The coordinates referred to in the text are indicated in (a). The magnitude of the transition matrix element is found by multiplying the relative numbers by $|M|^2$. For example, with light polarized along the wire direction of (c), the band edge C–LH transition strength $|M_{\text{T}}|^2 = (1/3) \times |M|^2$.

strength for band edge transitions is always conserved. The difference is that in quantum-confined structures a *redistribution* of the transition strength among the three polarizations occurs due to the nonuniform distribution of **k** vector directions.

To treat arbitrary polarizations, we can always break the field up into the three orthogonal components shown in Fig. 11. The transition matrix element is then simply given by the trigonometric sum of the three components, or

$$|M_T|_v^2 = |M|^2 \sum_i e_i^2 S_i^v, \qquad i = x, y, z, \tag{59}$$

where the $S_i^v$ are the transition strengths determined from Fig. 11. As an example of the use of (59), we examine the polarization-dependent characteristics of a quantum wire structure. From Fig. 11c, the *ratio* of the transition matrix element between the C–LH and C–HH transitions is 1/3 when $\hat{\mathbf{e}} \parallel$ wire and 5/3 when $\hat{\mathbf{e}} \perp$ wire. In general, from Fig. 11c and (59), we have

$$\frac{|M_T|_{LH}^2}{|M_T|_{HH}^2} = \frac{\frac{1}{3}\cos^2\theta + \frac{5}{6}\sin^2\theta}{\cos^2\theta + \frac{1}{2}\sin^2\theta}, \tag{60}$$

where $\theta$ is the angle between $\hat{\mathbf{e}}$ and the axis of the wire. Figure 12 compares the polarization dependence of the LH-to-HH matrix element ratio given by Eq. (60) to that calculated using a sophisticated finite element method (FEM) developed by Yi, *et al.* [63] which, for this example, numerically computes the envelope functions and transition matrix elements in a two dimensional potential well created by a 100 Å × 100 Å square GaAs/AlAs quantum wire (the solutions are obtained by numerically solving the valence band Hamiltonian to be introduced later in Section 7). As can be seen, the simplified treatment of transition strengths in quantized structures presented above is capable of explaining the basic features of the polarization dependence found using the more sophisticated approach.

The polarization dependence of a quantum well structure is also shown in Fig. 12. Because the electric field remains in the plane of the quantum well for all angles, no polarization dependence of the LH-to-HH matrix element ratio is expected for this case. Therefore, by experimentally measuring the absorption characteristics of a given sample as a function of input electric field polarization, we should be able to clearly distinguish between a quantum well and a quantum wire structure. This type of experiment has in fact been used to verify the existence of two dimensional confinement in tilted superlattice structures grown by molecular beam epitaxy [13, 14].

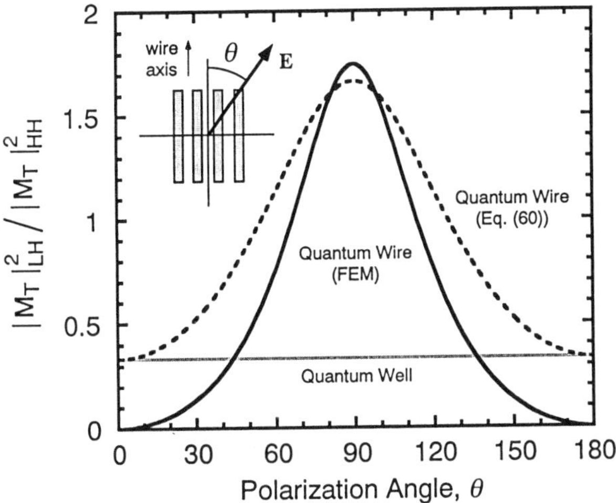


**Fig. 12.** Theoretical LH-to-HH transition matrix element ratio in a quantum wire and quantum well, as a function of the polarization angle defined in the inset. The dashed curve represents Eq. (60) in the text which estimates the ratio for a square quantum wire. The solid curve represents more exact calculations performed using a finite element method on a 100 Å × 100 Å square GaAs/AlAs quantum wire. The gray solid curve represents the ratio in a quantum well and is flat because the electric field remains in the plane of the well for all angles defined in the plot.

In the above discussion, we have only considered electron states that are near the band edge, where the **k** vectors consist of only quantized components. What about electron states further up in the band? As the energy of the electrons increase above the band edge, the **k** vectors begin to tilt away from the confinement axis (axes), and in general, we should expect the transition strengths in Fig. 11 to change [4, 5, 15]. However, significant mixing of the valence bands also begins to occur at energies away from the band edge in quantum wells, and we must unfortunately resort to evaluating (55) numerically. We will return to this point in Section 7, where band-mixing effects in quantum wells will be treated in detail. In quantum wires, the situation is even worse, because, the two dimensional confinement actually mixes the valence bands at the band edge [63], implying that the above treatment is really not valid, even at the band edge. However, the extent of band mixing at the band edge is expected to be small, and the above analysis should hold true approximately as the comparison in Fig. 12 reveals.

# 6. CONDUCTION BAND IN QUANTUM WELLS

## 6.1. Nondegenerate Effective Mass Equation

With the description of the transition matrix element given in Section 5, we now need to concentrate on determining how the band structure of the semiconductor is modified by the quantum well potential. We will also need to determine the subband gaps $E_g'$ (and hence the energy levels) of each subband transition pair, and the corresponding envelope functions $F$.

In our analysis, we will make the simplifying assumption that the conduction band is parabolic for all energies. In reality, nonparabolicity does exist at higher energies in the conduction band of bulk GaAs [38] (and most other III–V materials). In addition, this nonparabolicity is expected to be modified by quantum confinement. However, a full treatment is beyond the intent of the present chapter. We choose instead to use the simplified conduction band treatment as a guide to treating the more complex valence band structure in Section 7, where we will find that the method of solution closely parallels the conduction band treatment outlined in what follows.

For a nondegenerate energy band (excluding spin), such as the conduction band, it has been shown most notably by Luttinger and Kohn [16] (using a $k \cdot p$ formalism) that an "effective mass equation" or Schroedinger-like equation for the envelope function $F_e$, as defined in (3), can be derived. It can be given as

$$(H_c + V)F_e = E_c F_e, \tag{61}$$

where the Hamiltonian for the conduction band is simply given by

$$H_c = -\frac{\hbar^2}{2m_c}\nabla^2 \rightarrow \frac{\hbar^2}{2m_c}(k_x^2 + k_y^2 + k_z^2). \tag{62}$$

The arrow indicates the form of the Hamiltonian for plane wave solutions (when $V$ is constant). The potential $V$ now corresponds to the variation in the material band edge, and the total energy of the electron, $E_c$, is measured relative to the bottom of the conduction band. What is extremely appealing about the effective mass equation as compared with Eq. (1) is that the Bloch functions have been removed from the equation, and the effect of the periodic potential arising from the crystal lattice (and, hence, the coupling to other energy bands) is now replaced by a conduction band "effective" mass, $m_c$. In this approximation, the quantum well created by the interfacing of three materials of different bandgap truly becomes a textbook "particle in a box" problem with $F$ as the wavefunction, and the material band edges as the potential $V$.

### 6.2. General Solutions

For a one dimensional potential variation along $z$ such that $V = V(z)$, as in Fig. 13, we can assume the envelope function solutions to be of the form

$$F_{\mathrm{e}} = F_z(z)\exp[i(k_x x + k_y y)]. \tag{63}$$

Using this in (61), we obtain

$$-\frac{\hbar^2}{2m_i}\frac{d^2F_z}{dz^2} + (V_i + V_{k,i})F_z = E_{\mathrm{c}}F_z, \qquad \text{where } V_{k,i} = \frac{\hbar^2 k_{xy}^2}{2m_i}. \tag{64}$$

Here, the index $i = \mathrm{w}, \mathrm{b}$, standing for the well and barrier regions, respectively. For example, $V_{\mathrm{w}} = 0$ and $V_{\mathrm{b}} = V_0$, as shown in Fig. 13. We have also introduced an in-plane **k** vector such that $k_{xy}^2 \equiv k_x^2 + k_y^2$. The energy term $V_{k,i}$ is the contribution to the total energy of the electron from its momentum within the plane of the well. It is written here as a *potential* energy for reasons that will become clear later on. For now, we will assume that the electron is at rest and has no in-plane momentum, so that we can set $V_{k,i} = 0$.

Our first step in solving the quantum well problem will be to determine the general "bulk" solutions within the well and barrier regions. In the well region, where $V_{\mathrm{w}} = 0$, Eq. (64) yields the following general solutions:

$$F_{z,w} = A\exp(ik_z z) + B\exp(-ik_z z), \qquad k_z^2 = \frac{2m_{\mathrm{w}}}{\hbar^2}E_{\mathrm{c}}. \tag{65}$$

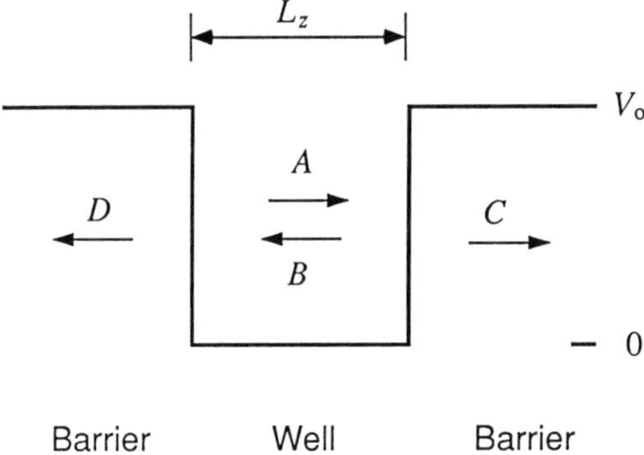


**Fig. 13.** Schematic of a typical quantum well potential with barrier height $V_0$ in the conduction band. We can define three distinct regions, each containing two counterpropagating plane wave solutions. In the barrier regions, the exponentially increasing solutions can be thrown away, and we are left with a total four plane waves. The directions and corresponding arbitrary coefficients $A, B, C, D$ of each wave are indicated in the figure.

For the barrier regions on either side of the well, $V_b = V_0$, and the general solutions become

$$\begin{aligned} F_{z,b+} &= C_1 \exp(\alpha_z z) + C_2 \exp(-\alpha_z z), \\ F_{z,b-} &= D_1 \exp(\alpha_z z) + D_2 \exp(-\alpha_z z), \end{aligned} \tag{66}$$

where $\alpha_z^2 = (2m_b/\hbar^2)(V_0 - E_c)$. To solve the quantum well problem, we need to determine the six unknown coefficients $A$, $B$, $C_i$, and $D_i$. By requiring the envelope functions to go to zero at infinity, we can set the exponentially growing solution in each barrier region equal to zero (i.e., $C_1 = D_2 = 0$), leaving us with four undetermined coefficients, as indicated in Fig. 13. These can be uniquely determined by matching the following two quantities across each well–barrier interface:

$$F_z \quad \text{and} \quad \frac{1}{m_i}\frac{dF_z}{dz}. \tag{67}$$

The appearance of the effective mass in the slope continuity boundary condition can be derived by integrating the effective mass equation (64) across the well–barrier interface, as discussed by Schuurmans and 't Hooft [64], and Bastard [65]. For the derivation to work properly, we must modify or "symmetrize" the Hamiltonian operator [66] so that $m^{-1}(z)\nabla^2 \rightarrow \nabla m^{-1}(z)\nabla$ (without a modification of this sort, the Hamiltonian would not be Hermitian for a spatially varying effective mass, as one can easily verify [1]. [*Note*: Some workers [21, 22, 27, 65, 67] have argued that the preceding boundary conditions arise from considerations of conservation of probability current density across the interface. However, as pointed out by Schuurmans and 't Hooft [64], the probability current density of real wavefunctions (as is the case in our quantum well problem) will always be zero on both sides of the interface regardless of how we choose to match the envelope functions.]

For a symmetric quantum well as in Fig. 13, symmetry requires that $B = \pm A$ and $D = \pm C$. Applying the boundary conditions in (67) at either one of the well–barrier interfaces, we can obtain the characteristic equation for energy for the even ($B = A$, $D = C$) and odd ($B = -A$, $D = -C$) solutions. They are given by

$$\tan\left(k_z \frac{L_z}{2}\right) = \frac{m_w}{m_b}\frac{\alpha_z}{k_z} \quad \text{(for even solutions)}, \tag{68a}$$

$$\cot\left(k_z \frac{L_z}{2}\right) = -\frac{m_w}{m_b}\frac{\alpha_z}{k_z} \quad \text{(for odd solutions)}, \tag{68b}$$

where $L_z$ is the quantum well width. Equations (68a, b) can only be satisfied for a discrete set of energies $E_{cn}$. The set of quantized energy levels $E_{cn}$ and corresponding envelope functions $F_{zn}$ represent the bound solutions of the quantum well [1].

In GaAs quantum wells, typically only two or three bound solutions exist in the conduction band (depending on the well width and barrier height). In Fig. 14a, we have plotted the lowest even and odd solutions of (65) and (66) using the energies found from Eqs. (68a, b). The index $n$ corresponds to the index $n_z$ introduced earlier in Section 2.2. We mentioned there that in a quantum well a set of two dimensional subbands is generated, one for each

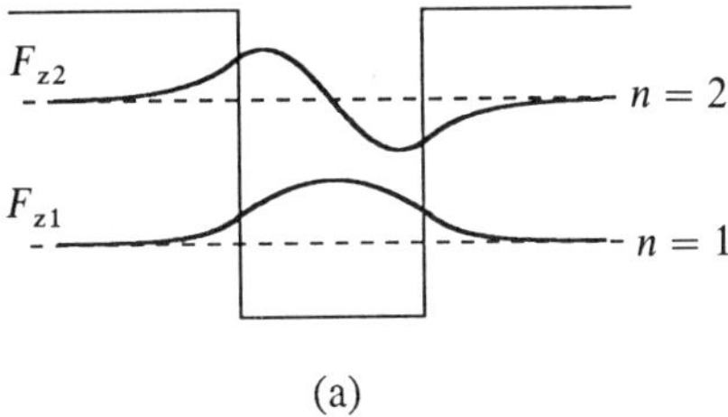


(a)

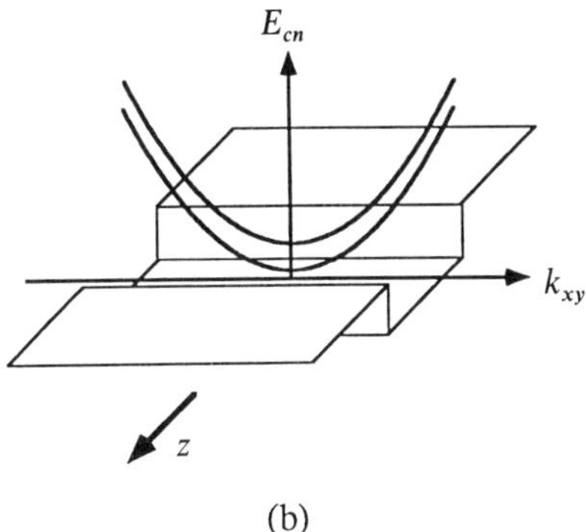


(b)

**Fig. 14.** Conduction band quantum well solutions. Part (a) illustrates the quantized envelope functions $F_{zn}$ for the first and second quantized energy levels (the zero levels of the two envelope functions have been shifted to correspond to the energy levels in the well). Part (b) illustrates the "in-plane" subband structure of the quantum well in the conduction band. Within the plane of the well, the electron still behaves like a "free" electron. Thus, for each quantized level, a parabolic energy subband exists as shown.

value of $n_z$. The subbands are a result of the freedom of motion that still exists for the electron in the plane of the well. The $E_{cn}$ solutions found earlier correspond to the $k_{xy} = 0$ solutions. To find the more general $E_{cn}(k_{xy})$ solutions, we must solve (64) for the more general case of $V_{k,i} \neq 0$.

The new potential profile when we include the in-plane momentum of the electron is given by $V_i + V_{k,i}$. Note that if $m_w = m_b$, $V_{k,i}$ becomes independent of $z$ and simply changes the reference potential of the problem (the entire quantum well simply floats at a higher potential). For the more general case of $m_w \neq m_b$, a finite in-plane momentum $k_{xy}$ will produce $V_{k,w} \neq V_{k,b}$ and will therefore change the overall potential profile $V_i + V_{k,i}$, altering the solutions found previously. In the extreme case that $V_{k,w} - V_{k,b} = V_0$, the overall potential profile will be flat, and no bound solutions will exist. Fortunately, $|V_{k,w} - V_{k,b}| \ll V_0$ in most cases, even for large $k_{xy}$, and it will suffice to treat $V_{k,i}$ as a small perturbation to the potential well shown in Fig. 13. The general procedure involves assuming that the new solutions to (64) are only slightly different than the "unperturbed" solutions found with $V_{k,i} = 0$. Therefore, we assume the new solutions to be of the form $F_{zn} = F_{zn0} + dF$, and $E_{cn}(k_{xy}) = E_{cn}(0) + dE$, where $F_{zn0}$ and $E_{cn}(0)$ represent the unperturbed solutions. If we then (1) substitute these expressions into (64); (2) multiply the entire equation by $(F_{zn0})^*$; (3) integrate from $z = -\infty$ to $z = +\infty$; and (4) perform some standard manipulations common to perturbation theory [1], we can obtain an approximate expression for the perturbed energy in terms of the unperturbed solutions. The final result can be expressed as

$$E_{cn}(k_{xy}) = E_{cn}(0) + \frac{\hbar^2 k_{xy}^2}{2m_{\text{eff}}}, \qquad \text{where } \frac{1}{m_{\text{eff}}} = \frac{\Gamma}{m_w} + \frac{1-\Gamma}{m_b}. \tag{69}$$

The confinement factor $\Gamma$ is defined as

$$\Gamma = \int_{-L_z/2}^{L_z/2} |F_{zn0}|^2\, dz \Bigg/ \int_{-\infty}^{\infty} |F_{zn0}|^2\, dz. \tag{70}$$

From (69), we now see that the energy as a function of the in-plane **k** vector is parabolic with an "in-plane effective mass" that is the weighted average of the bulk effective masses in and out of the well. Note also that each quantized level (each value of $n$) results in its own two dimensional subband in the $k_x$–$k_y$ plane. The resulting subband structure for the quantized levels shown in Fig. 14a are illustrated in Fig. 14b.

The $E_{cn}(k_{xy})$ and corresponding envelope functions $F_{en}$ in the conduction band are now completely specified with Eqs. (65)–(70), and we can move on to the description of the in-plane valence subband structure.

## 7. VALENCE BANDS IN QUANTUM WELLS

### 7.1. Degenerate Effective Mass Equation

The simplicity of the band structure in the conduction band of a quantum well relies on the assumption that the interaction with other energy bands is weak enough that we can treat it perturbatively by replacing that interaction with a conduction band effective mass. However, for bands degenerate in energy, the assumption of weak interaction is a poor one and cannot be used. Therefore, the effective mass equation (61) must be modified to include the strong degenerate band interaction explicitly.

A modified derivation for the degenerate band effective mass equation has also been treated by Luttinger and Kohn [16]. In this case, we still obtain an effective mass equation for each degenerate band similar to (61); however, as a result of the degeneracy, a coupling term is introduced, which couples the equations together. For the degenerate HH and LH bands near the band edge (see Fig. 7), this implies that we must work with *four* coupled effective mass equations (we must include the spin degeneracy)! [*Note*: We can actually include as many energy bands in the coupled set of equations as we desire (two equations for each spin-degenerate band). For example, in addition to the HH and LH bands, Eppenga *et al.* [19, 20] have included the SO band as well as the conduction band in their four band model, leading to eight coupled equations! However, interaction between the HH and LH bands is by far the strongest, and we do not pay a large penalty by neglecting the other bands. Only when we consider energy levels deep into the valence band (energies comparable to the spin–orbit splitting energy $\Delta$) do we need to include the SO band explicitly. In GaAs and InGaAs, $\Delta > 300$ meV [68, 69], implying that for most gain calculations with these materials we can neglect the SO band entirely. In InP, $\Delta$ is equal to about 100 meV [68]. However, well material typically used in this system is closer to lattice-matched InGaAs, where again $\Delta$ is closer to 300 meV.

The four coupled effective mass equations can be greatly simplified using a method first suggested by Kane [2] and later refined by Broido and Sham [23–26, 28, 29]. They pointed out that an appropriate linear combination of the four Bloch functions ($u_{\mathrm{hh}}, u_{\mathrm{lh}}, \bar{u}_{\mathrm{hh}}, \bar{u}_{\mathrm{lh}}$) into four new Bloch functions ($u_{\mathrm{A}}, u_{\mathrm{B}}, u_{\mathrm{C}}, u_{\mathrm{D}}$) decouples the four equations into two identical sets of two coupled equations. Thus, we actually only need to consider two coupled equations in our analysis. However, a price must be paid for this luxury. We must now restrict our attention to analyzing $E_{\mathrm{v}}(k_{xy})$ in a given *plane* in the crystal (the direction of $k_{xy}$ must be specified). Furthermore, the equations remain completely general only for the {100} and {110} planes [29]. However, for the present purposes, these represent only minor restrictions.

The two coupled effective mass equations for the degenerate bands can be expressed as

$$(H_{\mathrm{h}} + V)F_{\mathrm{h}} + WF_{\mathrm{l}} = F_{\mathrm{v}}F_{\mathrm{h}}, \tag{71}$$

$$(H_{l} + V)F_{\mathrm{l}} + W^{\dagger}F_{\mathrm{h}} = E_{\mathrm{v}}F_{\mathrm{l}}, \tag{72}$$

where $F_{\mathrm{h,l}}$ are the heavy- and light-hole envelope functions corresponding to two new Bloch functions $u_{\mathrm{A,B}}$, to be defined in the next subsection, and $W^{\dagger}$ is the Hermitian conjugate of $W$. The main difference between the degenerate (71)–(72) and nondegenerate (61) effective mass equations lies in the coupling term $W$. The energy $E_{\mathrm{v}}$ is what we would like to solve for, but because of $W$ we must now solve two equations simultaneously to find it. The form of the Hamiltonians, $H_{\mathrm{h,l}}$, are also slightly different from (62). Let us define $k_z$ and $k_t$ to be two perpendicular **k** vector components, with $k_z$ directed along a ⟨100⟩ direction and $k_t$ directed either along a ⟨100⟩ direction or a ⟨110⟩ direction within the $k_x$–$k_y$ plane (we use a transverse **k** vector, $k_t$, and not $k_{xy}$, for the in-plane **k** vector as was done in earlier sections, because $k_t$ represents a particular direction within the $k_x$–$k_y$ plane). With these, we can write (assuming plane wave solutions) [16–29]

$$H_{\mathrm{h}} = (\gamma_1 - 2\gamma_2)k_z^2 + (\gamma_1 + \gamma_2)k_t^2, \tag{73}$$

$$H_{\mathrm{l}} = (\gamma_1 + 2\gamma_2)k_z^2 + (\gamma_1 - \gamma_2)k_t^2, \tag{74}$$

where

$$\gamma_1 - 2\gamma_2 \equiv \frac{\hbar^2}{2m_{\mathrm{hh}}}, \qquad \gamma_1 + 2\gamma_2 \equiv \frac{\hbar^2}{2m_{\mathrm{lh}}}. \tag{75}$$

The form of the (73) and (74) is very similar to (62). The only surprising feature is the different effective masses used along $k_z$ and $k_t$. As we shall see in the next subsection, inclusion of the coupling term compensates for this apparent asymmetry in bulk material. However, in quantum well material, this asymmetry produces a much lighter HH mass in the plane of the well than along the confinement axis [27]. The material constants $\gamma_{1,2}$ are referred to as the Luttinger parameters [17], and are easily related to the HH and LH effective masses, $m_{\mathrm{hh}}$ and $m_{\mathrm{lh}}$, through (75). A third Luttinger parameter $\gamma_3$ exists in the coupling term $W$. The coupling term takes a slightly different form when $k_t$ is directed along ⟨100⟩ and ⟨110⟩ directions [29]. We can define the two forms as

$$W = \sqrt{3}k_t(\gamma_2 k_t - i\,2\gamma_3 k_z) \qquad \text{for } \{100\} \text{ planes,} \tag{76a}$$

$$W = \sqrt{3}k_t(\gamma_3 k_t - i\,2\gamma_3 k_z) \qquad \text{for } \{110\} \text{ planes.} \tag{76b}$$

The Hermitian conjugate is given by

$$W^\dagger = \sqrt{3}k_t(\gamma_{2(3)} k_t + i\,2\gamma_3 k_z), \tag{77}$$

independent of whether or not $k_z$ is complex (i.e., we do not take the complex conjugate of the Hermitian operator $k_z$ in defining $W^\dagger$). For either of the preceding forms, it is interesting to note that with $k_t = 0$ the coupling term disappears, and the effective mass equations decouple! Thus, in a quantum well, we can independently solve each of the effective mass equations (71) and (72) for the quantized $E_{vn}(0)$ of the HH and LH valence bands, just as was done for the conduction band in the previous section. However, for finite $k_t$ (i.e., away from the band edge), the equations become coupled and the solutions for $E_{vn}(k_t)$ become more complicated.

[*Note*: In this and the next section, the energy axis will be flipped upside-down to avoid the appearance of minus signs in every equation (i.e., the equations will be written in terms of the *hole* energy axis). However, most plots to be presented will still use the electron energy axis for drawing calculated band structures. Furthermore, it should be noted that all equations in Sections 1–5 are written relative to the electron energy axis exclusively.]

### 7.2. Bulk Solutions

As in Section 6, our first step in solving the quantum well problem is to find the bulk solutions within each material. To find a general relation for $E_v(\mathbf{k})$ in bulk material, it is convenient to cast the effective mass equations (71) and (72) into matrix form [16–29]:

$$\begin{bmatrix} H_h + V & W \\ W^\dagger & H_l + V \end{bmatrix} \begin{bmatrix} F_h \\ F_l \end{bmatrix} = E_v \begin{bmatrix} F_h \\ F_l \end{bmatrix}. \tag{78}$$

For the bulk solutions, we take $V$ to be a constant, $V_0$. Then, in looking at Eqs. (73)–(77), we find that for a given $\mathbf{k}$ vector the $2 \times 2$ Hamiltonian matrix consists of simple constants. The eigenvalue problem in this case is easily solved for the eigen-energies. In general form the bulk $E_v(\mathbf{k})$ relations for the HH and LH bands are given by

$$E_v(\mathbf{k}) - V_0 = \tfrac{1}{2}(H_h + H_l) \pm \tfrac{1}{2}[(H_h - H_l)^2 + 4W^\dagger W]^{1/2}. \tag{79}$$

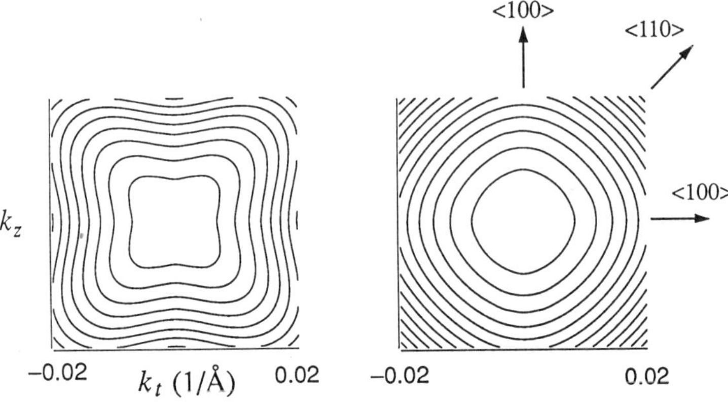


**Fig. 15.** Contours of constant energy within any $\{100\}$ plane of $k$ space for the HH and LH bands in bulk GaAs. The energy spacing between each contour level is 0.5 meV for the HH band and 3 meV for the LH band. The effective HH mass is much larger along the $\langle 110 \rangle$ direction than along the $\langle 100 \rangle$ direction, as indicated by the larger contour spacing. The effective LH mass is seen to be much more isotropic.

Using Eqs. (73)–(77), $E_v(\mathbf{k})$ can be given explicitly in terms of $\mathbf{k}$ within any $\{100\}$ plane by

$$E_v(\mathbf{k}) = \gamma_1(k_z^2 + k_t^2) \pm [4\gamma_2^2(k_z^2 + k_t^2)^2 + 12(\gamma_3^2 - \gamma_2^2)k_z^2 k_t^2]^{1/2} \tag{80}$$

(where we have set $V_0 = 0$ for simplicity). A similar equation can be found for $E_v(\mathbf{k})$ in any $\{110\}$ plane. Figure 15 illustrates the constant energy contour curves of (80) for both the HH band solution (negative root) and the LH band solution (positive root) [70, 71]. From these curves, we see that the third Luttinger parameter $\gamma_3$ can be related to the effective mass anisotropy along the $\langle 100 \rangle$ and $\langle 110 \rangle$ directions (because with $\gamma_3 = \gamma_2$, the contour curves would become circles, from (80)). In addition, looking along either $k_z$ or $k_t$ (along a $\langle 100 \rangle$ direction), the cross-term in Eq. (80) disappears, and we simply have

$$E_v(\mathbf{k}) = (\gamma_1 \pm 2\gamma_2)k^2, \qquad \mathbf{k} \text{ directed along any } \langle 100 \rangle \text{ direction.} \tag{81}$$

From the definition of the Luttinger parameters in (75), we see that the standard HH and LH $E_v(\mathbf{k})$ relations are obtained from (81). Thus, contrary to what (73) and (74) may suggest, the effective masses along $k_z$ and $k_t$ are identical in bulk material (within $\{100\}$ planes).

To completely specify the bulk solutions, we also need to find the eigenvectors of (78). With $E_h(\mathbf{k})$ and $E_l(\mathbf{k})$ given by the two roots of (79), the eigenvectors, apart from a normalization constant, are found to be

$$\psi_1(\mathbf{k}, \mathbf{r}) = \begin{bmatrix} F_h \\ F_l \end{bmatrix} = e^{i\mathbf{k}\cdot\mathbf{r}} \begin{bmatrix} H_l + V_0 - E_h \\ -W^\dagger \end{bmatrix} \equiv e^{i\mathbf{k}\cdot\mathbf{r}} \begin{bmatrix} \Delta_{1h}(\mathbf{k}) \\ \Delta_{1l}(\mathbf{k}) \end{bmatrix}, \tag{82}$$

$$\psi_2(\mathbf{k}, \mathbf{r}) = \begin{bmatrix} F_h \\ F_l \end{bmatrix} = e^{i\mathbf{k}\cdot\mathbf{r}} \begin{bmatrix} H_l + V_0 - E_l \\ -W^\dagger \end{bmatrix} \equiv e^{i\mathbf{k}\cdot\mathbf{r}} \begin{bmatrix} \Delta_{2h}(\mathbf{k}) \\ \Delta_{2l}(\mathbf{k}) \end{bmatrix}, \tag{83}$$

where, for either solution, the matrix notation implies

$$\psi = F_h u_A + F_l u_B. \tag{84}$$

The validity of (82) and (83) can be checked by substituting them into (72). In addition, the Hamiltonians in Eqs. (73)–(77) implicitly assumed plane wave solutions; thus, their inclusion in (82) and (83) is mandatory. Equation (84) gives the wavefunctions in vector notation. The Bloch functions $u_{A,B}$ are orthogonal to each other (analogous to two orthogonal unit position vectors in real space) and are given by linear combinations of the valence band Bloch functions defined in Section 5. We can write them as [24]

$$u_A = \frac{1}{\sqrt{2}}(\alpha u_{hh} - \alpha^* \bar{u}_{hh}), \tag{85}$$

$$u_B = \frac{1}{\sqrt{2}}(\beta \bar{u}_{lh} - \beta^* u_{lh}). \tag{86}$$

For $\{100\}$ planes [29], $\alpha = \beta = 1$. For $\{110\}$ planes [29], $\alpha = \exp[i\,3\pi/8]$, and $\beta = \exp[-i\,\pi/8]$.

### 7.3. Quantum Well Solutions

To solve the quantum well problem, we again choose the quantum well direction to be along $z$ (this is not mandatory, but the effective mass equations would not decouple at the band edge otherwise [see (76) and (77)]). In this case, $k_z$ is directed along the confinement axis, and $k_t$ is the transverse **k** vector component and lies in the plane of the well, as shown in Fig. 16a. We now construct a general solution out of all bulk solutions that exist at a given energy within each material. From Fig. 16b, we see that in general four plane wave solutions exist at a given energy [22]. The general solution in each region of Fig. 13 is then given by a linear combination of these four waves, or

$$\Psi = \sum A_\pm \psi_1(\pm k_h, k_t, \mathbf{r}) + \sum B_\pm \psi_2(\pm k_l, k_t, \mathbf{r}). \tag{87}$$

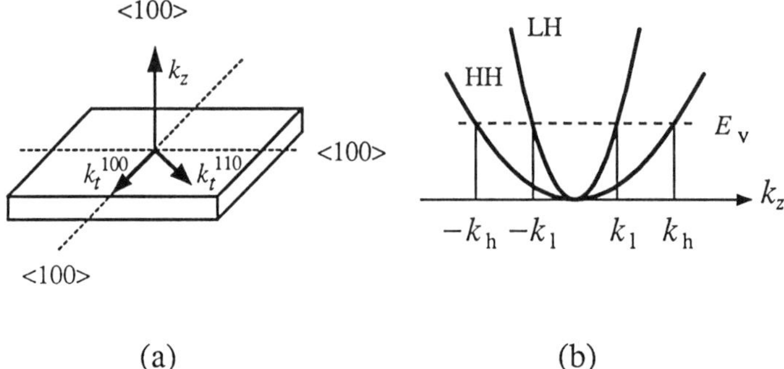


**Fig. 16.** (a) Coordinate system to be used in the valence band model. The transverse, or in-plane, **k** vector $k_t$ can be directed along either a $\langle 100 \rangle$ or $\langle 110 \rangle$ direction, whereas the confinement axis must be along a $\langle 100 \rangle$ direction. (b) Illustration of the four plane wave states that exist at a given energy in the bulk valence band structure with $k_t = 0$.

The sums are over the plus and minus going waves, and the $k_{\mathrm{h,l}}$ are defined in Fig. 16b. The four coefficients $A_\pm$, $B_\pm$ are unknown constants. [*Note*: Equation (87) is the valence band analog of (65) and (66) in the conduction band. In this sense, Eq. (79) is the more complicated valence band analog of the equations to the right of (65) and (66).] Both $\psi_1$ and $\psi_2$ are two-component vectors, from their definitions in (82) and (83). The general solution $\Psi$ is also a two-component vector. The components can be written as

$$F_{\mathrm{h}} = e^{ik_t r_t}\left[\sum A_\pm \Delta_{1\mathrm{h}}(\pm k_{\mathrm{h}}, k_t)e^{\pm ik_{\mathrm{h}}z} + \sum B_\pm \Delta_{2\mathrm{h}}(\pm k_{\mathrm{l}}, k_t)e^{\pm ik_{\mathrm{l}}z}\right], \tag{88}$$

$$F_{\mathrm{l}} = e^{ik_t r_t}\left[\sum A_\pm \Delta_{1\mathrm{l}}(\pm k_{\mathrm{h}}, k_t)e^{\pm ik_{\mathrm{h}}z} + \sum B_\pm \Delta_{2\mathrm{l}}(\pm k_{\mathrm{l}}, k_t)e^{\pm ik_{\mathrm{l}}z}\right], \tag{89}$$

where we have pulled out the common transverse plane wave component. Thus, in the case of the valence band, there are four unknown coefficients, $A_\pm$, $B_\pm$, in each region as opposed to just two, as was the case in the conduction band. To find the coefficients, we match the general solutions within each region at both well–barrier interfaces. The boundary conditions for the degenerate effective mass equation involve matching the following four quantities [21, 22, 27] across the interface:

$$F_{\mathrm{h}} \quad \text{and} \quad (\gamma_1 - 2\gamma_2)\frac{dF_{\mathrm{h}}}{dz} + \sqrt{3}\,\gamma_3 k_t F_{\mathrm{l}}; \tag{90}$$

$$F_{\mathrm{l}} \quad \text{and} \quad (\gamma_1 + 2\gamma_2)\frac{dF_{\mathrm{l}}}{dz} - \sqrt{3}\,\gamma_3 k_t F_{\mathrm{h}}. \tag{91}$$

With $k_t = 0$, these boundary conditions are identical to those given in Section 6 for the nondegenerate case. In addition, for any $k_t$, if the Luttinger parameters are the same on both sides of the interface, the second boundary conditions reduce to the simple slope continuity conditions. The generalized slope continuity conditions (90) and (91) can be derived by integrating Eq. (78) over an infinitesimal thickness which straddles the interface [21, 22, 27, 64, 65, 67] in a similar manner to that discussed in Section 6. However, in this case symmetrizing the Hamiltonian involves setting $(\gamma_1 \pm 2\gamma_2)k_z^2 \rightarrow k_z(\gamma_1 \pm 2\gamma_2)k_z$ in Eqs. (73–74), *and* setting $\gamma_3 k_z \rightarrow (\gamma_3 k_z + k_z\gamma_3)/2$ in Eqs. (76, 77), before setting $k_z \rightarrow -i(\partial/\partial z)$, since $\gamma_1, \gamma_2, \gamma_3$ all depend on $z$. We leave it as an exercise for the reader to verify that these symmetrizing substitutions guarantee that the Hamiltonian in Eq. (78) remains hermitian for any arbitrary $z$-dependence of $\gamma_1, \gamma_2, \gamma_3$. It should be noted that the preceding boundary conditions hold for both $\{100\}$ and $\{110\}$ planes.

Applying the four boundary conditions at each interface gives us a total of eight equations. There are four unknown coefficients in each of the three regions, or twelve total. However, requiring the envelope functions to go to zero at infinity leaves us with a total of eight unknown coefficients. Thus, our problem is now completely specified and we can solve the eight homogeneous equations by numerically finding the roots of the $8 \times 8$ determinant.

The general procedure for obtaining $E_v(k_t)$ is then as follows: (1) Find the $E_{vn}(0)$ of a particular HH or LH band edge energy level using the conventional method of Section 6; (2) Increment $k_t$ and guess at the new energy of the state; (3) Find $k_{h,l}$ from the two $E_v(\mathbf{k})$ relations given in (79) within each material (each material having its own $V_0$ and its own set of Luttinger parameters); (4) Evaluate the Δ's within each material from their definitions in (82) and (83); (5) Evaluate the $8 \times 8$ coefficient determinant; (6) If it is not equal to zero, use Newton's method to repeat the process until the energy root is found for that given $k_t$; (7) Increment $k_t$ and repeat the entire process to find the new energy root, using an educated initial guess. The entire $E_{vn}(k_t)$ can then be traced in this way. The rate of convergence is very good for this type of problem. For example, for each $k_t$, the energy root can be found in typically 2–3 iterations (Figure 17 was generated in less than 3 minutes on a Macintosh IIci).

The coefficients $A_\pm$, $B_\pm$ in each material can be found at each energy by Gaussian reduction of the eight equations, allowing us to reconstruct the total wavefunction in terms of the original valence band Bloch functions of Section 5 using Eqs. (82)–(89). From these, we can then evaluate the transition

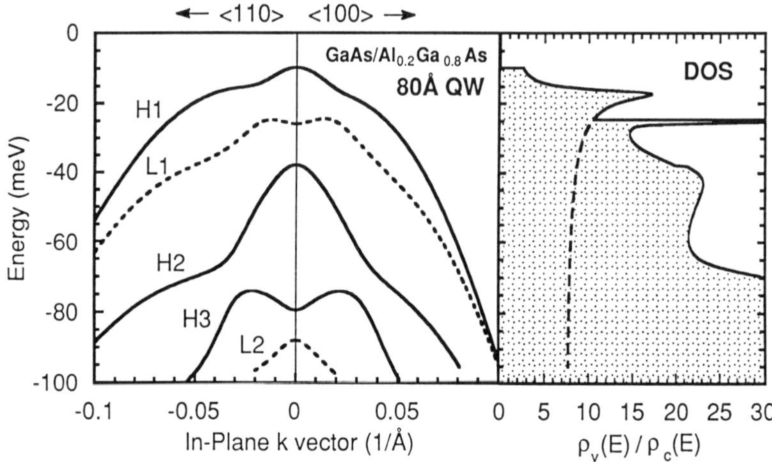


**Fig. 17.** The valence subband structure of an 80-Å $GaAs/Al_{0.2}Ga_{0.8}As$ quantum well ($V_0 \approx 95$ meV) with the inclusion of coupling between the HH and LH bands is plotted at left. The coupling between the HH (solid curves) and LH (dashed curves) subbands is very dramatic. The $\langle 110 \rangle$ and $\langle 100 \rangle$ dispersion characteristics are also seen to be slightly different, with a "heavier" band structure observed along the $\langle 110 \rangle$ direction, consistent with the bulk HH bands illustrated in Fig. 15. At right are the total (solid curve) and H1 subband (dashed curve) densities of states plotted relative to the density of states in the first conduction (C1) subband. The zero slope in the L1 subband away from $k = 0$ is seen to result in a pole in the DOS. The calculated DOS actually corresponds to a band structure that assumes a coupling term of the form $W_{ave} = (W_{100} + W_{110})/2$ (basically, the average between the $\langle 100 \rangle$ and $\langle 110 \rangle$ dispersions).

matrix element. However, because the general valence band wavefunction in (84) consists of both HH and LH envelope function components, the transition matrix element of Eq. (22) must be modified, and it is now expressed as

$$|M_T|_v^2 = 2 \sum_{u_c, \bar{u}_c} |\langle u_A|\hat{\mathbf{e}} \cdot \mathbf{p}|u_c\rangle\langle F_h|F_e\rangle + \langle u_B|\hat{\mathbf{e}} \cdot \mathbf{p}|u_c\rangle\langle F_l|F_e\rangle|^2. \quad (92)$$

The summation accounts for spin degeneracy in the conduction band. The factor of two accounts for transitions involving the spin degenerate counterparts of $u_A$ and $u_B$ (the $u_C$ and $u_D$ Bloch functions).

The transition matrix element in (92) can be placed in a more elegant form by following a procedure similar to that outlined in Section 5.2. However, in the present case we must average the transition matrix element over all in-plane **k** vector directions to remove the cross-term that results from squaring Eq. (92) [25]. With the help of Eqs. (85), (86), and (51), the transition matrix element for transverse magnetic (TM) and transverse electric (TE) polarizations of the optical field reduces to

$$|M_{\mathrm{T}}|_{\mathrm{v}}^{2} = |M|^{2}[\tfrac{4}{3}|\langle F_{\mathrm{l}}|F_{\mathrm{e}}\rangle|^{2}] \qquad \text{TM polarizations } (\hat{\mathbf{e}} \parallel \hat{\mathbf{z}}), \quad (93)$$

$$|M_{\mathrm{T}}|_{\mathrm{v}}^{2} = |M|^{2}[|\langle F_{\mathrm{h}}|F_{\mathrm{e}}\rangle|^{2} + \tfrac{1}{3}|\langle F_{\mathrm{l}}|F_{\mathrm{e}}\rangle|^{2}] \qquad \text{TE polarizations } (\hat{\mathbf{e}} \perp \hat{\mathbf{z}}). \quad (94)$$

At the band edge (where $F_{\mathrm{l}} = 0$ for HH states, and $F_{\mathrm{h}} = 0$ for LH states), these expressions give the same transition strengths as those in Fig. 11b assuming the overlap integrals are close to unity (which is true to within 5–10% typically, and would be true exactly if the effective mass in the C and V bands were identical). However, as we move away from the band edge, band-mixing occurs, such that both $F_{\mathrm{h,l}}$ are present in any one wavefunction, altering the transition strengths from those shown in Fig. 11b. In Section 9, we will consider the transition matrix element quantitatively as a function of the transverse **k** vector, $k_t$.

In finding the envelope functions and evaluating the overlap integrals numerically, we must make sure they are properly normalized. Normalization of the wavefunctions is obtained through the following relations:

$$F_{i,\mathrm{norm}} = \frac{F_i}{\sqrt{N_i}}, \qquad \text{where } N_{\mathrm{h,l}} = \langle F_{\mathrm{h}}|F_{\mathrm{h}}\rangle + \langle F_{\mathrm{l}}|F_{\mathrm{l}}\rangle,\ N_{\mathrm{e}} = \langle F_{\mathrm{e}}|F_{\mathrm{e}}\rangle. \quad (95)$$

These envelope functions refer to the functions along the confinement direction, and hence the brackets indicate integration along $z$ as defined in Eq. (21). The in-plane envelope functions are simple plane waves. Thus, we require **k** conservation in the plane of the well, which then yields an in-plane overlap integral of unity, as discussed near the end of Section 3.1.

Our description of the valence subband structure is now complete, and we can move on to an example. We saw in the previous section that the one dimensional quantum confinement in the conduction band gives rise to a set of parabolic subbands in the plane of the well, as illustrated in Fig. 14b. In the valence band, coupling between the HH and LH subbands changes the situation drastically, giving rise to much more interesting band structure than that shown in Fig. 14b. Figure 17 shows the valence subband structure for an 80-Å GasAs/$Al_xGa_{1-x}As$ quantum well with $x = 0.2$ in the barrier regions, calculated using the procedure outlined above [29]. The subband structure is seen to be far from parabolic, and in some regions the band curvature is even inverted, leading to a negative "local" hole mass [18–29].

With respect to predicting the optical gain achievable in a material with the subband structure shown in Fig. 17, we will be particularly interested in the density of states of the subbands, also shown in Fig. 17 (calculated using the general definition of Eq. (12)). The density of states is important because it determines the relationship between carrier density and the quasi-Fermi level of the band as discussed in Section 4.1. From Fig. 17, we see that $\rho_v$ is roughly $2.5 \times \rho_c$ near the band edge, but rapidly becomes very large as mixing between the bands starts to become significant (compare with the parabolic subbands in Fig. 2). From the discussion of Section 4.1, the mismatch between $\rho_c$ and $\rho_v$ as well as the overall large $\rho_v$ reduces the performance of the quantum well, increasing transparency levels and reducing the differential gain (ideally the DOS curve in Fig. 17 would be a straight line of magnitude one).

In the next section, we consider ways in which $\rho_v$ can be modified, in beneficial ways, by the introduction of strain.

## 8. STRAINED QUANTUM WELLS

It was originally suggested by Yablanovitch and Kane [30, 31], and independently by Adams [32], that the introduction of compressive strain into the crystal lattice of a semiconductor could lead to enhanced performance in semiconductor lasers. To understand why, we need to examine the effects of strain, particularly in a quantum well. Introducing compressive strain into a quantum well configuration is particularly simple: just grow the well layer out of a material with a *larger* native lattice constant than the barrier layers. Because the quantum well layer is typically very thin, instead of forming misfit dislocations the lattice actually compresses in the plane of the well to match that of the barrier layers [72, 73]. In addition, the lattice constant in the direction normal to the plane becomes elongated (in an effort to keep the volume of each unit cell the same), as shown in Fig. 18a.

Because the energy gap of a semiconductor is related to its lattice spacing, we might expect that distortions in the crystal lattice should lead to alterations in the bandgap of the strained layer (putting aside for the moment the changes created simply by quantum confinement). In fact, there are two types of modifications that occur [72–75]. The first effect produces an upward shift in the conduction band as well as a downward shift in both valence bands, increasing the overall bandgap by an amount $\delta\varepsilon_H$ (which is positive for compressive strain and negative for tensile strain). The H subscript indicates that this shift originates from the *hydrostatic* component

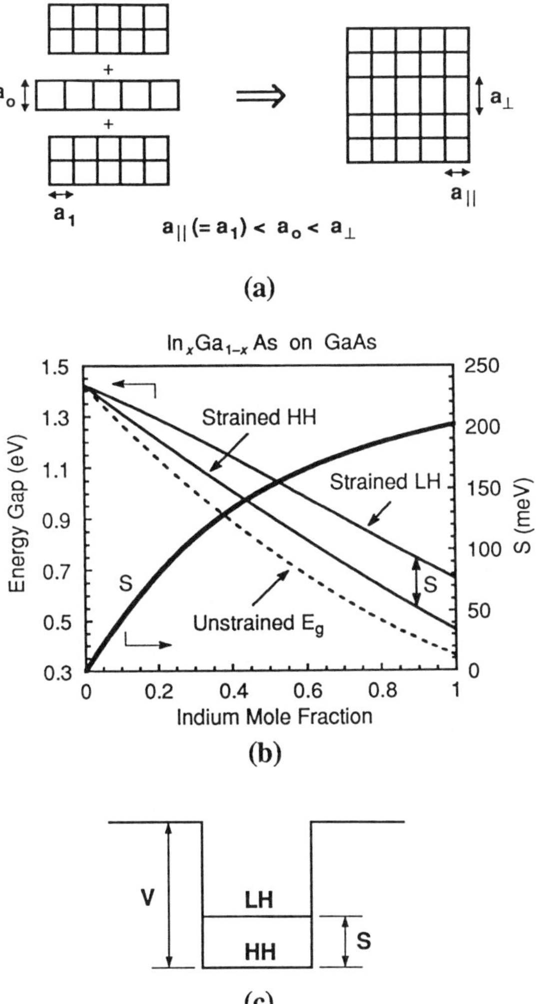


**Fig. 18.** (a) Illustration of the crystal lattice deformation resulting from the epitaxial growth of a thin layer of material with a native lattice constant $a_0$ between two thick layers with lattice constant $a_1 < a_0$. The lattice constant of the thin layer parallel to the plane is compressed to match that of the thick layers, while perpendicular to the plane it is elongated. (b) Effect of the lattice deformation on the bulk bandgap of $In_xGa_{1-x}As$ as a function of indium mole fraction, when the in-plane lattice constant is compressed to that of GaAs. The entire bandgap is noticeably increased. However, the more striking feature is the splitting of the HH and LH bands, defined by a splitting energy $S$. (c) The modified potential barrier(s) of a quantum well, when the well material is under compressive strain. The increase in the LH bandgap effectively pushes all LH quantized energy levels further up in the well.

of the strain. The second, more important effect separates the HH and LH bands, each being pushed in opposite directions from the center by an amount $\delta\varepsilon_s$. The S subscript indicates that this shift originates from the *shear* component of the strain. Thus, the band edge degeneracy of the two valence bands is removed and two energy gaps must now be defined. The total strained bandgap can be written as $E_g + \delta\varepsilon_H \pm \delta\varepsilon_S$, where the upper sign refers to the C–LH bandgap $E_g$(LH), and the lower sign refers to the C–HH bandgap $E_g$(HH). In Fig. 18b we have plotted the unstrained bulk bandgap as well as the two compressively strained bulk bandgaps of InGaAs on a GaAs substrate (which has a smaller lattice constant than InGaAs). Note that $\delta\varepsilon_S$ as defined is positive for compressive strain, since $E_g(\mathrm{LH}) > E_g(\mathrm{HH})$. For tensile strain, $\delta\varepsilon_S$ would be negative, and the bandgap ordering would be reversed.

The two energy shifts, $\delta\varepsilon_H$ and $\delta\varepsilon_S$, increase linearly with the lattice constant mismatch, which in turn increases linearly with indium mole fraction. We should expect then that the bandgap difference, $E_g(\mathrm{LH}) - E_g(\mathrm{HH})$, defined in the plot as $S$, should be a linearly increasing function of indium mole fraction, since, from the preceding discussion, $S = 2\delta\varepsilon_S$. For small indium mole fractions, we see from the plot that this is true. However, as the indium mole fraction increases, $S$ begins to saturate. This is a result of the interaction between the LH band and the SO band. When taken into account, this interaction introduces a correction term into the expression for the strained LH band gap such that, to second order, $S = 2\,\delta\varepsilon_S(1 - \delta\varepsilon_S/\Delta)$, where $\Delta$ is the spin–orbit splitting energy introduced earlier in Section 5 (the exact expression for $S$ used to calculate Fig. 18b is the solution to a quadratic equation [74, 75]).

In a quantum well, the splitting of the HH and LH bands can have dramatic consequences [29–35], since the large nonparabolicity of the subband structure in Fig. 17 is a direct result of the HH and LH band mixing. If we place the strained bandgaps shown in Fig. 18b into a quantum well, the situation becomes as shown in Fig. 18c, where the depth of the quantum well as seen by light holes is reduced by the splitting energy $S$. To predict the valence subband structure of the strained quantum well in Fig. 18c, we simply need to add a potential offset to the effective mass equation describing the LH band in the well [26–29]. Equation (78) in the previous section now simply becomes

$$\begin{bmatrix} H_h + V & W \\ W^\dagger & H_l + V + S \end{bmatrix} \begin{bmatrix} F_h \\ F_l \end{bmatrix} = E_v \begin{bmatrix} F_h \\ F_l \end{bmatrix}, \tag{96}$$

where $V$ is zero inside the well, and $S$ is zero outside the well. The procedure for solving (96) in a strained quantum well is entirely analogous to the procedure presented in Section 7 for an unstrained quantum well. Thus, we can immediately turn to an example calculation using Eq. (96).

We choose for our example the GaAs/InGaAs strained layer system. This system has received much attention in recent years [73, 75–80] because of its compatibility with MBE growth technology as well as its proven ability to yield high quality material comparable with, if not better than, its unstrained counterpart [81–85]. For direct comparison, we take the GaAs/AlGaAs 80-Å quantum well used in the example of Section 7 and simply add a bit of indium to the well layer. InGaAs has a larger native lattice constant than GaAs; thus, sandwiched between two AlGaAs layers, it will be compressed in the plane of the well. For an indium mole fraction of 20%, the resulting HH–LH splitting energy $S$ is approximately 80 meV (from Fig. 18b).

Figure 19 shows both the valence subband structure and the density of

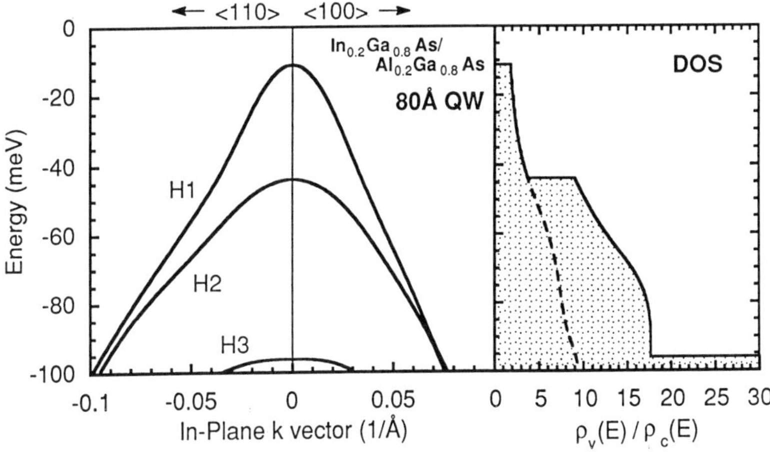


**Fig. 19.** The valence subband structure of an 80-Å $In_{0.2}Ga_{0.8}As/Al_{0.2}Ga_{0.8}As$ strained quantum well ($V_0 \approx 175$ meV) with the inclusion of coupling between the HH and LH bands is plotted at left. The LH bands have been pushed further out of the well as a result of the strain and cannot be seen with the energy scale shown. The effective removal of the LH subbands drastically reduces the band-mixing effects, as seen by comparison with Fig. 17. The result is an overall "lighter" band structure, as vividly displayed by the density of states shown on the right, where the total (solid curve) and H1 subband (dashed curve) densities of states are plotted relative to the density of states in the first conduction (C1) subband.

states of the 80-Å $In_{0.2}Ga_{0.8}As/Al_{0.2}Ga_{0.8}As$ strained quantum well calculated from Eq. (96) [35]. Immediately, we see that the LH bands have been pushed deep into the band (the full depth of the well is not included in the figure to keep the energy scale equal to Fig. 17; thus, we cannot even see the confined LH subbands). As a result, the band warping has been greatly reduced [29, 33, 35]. Comparison with Fig. 17 reveals that the density of states in the strained quantum well is reduced significantly, and matching between $\rho_c$ and $\rho_v$ is greatly improved. Both of these features should lead to lower transparency levels as well as higher differential gain than the unstrained quantum well, from the discussion of Section 4.1 (see Fig. 6). In the next section, we will find quantitatively that both of these improvements are indeed expected theoretically.

Equation (96) not only applies to quantum wells, but can also be applied to bulk strained material. The $E(k)$ relations in bulk material are obtained by finding the eigen-energies of Eq. (96) as we did for Eq. (78) in Section 7.2 (see (79) and (80)) (the definitions for $H$ and $W$ given in Section 7 apply here as well). The general bulk solutions for the eigen-energies of (96) can be obtained in closed form, but are somewhat messier than Eq. (80) [30]. We leave it as an exercise for the reader to show that, to first order in $\gamma_2 k^2/S$ (i.e., in the large strain regime where $S \gg \gamma_2 k^2$), the eigen-energies of (96) can be expressed as [71, 86]

$$E(k_{\|}) = (\gamma_1 \mp \gamma_2)k_t^2 \pm S/2 \qquad (k_z = 0), \tag{97}$$

$$E(k_{\perp}) = (\gamma_1 \pm 2\gamma_2)k_z^2 \pm S/2 \qquad (k_t = 0). \tag{98}$$

The upper signs refer to the LH band solutions, while the lower signs refer to the HH band solutions (we have thrown away any common shifts in the band edges to concentrate on the *difference* between the LH and HH bandgap energies). The parallel (in-plane) and perpendicular (normal to the plane) **k** vectors refer to the notation used in Fig. 18a. Note that the preceding relations are also obtained by setting $W = 0$ (as seen by combining (73) and (74) with (96)). This makes sense, because in the large strain regime we are basically saying that $S \gg W$, allowing us to neglect the coupling between the bands altogether. The modified band structure given by (97) and (98) is shown to scale in Fig. 20. Thus, even in bulk material, strain serves to reduce the effective mass of the HH band dramatically [86] within the "plane of compression." Perpendicular to the plane, it is interesting to note, however, that apart from the splitting of the HH and LH bands, the dispersion relation (98) remains unchanged compared with (81).

We have thus far not mentioned the effects of strain on the conduction

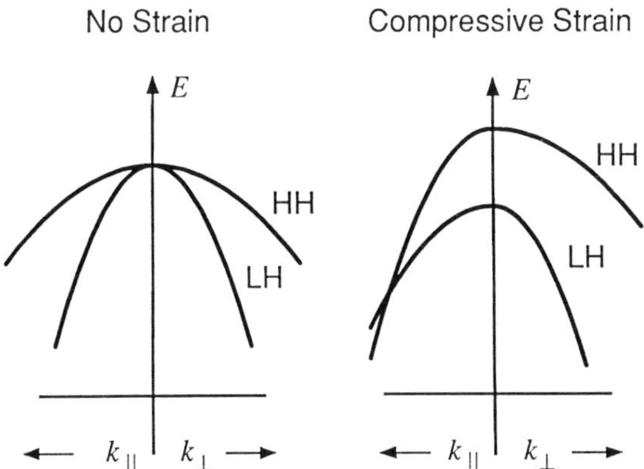


**Fig. 20.** Modification to the bulk valence band structure by the application of compressive strain, estimated using the approximate eigenvalues (97) and (98), under the assumption of large strain, such that $S \gg \gamma_2 k^2$. The bands split apart and the HH band becomes "light" parallel to the plane of compression, while the LH band becomes "heavier." Perpendicular to the plane, however, the bands do not change their shape. For large values of $k$, our assumption of $S \gg \gamma_2 k^2$ breaks down, and the apparent crossing of the HH and LH bands in the plane would be removed by use of the exact eigenvalues of Eq. (96). [*Note*: The relative band curvature of each band is drawn to scale to illustrate the relative changes in band shapes. In addition, the constant energy shift of both bands due to the overall increase in the bandgap is not shown in the figure.]

band. The reason for this is that due to its relative isolation from other bands, the conduction band curvature remains relatively unaffected by the shifting energy gaps. Equation (53) in Section 5 would suggest that the increase in the bandgap should increase the conduction band effective mass slightly. However, we must be careful here because the third parameter that ties into this equation, $|M|^2$, does not necessarily remain constant (for example, it is conceivable that $|M|^2$ could increase proportionally with the bandgap, leaving $m^*$ unaffected). In any case, the change in the conduction band curvature should be slight. In the examples to be given in the next section, we assume that strain has no effect on the conduction band effective mass.

In the preceding discussion of strain, we have attempted to extract the vital information necessary to understand how strain modifies the valence band structure in quantum wells and in bulk material. All that was really needed to do this was one phenomenological parameter, $S$. We have intentionally glossed over the specific details of how $S$ can be calculated

[72–75], for example, as a function of lattice mismatch or strain. Nor have we considered the very important concept of critical thickness [73] of the strained layer (beyond which the strained lattice becomes unstable). The interested reader will find that later chapters of this book discuss these issues in more detail.

## 9. GAIN IN QUANTUM WELLS — AN EXAMPLE

We now present some example gain calculations using the methods previously presented. We will consider the following: (a) an 80-Å GaAs/$Al_{0.2}Ga_{0.8}As$ quantum well; (b) an 80-Å $In_{0.2}Ga_{0.8}As/Al_{0.2}Ga_{0.8}As$ strained quantum well; and (c) a fictitious 80-Å "quasi-bulk" GaAs well using all bulk parameters (such as the bulk density of states, the bulk transition matrix element, and no quantized energy levels). [*Note*: Because the energy bands are different in the {110} and {100} planes, and because electrons will of course exist with **k** vectors in all planes, we use an average coupling term, $W_{ave} = (W_{100} + W_{110})/2$, to calculate the quantum well energy band structure. This approach is commonly known as the *axial* approximation [21]. For the "bulk" well, the energy bands (the HH and LH effective masses) are found using a "spherical average" of the effective masses along all directions in the crystal (see Fig. 15). It basically involves replacing $\gamma_2$ with $(2\gamma_2 + 3\gamma_3)/5$ in Eq. (75), and is known as the *spherical* approximation [87].]

The material parameters used in the calculations are given in Table III. Most bulk parameters were taken from Ref. 68, with the exception of the AlGaAs bandgap, which was taken from Casey and Panish [37]. The InGaAs bulk bandgap is a quadratic fit to the three well known GaAs, InAs, and $In_{0.53}Ga_{0.47}As$ bandgaps [68]. The strained bandgap and splitting energy $S$ were found using the procedure outlined by Ji *et al.* [75], and the material deformation potentials given by Adachi [69]. A quadratic curve fit to the calculated strained bandgap and splitting energy was then found for indium mole fractions less than 50%. [*Note*: $S$ would be a linearly increasing function of $x$; however, the LH band interacts with the SO band when the strain is large, leading to a correction term in $S$ that is dependent on the spin–orbit splitting energy [74, 75] (we assumed that $\Delta\,(\text{eV}) = 0.34(1 - x) + 0.371x$ in the GaAs/$In_xGa_{1-x}As$ system [68]).]

The conduction band offset in the GaAs/AlGaAs system was assumed to be 0.62, taken from Watanabe [88, 89]. The conduction band offset in the strained InGaAs/GaAs system is much more uncertain at present. Various

**Table III.**

Material Parameters Used for the InGaAs/AlGaAs System Considered in this Chapter

| $Al_xGa_{1-x}As$ Bulk Parameters | | $In_xGa_{1-x}As$ Bulk Parameters | |
|---|---|---|---|
| $E_g\,(eV) = 1.424 + 1.247x,$ | $x < 0.45$ | $E_g\,(eV) = 1.424 - 1.619x + 0.555x^2$ | |
| $E_g\,(eV) = 1.424 + 1.247x + 1.147(x - 0.45)^2,$ | $x > 0.45$ | | |
| $a\,(Å) = 5.6533 + 0.0078x$ | | $a\,(Å) = 5.6533 + 0.405x$ | |
| $m_c = 0.067 + 0.083x$ | $(\times m_0)$ | $m_c = 0.067 - 0.04x$ | $(\times m_0)$ |
| $\gamma_1 = 6.85(1 - x) + 3.45x$ | $\left(\times \frac{\hbar^2}{2m_0}\right)$ | $\gamma_1 = 6.85(1 - x) + 19.67x$ | $\left(\times \frac{\hbar^2}{2m_0}\right)$ |
| $\gamma_2 = 2.1(1 - x) + 0.68x$ | | $\gamma_2 = 2.1(1 - x) + 8.37x$ | |
| $\gamma_3 = 2.9(1 - x) + 1.29x$ | | $\gamma_3 = 2.9(1 - x) + 9.29x$ | |

| $In_xGa_{1-x}As/GaAs$ Strained Parameters | | | |
|---|---|---|---|
| Strained $E_g(HH) = 1.424 - 1.06x + 0.08x^2$ | $(x < 0.5)$ | $= 1.424 - 1.061x + 0.07x^2 + 0.03x^3$ | $(0 < x < 1)$ |
| $S \equiv E_g(LH) - E_g(HH) = 0.465x - 0.33x^2$ | | $= 0.48x - 0.43x^2 + 0.152x^3$ | |
| $Q_c \equiv \Delta E_c/\Delta E_g = 0.4 - 0.8$ | $[\Delta E_g \equiv E_g\,(\text{barrier}) - E_g(HH)]$ | | |

reports place $Q_c$ anywhere in the range 0.4–0.8 [75–80]. Joyce *et al.* [78] have even suggested that it is a function of indium composition. In the present examples, we will assume that it is the same as the GaAs/AlGaAs system, or 0.62 [90]. [*Note*: For $Q_c$ in the strained system, $\Delta E_g$ refers to the difference between the barrier material bandgap and the *strained* HH bandgap.]

In Section 5.3, we discussed polarization-dependent effects in quantum wells. We provided a general method for predicting band edge transition strengths, but mentioned that the transition strengths above the band edge change considerably. We can find the TM and TE transition strengths from (93) or (94) for each subband transition as a function of in-plane **k** vector. Figures 21a,b display the transition strengths calculated for the unstrained quantum well (case a), for both TM and TE modes, for the two lowest subband transitions [29]. The dashed curves represent what one would calculate if the valence subbands were parabolic [4, 5, 15]. At the band edge ($k_t = 0$), the calculated transition strengths agree well with the predictions for the band edge transition strengths given in Fig. 11b. However, away from the band edge, the strengths change dramatically. In the parabolic model, the change is attributed to the tilting of the electron **k** vectors away from the confinement axis [4, 5, 15], and these changes can be predicted using the method of Section 5.3 [15]. However, mixing between the HH and LH bands produces additional changes, which are not accounted for in the simple parabolic model [20, 25, 29].

Perhaps the most interesting feature of Figs. 21a,b is that, in contrast to Fig. 11, the total transition strength, $2 \times \text{TE} + \text{TM}$, is no longer conserved in a given subband transition, decreasing as the in-plane **k** vector increases [29], especially in the C1–LH1 transition (note that the average transition strength is always conserved in the parabolic model). From (44) and (45), this implies that for a given quasi-Fermi level separation (or carrier density), the radiative spontaneous emission is reduced from what one would calculate using the parabolic model (the band-mixing model predicts a longer carrier lifetime than the parabolic model [20]).

It might seem disturbing that the transition strength averaged over all polarizations is not conserved within a given subband transition. However, we must remember that the model includes mixing between the various subbands, and hence we might expect a redistribution of the transition strength among the various subband transitions. This is indeed the case, and in fact transitions that are strictly forbidden in the parabolic model (such as transitions between states of opposite parity, mentioned at the end of Section 3.1) are allowed once significant band mixing begins to occur [20, 29]. For

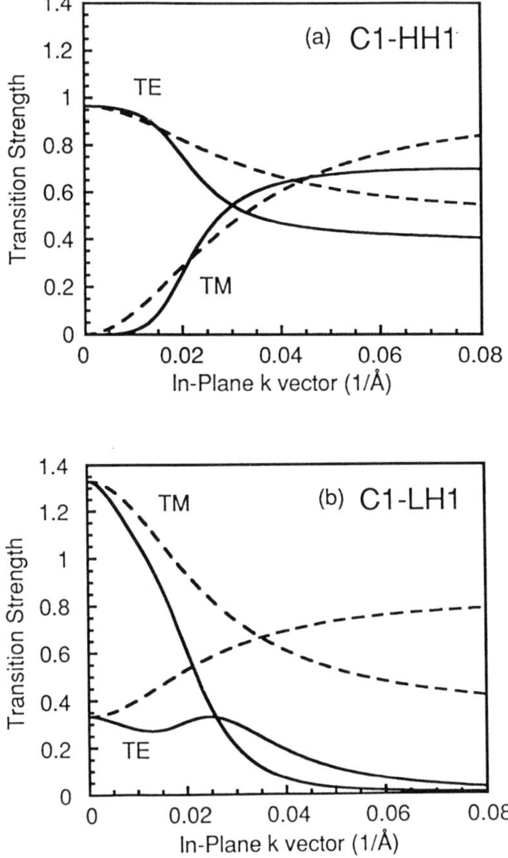


**Fig. 21.** Relative transition strengths for both TE and TM light polarization for the two lowest subband transitions in the unstrained 80-Å $GaAs/Al_{0.2}Ga_{0.8}As$ quantum well. The dashed curves represent what one would calculate assuming the valence subbands were parabolic. The transition matrix element $|M_T|^2$ is found by multiplying the transition strengths in the figure by $|M|^2$ from Table II.

example, this is dramatically illustrated in Fig. 22, where we show the TE transition strength of the C1–HH3 and C2-HH3 "forbidden" transitions in the unstrained quantum well (case a). At the band edge, where the subbands are decoupled, the transition strength is indeed zero for the odd parity transition, C2–HH3, but it immediately picks up some finite transition probability away from the band edge. Other forbidden transitions also

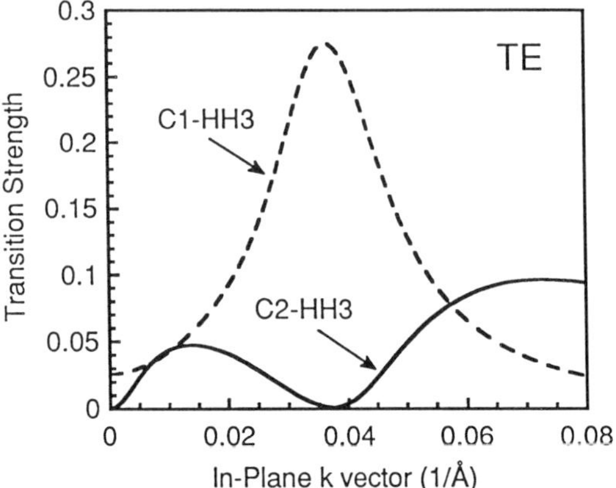


**Fig. 22.** "Forbidden" transition strengths in the unstrained 80-Å GaAs/$Al_{0.2}Ga_{0.8}As$ quantum well. The even-parity transition (C1–HH3) is not truly forbidden, in the sense that a model without band mixing would also predict a finite overlap between the envelope functions. However, the band-mixing model predicts that the strength of the transition becomes quite strong for in-plane **k** vectors away from zero. The odd-parity transition (C2–HH3) should be strictly forbidden in the absence of band mixing. The inclusion of band mixing reveals that for finite in-plane **k** vectors, when subband mixing begins to occur, a redistribution of the transition strength gives the C2–HH3 transition some finite probability of occurring.

display similar character. Thus, the band-mixing model predicts that certain forbidden transitions should be observable, due to a redistribution of the transition strength among the various subband transitions.

The above transition strengths along with the band structure of the previous two sections can now be used in (35) and (44) to obtain the spectral gain and spontaneous emission as a function of either carrier density or radiative current density. In Figs. 23a–d, we show the TE gain spectra for all cases a–c mentioned above as a function of increasing carrier density. In addition, we have also included case a assuming parabolic valence subbands [5, 48, 91] (no band mixing) using the dispersion relations given in (81). The dashed curves are the as-calculated spectra from (36), whereas the solid curves include the smoothing effect of the intraband scattering of the carriers as given in (38) (we have used the lineshape function approximated by Chinn *et al.* [48], since it is particularly simple to implement using fast Fourier transform techniques [92]).

The changes in the absorption characteristics of the material band edge from high absorption to high gain as a function of carrier density are clearly seen in Figs. 23a–d. The difference between the smooth, flat, bulk gain spectra and the more jagged quantum well spectra is also dramatically illustrated. The peaks in the gain/absorption spectra of the quantum well examples arise from the various subband transitions that are possible, as illustrated earlier in Fig. 3 of Section 3. The different subband transitions can be identified in Fig. 23 by the markers placed across the bottom of each quantum well plot. All subband transitions, forbidden or not, are indicated by the markers for both C–HH and C–LH transitions. Thus, forbidden transitions can easily be identified. For example, the C2–HH3 transition appears in Fig. 23a, whereas it is not observable in the parabolic model (Fig. 23c). In all quantum well cases, the peak gain is dominated by the first quantized energy level transitions (C1–HH1 and C1–LH1) for small carrier densities. However, for very high carrier densities contributions to the gain from the second quantized level transitions become significant. The two peaks in the gain spectra are characteristic of quantum well material and can actually cause a discrete change in the lasing wavelength of quantum well lasers (for example, as the length of the cavity is changed [93]).

The peak TE gain of the (smoothed) spectra shown in Figs. 23a–d as a function of both carrier density and radiative current density is shown in Figs. 24a,b [35]. For comparison purposes in Figs. 24a,b, we have included the simplified parabolic band approximations [5, 48 91] (dashed curves) for both the strained and unstrained quantum wells, using the bulk parabolic valence band dispersion relations (97)–(98) and (81), respectively. Features of Fig. 24 that are of practical interest include: (1) the amount of current or carriers necessary to reach zero loss and gain in the material, the so-called transparency level, $N_{tr}$ and $J_{tr}$; and (2) the differential gain $dg/dN$, which is the rate of increase in the gain as a function of carrier or current density. In Figs. 24a,b, the quasi-bulk GaAs well actually has a lower $N_{tr}$ and $J_{tr}$ than the unstrained quantum well; however, the gain increases very slowly; i.e., the differential gain is very low. The soft band edge resulting from using the bulk density of states is largely responsible for this "sluggish" behavior, as discussed in Section 4.1. The unstrained quantum well achieves higher differential gain mostly due to the step-like density of states [91], but it has a higher transparency than the "bulk" well. The advantages of the strained quantum well are clearly evident from Fig. 24, where the best of both worlds

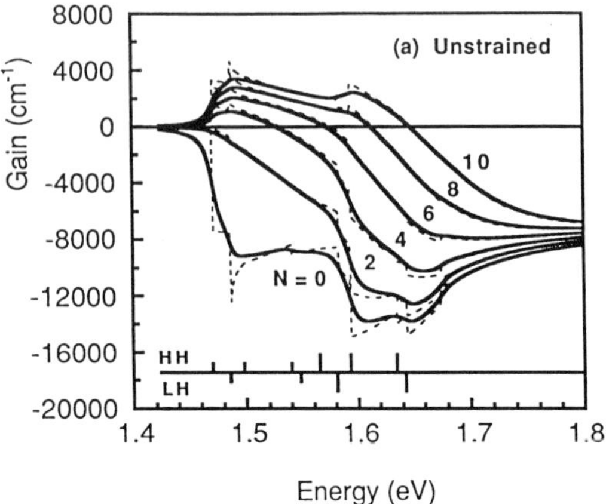


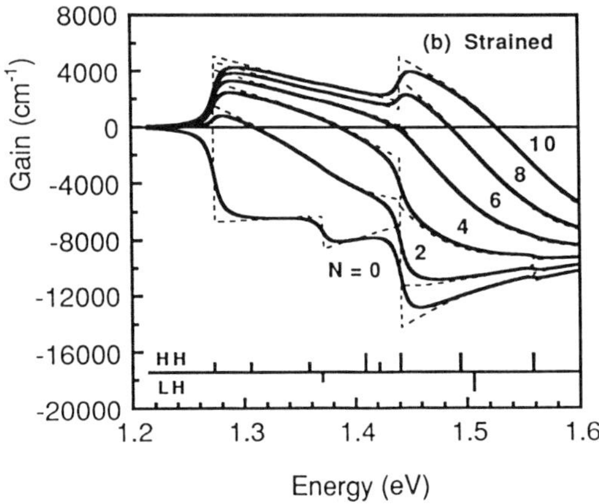


**Fig. 23.** TE gain (absorption) spectra at various indicated carrier densities (in units of $10^{18}$ $cm^{-3}$) for four cases: (a) the unstrained 80-Å $GaAs/Al_{0.2}Ga_{0.8}As$ quantum well; (b) the strained 80-Å $In_{0.2}Ga_{0.8}As/Al_{0.2}Ga_{0.8}As$ quantum well; (c) same as (a), assuming parabolic valence subbands; and (d) bulk material. In all cases, undoped material was assumed. The dashed curves are the as-calculated spectra, while the solid curves have been convolved to take into account the spectral broadening (using a lineshape function defined by Chinn *et al.* [48]). The markers at the bottom of each graph indicate all of the various subband transitions. For example, in (a), the first three short ticks above the line indicate the C1–HH(1–3) subband transitions, whereas the second set of three ticks (twice as tall) indicate the C2–HH(1–3) subband transitions.

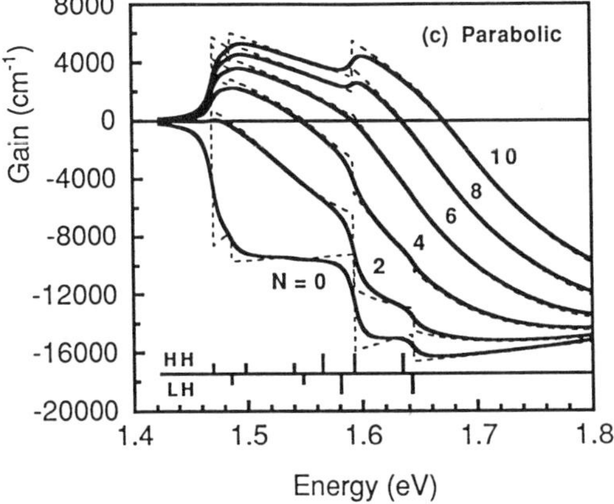


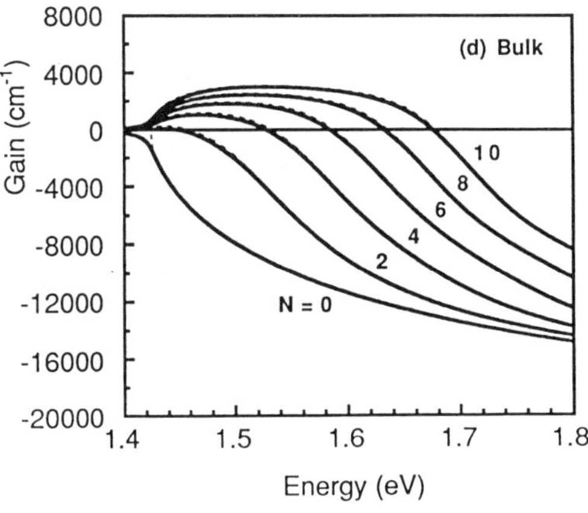


**Fig. 23.** (*Continued*)

is achieved, with both the lowest transparency level and the highest differential gain. This is again mainly due to close matching of the conduction and valence band density of states, as mentioned in the previous section and Section 4.1.

In comparing the parabolic models with the band-mixing models, we find that the parabolic model dramatically underestimates the carrier density required to achieve a given gain in both strained and unstrained cases.

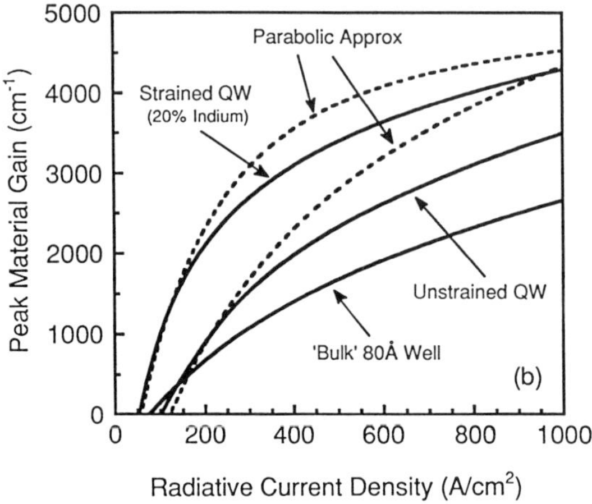


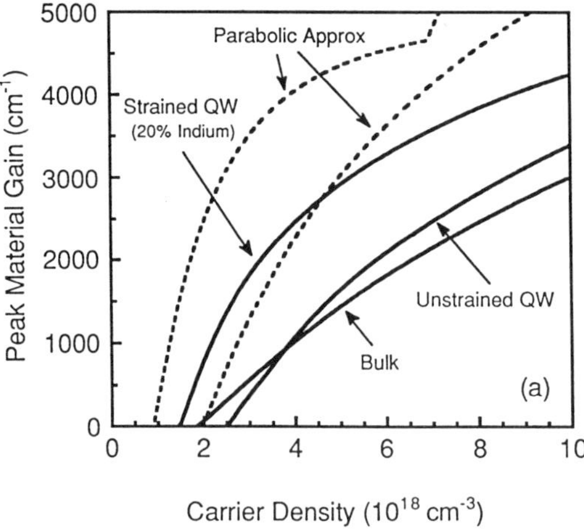


**Fig. 24.** Comparison of the peak material TE gain (from the convolved spectra) as a function of (a) carrier density, and (b) radiative current density, after Ref. [35]. Auger recombination and other nonradiative mechanisms have not been included; thus, the actual material gain is expected to be suppressed by 10–30% of that shown, under high injection conditions.

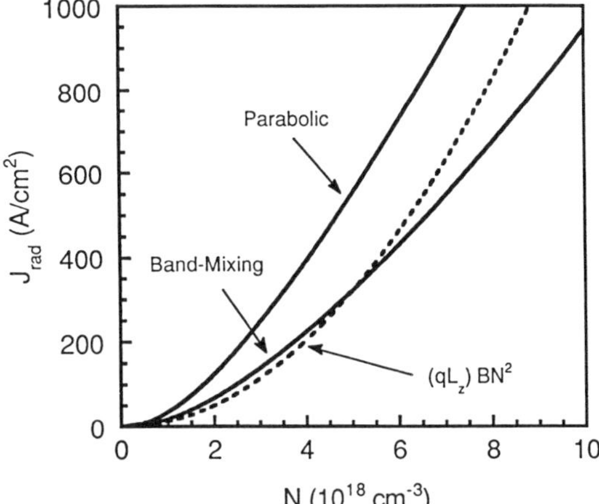


**Fig. 25.** Radiative current density as a function of carrier density for three different models applied to the unstrained 80-Å GaAs/$Al_{0.2}Ga_{0.8}As$ undoped quantum well. For the dashed curve, we used a bimolecular recombination rate constant $B = 10^{-10}$ cm$^3$/s, which is typical for GaAs [94].

Curiously, however, its overestimation of the radiative current density (as a result of the difference in average transition strengths mentioned earlier) compensates for this, and the end result is that the parabolic model predicts a radiative current density similar to the band-mixing model for low current densities [20, 29]. It is interesting to plot the current density directly as a function of the carrier density to see this effect more clearly. Figure 25 compares $J$ versus $N$ for the parabolic band model and the band-mixing model. Also included is the standard bimolecular recombination rate [94] ($J = (eL_z)BN^2$, with $B = 10^{-10}$ cm$^3$/s in GaAs [94]). The overestimation of current density in the parabolic model is very evident from this figure. It is also interesting that neither model predicts an exact square-law dependence, shown by the dashed curve. However, the band-mixing model does yield a pretty good match for low carrier densities.

We have so far been discussing TE gain. Another benefit of the strained quantum well is seen in Fig. 26, where we have plotted the TE and TM peak gain as a function of current density for the strained and unstrained quantum well. The TM gain in the strained quantum well is drastically suppressed due to the effective removal of the LH subband transitions (see Fig. 19), which are largely responsible for TM gain (see Fig. 21). In the conventional

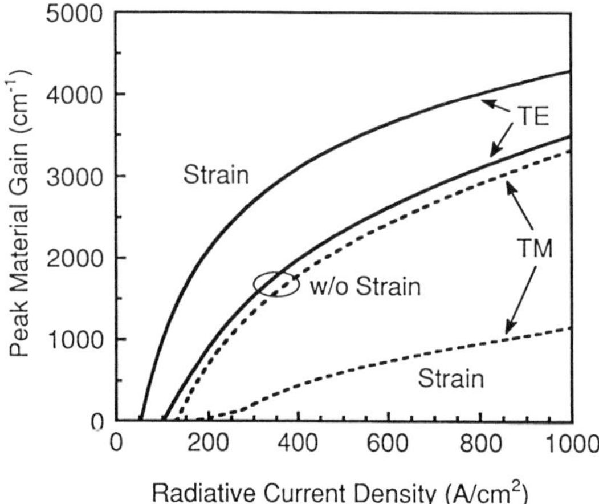


**Fig. 26.** Comparison of TE and TM peak gain in the strained and unstrained quantum wells as a function of radiative current density.

quantum well, the TM gain could compete with the TE gain, possibly reducing the TE to TM mode discrimination in a laser.

Figure 27 displays $J_{tr}$, $N_{tr}$, and $dg/dN$ as a function of indium mole fraction in the 80-Å strained quantum well [35]. We see that large reductions in transparency levels are theoretically predicted for strained quantum wells. In addition, the differential gain, which is related to the ultimate modulation speed of quantum well lasers, is improved by as much as a factor of two or three, implying that strained quantum well lasers should be faster [33–35] as well as having lower threshold currents than their unstrained counterparts. Later chapters in this book will cover both of these exciting prospects in greater detail.

Chen *et al.* [95] have performed a detailed experimental study of the transparency current density $J_{tr}$ in GaAs/AlGaAs quantum well lasers. Their best result (the best reported to date) was a threshold current density of 80 A/cm$^2$. However, Fig. 1 of their paper reveals more typical long cavity threshold current densities in the range of 110–150 A/cm$^2$, with little or no correlation between quantum well width and threshold for $L_z \sim 65$–125 Å. From Fig. 24b, the band-mixing model presented in this chapter predicts a transparency level of about 100 A/cm$^2$ for the unstrained quantum well (the parabolic model predicts a higher value, in the range of $\sim 125$ A/cm$^2$). [*Note*: The parabolic model of Chinn *et al.* [48] indicates a value closer to

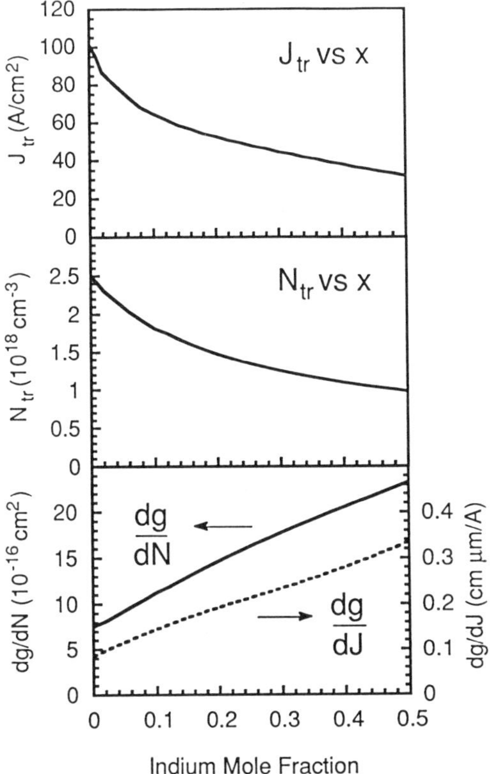


**Fig. 27.** Gain characteristics of the strained 80-Å $In_xGa_{1-x}As/Al_{0.2}Ga_{0.8}As$ quantum well as a function of indium mole fraction $x$. (The differential gain $dg/dJ$ has been multiplied by the well width of 80 Å to obtain units of cm $\mu$m/A.)

100 A/cm$^2$; however, the matrix element they used was 25% too small.] However, the present model does not include nonradiative currents or leakage currents, which have been shown theoretically by Chinn *et al.* [48] to increase the predicted transparency level by 10–20%. In addition, unbound radiative transitions at energies above the well barrier have also been neglected in the present model [52] (these contributions to the radiative current, however, should be small, especially near transparency). Thus, numbers in the range of 110–120 A/cm$^2$ are to be expected theoretically, in reasonable agreement with Chen's experimental findings.

Welch *et al.* [84] have performed an experimental study of the transparency current density as a function of indium mole fraction in strained quantum wells. Their experimental findings are in very good qualitative agreement with Fig. 27. Their *measured* threshold currents densities vary from 160 to 115 A/cm$^2$ as the indium mole fraction increases from 0 to 0.2. Extrapolated transparency levels found by varying the laser cavity lengths indicate corresponding transparency levels that vary from 70 to 50 A/cm$^2$; however, these are not directly measured current densities. Recently, Choi and Wang [96] have reported a record low threshold current density of 65 A/cm$^2$ in strained InGaAs/AlGaAs quantum wells emitting at 1.02 $\mu$m, which is in more quantitative agreement with the theoretical trend expected from Fig. 27.

In all of the gain calculations presented thus far, we have assumed that the quantum well material was undoped. If we were to dope the quantum well n-type, for example, the coulomb potential created by the ionized donor impurities would create donor energy levels just below the conduction band edge (corresponding to the "binding energy" of the electron to the positively charged ion). This changes our quantum well model in two ways. First of all, if the concentration of ionized donors is high enough, the isolated donor energy levels will split into many finely separated levels, creating band tail states and thus altering the band structure near the band edge, which ultimately changes the density of states. The second important change is that these donor levels correspond to an electron bound to an ionized donor, and hence the electron's envelope function is localized around the impurity. Thus, the electron no longer has a sharp momentum, and the **k**-selection rule can no longer be used (see Section 3.1). The use of a reduced density of states (as used in (35)), implicitly assumes **k**-selection. Thus, the expression for gain must be rederived for the case of band-to-bound and bound-to-bound transitions. We will not pursue such treatments in the present chapter; however, interested readers are referred to Casey and Panish [37], where these issues are discussed in great detail.

While the present model cannot take into account the doping effects previously mentioned, it is possible to model the changes required by charge neutrality in the well when ionized impurities are introduced. Charge neutrality will force the quasi-Fermi levels to shift either further into the conduction band (for n-type doping) or further into the valence band (for p-type doping). This shift has dramatic effects on the gain characteristics of the material. To demonstrate this, in Fig. 28 we show how the electron and hole densities, the current density, and the differential gain, all taken at

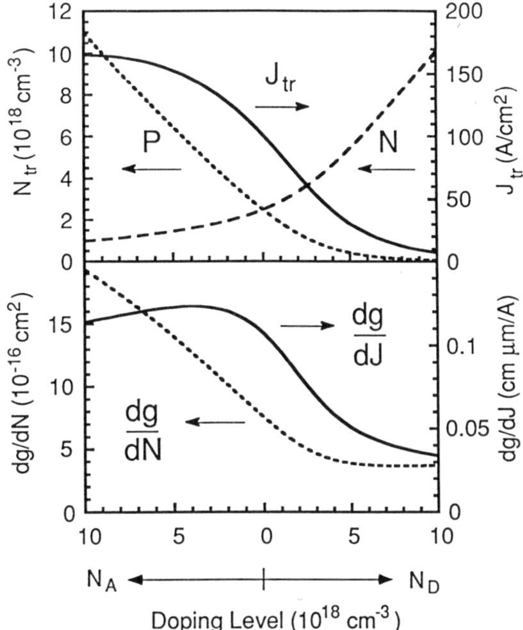


**Fig. 28.** Illustration of the doping dependence of the transparency electron (large dashed line) and hole (small dashed line) density, the transparency current density (solid line), as well as the differential gain at transparency for the case a quantum well. (The differential gain $dg/dJ$ has been multiplied by the well width of 80 Å to obtain units of cm $\mu$m/A). N-type doping is to the right and p-type doping is to the left. Note that, for no doping, the transparency electron and hole densities are equal, as required by charge neutrality. One should keep in mind that these calculations do not take into account band tail states [37] or band-to-bound transitions [37], as mentioned in the text.

transparency, are affected as either p-type acceptor impurities (to the left) or n-type donor impurities (to the right) are introduced into the well of case a defined earlier. Note that the transparency current density can be dramatically reduced when the well is doped n-type. Note also, however, that there is a tradeoff involved, since the differential gain is also reduced in this case. Doping the well p-type should increase the differential gain, but at the expense of a higher transparency current density. In high speed laser applications, where the modulation speed of the laser is perhaps more important than its threshold current, p-type doping would be the most desirable (since increased differential gain should lead to higher speed lasers). The transparency hole and electron densities are also interesting to observe. Note that the hole density in the n-type case is suppressed quite a bit more than the electron density is in the p-type case. This is a result of the asymmetry in the density of states functions in the two bands. We leave it as an interesting exercise for the reader to qualitatively verify the trends observed in Fig. 28, using simple arguments and diagrams similar to those used in Section 4.1 and Fig. 6.

Before closing this section, we would like to point out that one large uncertainty in *all* semiconductor quantum well gain models involves the intraband scattering lineshape function discussed in Section 4.1. Figure 29 shows the effect of the assumed lineshape function on transparency level for the unstrained quantum well. As can be seen, the assumed lineshape can have dramatic effects on the predictions of the theory. Thus, the accuracy of the theoretical predictions of *any* quantum well gain model is largely dependent on a good description of the energy broadening lineshape function. For this reason, M. Asada has contributed an entire chapter of this

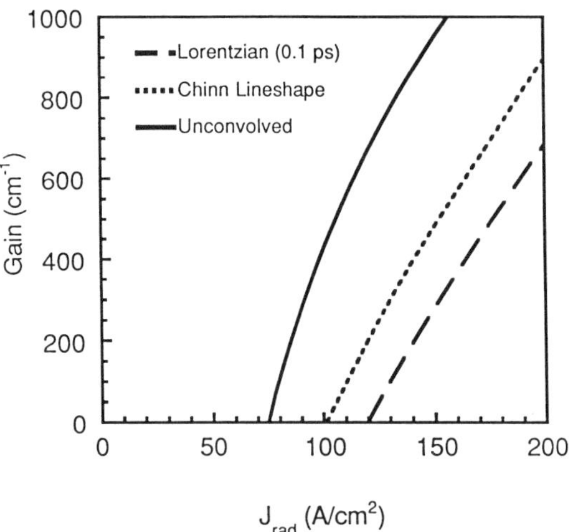


**Fig. 29.** Demonstration of the effects of the assumed lineshape function on theoretically calculated peak gain as a function of radiative current density. The transparency condition can vary by as much as 50% depending on how one theoretically treats the spectral broadening of electron states.

book (Chapter 2) devoted entirely to the important discussion of energy broadening mechanisms in quantum wells.

## 10. CONCLUSION

In the present chapter, we have tried to present some of the current methods used to understand and predict the behavior of optical gain in III-V semiconductors. While it has not been possible to cover all aspects of gain theory in this one chapter, we have attempted to highlight some of the more recent developments not currently covered in standard textbook treatments on the subject. These include, among other things, a detailed analysis of band-mixing effects in the valence bands of quantum wells, the effects of strain on the valence subband structure, and a simple description of polarization-dependent phenomena observable in quantum wells and quantum wires. The following chapters should provide good complementary reading to the present chapter, where strained quantum wells, long wavelength semiconductors, Auger recombination processes, intraband scattering mechanisms in quantum wells, and quantum wire lasers are discussed extensively.

## REFERENCES


1. Liboff, R. L. (1980). *Introductory quantum mechanics.* Holden-Day, Oakland.
2. Kane, E. O. (1956). *J. Phys. Chem. Solids* **1**, 82.
3. Kane, E. O. (1957). *J. Phys. Chem. Solids* **1**, 249.
4. Yamanishi, M., and Suemune, I. (1984). *Japan. J. Appl. Phys.* **23**, L35.
5. Asada, M., Kameyama, A., and Suematsu, Y. (1984). *IEEE J. Quantum Electron.* **QE-20**, 745.
6. Asada, M., Miyamoto, Y., and Suematsu, Y. (1985). *Japan. J. Appl. Phys.* **24**, L95.
7. Asada, M., Miyamoto, Y., and Suematsu, Y. (1986). *IEEE J. Quantum Electron.* **QE-22**, 1915.
8. Iwamura, H., Saku, T., Kobayashi, H., and Horikoshi, Y. (1983). *J. Appl. Phys.* **54**, 2692.
9. Yamada, M., Ogita, S., Yamagishi, M., Tabata, K., Nakaya, N., Asada, M., and Suematsu, Y. (1984). *Appl. Phys. Lett.* **45**, 324.
10. Yamada, M., Ogita, S., Yamagishi, M., and Tabata, K., (1985). *IEEE J. Quantum Electron.* **QE-21**, 640.
11. Weiner, J. S., Chemla, D. S., Miller, D. A. B., Haus, H. A., Gossard, A. C., Wiegmann, W., and Burrus, C. A. (1985). *Appl. Phys. Lett.* **47**, 664.

12. WEINER, J. S., MILLER, D. A. B., CHEMLA, D. S., DAMEN, T. C., BURRUS, C. A., WOOD, T. H., GOSSARD, A. C., and WIEGMANN, W. (1985). *Appl. Phys. Lett.* **47**, 1148.
13. TSUCHIYA, M., GAINES, J. M., YAN, R. H., SIMES, R. J., HOLTZ, P. O., COLDREN, L. A., and PETROFF, P. M. (1989). *Phys. Rev. Lett.* **62**, 466.
14. TANAKA, M., and SAKAKI, H. (1989). *Appl. Phys. Lett.* **54**, 1326.
15. CORZINE, S. W., YAN, R. H., and COLDREN, L. A., unpublished.
16. LUTTINGER, J. M., and KOHN, W. (1955). *Phys. Rev.* **97**, 869.
17. LUTTINGER, J. M. (1956). *Phys. Rev.* **102**, 1030.
18. SUEMUNE, I., and COLDREN, L. A. (1988). *IEEE J. Quantum Electron.* **QE-24**, 1778.
19. EPPENGA, R., SCHUURMANS, M. F. H., and COLAK, S. (1987). *Phys. Rev. B* **36**, 1554.
20. COLAK, S., EPPENGA, R., and SCHUURMANS, M. F. H. (1987). *IEEE J. Quantum Electron.* **QE-23**, 960.
21. ALTARELLI, M., EKENBERG, U., and FASOLINO, A. (1985). *Phys. Rev. B* **32**, 5138.
22. ANDREANI, L. C., PASQUARELLO, A., and BASSANI, F. (1987). *Phys. Rev. B* **36**, 5887.
23. BROIDO, D. A., and SHAM, L. J. (1985). *Phys. Rev. B* **31**, 888.
24. CHUANG, S. L. (1989). *Phys. Rev. B* **40**, 10379.
25. AHN, D., and CHUANG, S. L. (1990). *IEEE J. Quantum Electron.* **QE-26**, 13.
26. AHN, D., and CHUANG, S. L. (1988). *IEEE J. Quantum Electron.* **QE-24**, 2400.
27. BASTARD, G., and BRUM, J. A. (1986). *IEEE J. Quantum Electron.* **QE-22**, 1625. See also BASTARD, G. (1988). *Wave mechanics applied to semiconductor heterostructures.* Editions de physique, Les Ulis, France.
28. O'REILLY, E. P. (1989). *Semicond. Sci. Technol.* **4**, 121.
29. CORZINE, S. W., YAN, R. H., and COLDREN, L. A., unpublished.
30. YABLANOVITCH, E., and KANE, E. O. (1986). *IEEE J. Lightwave Technol.* **LT-4**, 504.
31. YABLANOVITCH, E., and KANE, E. O. (1986). *IEEE J. Lightwave Technol.* **LT-4**, 961.
32. ADAMS, A. R. (1986). *Electron. Lett.* **22**, 249.
33. SUEMUNE, I., COLDREN, L. A., YAMANISHI, M., and KAN, Y. (1988). *Appl. Phys. Lett.* **53**, 1378.
34. LAU, K. Y., XIN, S., WANG, W. I., BAR-CHAIM, N., and MITTELSTEIN, M. (1989). *Appl. Phys. Lett.* **55**, 1173.
35. CORZINE, S. W., YAN, R. H., and COLDREN, L. A. (1990). *Appl. Phys. Lett.* **57**, 2835.
36. MADELUNG, O. (1978). *Introduction to solid-state theory.* Springer-Verlag, Berlin.
37. CASEY, H. C., Jr., and PANISH, M. B. (1978). *Heterostructure lasers; Part A: Fundamental principles.* Academic Press, Orlando, Florida.
38. BLAKEMORE, J. S. (1982). *J. Appl. Phys.* **53**, R123.
39. FEYNMAN, R. P., LEIGHTON, R. B., and SANDS, M. (1963). *The Feynman lectures on physics*, vols. II and III. Addison-Wesley, Reading, Massachusetts.
40. WEST, L. C., and EGLASH, S. J. (1985). *Appl. Phys. Lett.* **46**, 1156.
41. GOWER, J. (1984). *Optical communication systems.* Prentice-Hall, London.
42. NISHIMURA, Y., (1974). *Japan. J. Appl. Phys.* **13**, 109.
43. ZEE, B. (1978). *IEEE J. Quantum Electron.* **QE-14**, 727.
44. YAMADA, M., ISHIGURO, H., and NAGATO, H. (1980). *Japan. J. Appl. Phys.* **19**, 135.

45. Yamada, M., and Ishiguro, H. (1981). *Japan. J. Appl. Phys.* **20**, 1279.
46. Landsberg, P. T., and Robbins, D. J. (1985). *Solid State Electron.* **28**, 137.
47. Yamanishi, M., and Lee, Y. (1987). *IEEE J. Quantum Electron.* **QE-23**, 367.
48. Chinn, S. R., Zory, P., and Reisinger, A. R. (1988). *IEEE J. Quantum Electron.* **QE-24**, 2191.
49. Asada, M. (1989). *IEEE J. Quantum Electron.* **QE-25**, 2019.
50. Kucharska, A. I., and Robbins, D. J. (1990). *IEEE J. Quantum Electron.* **QE-26**, 443.
51. Agrawal, G. P., and Dutta, N. K. (1986). *Long-wavelength semiconductor lasers.* Van Nostrand Reinhold, New York.
52. Nagarajan, R., Kamiya, T., and Kurobe, A. (1989). *IEEE J. Quantum Electron.* **QE-25**, 1161.
53. Long, D. (1968). *Energy bands in semiconductors.* Interscience, New York.
54. Cohen-Tannoudji, C., Diu, B., and Laloe, F. (1977). *Quantum mechanics, vol. I*, p. 842. Wiley-Interscience, Toronto.
55. Ehrenreich, H. (1961). *Suppl. J. Appl. Phys.* **32**, 2155.
56. Cardona, M. (1963). *Phys. Rev.* **131**, 98.
57. Chadi, D. J., Clark, A. H., and Burnham, R. D. (1976). *Phys. Rev. B* **13**, 4466.
58. Hermann, C., and Weisbuch, C. (1977). *Phys. Rev. B* **15**, 823.
59. See Agranovich, V. M., and Maradudin, A. A., eds. (1984). *Modern problems in condensed matter sciences; vol. 8: Optical orientation*, pp. 463–508. North-Holland Physics Publishing Co.
60. Jani, B., Gilbart, P., Portal, J. C., and Aulombard, R. L. (1985). *J. Appl. Phys.* **58**, 3481.
61. Nicholas, R. J., Portal, J. C., Houlbert, C., Perrier, P., and Pearsall, T. P. (1979). *Appl. Phys. Lett.* **34**, 492.
62. Yan, R. H., Corzine, S. W., and Coldren, L. A. (1990). *IEEE J. Quantum Electron.* **QE-26**, 213.
63. Yi, J. L., and Dagli, N., submitted to *IEEE J. Quantum Electron.*
64. Schuurmans, M. F. H., and 't Hooft, G. W. (1985). *Phys. Rev. B* **31**, 8041.
65. Bastard, G. (1981). *Phys. Rev. B* **24**, 5693.
66. Einevoll, G. T., Hemmer, P. C., and Thomsen, J. (1990). *Phys. Rev. B* **42**, 3485.
67. Alterelli, M. (1983). *Phys. Rev. B* **28**, 842.
68. Landolt-Bornstein (1982). In *Numerical data and functionships in science and technology* (Madelung, O., ed.), New Series III, Vols. 17 and 22a. Springer, Berlin.
69. Adachi, S. (1985). *J. Appl. Phys.* **58**, R1.
70. Dresselhauss, G., Kip, A. F., and Kittel, C. (1955). *Phys. Rev.* **98**, 368.
71. Hensel, J. C., and Feher, G. (1963). *Phys. Rev.* **129**, 1041.
72. Asai, H., and Oe, K. (1983). *J. Appl. Phys.* **54**, 2052.
73. Kolbas, R. M., Anderson, N. G., Laidig, W. D., Sin, Y. K., Lo, Y. C., Hsieh, K. Y., and Yang, Y. J. (1988). *IEEE J. Quantum Electron.* **QE-24**, 1605.
74. Lu, X. Z., Garuthara, R., Lee, S., and Alfano, R. R. (1988). *Appl. Phys. Lett.* **52**, 93.

75. Ji, G., Huang, D., Reddy, U. K., Henderson, T. S., Houdre, R., and Morkoc, H. (1987). *J. Appl. Phys.* **62**, 3366.
76. Marzin, J. Y., Charasse, M. N., and Sermage, B. (1985). *Phys. Rev. B* **31**, 8298.
77. Menendez, J., Pinczuk, A., Werder, D. J., Sputz, S. K., Miller, R. C., Sivco, D. L., and Cho, A. Y. (1987). *Phys. Rev. B* **36**, 8165.
78. Joyce, M. J., Johnson, M. J., Gal, M., and Usher, B. F. (1988). *Phys. Rev. B* **38**, 10978.
79. Gershoni, D., Vandenberg, J. M., Chu, S. N. G., Temkin, H., Tanbun-Ek, T., and Logan, R. A. (1989). *Phys. Rev. B* **40**, 10017.
80. Moore, K. J., Duggan, G., Woodbridge, K., and Roberts, C. (1990). *Phys. Rev. B* **41**, 1090.
81. Beernink, K. J., York, P. K., Coleman, J. J., Waters, R. G., Kim, J., and Wayman, C. M. (1989). *Appl. Phys. Lett.* **55**, 2167.
82. Larsson, A., Cody, J., and Lang, R. J. (1989). *Appl. Phys. Lett.* **55**, 2268.
83. Beernink, K. J., York, P. K., and Coleman, J. J. (1989). *Appl. Phys. Lett.* **55**, 2585.
84. Welch, D. F., Streifer, W., Schaus, C. F., Sun, S., and Gourley, P. L. (1990). *Appl. Phys. Lett.* **56**, 10.
85. Bour, D. P., Martinelli, R. U., Hawrylo, F. Z., Evans, G. A., Carlson, N. W., and Gilbert, D. B. (1990). *Appl. Phys. Lett.* **56**, 318.
86. Chong, T. C., and Fonstad, C. G. (1989). *IEEE J. Quantum Electron.* **QE-25**, 171.
87. Baldereschi, A., and Lipari, N. O. (1973). *Phys. Rev. B* **8**, 2697.
88. Watanabe, M. O., Yoshida, J., Mashita, M., Nakanisi, T., and Hojo, A. (1985). *J. Appl. Phys.* **57**, 5340.
89. Kroemer, H. (1986). *Surface Sci.* **174**, 299.
90. Reithmaier, J. P., Hoger, R., and Riechert, H. (1990). *Appl. Phys. Lett.* **56**, 536.
91. Arakawa, Y., and Yariv, A. (1986). *IEEE J. Quantum Electron.* **QE-22**, 1887.
92. Cochran, W. T., Cooley, J. W., Favin, D. L., Helms, H. D., Kaenel, R. A., Lang, W. W., Maling, G. C., Nelson, D. E., Rader, C. M., and Welch, P. D. (1967). *IEEE Trans. Audio Electroacoust.* **AU-15**, 45.
93. Zory, P. S., Reisinger, A. R., Waters, R. G., Mawst, L. J., Zmudzinski, C. A., Emanuel, M. A., Givens, M. E., and Coleman, J. J. (1986). *Appl. Phys. Lett.* **49**, 16.
94. Kinoshita, S., and Iga, K. (1987). *IEEE J. Quantum Electron.* **QE-23**, 882.
95. Chen, H. Z., Ghaffari, A., Morkoc, H., and Yariv, A. (1987). *Appl. Phys. Lett.* **51**, 2094.
96. Choi, H. K., and Wang, C. A. (1990). *Appl. Phys. Lett.* **57**, 321.

# Chapter 2

# INTRABAND RELAXATION EFFECT ON OPTICAL SPECTRA

**Masahiro Asada**

*Department of Electrical and Electronic Engineering*
*Tokyo Institute of Technology*
*Tokyo, Japan*



## 1. INTRODUCTION

Optical spectra, e.g., gain, spontaneous emission, etc., are the most basic properties of semiconductor lasers. Main factors determining an optical spectrum are (i) carrier distribution in the energy bands, (ii) transition



ISBN 0-12-781890-1

matrix elements between electrons and holes, and (iii) intraband relaxation of carriers due to various scattering processes.

Carrier distribution is the most important factor that determines the overall profile of a spectrum and the relation between the peak intensity and carrier density. Change of the density of states from the parabola of bulk semiconductors to the step-like shape of quantum wells brings about various superior characteristics, such as the increase of peak gain and the threshold reduction [27–30]. The transition matrix element determines the intensity of a spectrum and, also, its anisotropy results in the polarization dependence in quantum well lasers [3, 23, 24].

The intraband relaxation causes broadening of an optical spectrum. Deformation and reduction of the peak values take place in the gain and emission spectra [1, 2]. This effect is quite remarkable in quantum well lasers [3, 4, 8, 17]. The spectral shape becomes smooth and broad in spite of the sharp step-like density of states. The intraband relaxation plays an important role also in oscillation mode behavior [1, 15, 31], where the relaxation is the main factor determining the magnitude of the gain suppression coefficient.

In this chapter, theoretical analysis is described for the effect of intraband relaxation on optical spectra in quantum well lasers. In Section 2, optical spectra affected by intraband relaxation are analyzed, with the gain spectra as an example. The effect of intraband relaxation is expressed by the line broadening function, which is the Fourier transform of the response of the polarization to an impulse electric field. The Lorentzian shape is obtained for the broadening function as an approximation, the half width of which is given by the reciprocal scattering probability, i.e., the intraband relaxation time. The analysis of the intraband relaxation time [9] in quantum well lasers is shown in Section 3. Carrier–carrier and carrier–longitudinal optical (LO) phonon scattering processes are taken into account. In Section 4, spectral lineshape is calculated without the preceding Lorentzian approximation. Deviation of the lineshape from the Lorentzian is discussed. It is shown that the lineshape is narrower than the Lorentzian with the intraband relaxation time obtained in Section 3, and the low-energy tail is much steeper than the Lorentzian. An approximate formula for the line broadening function is also given with the parameters used in the Lorentzian, and the calculation process is summarized in this section. The conclusion is given in Section 5.

Discussion is restricted to only spectral broadening due to intraband relaxation, which is one of the main results of many-body theory in an electron–hole system. Other phenomena related to many-body theory [11] are not considered here.

## 2. OPTICAL TRANSITION WITH INTRABAND RELAXATION

### 2.1. Introduction

The effect of intraband relaxation on optical spectra is schematically shown in Fig. 1 for a quantum well structure (see also Fig. 29 of Chapter 1). Due to the intraband relaxation, broadening and reduction of the peak values take place in the gain and emission spectra.

The intraband relaxation has been taken into account in a gain calculation by the following equation [1–4, 15, 17, 31]:

$$G(\hbar\omega) = \int_{E'_g}^{\infty} g(E_{cv})L(\hbar\omega - E_{cv})\, dE_{cv}, \tag{1}$$

here $G(\hbar\omega)$ is the gain coefficient as a function of photon energy $\hbar\omega$ including the effect of the intraband relaxation, which corresponds to the solid curve in Fig. 1a, and $g(E_{cv})$ is the gain coefficient without the intraband relaxation; i.e., $g(\hbar\omega)$ corresponds to the dashed curve in Fig. 1a. $g(\hbar\omega)$ is determined by the carrier distribution in the energy band and the transition matrix element and has been discussed in detail in Chapter 1. $L(\hbar\omega - E_{cv})$ is the line broadening function characterized by the intraband relaxation phenomenon, which is the central part of the discussion in this chapter.

Equation (1) means that the gain $g$ at the transition energy $E_{cv}$ ($\neq$ the photon energy $\hbar\omega$) is added to the gain at the transition energy just equal to $\hbar\omega$ with the weight $L(\hbar\omega - E_{cv})$. Since negative $g$ at large $E_{cv}$ affects the

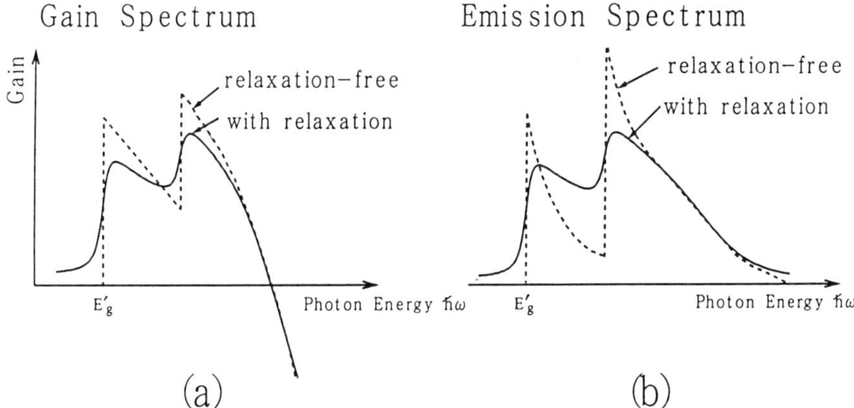


**Fig. 1.** Schematic spectra of (a) optical gain and (b) spontaneous emission under carrier injection in quantum well structures. Spectra become smooth and broad due to the relaxation by scattering.

gain peak through $L(\hbar\omega - E_{cv})$, $G$ is usually smaller than $g$ around the gain peak.

If we employ density matrix theory with the phenomenologically introduced intraband relaxation time [1], then the following Lorentzian lineshape often used for the gain analysis [1–4, 15, 17, 31] is obtained for the line broadening function $L(\hbar\omega - E_{cv})$:

$$L(\hbar\omega - E_{cv}) = \frac{1}{\pi} \frac{\hbar/\tau_{in}}{(\hbar\omega - E_{cv})^2 + (\hbar/\tau_{in})^2}, \tag{2}$$

where $\tau_{in}$ is called the intraband relaxation time, which is regarded as the reciprocal intraband scattering probability averaged between electrons and holes [1]. Theoretical estimation has been made for intraband relaxation time in bulk semiconductors [5–7] and quantum well structures [9] by taking into account various scattering mechanisms.

The overall gain profile and its magnitude depend considerably on the shape of the line broadening function via Eq. (1). Since the preceding Lorentzian shape is a result of phenomenological consideration, it is desirable to deduce the line broadening function analytically.

In this section, we derive the gain formula Eq. (1) without introducing any phenomenological relaxation times. General expression of the line broadening function in Eq. (1) is obtained by this derivation. This line broadening function is of non-Lorentzian shape [8, 9, 26], and it is shown that the Lorentzian shape is obtained from this general expression *as an approximation.* The intraband relaxation time is related to the scattering probability through this approximation. General expression of the broadening (non-Lorentzian shape) and its effect on the gain is discussed in detail in Section 4.

## 2.2. Theoretical Analysis of Optical Transition with Intraband Relaxation

The effect of intraband relaxation is discussed, with the gain spectra as an example. Optical gain results from the polarization induced by a lightwave train in an excited laser medium. The effect of intraband relaxation is to disturb this polarization by various carrier scattering processes. Scattering acts as frictional resistance on oscillation of the polarization, and the resonance spectrum of the oscillation to the lightwave train becomes broad. Here, we use the semiclassical theory; i.e., the relation between the gain and the polarization is obtained classically, and the polarization and its response to a lightwave is treated quantum mechanically.

The gain coefficient $G$ is related to the polarization by the following equation, a result of the Maxwell's equation [1, 21]:

$$G = \frac{\omega}{n_r}\sqrt{\frac{\mu_0}{\varepsilon_0}}\frac{1}{E_0 T_0}\,\mathrm{Im}\int_0^{T_0} P(t)e^{i\omega t}\,dt, \tag{3}$$

where $\omega$ is the angular frequency of light, $n_r$ is the refractive index, $\mu_0$ and $\varepsilon_0$ are the permeability and dielectric constant of the vacuum, respectively, $T_0$ is the period of the lightwave ($T_0 = 2\pi/\omega$), the notation Im refers to the imaginary part, and the polarization $P(t)$ is assumed to be induced by an incident light with the electric field $E(t) = E_0 \exp(-i\omega t)$ + complex conjugate.

The polarization $P(t)$ is given as follows by the *one particle* density operator $\rho$ [1]:

$$P(t) = N\,\mathrm{Tr}(\rho R) = N \sum_{m,n} [\rho_{mn}(t)R_{nm} + \rho_{nm}(t)R_{mn}], \tag{4}$$

where $N$ is the electron density, $R$ is the dipole moment operator [1, 3, 15], and the subscripts $m$ and $n$ refer to the energy levels in conduction and valence bands, respectively. $\rho_{mn}(t)$ is usually calculated by the equation of motion of the density operator with the phenomenologically introduced intraband relaxation time [1, 31]. Rather than use this method, many particle quantum theory is used here in order to take into account various scattering processes. $\rho_{mn}(t)$ is written with the second quantization representation as

$$\begin{aligned}\rho_{mn}(t) &= \langle a_m a_n^\dagger \rangle / N \\ &= \frac{\mathrm{Tr}[e^{-(H_0+H_r)/kT} a_m(t) a_n^\dagger(t)]}{N\,\mathrm{Tr}[e^{-(H_0+H_r)/kT}]},\end{aligned} \tag{5}$$

where $\langle \cdots \rangle$ denotes the ensemble average, $a_m$ and $a_n^\dagger$ are the annihilation and creation operators of electrons in the Schrödinger representation, respectively, and $a_m(t)$ and $a_n^\dagger(t)$ are those in the Heisenberg representation. $a_m(t)$ and $a_n^\dagger(t)$ satisfy the following equation of motion:

$$a_m(t) = U^\dagger(t) a_m U(t), \tag{6}$$

$$a_n^\dagger(t) = U(t) a_n^\dagger U^\dagger(t), \tag{7}$$

with

$$i\hbar \frac{\partial U(t)}{\partial t} = H U(t). \tag{8}$$

The total Hamiltonian $H$ of the system is given by

$$H = H_0 + H_r + H_c \tag{9}$$

where $H_0$ is the Hamiltonian of the many electron system without any perturbation; $H_r$ is that of various scattering processes; and $H_c$ is the coherent interaction of electrons with light, given by

$$H_c(t) = -\sum_{m,n} [a_m^\dagger a_n R_{nm} + a_n^\dagger a_m R_{mn}]E(t), \tag{10}$$

where the electric field of light $E(t)$ is treated classically, as before.

Calculating $a_m(t)a_n^\dagger(t)$ in Eq. (5) up to the first order perturbation with respect to $H_c$ by Eqs. (6)–(10) (see, for example, [10]), and using the rotating wave approximation [21], $\rho_{mn}(t)$ is obtained as

$$\rho_{mn}(t) = R_{mn} \int_{-\infty}^{\infty} Q^R(t')e^{i\omega t'}\, dt'\, E_0 e^{-i\omega t}/N, \tag{11}$$

where

$$\begin{aligned} Q^R(t) &= (1/i\hbar)\langle[a_m^{\dagger\prime}(t)a_n'(t),\, a_m'(0)a_n^{\dagger\prime}(0)]\rangle u(t) \\ &= \frac{\operatorname{Tr}(e^{-(H_0+H_r)/kT}[a_m^{\dagger\prime}(t)a_n'(t),\, a_m'(0)a_m^{\dagger\prime}(0)])}{i\hbar \operatorname{Tr}(e^{-\mathrm{H}_0+H_r)/kT})} u(t), \end{aligned} \tag{12}$$

where [ , ] denotes the commutator bracket; $u(t)$ is the unit step function: i.e., $u(t) = 1$ if $t > 0$, and otherwise $u(t) = 0$; and prime means the interaction representation, with $H_0 + H_r$ being the unperturbed Hamiltonian. Equation (11) gives the response of $\rho_{mn}$ to the incident lightwave $E_0 e^{-i\omega t}$. If the incident light is an impulse $E_0\delta(t)$ $(=E_0 \int_{-\infty}^{\infty} e^{-i\omega t}\, d\omega)$, then $\rho_{mn}$ is obtained by integrating Eq. (11) with respect to $\omega$, and it is just proportional to $Q^R(t)$. From this result and Eq. (4), $Q^R(t)$ means the response of the polarization to an impulse light. This function $Q^R(t)$ is a kind of retarded Green's function (see, for example, [10]). The preceding treatment is an application of the linear-response theory of Kubo [19].

From Eqs. (3), (4), and (11), the gain coefficient is written as

$$G = \frac{\omega}{n_r}\sqrt{\frac{\mu_0}{\varepsilon_0}} \sum_{m,n} |R_{mn}|^2 \operatorname{Im} \int_{-\infty}^{\infty} Q^R(t)e^{i\omega t}\, dt. \tag{13}$$

Thus, the gain coefficient is expressed by the Fourier transform of the impulse response of the polarization $Q^R(t)$. If $Q^R(t)$ decays exponentially, then the Fourier transform gives the Lorentzian shape.

For calculation of $Q^R(t)$, there are two ways. One is to calculate Eq. (12) directly with Eqs. (6)–(9) using the perturbation method, as is done by Yamanishi and Lee [16, 20]. The other is the method employing the temperature Green's function, which is usually used in the many particle theory (see, for example, [10, 11]).

We use the latter method. This method gives the Fourier transform of $Q^{R}(t)$ directly without the derivation of temporal behavior. The calculation process is mentioned here briefly. Comparison between these two methods is discussed at the end of Section 4. First, the temperature Green's function corresponding to $Q^{R}(t)$ is considered [10]. Its Fourier expansion coefficient is calculated in the form of the perturbation power series with respect to the Hamiltonian $H_r$. The carrier–carrier and carrier–LO phonon scattering processes are considered here for $H_r$. In this calculation process, the temperature Green's function is written as a product of the one particle temperature Green's functions of electron and hole, if the direct scattering (the pair states [11]) is neglected between electron and hole constructing the dipole moment. Next, the Fourier transform of $Q^{R}(t)$ is obtained from the Fourier coefficient of the temperature Green's function by analytic continuation in the complex plane.

By this calculation process, the Fourier transform of $Q^{R}(t)$ is obtained as

$$\int_{-\infty}^{\infty} Q^{R}(t)e^{i\omega t}\,dt = -\int_{-\infty}^{\infty}\int_{-\infty}^{\infty}[f_c(E)+f_v(E')-1] \times \frac{D_c(E)D_v(E')}{(\hbar\omega - E - E' - E_g) + i0^{\dagger}}\,dE\,dE' \tag{14}$$

$$\simeq -[f_c(E_m)+f_v(E_n)-1] \times \int_{-\infty}^{\infty}\int_{-\infty}^{\infty}\frac{D_c(E)D_v(E')}{(\hbar\omega - E - E' - E_g) + i0^{+}}\,dE\,dE' \tag{15}$$

and

$$\operatorname{Im}\int_{-\infty}^{\infty} Q^{R}(t)e^{i\omega t}\,dt \simeq [f_c(E_m)+f_v(E_n)-1] \times \int_{-\infty}^{\infty} D_c(E)D_v(\hbar\omega - E - E_g)\,dE, \tag{16}$$

where $0^{+}$ is a positive small quantity, $f_c$ and $f_v$ are the Fermi functions of electrons and holes given later, $E_g$ is the bulk bandgap energy, and $D_c$ and $D_v$ are the energy spectral functions of the levels $m$ and $n$ obtained from the imaginary parts of the one particle Green's functions [10]. Detailed formulae are given in the following sections. These functions appear in the preceding equations since the original Green's function is separated into two one-particle Green's functions as mentioned in the preceding. In the calculation from Eq. (14) to Eq. (15), we have assumed that $D_c$ and $D_v$ have sharp peaks around $E_m$ and $E_n$, respectively, and that their distributions are much narrower than those of $f_c$ and $f_v$.

By substituting Eq. (16) into Eq. (13), the gain coefficient including the line broadening function is obtained. This is discussed in the next section.

### 2.3. Gain Spectra with Relaxation Broadening

Using Eqs. (13) and (15), the gain coefficient in a quantum well structure is determined as follows. First, the subscripts $m$ and $n$ are rewritten as c$j\mathbf{k}_{\parallel}$ and v$j'\mathbf{k}'_{\parallel}$, where c and v refer to conduction and valence bands, respectively, $j$ and $j'$ are the subband numbers of a quantum well structure, and $\mathbf{k}_{\parallel}$ and $\mathbf{k}'_{\parallel}$ are the wave vectors parallel to the well interface. In the matrix element $R_{mn}$, the selection rule $j = j'$ and $\mathbf{k}_{\parallel} = \mathbf{k}'_{\parallel}$ holds between the levels $m$ and $n$ (see, for example, [3]). Substituting Eq. (16) into Eq. (13) and transforming the summation in Eq. (13) to the integral, the final formula for the gain spectrum is obtained as

$$G = \frac{\pi\omega}{n_r}\sqrt{\frac{\mu_o}{\varepsilon_0}}\sum_j \int_{E_g+E_{cj}+E_{vj}}^{\infty} \rho_{cvj}(E_{cv})\langle R_{cv}^2(E_{cv})\rangle \times [f_c(E_{cj\mathbf{k}_{\parallel}}) + f_v(E_{vj\mathbf{k}_{\parallel}}) - 1]L(\hbar\omega - E_{cv})\, dE_{cv}, \tag{17}$$

where $\langle R_{cv}^2(E_{cv})\rangle$ is the averaged square of the dipole moment [3], and $E_{cj\mathbf{k}_{\parallel}}$ and $E_{vj\mathbf{k}_{\parallel}}$ are given by

$$E_{cj\mathbf{k}_{\parallel}} = E_{cj} + \hbar^2 k_{\parallel}^2/2m_c, \tag{18}$$

$$E_{vj\mathbf{k}_{\parallel}} = E_{vj} + \hbar^2 k_{\parallel}^2/2m_v, \tag{19}$$

with the quantized energy levels $E_{cj}$ and $E_{vj}$ of the quantum well and effective masses $m_c$ and $m_v$. $E_{cv}$ is the energy difference between electron and hole in the optical transition given by

$$E_{cv} = E_g + E_{cj\mathbf{k}_{\parallel}} + E_{vj\mathbf{k}_{\parallel}}, \tag{20}$$

where $E_g$ is the bulk bandgap energy. Positions of these energy levels are schematically shown in Fig. 2. $\rho_{cvj}$, $f_c$, and $f_v$ are the density of states of subband $j$ and the Fermi distribution functions of the electron in the conduction band and the hole (*not electron*) in valence band, respectively.

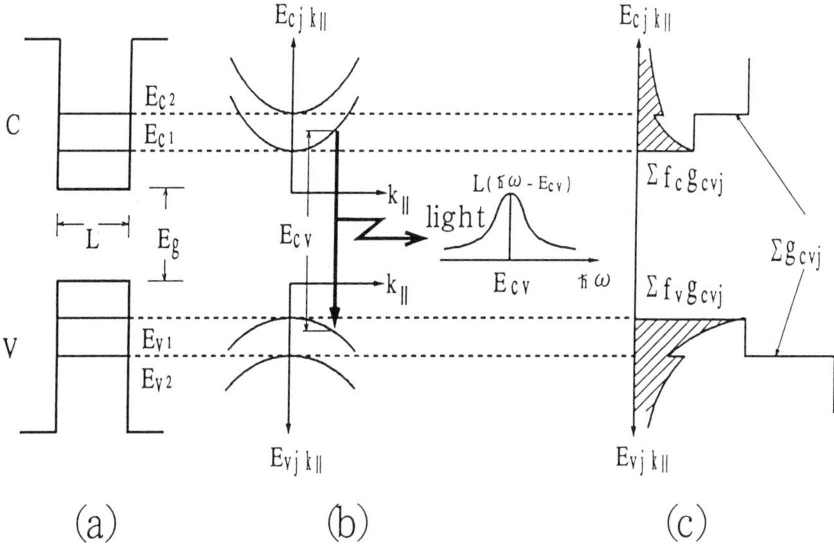


**Fig. 2.** Subbands and carrier distribution in a quantum well. Upper and lower parts are conduction and valence bands, respectively. (a) Wells and quantized energy levels; (b) subband structures and optical transition; (c) carrier distribution.

We have

$$\rho_{\mathrm{cv}j}(E_{\mathrm{cv}}) = \frac{1}{\pi\hbar^2 L}\frac{m_{\mathrm{c}}m_{\mathrm{v}}}{m_{\mathrm{v}}+m_{\mathrm{v}}}\,u(E_{\mathrm{cv}}+E_{\mathrm{g}}-E_{\mathrm{c}j}-E_{\mathrm{v}j}), \tag{21}$$

$$f_{\mathrm{c}}(E_{\mathrm{c}j\mathbf{k}_\parallel}) = 1/[1+\exp\{((E_{\mathrm{c}j\mathbf{k}_\parallel}-E_{\mathrm{fc}})/kT\}], \tag{22}$$

$$f_{\mathrm{v}}(E_{\mathrm{v}j\mathbf{k}_\parallel}) = 1/[1+\exp\{(E_{\mathrm{v}j\mathbf{k}_\parallel}-E_{\mathrm{fv}})/kT\}], \tag{23}$$

where $L$ is the well width, $u(E)$ is the unit step function, and $E_{\mathrm{fc}}$ and $E_{\mathrm{fv}}$ are the quasi-Fermi levels of conduction and valence bands, respectively.

$L(\hbar\omega - E_{\mathrm{cv}})$ in Eq. (17) is the line broadening function given from the last factor in Eq. (16) as

$$L(\hbar\omega - E_{\mathrm{cv}}) = \int_{-\infty}^{\infty} D_{\mathrm{c}}(E)D_{\mathrm{v}}(\hbar\omega - E - E_{\mathrm{g}})\,dE. \tag{24}$$

Thus, the gain coefficient is expressed in the form of Eq. (1) including the intraband relaxation effect in the line broadening function $L(\hbar\omega - E_{\mathrm{cv}})$. $L(\hbar\omega - E_{\mathrm{cv}})$ in Eq. (24) is discussed in the following section by giving the explicit formulae for $D_{\mathrm{c}}$ and $D_{\mathrm{v}}$.

### 2.4. Line Broadening Function

As previously mentioned, the effect of intraband relaxation on the gain spectra is expressed by the line broadening function $L(\hbar\omega - E_{cv})$ given by Eq. (24). $D_c(E)$ and $D_v(E)$ are the energy spectral functions of electron and hole, which respectively have energy $E_{cj\mathbf{k}_\parallel}$ and $E_{vj\mathbf{k}_\parallel}$ if there exists no scattering, given as follows by the imaginary parts of the one particle Green's functions:

$$D_c(E) = \frac{1}{\pi} \frac{\Gamma_{cj\mathbf{k}_\parallel}(E)}{(E - E_{cj\mathbf{k}_\parallel} - \Delta_{cj\mathbf{k}_\parallel})^2 + \Gamma^2_{cj\mathbf{k}_\parallel}(E)}, \tag{25}$$

$$D_v(E) = \frac{1}{\pi} \frac{\Gamma_{vj\mathbf{k}_\parallel}(E)}{(E - E_{vj\mathbf{k}_\parallel} - \Delta_{vj\mathbf{k}_\parallel})^2 + \Gamma^2_{vj\mathbf{k}_\parallel}(E)}, \tag{26}$$

where the $\Delta$ and $\Gamma$ are the factors for energy shift and broadening due to the intraband relaxation, and are the real and imaginary parts of the self-energy, respectively [10]. These factors characterize the effect of the intraband relaxation. If the intraband relaxation is negligibly small, the $\Delta$ and the $\Gamma$ approach zero, and $D_c(E)$ and $D_v(E)$ approach $\delta(E - E_{cj\mathbf{k}_\parallel})$ and $\delta(E - E_{vj\mathbf{k}_\parallel})$, respectively. In this case, $L(\hbar\omega - E_{cv})$, given by Eq. (24), is equal to $\pi\, \delta(\hbar\omega - E_{cv})$, and therefore the gain coefficient in Eq. (17) is determined only by the carrier distribution and dipole moment ($G = g$ in Eq. (1)). Finite values of the $\Delta$ and the $\Gamma$ result in the relaxation broadening.

Note that $D_c(E)$ and $D_v(E)$ are *not* the Lorentzian, because the $\Gamma$ depend on energy $E$, and therefore $L(\hbar\omega - E_{cv})$ generally deviates from the Lorentzian. If the $\Gamma$ are constant for $E$, $D_c(E)$, $D_v(E)$, and $L(\hbar\omega - E_{cv})$ become the Lorentzian. This condition is discussed in the following section. The $\Delta$ in Eqs. (25) and (26) are constant for $E$ if we consider only the first order perturbation term in the energy shift, as mentioned later.

The $\Delta$ and $\Gamma$ are obtained from the perturbation expansion of the one particle Green's functions with respect to various scattering processes in the Hamiltonian $H_r$ in Eq. (9) [10]. Carrier–carrier scattering and carrier–LO phonon scattering are considered here. For carrier–carrier scattering,

$$\begin{aligned}\Gamma^{c-c}_{cj\mathbf{k}_\parallel}(E) = \pi \sum_{n=c,v} \sum_{\mathbf{k}'_\parallel,\mathbf{p}'_\parallel} \sum_{i,i',j}{}^{2} \,|V_{cn}(\mathbf{k}_\parallel \mathbf{k}'_\parallel, ii'jj')|^2 \\ \times\, \delta(E + E_{ni\mathbf{p}_\parallel} - E_{cj'\mathbf{k}'_\parallel} - E_{ni'\mathbf{p}'_\parallel}) \\ \times\, [f_c(E_{cj'\mathbf{k}'_\parallel}) f_n(E_{ni'\mathbf{p}'_\parallel})\{1 - f_n(E_{ni\mathbf{p}_\parallel})\} \\ + \{1 - f_c(E_{cj\mathbf{k}'_\parallel})\}\{1 - f_n(E_{ni'\mathbf{p}'_\parallel})\} f_n(E_{ni\mathbf{p}_\parallel})],\end{aligned} \tag{27}$$

where $V_{cn}$ is the scattering matrix element due to the screened Coulomb potential, discussed in the next section; only the ring diagram [10] is considered; and the imaginary part of the potential is assumed to be much smaller than the real part.

The scattering processes given by the first and second terms of square brackets in Eq. (27) are schematically shown in Figs. 3a and b, respectively, for $n = \mathrm{c}$. In Fig. 3a, two electrons at the levels $\mathrm{c}j'\mathbf{k}'_{\|}$ and $\mathrm{c}i'\mathbf{p}'_{\|}$ collide with each other, and they are scattered into two holes at $\mathrm{c}j\mathbf{k}_{\|}$ and $\mathrm{c}i\mathbf{p}_{\|}$. Formation of a hole at $\mathrm{c}j\mathbf{k}_{\|}$ during optical transition is intercepted by this process in a finite time, resulting in the energy uncertainty and spectral broadening. In Fig. 3b, optical transition of an electron at $\mathrm{c}j\mathbf{k}_{\|}$ is intercepted by the collision of two holes at $\mathrm{c}j'\mathbf{k}'_{\|}$ and $\mathrm{c}i'\mathbf{p}'_{\|}$. The spin degeneracy at $\mathrm{c}i\mathbf{p}_{\|}$ and $\mathrm{c}i'\mathbf{p}'_{\|}$ results in the factor of two in Eq. (27). The spin at $\mathrm{c}j\mathbf{k}_{\|}$ is specified, and its degeneracy is taken into account in $\rho_{\mathrm{cv}j}$ in Eq. (17) for the calculation of the total optical spectra.

The process for $n = \mathrm{v}$ in Eq. (27) is similarly given. Electron–electron scattering and electron–hole scattering correspond to $n = \mathrm{c}$ and $n = \mathrm{v}$, respectively; $\Gamma^{\mathrm{c}-\mathrm{c}}_{\mathrm{v}jk_{\|}}$ is given by replacing the subscript c with v in Eq. (27); $n = \mathrm{c}$ and $n = \mathrm{v}$ correspond to hole–electron scattering and hole–hole scattering in this case, respectively.

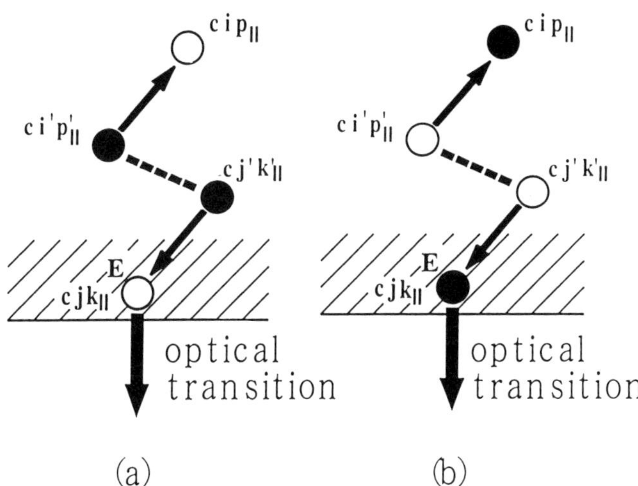


**Fig. 3.** Schematic illustration of electron–electron scattering in conduction band. (a) Formation of a hole at level $\mathrm{c}j\mathbf{k}_{\|}$ under the optical transition is intercepted due to collision of two electrons at $\mathrm{c}j'\mathbf{k}'_{\|}$ and $\mathrm{c}i'\mathbf{p}'_{\|}$, resulting in the spectral broadening. (b) An electron under the optical transition is scattered due to collision of two holes. (From Ref. [9], © 1989 IEEE.)

For carrier–LO phonon scattering,

$$\Gamma_{cj\mathbf{k}_{\parallel}}^{c-LO}(E) = \pi \sum_{\mathbf{k}'_{\parallel}} \sum_{j'} |V_c(\mathbf{k}_{\parallel}\mathbf{k}'_{\parallel}, jj')|^2 \times [\delta(E - E_{cj'\mathbf{k}'_{\parallel}} \pm \hbar\omega_{LO}) f_c(E_{cj'\mathbf{k}'_{\parallel}}) + \delta(E_{cj'\mathbf{k}'_{\parallel}} - E \pm \hbar\omega_{LO})\{1 - f_c(E_{cj'\mathbf{k}'_{\parallel}})\}], \tag{28}$$

where $V_c$ is the scattering matrix element, and $\hbar\omega_{LO}$ is the energy of LO phonon. The scattering processes given by the first and second terms in the square brackets of Eq. (28) are schematically shown in Figs. 4a and b, respectively, which correspond to Figs. 3a and b of carrier–carrier scattering.

$\Gamma_{cj\mathbf{k}_{\parallel}}(E)$ is given by the sum of Eqs. (27) and (28). $\Delta_{cj\mathbf{k}_{\parallel}}$ also includes the carrier–carrier and carrier–LO phonon scatterings. For carrier–carrier scattering, $\Delta_{cj\mathbf{k}_{\parallel}}$ is given by the exchange energy (the first order perturbation term), with the screened Coulomb potential if the imaginary part of the potential is much smaller than the real part. For carrier–LO phonon scattering, $\Delta_{cj\mathbf{k}_{\parallel}}$ is given by the correlation energy (the second-order perturbation term), related by the Kramers–Kronig relation to Eq. (28). $\Delta_{cj\mathbf{k}_{\parallel}}$ is neglected in Section 3, and $\Delta_{cj\mathbf{k}_{\parallel}}$ due to hole–hole scattering is considered in Section 4. For carrier–carrier scattering, $\Delta_{cj\mathbf{k}_{\parallel}}$ is given using the static screened Coulomb potential by

$$\Delta_{cj\mathbf{k}_{\parallel}} = -\sum_{j'} \sum_{\mathbf{k}_{\parallel}} V_{cc}(\mathbf{k}_{\parallel}\mathbf{k}'_{\parallel}, j'jjj') f_c(E_{cj'\mathbf{k}'_{\parallel}}), \tag{29}$$

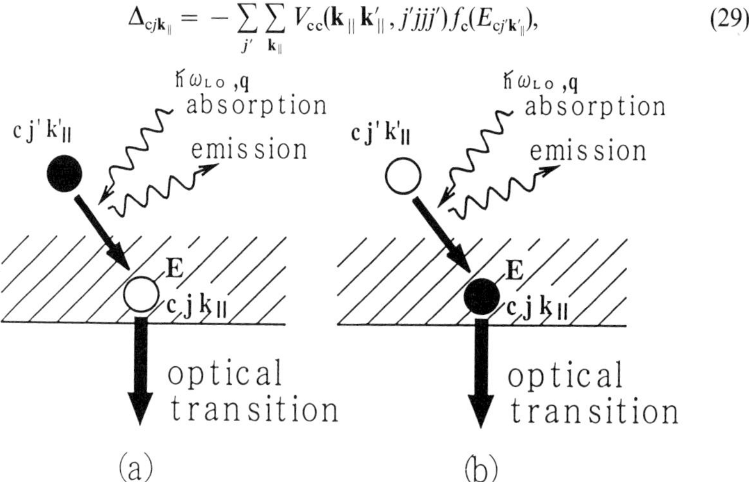


**Fig. 4.** Schematic illustration of electron–LO phonon scattering in conduction band. (a) Formation of a hole at level $cj\mathbf{k}_{\parallel}$ under the optical transition is intercepted by an electron at $cj'\mathbf{k}'_{\parallel}$, which emits or absorbs a phonon. (b) An electron under the optical transition is scattered by a hole, which emits or absorbs a phonon. (From Ref. [9], © 1989 IEEE.)

which becomes constant for energy $E$. $\Delta_{\mathrm{v}j\mathbf{k}_\parallel}$ is given by replacing the subscript c with v in Eq. (29). These factors give the bandgap shrinkage at high injection.

If the $\Delta$ and the $\Gamma$ include all the orders of the perturbation terms, $L(\hbar\omega - E_{\mathrm{cv}})$ is necessarily normalized so that the integal over entire range of $E_{\mathrm{cv}}$ is equal to unity. However, since only limited perturbation terms are included in Eqs. (27) and (28), this normalization is not satisfied, and an additional renormalization coefficient must multiply $L(\hbar\omega - E_{\mathrm{cv}})$.

## 2.5 Lorentzian Approximation and Intraband Relaxation Time

Usually, the broadening function $L(\hbar\omega - E_{\mathrm{cv}})$ in Eq. (24) is approximated by the Lorentzian shape with the intraband relaxation time, i.e., reciprocal scattering probability per unit time [3–7]. This shape is obtained from Eqs. (24)–(26) as an approximation as follows. $D_{\mathrm{c}}$ and $D_{\mathrm{v}}$ have sharp peaks around $E \simeq E_{\mathrm{c}j\mathbf{k}_\parallel}$ and $E \simeq E_{\mathrm{v}j\mathbf{k}_\parallel}$, respectively. (The energy shifts $\Delta$ are neglected.) Therefore, if $\Gamma_{\mathrm{c}j\mathbf{k}_\parallel}$ and $\Gamma_{\mathrm{v}j\mathbf{k}_\parallel}$ do not vary rapidly round these peaks, the integral in Eq. (24) can be calculated by putting $E = E_{\mathrm{c}j\mathbf{k}_\parallel}$ and $E = E_{\mathrm{v}j\mathbf{k}_\parallel}$ in $\Gamma_{\mathrm{c}j\mathbf{k}_\parallel}$ in Eq. (25) and in $\Gamma_{\mathrm{v}j\mathbf{k}_\parallel}$ in Eq. (26), respectively. By this approximation, the line broadening function is given, neglecting the factors of the energy shift ($\Delta_{\mathrm{c}j\mathbf{k}_\parallel}$and $\Delta_{\mathrm{v}j\mathbf{k}_\parallel}$), by

$$L(\hbar\omega - E_{\mathrm{cv}}) = \frac{1}{\pi} \frac{\Gamma_{\mathrm{cv}}}{(\hbar\omega - E_{\mathrm{cv}})^2 + \Gamma_{\mathrm{cv}}^2}, \tag{30}$$

where

$$\Gamma_{\mathrm{cv}} = \Gamma_{\mathrm{c}j\mathbf{k}_\parallel}(E_{\mathrm{c}j\mathbf{k}_\parallel}) + \Gamma_{\mathrm{v}j\mathbf{k}_\parallel}(E_{\mathrm{v}j\mathbf{k}_\parallel}) \tag{31}$$

Thus, lineshape is approximated by the Lorentzian with the half width of the half maximum $\Gamma_{\mathrm{cv}}$. ("Lorentzian" means that Eq. (30) is Lorentzian with respect to the photon energy $\hbar\omega$; i.e., $\Gamma_{\mathrm{cv}}$ is constant for $\hbar\omega$, although it is a function of $E_{\mathrm{cv}}$ via Eqs. (31) and (20).)

In this Lorentzian approximation, $L$ approaches $(1/\pi)\Gamma_{\mathrm{cv}}/(\hbar\omega - E_{\mathrm{cv}})^2$ as $|\hbar\omega - E_{\mathrm{cv}}| \to \infty$, and thus the effect of tail region of $L$ on the gain profile is quite large. This results in underestimation of the gain magnitude (see Fig. 29 of Chapter 1). The line broadening function without this approximation, i.e., non-Lorentzian shape, and its effect on the gain are discussed in Section 4.

This Lorentzian approximation can be explained qualitatively in the temporal behavior as follows. The preceding approximation $E \simeq E_{\mathrm{c}j\mathbf{k}_\parallel}$ assumes that the uncertainty in electron energy at optical transition is very small. From the uncertainty relation, this means that the condition $\hbar/t \ll \Delta E$

holds, where $t$ is the observation time, nearly equal to the relaxation time in the present case, and $\Delta E$ is the distribution width of carriers determined by the Fermi function and the density of states. It is known from time-dependent perturbation theory (see, for example, [20, 22]) that the transition (scattering) probability is given as $1 - \gamma t + \cdots \simeq \exp(-\gamma t)$ in the limit $\hbar \ll \Delta E \times$ (the observation time $t$), where $\gamma = 1/\tau$ (scattering probability per unit time), the observation time $t \simeq \tau$, and energy conservation holds ($E = E_{cj\mathbf{k}_\parallel}$ in the present case). This exponential decay results in the Lorentzian energy distribution. In the other limit $\Delta Et \ll \hbar$, the scattering probability is given by $1 - \gamma' t^2 + \cdots + \simeq \exp(-\gamma' t^2)$, with a constant $\gamma'$.

Consequently, the spectral lineshape is the Lorentzian if $\Delta E\tau \gg \hbar$, while it is the Gaussian if $\Delta E\tau \ll \hbar$. In a semiconductor, the situation is located between these two limits. The lineshape approaches the Lorentzian at high injection, since the carrier distribution width $\Delta E$ becomes larger. Also, in lower dimensional quantum wells, i.e., quantum wire and quantum box, the shape approaches the Gaussian [26], since $\Delta E$ becomes smaller. The case $\Delta E\tau \ll \hbar$ corresponds to the non-Markovian process [16] with the averaged correlation time given by $\hbar/\Delta E$.

The preceding approximation also corresponds to that discussed for the temporal behavior of the Green's function [10]. If $|E_{cj\mathbf{k}_\parallel} - E_{fc}| \gg \hbar/t > \Gamma_{cj\mathbf{k}_\parallel}$ and $|E_{vj\mathbf{k}_\parallel} - E_{fv}| \gg \hbar/t > \Gamma_{vj\mathbf{k}_\parallel}$, where $t$ is the observation time, then the one particle Green's functions $G_c(t)$ and $G_v(t)$ are nearly proportional to $\exp(-\Gamma_{cj\mathbf{k}_\parallel} t/\hbar)$ and $\exp(-\Gamma_{vj\mathbf{k}_\parallel} t/\hbar)$, respectively [10]. $D_c$ and $D_v$ are given by the imaginary parts of the Fourier transform of $G_c(t)$ and $G_v(t)$, respectively, and become the Lorentzian shape.

One particle Green's functions are defined as the expectation values of the electron wave amplitude. Since the electron number is proportional to the absolute square of the amplitude, the relaxation times $\tau_c$ and $\tau_v$ at the levels $cj\mathbf{k}_\parallel$ and $vj\mathbf{k}_\parallel$ are defined from the exponential decay of $G_c(t)$ and $G_v(t)$ as

$$1/\tau_c = 2\Gamma_{cj\mathbf{k}_\parallel}(E_{cj\mathbf{k}_\parallel})/\hbar, \tag{32}$$

$$1/\tau_v = 2\Gamma_{vj\mathbf{k}_\parallel}(E_{vj\mathbf{k}_\parallel})/\hbar. \tag{33}$$

Substituting Eqs. (27) and (28) into these equations, it is seen that $1/\tau_c$ and $1/\tau_v$ are just equal to the scattering probability per unit time given by Fermi's golden rule.

The polarization formed between electrons and holes is proportional to $G_c(t)G_v^*(t)$. Thus, the intraband relaxation time determining the spectral broadening, which is the relaxation time of the dipole moment (transverse relaxation time), is obtained from Eqs. (30)–(33) as

$$1/\tau_{\text{in}} = \Gamma_{\text{cv}}/\hbar = (1/\tau_{\text{c}} + 1/\tau_{\text{v}})/2. \tag{34}$$

The adiabatic term [18] and excitonic effect have been neglected in the preceding discussion.

## 3. INTRABAND RELAXATION TIME IN QUANTUM WELL LASERS

### 3.1. Carrier–Carrier Scattering

In the Lorentzian approximation discussed in Section 2.5, the intraband relaxation time determines the width of the line broadening function and is related to the scattering probability.

Calculation of the intraband relaxation time for quantum well lasers [9] is shown here, based on the preceding section. The results in this section are also used for the discussion of the non-Lorentzian shape in the next section.

$\tau_{\text{c}}$ and $\tau_{\text{v}}$ due to carrier–carrier scattering is calculated using Eqs. (27), (32), and (33). The scattering process is shown in Fig. 3 ($E = E_{cj\mathbf{k}_\parallel}$, as discussed in Section 2.5). As discussed in the previous section, $1/\tau_{\text{c}}$ and $1/\tau_{\text{v}}$ are just equal to the scattering probability obtained by Fermi's golden rule.

For the interaction matrix element $V_{\text{c}n}$ in Eq. (27), the screened Coulomb potential is used. Deviation from the exact calculation is discussed later. We have

$$\begin{aligned} V_{\text{c}n}(\mathbf{k}_\parallel \mathbf{k}'_\parallel, ii'jj') = \iint \Psi^*_{\text{c}j'\mathbf{k}'_\parallel}(\mathbf{r}_1)\Psi^*_{ni'\mathbf{p}_\parallel}(\mathbf{r}_2)\,\frac{e^2 \exp(-\lambda_{\text{s}} r)}{4\pi\varepsilon r} \\ \times\, \Psi_{\text{c}j\mathbf{k}_\parallel}(\mathbf{r}_1)\Psi_{nip_\parallel}(\mathbf{r}_2)\, d^3\mathbf{r}_1\, d^3\mathbf{r}_2, \end{aligned} \tag{35}$$

where $e$ is the electron charge; $\varepsilon$ is the static dielectric constant; $r = |\mathbf{r}_1 - \mathbf{r}_2|$; $\lambda_{\text{s}}$ is the inverse screening length given in the following; the $\Psi$ are the electron wavefunctions in a quantum well approximated as

$$\psi_{nj\mathbf{k}_\parallel}(\mathbf{r}) \simeq u_{n\mathbf{k}}(\mathbf{r})\Phi_{nj}(z)\exp(i\mathbf{k}_\parallel \cdot \mathbf{r}_\parallel)/\sqrt{S}, \tag{36}$$

with the envelope function given by

$$\Phi_{nj}(z) = \sqrt{2/L_{nj}}\,\sin(j\pi z/L_{nj}), \tag{37}$$

where the $z$ axis is perpendicular to the well interface, $S$ is the interface area of the sample, $u$ is the periodic part of the bulk Bloch function, $\mathbf{r}_\parallel$ is the component of position vector parallel to the interface, and $L_{nj}$ is the effective well width defined by

$$L_{nj} = \pi\hbar/\sqrt{2m_n E_{nj}} = \pi/k_{nj\perp} \tag{38}$$

where $k_{nj\perp}$ is the equivalent $z$ component of the wavevector related to the quantized energy level by this equation. The wavefunction using this effective well width approximation is schematically illustrated in Fig. 5.

The inverse screening length in Eq. (35) is given as follows for the carrier injection case, where electrons and holes simultaneously exist:

$$\begin{aligned}\lambda_s^2 &= -\frac{e^2}{\varepsilon}\sum_j\left[\int_{E_{cj}}^{\infty}\frac{\partial f_c}{\partial E_c}\rho_{cj}\,dE_c + \int_{E_{vj}}^{\infty}\frac{\partial f_v}{\partial E_v}\rho_{vj}\,dE_v\right] \\ &= \frac{e^2}{\pi\hbar^2\varepsilon}\sum_j\,[m_c f_c(E_{cj})/L_{cj} + m_v f_v(E_{vj})/L_{vj}],\end{aligned} \tag{39}$$

where $\rho_{cj}$ and $\rho_{vj}$ are the step-like density of states of subband $j$ in conduction and valence bands, respectively.

Substituting Eq. (36) into Eq. (35), the matrix element is calculated as

$$\begin{aligned}V_{cn}(\mathbf{k}_{\parallel}\,\mathbf{k}'_{\parallel},\,ii'jj') &= \frac{e^2}{2\varepsilon S}\frac{\delta(\mathbf{k}_{\parallel} - \mathbf{k}'_{\parallel},\,\mathbf{p}'_{\parallel} - \mathbf{p}_{\parallel})}{\sqrt{g_{\parallel}^2 + \lambda_s^2}} \\ &\quad\times\iint\Phi^*_{cj'}(z_1)\Phi_{cj}(z_1)\Phi^*_{ni'}(z_2)\Phi_{ni}(z_2) \\ &\quad\times\exp(-|z_1 - z_2|\sqrt{g_{\parallel}^2 + \lambda_s^2})\,dz_1\,dz_2,\end{aligned} \tag{40}$$

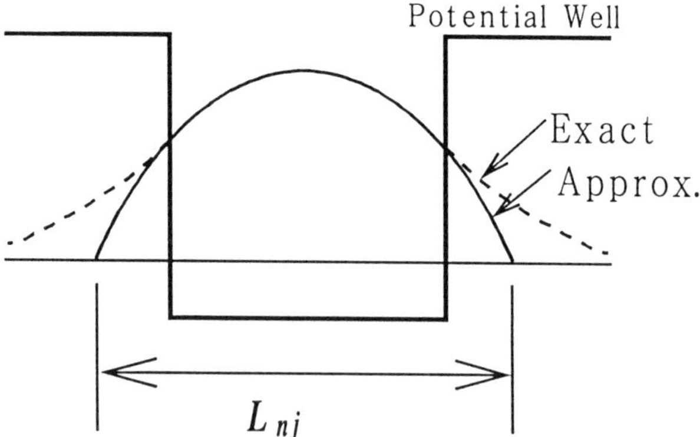


**Fig. 5.** The envelope function $\Phi_{nj}(z)$ of Eq. (36). Dashed curve is exact, and solid curve is an approximation given by Eq. (37), which is an extension of the exact curve in the well to the outside.

where the $\delta$ notation ($\delta(\alpha, \beta) = 1$ if $\alpha = \beta$; $\varepsilon(\alpha, \beta) = 0$ otherwise) represents the momentum conservation within the plane parallel to the interface, and $g_{\parallel} = \mathbf{k}_{\parallel} - \mathbf{k}'_{\parallel}$.

Assuming only the lowest subband both in conduction and valence bands ($i = i' = 1, j = j' = 1$), the absolute square of the matrix element is calculated as

$$|V_{cn}(\mathbf{k}_{\parallel}\mathbf{k}'_{\parallel}, 1111)|^2 = \left(\frac{e^2}{2\varepsilon S L_e}\right)^2 \delta(\mathbf{k}_{\parallel} - \mathbf{k}'_{\parallel}, \mathbf{p}'_{\parallel} - \mathbf{p}_{\parallel})\, T(\mathbf{g}_{\parallel}, k_{c1\perp}, k_{n1\perp}), \quad (41)$$

where $L_e$ is the minimum of the effective well widths of the four envelope functions in Eq. (40), and

$$\begin{aligned} T(g_{\parallel}, k_{c1\perp}, k_{n1\perp}) &= \left[\frac{2}{\alpha} + \frac{\delta(k_{c1\perp}, k_{n1\perp})}{\alpha + 4k_{c1\perp}^2} - \frac{2}{L_e\sqrt{\alpha}}\{1 - \exp(-L_e\sqrt{\alpha})\}\right. \\ &\quad \times \left\{\frac{1}{\alpha} - \frac{\alpha + 4k_{c1\perp}^2 + 4k_{n1\perp}^2}{(\alpha + 4k_{n1\perp}^2)(\alpha + 4k_{n1\perp}^2)}\right. \\ &\quad \left.\left. + \frac{L_e k_{c1\perp}(L_e k_{c1\perp} - \pi)(\alpha - 4k_{n1\perp}^2)}{4(\alpha + 4k_{c1\perp}^2)(\alpha + 4k_{n1\perp}^2)\{\exp(L_e\sqrt{\alpha}) - 1\}}\right\}\right]^2, \end{aligned} \quad (42)$$

$$\alpha = g_{\parallel}^2 + \lambda_s^2. \quad (43)$$

The first and second terms in the square brackets of Eq. (42) are essentially the same as those in bulk semiconductors. These terms arise from the forward and backwad waves in the standing wave along the $z$ axis included in Eq. (37). The third term, which approaches zero at the limit $L \to \infty$, is peculiar to the quantum well structures, and results from the localization of the wavefunction. Equation (42) coincides with that for the perfect two dimensional case at the limit $L \to 0$ due to the existence of the third term.

In the calculation of Eq. (42), $\delta(k_{c1\perp}, k_{n1\perp})$ in the second term is unity for electron–electron and hole–hole scatterings and is zero for electron–hole and hole–electron scatterings. The last term is equal to zero for electron–electron and hole–hole scatterings.

In the derivation of Eq. (42), the screened Coulomb potential is used, which is an approximation using the static and long wavelength ($g_{\parallel} \to 0$) limits. The exact calculation for the perfect two dimensional case shows that the screening effect falls off rapidly at $g_{\parallel} > 2k_{f\parallel}$ with the increae of $g_{\parallel}$ [13] at $T = 0$ K, where $k_{f\parallel}$ is component of the Fermi wavevector parallel to the well interface. A similar result has also been obtained for bulk case [10].

Since the approximate result in Eq. (42) decreases with $g_{\parallel}$, the difference between the exact calculation and the present approximation is small if $2k_{f\parallel} > \lambda_s$. This condition is satisfied when the sheet carrier density $n_{2D}$ is in the order of $10^{12}$ cm$^{-2}$ (the threshold level of the laser), and well width is larger than about 3 nm at $T = 300$ K. For low carrier concentration and very small well width, the screening effect becomes weaker than that in the approximation in Eqs. (39) and (42), and thus deviation becomes large. Under the long wavelength limit, it has been shown that the screening effect becomes anisotropic for quantum wells [12]. Since this anisotropy is small for the range of well width in the present calculation, the isotropic screening given in Eqs. (35) and (39) is used here.

Transforming the summation for $\mathbf{k}'_{\parallel}$ and $\mathbf{p}'_{\parallel}$ in Eq. (27) into an integral, and doing a calculation similar to that for bulk semiconductors [6], $1/\tau_c$ is given as

$$\begin{aligned}1/\tau_c = \sum_n \frac{m_c e^4}{8\pi^5\hbar^3\varepsilon^2}\left(\frac{m_n}{m_c}\right)^2 \\ \times \int_0^{2\pi} d\phi \int_0^{\infty} dk'_{\parallel} \int_0^{\infty} du\, T(g_{\parallel}, k_{c1\perp}, k_{n1\perp})(k_{c1\perp}^2 k'_{\parallel}|\beta|/g_{\parallel}^2) \\ \times [f_c(E_{c1\mathbf{k}'_{\parallel}}) f_n(E_{n1\mathbf{p}'_{\parallel}})\{1 - f_n(E_{n1\mathbf{p}_{\parallel}})\} \\ + \{1 - f_c(E_{c1\mathbf{k}'_{\parallel}})\}\{1 - f_n(E_{n1\mathbf{p}'_{\parallel}})\} f_n(E_{n1\mathbf{p}_{\parallel}})],\end{aligned} \tag{44}$$

with

$$g_{\parallel}^2 = k_{\parallel}^2 + k_{\parallel}'^2 - 2k_{\parallel}k'_{\parallel}\cos\phi, \tag{45}$$

$$\begin{aligned}\beta = k_{\parallel}'^2 - k'_{\parallel}k_{\parallel}\cos\phi + (m_c/m_n - 1)g_{\parallel}^2/2 \\ - (m_c/\hbar^2)E,\end{aligned} \tag{46}$$

$$E_{c1\mathbf{k}'_{\parallel}} = (\hbar^2/2m_c)(k_{\parallel}'^2 + k_{c1\perp}^2), \tag{47}$$

$$\begin{aligned}E_{n1\mathbf{p}'_{\parallel}} = (\hbar^2/2m_n)\{(m_n/m_c)^2(\beta/g_{\parallel})^2(u^2 + 1) \\ + g_{\parallel}^2 - 2(m_n/m_c)\beta + k_{n1\perp}^2\},\end{aligned} \tag{48}$$

$$E_{n1\mathbf{p}_{\parallel}} = E_{c1\mathbf{k}'_{\parallel}} + E_{n1\mathbf{p}'_{\parallel}} - E, \tag{49}$$

where $E = E_{c1\mathbf{k}_{\parallel}}$ (the case $E \neq E_{c1\mathbf{k}_{\parallel}}$ is discussed in Section 4), the parabolic band structure is assumed, and the summation with respect to $n$ corresponds to electron–electron scattering and electron-hole scattering for $n = \mathrm{c}$ and $n = \mathrm{v}$, respectively. In the numerical calculation, temperature $T$, sheet carrier density $n_{2D}$, and the energy $E_{c1\mathbf{k}_{\parallel}}$ are specified at first, and the quasi-Fermi levels ($E_{fc}$ and $E_{fv}$) are determined with the charge neutrality of electrons and holes [1]. $k_{c1\perp}$ and $k_{\parallel}$ are obtained from Eqs. (18) and (38), and thus, using Eqs. (42), (43), and (45)–(49), $1/\tau_c$ in Eq. (44) is calculated. The intraband relaxation time for valence band $\tau_v$ is calculated by the same procedure as in the preceding, replacing the subscript c with v throughout Eq. (44)

### 3.2. Carrier–Phonon Scattering

The intraband relaxation time due to carrier–LO phonon scattering is calculated using Eqs. (28), (32), and (33). The scattering process is shown in Fig. 4 ($E = E_{cj\mathbf{k}_{\parallel}}$).

Assuming the electron wavefunction of Eq. (36), the square of the matrix element $|V_c(k_{\parallel} k'_{\parallel}, jj')|^2$ is given as follows, taking into account the screening effect [14]:

$$|V_c(k_{\parallel} k'_{\parallel}, jj')|^2 = \sum_q \frac{e^2\hbar\omega_{LO}}{2V}\left(\frac{1}{\varepsilon_\infty} - \frac{1}{\varepsilon}\right)\frac{q^2}{(q^2+\lambda_s^2)^2} \times \begin{Bmatrix} n_q + 1 \\ n_q \end{Bmatrix} \delta(\mathbf{k}_{\parallel} - \mathbf{k}'_{\parallel}, \mathbf{q}_{\parallel}) \left| \int \Phi^*_{cj}(z)\Phi_{cj'}(z)e^{-iq\perp z}\, dz \right|^2, \tag{50}$$

where $\mathbf{q}$, $\mathbf{q}_{\parallel}$, and $\mathbf{q}_{\perp}$ are the phonon wavevector and its components parallel and perpendicular to the well interface, respectively; $\varepsilon$ and $\varepsilon_\infty$ are the static and optical dielectric contants, respectively; $V$ is the volume of the system; and the factors $n_q + 1$ and $n_q$ correspond to the emission and absorption of phonons, respectively, where $n_q$ is the phonon number per mode $q$, given by

$$n_q = 1/[\exp(\hbar\omega_{LO}/kT) - 1]. \tag{51}$$

The intensity of the phonon mode confined in the well is assumed to be much smaller than that of the bulk mode, and it is neglected in Eq. (50).

Transforming the summation for $\mathbf{k}'_{\parallel}$ and $\mathbf{q}$ in Eqs. (28) and (50) into an integral, $1/\tau_c$ is calculated for the one subband case ($j' = 1$) as

$$1/\tau_c = \frac{m_c e^2}{2\pi^2\hbar^3}\left(\frac{1}{\varepsilon_\infty} - \frac{1}{\varepsilon}\right)\hbar\omega_{LO} \times \left\{[n_q + 1 - f_c(E_{c1\mathbf{k}_{\parallel}} - \hbar\omega_{LO})]\int_0^{\phi_{max}} \frac{[I(a_1) + I(a_2)]\, d\phi}{\sqrt{k_{\parallel}^2\cos^2\phi - 2m_c\omega_{LO}/\hbar}} + [n_q + f_c(E_{c1\mathbf{k}_{\parallel}} + \hbar\omega_{LO})]\int_0^{\pi} \frac{I(a_3)\, d\phi}{\sqrt{k_{\parallel}^2\cos^2\phi + 2m_c\omega_{LO}/\hbar}}\right\}, \tag{52}$$

with

$$\phi_{\max} = \cos^{-1}\sqrt{2m_c\omega_{LO}/\hbar k_{\|}^2}, \tag{53}$$

$$I(a) = \int_0^\infty \frac{ab(b^2x^2 + a^2)\sin^2 x}{(b^2x^2 + a^2 + 1)^2(x^2/\pi^2 - 1)^2}\,dx, \tag{54}$$

$$\simeq (\pi/4)a(2a^2 + 1)(a^2 + 1)^{-3/2} \qquad \text{for } L_{njq_\perp} \ll 1,$$

$$a_1 = \gamma - \sqrt{\gamma^2 - \delta^2}, \tag{55}$$

$$a_2 = \gamma + \sqrt{\gamma^2 - \delta^2}, \tag{56}$$

$$a_3 = \gamma + \sqrt{\gamma^2 + \delta^2}, \tag{57}$$

$$b = 2k_{c1\perp}/\pi\lambda_s, \tag{58}$$

$$\gamma = (k_{\|}/\lambda_s)\cos\phi, \tag{59}$$

$$\delta = \sqrt{2m_c\omega_{LO}/\hbar\lambda_s^2}, \tag{60}$$

where the first term of Eq. (52) expresses the probability of phonon emission, which is equal to zero if $\hbar\omega_{LO} > \hbar^2 k_{\|}^2/2m_c$.

The process of numerical calculation is similar to that of carrier–carrier scattering. Scattering of holes in the valence band by LO phonon is obtained by replacing the subscript c with v in the preceding equations.

### 3.3. Calculation Result of Intraband Relaxation Times

#### *3.3.1. Well Width Dependence*

Figure 6 shows the calculation result of the well width dependence of the broadening factors, i.e., reciprocal intraband relaxation times ($\hbar/\tau_c$ and $\hbar/\tau_v$) at subband edges ($k_{\|} = 0$) for $Ga_{0.47}In_{0.53}As/InP$ and $GaAs/Ga_{0.8}Al_{0.2}As$ quantum well structures [9]. The calculation was made for the lowest and highest subbands in conduction and valence bands, respectively. As can be seen, hole–hole, electron–hole, and hole–LO phonon scatterings are dominant. The difference of the values between the two material systems in Fig. 6 is mainly due to that of the electron effective mass. $\hbar/\tau_c$ and $\hbar/\tau_v$ are given by the sum of the curves in Figs. 6a and b, respectively. The broadening factor for optical spectra $\hbar/\tau_{in}$ is given by Eq. (34).

A slight decrease of the total broadening factor with the decrease of well width is caused mainly by $\lambda_s^2$, $k_{c1\perp}^2$, and $k_{v1\perp}^2$ in the first two terms in Eq. (42) of carrier–carrier scattering. The third term in Eq. (42) is arising from the localization of the wavefunction, and it enhances the scattering rate and broadening with the decrease of well width.

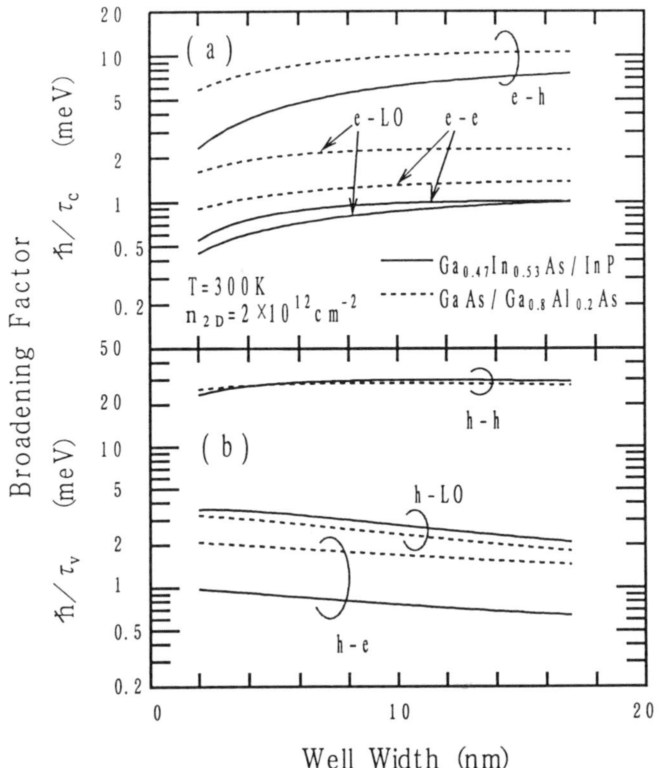


**Fig. 6.** Well width dependence of the broadening factor (reciprocal intraband relaxation time) in (a) conduction and (b) valence bands at subband edges. Solid and dashed lines are $Ga_{0.47}In_{0.53}As/InP$ and $GaAs/Ga_{0.8}Al_{0.2}As$ quantum wells, respectively. Hole–hole, electron–hole, and hole–LO phonon scattering dominate in the broadening. The broadening factor for optical spectra $\hbar/\tau_{in}$ is given by Eq. (34). (From Ref. [9], © 1989 IEEE.)

Hole-involved scattering, i.e., hole–hole, hole–LO, and electron–hole scattering, is the dominant scattering mechanism, because holes with large effective mass act as strong scatterers. This result is the same as that of bulk calculations [5, 7, 8]. In the compressively strained superlattice, spectral broadening can be small, because the effective mass parallel to the interface in the highest valence band can become small.

### *3.3.2. Dependence on Temperature, Carrier Density, and Energy*

Figure 7 shows the temperature dependence of broadening factors $\hbar/\tau_c$ and $\hbar/\tau_v$ [9]. The variation of broadening factors is caused by that of the

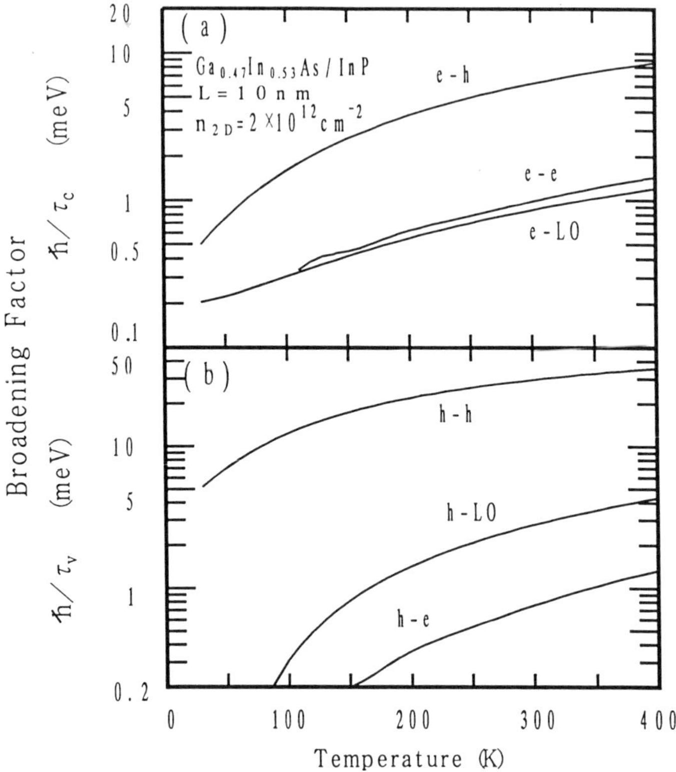


**Fig. 7.** Temperature dependence of the broadening factor (reciprocal intraband relaxation time) at subband edges in (a) conduction and (b) valence bands for $Ga_{0.47}In_{0.53}As/InP$. (From Ref. [9], © 1989 IEEE.)

thermal distribution of electrons and holes and the inverse screening length in Eq. (39). Around room temperature, the broadening factor increases slightly with temperature.

Figure 8 shows the sheet carrier density dependence of $\hbar/\tau_c$ and $\hbar/\tau_v$. The broadening factor is almost constant at carrier density of about $10^{12}$ cm$^{-2}$ (near the threshold level of the laser), because an increase of scatterers (carriers) is canceled out by the decrease of screening length with carrier density. Calculation is made for undoped semiconductors. At low carrier densities, carriers generated by impurities will be dominant in the actual case, and moreover carrier–impurity scattering must be considered. Carrier density dependence of spectral lineshape is discussed without relaxation time approximation in Section 4.

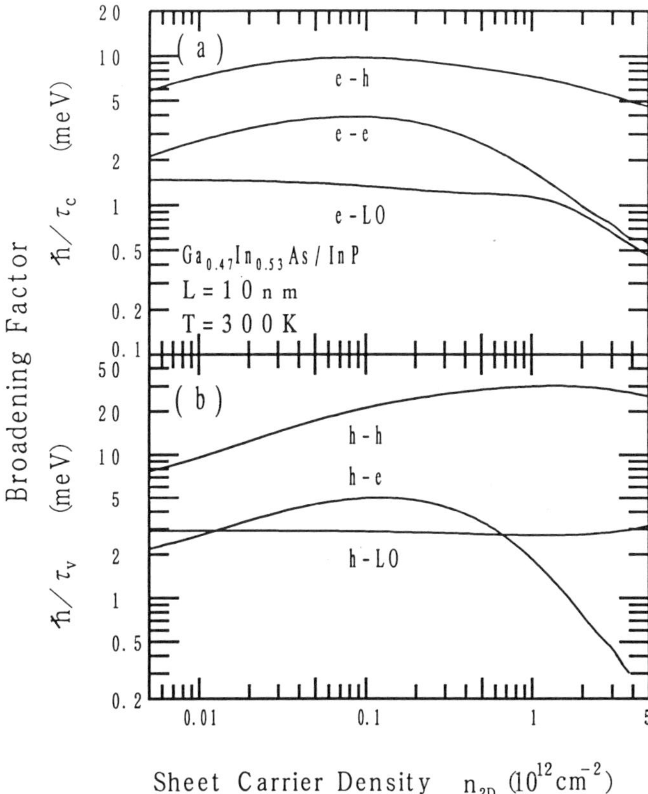


**Fig. 8.** Sheet carrier density dependence of the broadening factor (reciprocal intraband relaxation time) at subband edges in (a) conduction and (b) valence bands for $Ga_{0.47}In_{0.53}As/InP$.

Figure 9 shows the dependence of $\hbar/\tau_c$ and $\hbar/\tau_v$ on electron and hole energy measured from the subband edges [9]. For carrier–carrier scattering, the first term in the square bracket of Eq. (27) (Fig. 3a) decreases with energy [6], while the second term (Fig. 3b) increases. In total, the broadening factor decreases slowly with energy. The situation is the same for carrier–LO phonon scattering. In addition, there exists a discontinuous step at the phonon energy $\hbar\omega_{LO}$, because electrons with energy higher than $\hbar\omega_{LO}$ can be scattered by phonon emission. In bulk semiconductors, there is a kink at $\hbar\omega_{LO}$ instead of a discontinuous step, due to the difference in the density of states. If the laser oscillates near the lowest subband edge, then there is no broadening due to phonon emission.

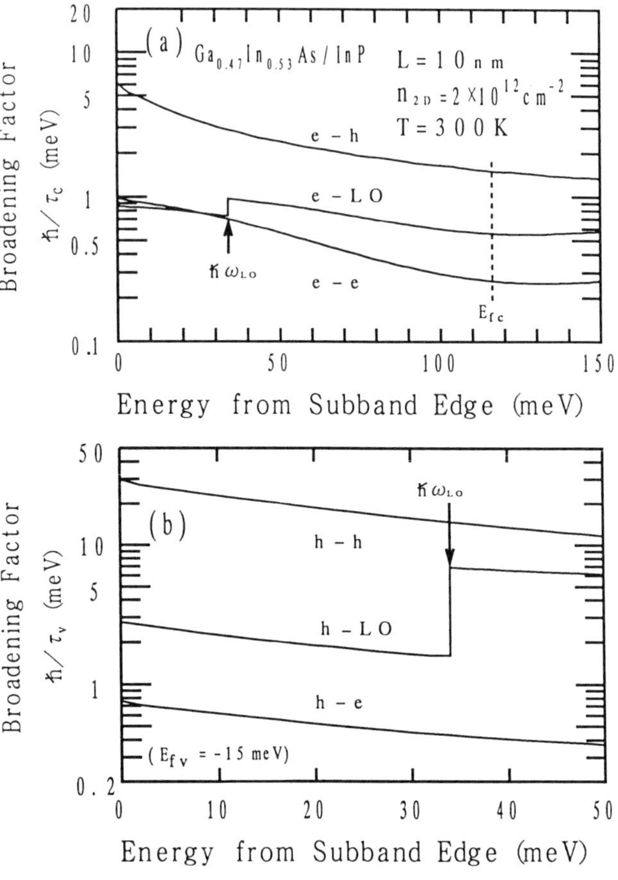


**Fig. 9.** Dependence of the broadening factor (reciprocal intraband relaxation time) on energy of electron and hole under the optical transition in (a) conduction and (b) valence bands, respectively, for $Ga_{0.47}In_{0.53}As/InP$. Energy is measured from the lowest subband edge into the band. (From Ref. [9], © 1989 IEEE.)

## 4. SPECTRAL LINESHAPE ANALYSIS

### 4.1. Derivation of Non-Lorentzian Lineshape

In Section 2.5, the Lorentzian shape in Eq. (30) was derived as an approximation, and the intraband relaxation time, which determines the width of the Lorentzian, was calculated in Section 3. However, calculation of optical spectra with the Lorentzian shape sometimes fails to explain experimental results. For example, fundamental absorption spectrum is known to decay exponentially at the band edge (the Urbach tail). This cannot be explained by the Lorentzian shape, because the Lorentzian spectrum decays as $(\hbar\omega)^{-2}$ at the tail region, which is much slower than the exponential decay.

From the qualitative discussion in Section 2.5 for the Lorentzian shape, deviation between the calculation and the observation is reasonable, and is understood as follows. Consider the hole–hole scattering that is dominant in the relaxation broadening. The energy distribution width of holes $\Delta E$ is nearly equal to $kT$, because the hole distribution is almost nondegenerate even at the laser threshold level. From the calculation result of $\hbar/\tau_v$ in Section 3, $\Delta E(\sim kT)$ is comparable with $\hbar/\tau_v$. This situation is located between Lorentzian and Gaussian shapes, as discussed in Section 2.5 with the uncertainty relation.

In this section, spectral lineshape is calculated using a general expression (non-Lorentzian shape) for the line broadening function obtained in Eqs. (24)–(29) of Section 2. The lineshape is expressed with the parameters used in the Lorentzian. Then, the effect of line broadening on gain spectra and its magnitude is discussed. In the last part of this section, we show the equivalence between our method and the method employed by Yamanishi and Lee [16], in which the temporal behavior of the polarization is traced.

The lineshape is calculated using Eqs. (24)–(29) without the approximation done in Section 2.5. Only the hole–hole scattering is considered here, because it is dominant between various processes, and it is the most important, as previously stated. In this case, $\Delta_{cj\mathbf{k}_\parallel}$ and $\Gamma_{cj\mathbf{k}_\parallel} \to 0$ in Eq. (25), and therefore $D_c(E) \simeq \delta(E - E_{cj\mathbf{k}_\parallel})$. Substituting this $D_c$ into Eq. (24) and using $D_v$ in Eq. (26), the line broadening function is obtained as

$$L(\hbar\omega - E_{cv}) \simeq \frac{1}{\pi} \frac{\Gamma_{vj\mathbf{k}_\parallel}(\hbar\omega - E_{cv} + E_{vj\mathbf{k}_\parallel})}{(\hbar\omega - E_{cv} - \Delta_{vj\mathbf{k}_\parallel})^2 + \Gamma^2_{vj\mathbf{k}_\parallel}(\hbar\omega - E_{cv} + E_{vj\mathbf{k}_\parallel})}. \tag{61}$$

Equation (61) is non-Lorentzian shape, because $\Gamma_{vj\mathbf{k}_\parallel}$ is a function of photon energy $\hbar\omega$, in contrast to Eqs. (30) and (31). The quantity $\Gamma_{vj\mathbf{k}_\parallel}(\hbar\omega - E_{cv} + E_{vj\mathbf{k}_\parallel})$ included in Eq. (61) is calculated for hole–hole scattering by transforming Eq. (27) into an integral, which is similar to the manner that we obtained Eq. (44) from Eq. (27). $\Delta_{vj\mathbf{k}_\parallel}$ in $D_v$ is obtained via Eq. (29). Only the highest subband of the valence band is considered here.

Figure 10 shows the calculated results for (a) $\Gamma_{vj\mathbf{k}_\parallel}(E)$ $(E = \hbar\omega - E_{cv} + E_{vj\mathbf{k}_\parallel})$, and (b) $L(\hbar\omega - E_{cv})$ at subband edge $(k_\parallel = 0)$. The Lorentzian shape obtained by the approximation in Section 2.5 and Section 3 is also shown, in Fig. 10b, for comparison. The dashed curves are the results of the approximate formula for the non-Lorentzian to be given in the following.

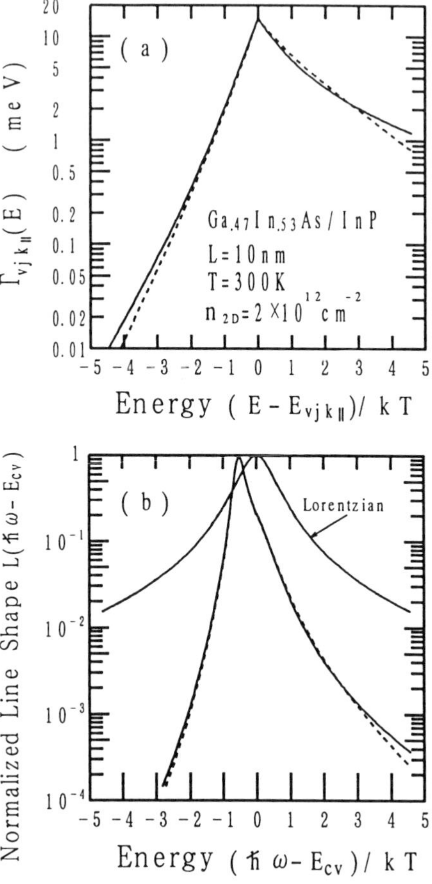


**Fig. 10.** Calculated results of (a) $\Gamma_{vj\mathbf{k}_{\parallel}}$ and (b) the line broadening function $L(\hbar\omega - E_{cv})$ at the subband edge. Only the hole–hole scattering is considered. The Lorentzian shape obtained with the approximation in Sections 2.5 and 3 is also shown in (b) for comparison. The dashed curves are the approximate calculation of the non-Lorentzian shape (Eqs. (62)–(64)). The peak values are normalized in (b).

Figure 11 shows the spectral lineshape at subband edges for different values of $E_{fv}/kT$.

As can be seen in Fig. 10b, the calculated lineshape is narrower than the Lorentzian with the intraband relaxation time obtained in Section 3, and the low-energy tail is much steeper than the Lorentzian; it decays exponentially, as will be shown later. Since the limited perturbation terms are included in the calculation of $\Gamma_{vj\mathbf{k}_{\parallel}}(E)$ in Section 2.4, there is a small kink on the high energy side, which will become smooth if higher order perturbations are considered. From Fig. 11, the half width does not change very much with

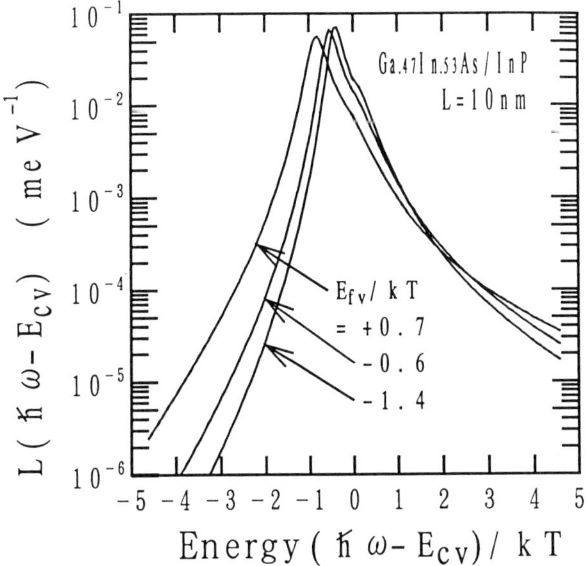


**Fig. 11.** Spectral line shapes at subband edges for normalized quasi-Fermi levels $E_{fv}/kT = -1.4, -0.6$, and $+0.7$, which correspond to sheet carrier densities $n_{2D} = 1, 2$, and $5 \times 10^{12}\ \mathrm{cm}^{-2}$ at 300 K, respectively, with the one subband assumption. These curves are normalized so that the integral over the entire range is equal to unity.

$E_{fv}/kT$, corresponding to the result in Fig. 8. The dependence of the lineshape on the energy $E_{vj\mathbf{k}_\parallel}$ is also calculated, and it is found to be almost the same as that at the subband edges.

## 4.2. Approximate Formula for the Line Broadening Function

### *4.2.1. Approximate Formula*

By using Eq. (61) or Fig. 11 in Eq. (17) of Section 2, we can now calculate the gain with intraband relaxation. However, since calculation of $\Gamma_{vj\mathbf{k}_\parallel}$ and $\Delta_{vj\mathbf{k}_\parallel}$ with Eqs. (27) and (29) needs numerical integration, we show an approximate formula for the non-Lorentzian shape here.

Assuming that the hole distribution is almost nondegenerate, $\Gamma_{vj\mathbf{k}_\parallel}(E)$ and $\Delta_{vj\mathbf{k}_\parallel}(E)$ in Eqs. (27) and (29) are approximated as follows:

$$\Gamma_{vj\mathbf{k}_\parallel}(E) \simeq \frac{\hbar}{2\tau_v} \frac{1 + e^{(E - E_{vj\mathbf{k}_\parallel} - E_{fv})/kT}}{1 + e^{-E_{fv}/kT}} e^{-K|E - E_{vj\mathbf{k}_\parallel}|/kT}, \tag{62}$$

$$\Delta_{vj\mathbf{k}_\parallel} \simeq -\frac{e^2}{4\varepsilon L_{v1}} T^{1/2}(0, k_{v1\perp}, k_{v1\perp}) n_{2D}, \tag{63}$$

with

$$K = \frac{e^4 \tau_v}{48\pi\varepsilon^2 \hbar k T L_{v1}^2} \frac{T(0, k_{v1\perp}, k_{v1\perp})}{1 + e^{-E_{fv}/kT}}, \tag{64}$$

$T(0, k_{v1\perp}, k_{v1\perp})$ in Eq. (42) ($g_{\parallel} = 0$), and $\tau_v$ obtained in Section 3 for hole–hole scattering. $L(\hbar\omega - E_{cv})$ is calculated using Eqs. (62)–(64) in Eq. (61). This approximation is close to the exact result if $E_{fv}/kT$ is close to zero. Approximate results are shown by the dashed curves in Fig. 10, where they are compared with the exact ones.

The half width of the half maximum of the dashed curve in Fig. 10 is given by

$$\frac{\hbar}{2\tau_v} \left[ \frac{1 + e^{(\Delta_{vj\mathbf{k}_{\parallel}} - E_{fv})/kT}}{1 + e^{-E_{fv}/kT}} e^{-K\Delta_{vj\mathbf{k}_{\parallel}}/kT} \right]. \tag{65}$$

Since the half width of the half maximum of the Lorentzian approximation is given by $\hbar/2\tau_v$ ($\tau_c \gg \tau_v$), the inverse of the factor in the brackets in Eq. (65) is considered to be a modification factor to the intraband relaxation time given in the Lorentzian approximation.

The low-energy tail in the line broadening function is approximated by the exponential curve as given by Eq. (62), which agrees with experimental observations. The asymmetric shape of the broadening arises from the factor $1 + \exp\{(E - E_{vj\mathbf{k}_{\parallel}} - E_{fv}/kT\}$ in Eq. (62), which reflects the Fermi distribution. In the calculation of opical spectra, the value of $L(\hbar\omega - E_{cv})$ obtained previously must be normalized so that the integral over the entire range of $E_{cv}$ is equal to unity.

### 4.2.2. *Summary of the Calculation Process*

The calculation process of the line broadening function is summarized here. For the non-Lorentzian shape, the broadening function is given by Eq. (61). $\Gamma_{vj\mathbf{k}_{\parallel}}$ and $\Delta_{vj\mathbf{k}_{\parallel}}$ in Eq. (61) is given by Eq. (62), Fig. 10a, and Eq. (63). ($E$ in Eq. (62) is replaced by $\hbar\omega - E_{cv} + E_{vj\mathbf{k}_{\parallel}}$.)

In Eq. (62) for $\Gamma_{vj\mathbf{k}_{\parallel}}$, $\tau_v$ is obtained with Figs. 6–9 in Section 3, $K$ is given by Eq. (64), the quasi-Fermi energy $E_{fv}$ is obtained by the hole density, and $E_{vj\mathbf{k}_{\parallel}}$ is related to the transition energy $E_{cv}$ by Eq. (20) in Section 2. In Eq. (63) for $\Delta_{vj\mathbf{k}_{\parallel}}$ and Eq. (64), $T$ is given by Eq. (42) and is calculated with $g_{\parallel} = 0$ and Eqs. (43), (39), and (38) in Section 3. The value of $\Delta_{vj\mathbf{k}_{\parallel}}$ is nearly equal to the shift of peak energy from the center seen in Fig. 10b. Finally, the calculated $L(\hbar\omega - E_{cv})$ must be normalized so that the integral over the entire range of $E_{cv}$ is equal to unity.

Using the line broadening function $L(\hbar\omega - E_{cv})$ obtained by this procedure in Eq. (17) or (1), the gain coefficient with intraband relaxation is calculated.

If the scattering probability is comparable between electrons and holes, or if hole density is so high that the quasi-Fermi level of holes is much deeper than the subband edge, then the difference between the preceding approximation and exact calculation of the non-Lorentzian shape becomes large. In this case, Eqs. (24)–(29) in Section 2 must be calculated in a similar manner as the calculation in Section 3.

### 4.3. Discussion of the Gain Coefficient

The gain with intraband relaxation is related to that without relaxation by Eq. (1) (and also Eq. (17)) in Section 2.

The gain without relaxation is negative for the transition energy $E_{cv} > E_{fc} + E_{fv} + E_g$, where $E_{fc}$ and $E_{fv}$ are the quasi-Fermi levels of electrons and holes measured from the bulk band edge into the bands, and $E_g$ is the bulk bandgap energy. This negative gain at large $E_{cv}$ reduces the positive gain near the band edge via the integration in Eq. (1), and therefore the magnitude of the gain with relaxation is usually smaller than that without relaxation. This reduction depends significantly on the low-energy tail shape of the line broadening function. If we use the Lorentzian for the gain calculation, then the low-energy tail varies as $(\hbar\omega - E_{cv})^{-2}$ at $\hbar\omega \ll E_{cv} - \hbar/\tau_{in}$, and decreases very slowly with the decrease of $\hbar\omega$. This results in the large reduction of the peak gain by the gain ($<0$) at large $E_{cv}$, and this reduction is larger for smaller intraband relaxation times. Since the low-energy tail of the more exact lineshape (non-Lorentzian) in Fig. 10 decreases much faster than that of the Lorentzian, the gain magnitude with the Lorentzian shape may be underestimated. This has been seen in Fig. 29 of Chapter 1. The transparency current density $J_{tr}$ ($J_{rad}$ at $G = 0$) is larger for the Lorentzian due to the just mentioned underestimation. Therefore, the lineshape, in particular its behavior at the low-energy side, is important in the gain analysis.

For the gain suppression coefficient [1], which is used in the analysis of the mode competition phenomena above threshold, only the shape around the peak rather than the overall lineshape is important, in contrast to the gain coefficient. In this case, the Lorentzian shape with the half width given by Eq. (65) may be useful.

### 4.4. The Line Broadening Function by Temporal Behavior Calculation of Polarization

In Section 2, the spectral broadening function has been calculated from the Fourier transform of the impulse response of the polarization $Q^R(t)$. Without tracing the temporal behavior of $Q^{R}(t)$ in Eq. (12), we have obtained the Fourier transform of $Q^{R}(t)$ using the temperature Green's function method.

Before closing this section, we introduce here the direct calculation of the impulse response of the polarization done by Yamanishi and Lee [16], and we show that this and our calculation are originally equivalent. Yamanishi and Lee have obtained the time-dependent polarization $P(t)$, which corresponds to $iQ^{R}(t) - iQ^{R*}(-t)$ in our notation. The line broadening function is obtained by the Fourier transform of this $P(t)$ without calculating the imaginary part as in the case of $Q^{R}(t)$ (see Eq. (13).)

$P(t)$ has been calculated using the second order perturbation term of the scattering Hamiltonian. Since $Q^{R}(t) \propto P(t)$ for $t > 0$, the calculation result is written using $Q^{R}(t)$ as

$$Q^{R}(t) \propto e^{-i(E_g + E_{cj\mathbf{k}_\parallel} + E_{cj\mathbf{k}_\parallel})t/\hbar}(1 - L_c(t) + \cdots)(1 - L_v(t) + \cdots)u(t) \tag{66}$$

$$\simeq e^{-iE_{cv}t/\hbar} \exp\{-[L_c(t) + L_v(t)]\}u(t), \tag{67}$$

where $u(t)$ is the unit step function, and $L_c(t)$ has the form [16]

$$L_c(t) = \sum_{E_l} A_c \int_0^t (t - \tau)e^{-i(E_{cj\mathbf{k}_\parallel} - E_l)\tau/\hbar}\, d\tau, \tag{68}$$

where $A_c$ includes the Fermi distribution function and the scattering matrix element, and $E_l$ and $A_c$ are given by considering detailed scattering mechanisms, for example, carrier–carrier and carrier–phonon scattering, etc. [16]. $L_v(t)$ is given similarly by changing the subscripts c and $m$ to v and $n$, respectively.

Hereafter, we express the factors $e^{-iE_{cj\mathbf{k}_\parallel}t/\hbar}(1 - L_c(t) + \cdots)$ and $e^{-iE_{cj\mathbf{k}_\parallel}t/\hbar}(1 - L_v(t) + \cdots)$ in Eq. (66) as $G_c(t)$ and $G_v^*(t)$, respectively. The imaginary part of the Fourier transform of $Q^{R}(t)$ is given by

$$\begin{aligned}
\mathrm{Im} \int_{-\infty}^{\infty} Q^{R}(t)e^{i\omega t}\, dt &= \frac{1}{2i}\int_{-\infty}^{\infty} [Q^{R}(t) - Q^{R*}(-t)]e^{i\omega t}\, dt \\
&\propto \int_{-\infty}^{\infty} [G_c(t)u(t) - G_c^*(-t)u(-t)] \\
&\quad \times [G_v^*(t)u(t) - G_v(-t)u(-t)]e^{-iE_g t/\hbar}e^{i\omega t}\, dt \qquad (69) \\
&= 4\int_{-\infty}^{\infty} \left[\mathrm{Im}\int_{-\infty}^{\infty} G_c(t)u(t)e^{i\omega' t}\, dt\right] \\
&\quad \times \left[\mathrm{Im}\int_{-\infty}^{\infty} G_v(t)u(t)^{i(\omega - \omega' - E_g/\hbar)t}\, dt\right] d\omega'.
\end{aligned}$$

The Fourier transform of $G_c(t)u(t)$ in this equation is calculated using Eq. (68) as follows:

$$\int_{-\infty}^{\infty} e^{-iE_{cj\mathbf{k}_\parallel}t/\hbar}(1 - L_c(t) + \cdots)e^{i\omega t}u(t)\,dt$$

$$= \frac{i\hbar}{\hbar\omega - E_{cj\mathbf{k}_\parallel}} - \sum_{E_l} A_c \frac{i\hbar}{(\hbar\omega - E_{cj\mathbf{k}_\parallel})^2(\hbar\omega - E_l + i0^+)} + \cdots \tag{70}$$

$$\simeq \frac{i\hbar}{\hbar\omega - E_{cj\mathbf{k}_\parallel} + \Sigma_{E_l} A_c/(\hbar\omega - E_l + i0^+)}.$$

Separating the third term in the denominator of Eq. (70) into the real and imaginary parts and comparing the explicit formula of $A_c$ in [16] with Eqs. (27) and (28) in this chapter, we obtain the following relation both for carrier–carrier and carrier–phonon scattering:

$$\begin{aligned}\mathrm{Im}\left(\sum_{E_l} A_c \frac{1}{\hbar\omega - E_l + i0^+}\right) &= -\sum_{E_l} \pi A_c\, \delta(\hbar\omega - E_l) \\ &= -\Gamma_{cj\mathbf{k}_\parallel}(\hbar\omega).\end{aligned} \tag{71}$$

Also, the real part of the third term in the denominator of Eq. (70) is equal to $-\Delta_{cj\mathbf{k}_\parallel}$ in our notation, if we consider the correlation energy (the second order perturbation term). The exchange energy (the first order perturbation term) is not included in this discussion, because [16] has treated only the second order perturbation term of the scattering Hamiltonian.

By these relations, Eq. (70) becomes

$$\frac{i\hbar}{\hbar\omega - E_{cj\mathbf{k}_\parallel} - \Delta_{cj\mathbf{k}_\parallel} - i\Gamma_{cj\mathbf{k}_\parallel}(\hbar\omega)}. \tag{72}$$

The Fourier transform of $G_v(t)u(t)$ in Eq. (69) is given similarly. Consequently, we obtain

$$\mathrm{Im} \int_{-\infty}^{\infty} Q^{\mathrm{R}}(t)e^{i\omega t}\,dt \propto \int_{-\infty}^{\infty} D_c(E)D_v(\hbar\omega - E - E_g)\,dE, \tag{73}$$

where $D_c(E)$ and $D_v(E)$ are just equal to the expressions given in Eqs. (25) and (26) in Section 2.

Thus, we see that the temporal behavior calculation in [16] and our calculation in Section 2 are equivalent. One different point is that, for the energy shift factor $\Delta_{cj\mathbf{k}_\parallel}$, [16] and [26] consider the second order perturbation terms (correlation energy) of carrier–carrier and carrier–phonon scattering, while we consider the first order perturbation term (exchange energy) with the screened Coulomb potential for hole–hole scattering in the calculation in Section 4. The exchange energy is constant for photon energy in Eqs. (29) and (63), while the correlation energy is dependent on photon energy and changes sign near the center of the line shape. Both of them make the lineshape narrow. The difference between these narrowing effects appears to be small, as seen in the lineshape results of [26] and Fig. 10 of this chapter, maybe because the magnitudes of the exchange energy and the peak correlation energy are comparable. (In the previous lineshape calculation in [9], the energy shift factor has been neglected, and the linewidth is larger than these two calculations.) The difference between the lineshapes in [26] and this section at large values of $|\hbar\omega - E_{cv}|$ on the low-energy side may be due to the scattering neglected in this section. The effect of this difference on the gain is small, because both of the lineshape values are low enough in this energy region. The peak energy shift is larger for the exchange energy.

In [16] and [26], the intraband relaxation time is defined by $\mathrm{Re}[L(t)]/t$ at a large value of $t$ for the case where $L(t)$ approaches a linear function of $t$ as $t \to \infty$, where $L(t) = L_c(t) + L_v(t)$ in the preceding notation. The inverse of this intraband relaxation time approximately gives the half width of the non-Lorentzian lineshape, because the behavior of $L(t)$ at large $t$ mainly determines the lineshape around the peak after the Fourier transform. Therefore, this intraband relaxation time just corresponds to the modified one discussed in Eq. (65) of this chapter. This modified intraband relaxation time is larger than the inverse scattering probability calculated in Section 3.

The temporal behavior calculation is convenient in discussing the response of materials to ultrahigh speed optical pulses, and the temperature Green's function method is convenient for spectral analysis.

## 5. CONCLUSION

In this chapter, theoretical analysis was described for the effect of intraband relaxation on optical spectra in quantum well lasers, with the gain spectra as an example. The intraband relaxation effect was expressed by the line broadening function, which was the Fourier transform of the response of the polarization to an impulse electric field. The Lorentzian shape often used in gain analyses was obtained from the line broadening function as an approximation, the half width of which was given by the reciprocal scattering probability, i.e., the intraband relaxation time. Generally, the lineshape approaches the Lorentzian, if the energy width of carrier distribution is much larger than $\hbar \times$ (the scattering probability). The intraband relaxation time in quantum well lasers was dominated by the hole-involved scattering mechanism, such as the hole–hole, hole–LO phonon, and electron–hole scattering. Spectral lineshape without the preceding approximation of the Lorentzian shape was calculated. The lineshape was narrower than the Lorentzian with the intraband relaxation time obtained from the scattering probability, and the low-energy tail of the lineshape was much steeper than the Lorentzian. An approximate formula for this non-Lorentzian lineshape was given. The low-energy tail was shown to decay exponentially. The half width of the non-Lorentzian shape was given by the modified intraband relaxation time, which was proportional to but smaller than the reciprocal scattering probability. The calculation process for obtaining the non-Lorentzian shape was summarized. It was shown that the lineshape calculation via the temporal behavior of the polarization is equivalent to the method used in this chapter.

The overall shape of the line broadening function is important in the gain analysis. In particular, the shape of the low-energy tail affects significantly the gain magnitude. The lineshape depends on the band structure and carrier distribution. For quantum wire and quantum box structures, deviation of the lineshape from the Lorentzian is remarkable because of the narrow density of states, which has been shown by the temporal behavior calculation of the polarization [26].

## REFERENCES


1. Yamada, M., and Suematsu, Y. (1981). *J. Appl. Phys.* **52**, 2653.
2. Yamada, M., Ishiguro, H., and Nagato, H. (1980). *Japan. J. Appl. Phys.* **19**, 135.
3. Asada, M., Kameyama, A., and Suematsu, Y., (1984). *IEEE J. Quantum Electron,* **QE-20**, 745.
4. Zielinski, E., Schweizer, H., Hausser, S., Stuber, R., Pilkuhn, M., and Weimann, G., (1987). *IEEE J. Quantum Electron.* **QE-23**, 969.
5. Yamanishi, M. (1981). *Tech. Report on Opt. Quantum Electron., IECE of Japan,* **OQE80-103**, 19 (in Japanese).
6. Landsberg, P. T., and Robbins, D. J. (1985). *Solid State Electron.* **28**, 137.
7. Sugimura, A., Pazak, E., and Meissner, P. (1986). *J. Phys. D* **19**, 7.
8. Takehima, M. (1987). *Phys. Rev. B* **36**, 8082.
9. Asada, M. (1989). *IEEE J. Quantum Electron.* **QE-25**, 2019.
10. Fetter, A. L., and Walecka, J. D. (1971). *Quantum theory of many-particle systems.* McGraw-Hill, New York.
11. Haug, H., and Schmitt-Rink, S. (1984). *Prog. Quantum Electron.* **9**, no. 1.
12. Takeshima, M. (1986). *Phys. Rev. B* **34**, 1041.

13. Ando, T., Fowler, A. B., and Stern, F. (1982). *Rev. Mod. Phys.* **54**, 437.
14. Ehrenreich, H. (1959). *J. Phys. Chem. Solids* **9**, 129.
15. Asada, M., and Suematsu, Y. (1985). *IEEE J. Quantum Electron.* **QE-21**, 434.
16. Yamanishi, M., and Lee, Y. (1987). *IEEE J. Quantum Electron.* **QE-23**, 367.
17. Yamada, M., Ogita, S., Yamaguchi, M., Tabata, K., Nakaya, N., Asada, M., and Suematsu, Y. (1984). *Appl. Phys. Lett.* **45**, 324.
18. Bloembergen, N. (1965). *Nonlinear optics.* Benjamin, New York.
19. Kubo, R. (1957). *J. Phys. Soc. Japan* **12**, 570.
20. Toyozawa, Y. (1958). *Prog. Theo. Phys.* **20**, 53.
21. Sargent, M., III, Scully, M. O., Lamb, W. E., Jr. (1974). *Laser physics.* Addison-Wesley, Reading, Massachusetts.
22. Schiff, L. I. (1968). *Quantum mechanics.* McGraw-Hill, New York.
23. Leburton, J. P., and Hess, K. (1983). *J. Vac. Sci. Technol.* **B1**, 415.
24. Yamanishi, M., and Suemune, I. (1984). *Japan. J. Appl. Phys.* **23**, L35.
25. Iwamura, H., Saku, T., Kobayashi, H., and Horikoshi, Y. (1983). *J. Appl. Phys.* **54**, 2692.
26. Ohtoshi, T., and Yamanishi, M. (1991). *IEEE J. Quantum Electron.* **QE-27**, 46.
27. Tsang, W. T. (1981). *Appl. Phys. Lett.* **39**, 786.
28. Dutta, N. K. (1982). *J. Appl. Phys.* **53**, 7211.
29. Sugimura, A. (1984). *IEEE J. Quantum Electron.* **QE-20**, 336.
30. Kasemset, D., Hong, C. S., Patel, N. B., and Dapkus, P. D. (1983). *IEEE J. Quantum Electron.* **QE-19**, 1025.
31. Nishimura, Y., and Nishimura, Y. (1973). *IEEE J. Quantum Electron.* **QE-9**, 1011.

# *Chapter 3*

# MULTIQUANTUM WELL LASERS: THRESHOLD CONSIDERATIONS

**Reinhart W. H. Engelmann**

*Department of Electrical Engineering and Applied Physics*
*Oregon Graduate Institute of Science and Technology*
*Beaverton, Oregon*

and

**Chan-Long Shieh**

*Phoenix Corporate Research Laboratories*
*Motorola Corporation*
*Tempe, Arizona*

and

**Chester Shu**

*Department of Electronic Engineering*
*The Chinese University of Hong Kong*
*Shatin, N.T., Hong Kong*






ISBN 0-12-781890-1

# 1. INTRODUCTION

Single (SQW) and multiquantum well (MQW) laser structures are promising candidates for use in high speed integrated optics and optical communications because of their significant superiority in performance over conventional double-hetero (DH) structures. Quantum well (QW) lasers possess the potential of lower threshold current density with lower temperature sensitivity; higher differential quantum efficiency; improved coherency with reduced lasing linewidth and superior mode stability; larger modulation bandwidth; and reduced chirping during modulation [1]. This chapter is devoted to the topic of utilizing the unique properties of QW gain to the best advantage in the design of MQW laser structures. Of particular interest are the performance advantages or disadvantages with respect to the simpler SQW versions. Based on model calculations, QW gain spectra will be analyzed as a function of injection rates, and the consequences of this functional relationship for the dependence of the laser threshold current on cavity length and number of quantum wells will be evaluated. As an important result, it will be concluded that the unique advantages of QW gain, being limited to relatively low carrier injection densities, can be best utilized with MQW, rather than SQW, lasers, particularly when short lasing cavities are required. On the other hand, MQW laser performance is more strongly degraded by interface recombination and, of course, is also affected by fluctuations of QW layer thickness. Thus, special attention has to be placed on uniformity and interface quality during epitaxial growth of the MQW layer sequence.

Many approaches of various complexity have been used in the literature to model gain in QW structures. We shall concentrate here on selecting an approach that is still structurally simple but adequate to describe the experimental observations on the threshold current in QW lasers. Thus, the main emphasis is on estimating the spectral gain maximum that determines laser threshold rather than modeling the details of the full gain spectrum. In analyzing the gain/current relationship, the influence of homogeneous broadening of the gain spectrum by intraband carrier scattering [2] is discussed, and the polarization dependence resulting from the anisotropy of the QW structure is included [3]. Broadening and gain enhancement of TE polarization[1] largely compensate each other in their influence on the gain maximum for commonly used QW sizes around 7.5 nm. This leads us to define a "simplified" model neglecting these two effects, useful for rough estimates.

[1] TE polarization refers to light polarized parallel to the plane of the QW as opposed to TM polarization, which refers to light polarized perpendicular to the plane of the QW. Semiconductor lasers commonly operate in the TE mode.

The more complex "accurate" model that includes both these effects forms the basis for the comparison with experimental threshold current data from GaAs/AlGaAs single and multiple QW lasers. Additionally, various nonideal contributions to the total current are estimated, and the conditions for their minimization are established. These contributions result from nonradiative and leakage processes, such as Auger recombination, carrier losses out of the QWs [4, 5], and, most importantly, interface recombination at the QW boundaries [6].

It is shown that threshold current in MQW lasers can be minimized near relatively moderate injection levels above the transparency condition by choosing an optimum number of QWs for a given cavity length [7, 8]. A slight increase of the number of QWs from this optimized value may be beneficial, since it tends to reduce the injection level, thereby maximizing the advantages of QW gain as well as reducing the nonideal current contributions in a MQW laser with only a small penalty in increased threshold current. Minimizing injection levels is particularly important for long wavelength InGaAsP/InP MQW lasers because of the large Auger contribution in these materials [9].

We start with a review of the gain and radiative current calculations in QW lasers discussing the simplified and more accurate models. Next, we analyze the nonideal contributions to the current, continue with the description of threshold current versus QW number and cavity length, and eventually conclude by comparing the modeled results with experimental data.

## 2. SPECTRAL GAIN MAXIMUM AND RADIATIVE QW CURRENT

In determining the threshold current of QW lasers, we are concerned primarily with the spectral gain maximum and its dependence on the injection level. For the purpose of analyzing this gain maximum in the context of threshold conditions in MQW lasers, the procedure of calculating the gain and spontaneous emission spectra, as well as the radiative injection current density generating those spectra, is reviewed. First, we discuss the "simplified" model that is based on the energy-independent momentum matrix element of the bulk semiconductor without broadening, which is useful for rough estimates. Subsequently, to provide more general validity, we proceed to the physically more "accurate" model that takes broadening and TE polarization enhancement into account [10]. In our calculations, we use the widely accepted assumptions that the radiative transitions are subject to: (1) momentum ($k$) conservation in the plane of the quantum well; and

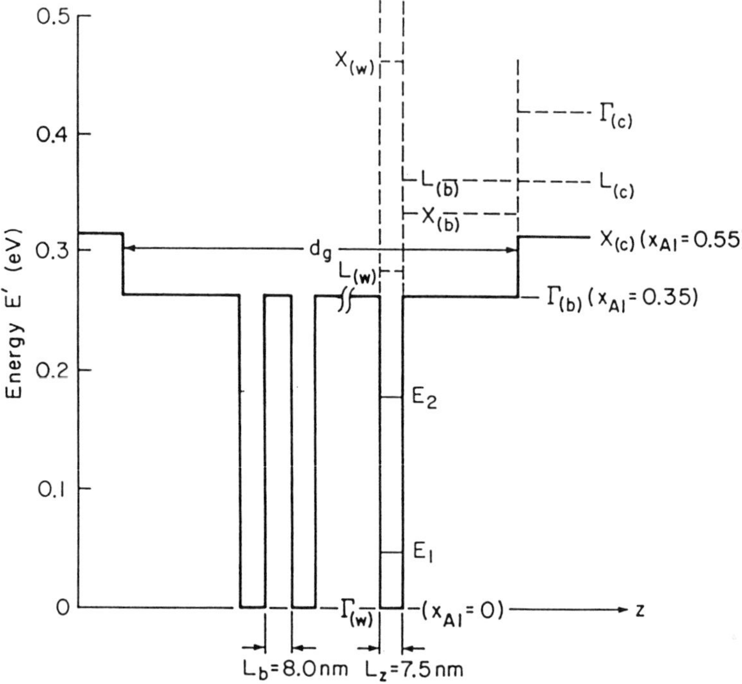


**Fig. 1.** Conduction band schematic (at the flat-band injection condition) of the representative GaAs/AlGaAs separate confinement MQW structure used in the numerical calculations (when choosing a gain confinement of $\Gamma = 0.023\ \nu_{QW}$, the guide thickness selected was $d_g = 250\ \mu m$; when optimizing $\Gamma$, it followed as $d_g = 258\ \mu m$ for $\nu_{QW} = 1$, $d_g = 219\ \mu m$ for $\nu_{QW} = 3$, and $d_g = 215\ \mu m$ for $\nu_{QW} = 6$). Important characteristic energy levels are also indicated: $\Gamma_{(i)}$, direct conduction band valleys in zone center ($k = 0$); $X_{(i)}$, indirect conduction band valleys at zone boundaries with $k$ along $\langle 100 \rangle$ crystal directions; $L_{(i)}$, indirect conduction band valleys at zone boundaries with $k$ along $\langle 111 \rangle$ crystal directions, for the well ($i =$ w), the barrier ($i =$ b), and the cladding ($i =$ c) materials; $E_n$ quantum levels in the well, with $n = 1, 2$.

(2) conservation of the well quantum number $n$ [1, 11]. It should be noted that other models without $k$ conservation have also been suggested to explain gain broadening in QW lasers [12, 13]. For details on the subject of QW gain, the reader is referred to Chapter 1.

Our numerical calculations were performed for a representative separate confinement MQW laser structure in the GaAs/AlGaAs system. Figure 1 illustrates this structure, showing a schematic of the conduction band potentials at the flat-band injection condition. We chose a QW size of

$L_z = 7.5$ nm, judged small enough to achieve minimum threshold current and/or threshold current density in optimized designs [7, 8; see also Chapter 4] without compromising the confinement of the electronic wave functions to the QW at smaller $L_z$. The waveguide region of thickness $d_g$ (typically about 250 nm) consists of a number ($\nu_{QW}$) of GaAs QWs symmetrically imbedded in the barrier material of 35% AlGaAs. A barrier size of 8 nm was selected between two neighboring QWs, large enough to ensure negligible coupling of the QW wavefunctions (negligible carrier tunneling across the barriers) but still small enough for maximum MQW packing density. The total waveguide region is considered to be undoped ("i-layer"), ensuring uniform injection of electrons and holes from the respective cladding layers into the QWs [14] without requiring any carrier tunneling [15]. This also ensures that the barrier and well potentials remain flat; i.e., no diffusion potentials develop at the well–barrier hetero-interfaces as a result of modulation doping effects [16, 17]. The 55% AlGaAs cladding layers doped p- or n-type, respectively, provide optical confinement in addition to the contacts for carrier injection. Table I lists the numerical material parameters used in the calculations.

**Table I.**

Numerical Values Used in the Calculations

| | |
|---|---|
| Temperature | 300 K |
| Band gaps [18] $E_g(x_{Al} = 0)$, $E_{g,b}(x_{Al} = 0.35)$ (eV) | |
| $Al_xGa_{1-x}As$ at $\Gamma(x_{Al} < 0.45)$ | $1.424 + 1.247x_{Al}$ |
| $\Gamma\ (x_{Al} > 0.45)$ | $1.424 + 1.247x_{Al} + 1.147(x_{Al} - 0.45)^2$ |
| L | $1.708 + 0.642x_{Al}$ |
| X | $1.900 + 0.125x_{Al} + 0.143x_{Al}^2$ |
| Band offset [19] $Al_xGa_{1-x}As/GaAs$ at $\Gamma$ (eV) | |
| $\Delta E_c\ (x_{Al} < 0.45)$ | $0.747x_{Al}\ (\Delta E_c{:}\Delta E_v = 0.6{:}0.4)$ |
| $\Delta E_c\ (x_{Al} > 0.45)$ | $0.747x_{Al} + 1.147\ (x_{Al} - 0.45)^2$ |
| $\Delta E_v$ | $0.50x_{Al}$ |
| Effective masses [20] $Al_xGa_{1-x}As$ at $\Gamma$, in units of $m_0$ | |
| electron $m_c$ | $0.0665 + 0.0835x_{Al}$ |
| heavy hole $m_{v,h}$ | $0.45 + 0.302x_{Al}$ |
| light hole $m_{v,l}$ | $0.08 + 0.057x_{Al}$ |
| Refractive index [18] $\bar{n}_{eq} \approx \bar{n}'_g \approx \bar{n}'$ (GaAs) | 3.6 |
| Matrix element prefactor [18] $\xi$ (GaAs) | 2.66 |
| Intraband scattering time [3, 21] $\tau_{in}$ (GaAs) | 0.1 ps |
| Auger coefficient [4] $C_A$ (GaAs) | $5 \times 10^{-30}\ cm^6\ s^{-1}$ |
| mode facet reflectivity [4] $R$ | 0.3 |
| Internal mode loss [4] $\alpha_i$ | $5\ cm^{-1}$ |

### 2.1. Simplified Model

In the simplified approach to calculate the gain spectra of QW semiconductor lasers, only the change in the density of states function from the bulk to the two dimensional (2D) quantum well is considered [22]. Additionally, the following assumptions are made regarding the electron–hole dipole transition in the QW: (1) There is a translational symmetry in the plane $(x, y)$ of the junction; therefore the momentum parallel to the junction plane is conserved, i.e., $\Delta k_{(x,y)} = 0$. (2) Because the quantized wavefunctions of the vertical dimension $z$ are orthogonal to each other, the vertical quantum number is conserved, i.e., $\Delta n = 0$. (3) Polarization effects for stimulated transitions (i.e., TE versus TM) due to the anisotropy of the quantum well structure are neglected and, thus, the energy independent transition matrix element $M_b$ of the bulk semiconductor is retained for the QW. (4) Broadening effects due to intraband scattering of the injected carriers are ignored. Assumptions (1) and (2) are reasonable because QW lasers are operated at higher injection levels, and therefore the band-tail states are negligible, and the translational symmetry is still preserved [23]; band-mixing effects that violate the strict $\Delta n = 0$ selection rule [24, 25] are neglected. Assumptions (3) and (4) are for simplification only, but results provide a rough estimate for the spectral TE gain maximum, since, with preferred well sizes $L_z$ around 7.5 nm, the reduction of the gain maximum caused by broadening tends to be largely compensated by the gain enhancement for TE polarization in the QW.

The QW constitutes a quasi two dimensional (2D) system for the movement of the charge carriers. In the (horizontal) plane $(x, y)$ of the QW, the movement remains unrestricted; in the vertical dimension $z$, it is confined. The density of states in such a system is obtained from the energy levels $E_n$ $(n = 1, 2, 3, \ldots)$ of the quantized dimension $z$, each forming a subband $n$ with respect to the free $x, y$ movement. The 2D density of states $D(E')$, i.e., the available states per unit area and unit energy, can be written as the sum over the step-like density of states functions of each individual subband [26]:

$$D(E') = \sum_n D_n(E'), \tag{1a}$$

$$D_n(E') = (m_n/2\pi\hbar^2)\, H(E' - E_n), \tag{1b}$$

where we have disregarded spin degeneracy (following the treatment in Chapter 1 we will include spin degeneracy in the transition matrix element later introduced by Eq. (7)). $H$ is the Heaviside unit step function, $m_n$ is the effective mass in the $n$th 2D subband with respect to the $x, y$ movement, and $\hbar$ is Planck's constant/$2\pi$. The energy $E'$ is measured from the band edge of the bulk semiconductor into the band. The effects of band mixing in QWs [24,25] lead to energy-dependent effective masses $m_n(E')$ in the subbands. As already mentioned, we ignore these effects; i.e., the subbands are treated within the parabolic band approximation, with $m_n$ becoming independent of $n$. The quantized energy levels $E_n$ are sometimes estimated from an infinite square well model [11]; however, we use here the more accurate approach of finite square wells with an effective mass discontinuity at the QW boundaries [27, 28], based on a band offset ratio for the AlGaAs/GaAs hetero-interface of $\Delta E_c{:}\Delta E_v = 0.6{:}\ 0.4$ [19].

From the density of states in both the conduction (c) and valence (v) bands, we are able to calculate the gain spectrum $g(E)$ as well as the spontaneous emission spectrum $R_{sp}(E)$ for any injection level by a proper summation over the respective electron transitions between the conduction and valence bands at an energy separation $E$ corresponding to the energy of the emitted photon, i.e., $E = \hbar\omega$. Strictly speaking, states above the edges of the QW should be included in this summation [5, 10]. In our approach, we neglect those bulk-like high energy transitions for calculating $g(E)$ and $R_{sp}(E)$, since these would affect only the high-energy tails of the spectra, whereas our interest is concentrated on the wavelength region near the gain maximum. However, as pointed out by Chinn *et al.* [21], their contribution to the current could be significant when integrating over the high-energy tail of $R_{sp}(E)$, see Eq. (13) following. In our treatment, we consider this contribution as part of the carrier "leakage" out of the well, which will be discussed in Section 3.

The injection level can be represented by the quasi-Fermi levels of the electrons (e) and the holes (h), assuming that the carriers relax within their respective bands immediately upon injection to attain the Fermi–Dirac distribution. This implies that the carrier relaxation times (order of 0.1 ps) are small compared with the carrier recombination times (order of ns) and, thus, there are no hot carriers. Additionally, we postulate flat quasi-Fermi levels both in the conduction band and the valence band across the entire waveguide core containing the QWs (Fig. 1). On the other hand, the quasi-Fermi levels for electrons and holes are related through the charge neutrality conditions. For, imposing this condition, we count carriers occupying QW energy states only; i.e., we restrict ourselves to injection levels at which any high-energy carriers that are thermally excited into the bulk-like states beyond the edges of the QWs can be neglected. It is advantageous to use the hole quasi-Fermi level as the measure of the injection level because there are two branches, namely, the heavy (h) and light (l) holes, in the valence band. This in turn leads to a straightforward calculation of the electron quasi-Fermi level from the hole quasi-Fermi level.

As a result of the conservation of in-plane momentum $k_{(x,y)}$ and confinement quantum number $n$ during a dipole transition of an injected electron from the conduction to the valence band, the 2D density of states functions of the two subbands with equal $n$ can be combined to form a reduced 2D density of states describing the dipoles of energy $E = \hbar\omega$ between the electron and the respective hole [9]; i.e.,

$$D_{\mathrm{r},jn}(E) = (m_{\mathrm{r},j}/2\pi\hbar^2)\, H(E - E_{\mathrm{g},jn}), \qquad j = \mathrm{h}, \mathrm{l}, \quad n = 1, 2, \ldots, \tag{2}$$

where the effective mass $m_n$ of Eq. (1) is replaced by the reduced effective mass

$$m_{\mathrm{r},j} = m_{\mathrm{c}} m_{\mathrm{v},j}/(m_{\mathrm{c}} + m_{\mathrm{v},j}), \tag{3}$$

and (see Fig. 2)

$$E_{\mathrm{g},jn} = E_{\mathrm{g}} + E_{\mathrm{c},n} + E_{\mathrm{v},jn} \tag{4}$$

is the gap between the subband pair e$n$/$j$h$n$. For a QW size of $L_z$, we then simply obtain the volume density of states

$$\rho_{\mathrm{r},jn}(E) = D_{\mathrm{r},jn}(E)/L_z, \tag{5}$$

and, taking all the subband pairs e$n$/$j$h$n$ into account by summing over $j$ and $n$, the gain spectrum is calculated from the following equation [29]:

$$g(E) = g_0(E)|M_{\mathrm{b}}|^2 \sum_{j,n} \rho_{\mathrm{r},jn}(E)[f_{\mathrm{c}} - f_{\mathrm{v}}]_{E,jn}; \tag{6a}$$

here, the gain prefactor is given by

$$g_0(E) = \pi\hbar e^2/m_0^2 \bar{n}_{\mathrm{eq}} E c \varepsilon_0, \tag{6b}$$

and $M_{\mathrm{b}}$ is the average, energy-independent, momentum matrix element for the dipole transition (*transition* matrix element) in the bulk semiconductor (spin degeneracy *included*), i.e.,

$$|M_b|^2 = (2/3)|M|^2 \tag{7a}$$

where

$$|M|^2 = (m_0/m_{\mathrm{c}} - 1)[(E_{\mathrm{g}} + \Delta)/2\{E_{\mathrm{g}} + (2/3)\Delta\}]m_0 E_{\mathrm{g}} \tag{7b}$$

is the absolute square of the "basis function" momentum matrix element (see Chapter 1; the factor 1/3 in Eq. (7a) results from averaging over all directions, the factor 2 from spin degeneracy). Introducing a prefactor $\xi$, following Dutta [11], we may also write

$$|M_{\mathrm{b}}|^2 = 2\xi m_0 E_{\mathrm{g}}. \tag{7c}$$

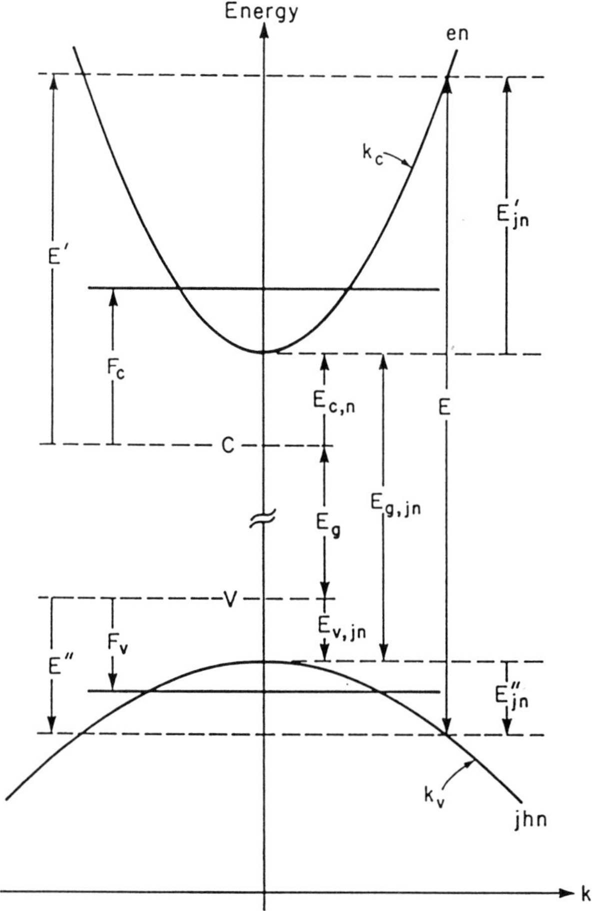


**Fig. 2.** Energy band diagram, $E$ versus $k$, for a particular subband pair $n$ ($n = 1, 2, \ldots$) based on the parabolic band approximation. The energy scales used in the text are defined for a $\Delta k = 0$ subband transition from the $k_c$ level e$n$ to the $k_v$ level $j$h$n$ ($j$ = h, 1), producing a photon of energy $E = \hbar\omega$.

The parameters appearing in Eqs. (6) and (7) are defined as follows: $e$, electron charge; $m_0$, free electron mass; $c$, speed of light in vacuum; and $\varepsilon_0$, permittivity of vacuum; $\bar{n}_{eq}$ is an equivalent refractive index of the propagating optical mode [10] and can be expressed in terms of the mode group index $\bar{n}_g$ and the QW phase index $\bar{n}'$ as $\bar{n}'^2/\bar{n}_g$ (see Chapter 1); $\Delta$ is the spin–orbit splitting of the valence band in the QW material; $f_c$ and $f_v$ are the Fermi–Dirac functions for the electrons in the conduction and valence bands, respectively.

In estimating the matrix-element prefactor $\xi$, temperature effects are usually neglected. For GaAs, Eq. (7) yields $\xi = 2.66$ (Table I) based on generally accepted room temperature parameters. In formulating Eq. (7b), we have ignored corrections caused by the contributions from higher lying energy bands to the effective mass of the electrons near the conduction band edge [30]. This contribution, which would increase the prefactor $\xi$ by about 27%, appears to be largely compensated, in the case of the preferred QW sizes near 7.5 nm, by the rather complex effects of QW subband mixing [24, 25], which we chose to ignore here also. For a discussion of band-mixing effects see Chapters 1 and 7.

The Fermi–Dirac functions $f_c$ and $f_v$ have to be evaluated for the proper conduction–valence subband combination e$n$/$j$h$n$ at energy separation $E$ under the condition of equal $k_{(x,y)}$, i.e.,

$$k_c(E'_{jn}) = k_v(E''_{jn}) \qquad (k\text{-selection rule}), \tag{8}$$

with the respective quasi-Fermi levels being fixed by the injection level as mentioned before. The procedure is sketched in Fig. 2, where we show a representative $E$ versus $k$ diagram for a particular subband pair e$n$/$j$h$n$ in the parabolic band approximation (constant effective masses); the energies

$$E'_{jn} = (E - E_{g,jn})m_{r,j}/m_c, \tag{9a}$$

$$E''_{jn} = (E - E_{g,jn})m_{r,j}/m_{v,j} \tag{9b}$$

are measured from the bottom of the conduction (upwards $E'_{jn}$) and the top of the valence subband (downwards $E''_{jn}$), respectively, satisfying the relation

$$E = E_{g,jn} + E'_{jn} + E''_{jn} \qquad \text{(energy conservation)}. \tag{9c}$$

Thus,

$$[f_c]_{E,jn} = [1 + \exp\{(E'_{jn} + E_{c,n} - F_c)/k_B T\}]^{-1}, \tag{10a}$$

$$[f_v]_{E,jn} = [1 + \exp\{-(E''_{jn} + E_{v,jn} - F_v)/k_B T\}]^{-1}, \tag{10b}$$

where $F_c$ and $F_v$ are the quasi-Fermi-level positions measured from the respective band edges of the bulk semiconductor (up and downwards, respectively), and $k_B$ is Boltzmann's constant; a choice of $F_v$ determines the hole injection level $P$, and $F_c$ is then obtained from the neutrality condition. Thus, we have, counting the energy states in the QWs only[2] (the factor 2

[2] In order to simplify the integration, we have actually extended the upper limit to $\infty$ by extrapolating the QW density of states functions $\rho_c$ and $\rho_v$ beyond $\Delta E_c$ and $\Delta E_v$, respectively. This provides analytical expressions for the carrier densities and the error introduced is negligible if the injection levels are not excessive.

accounts for spin degeneracy),

$$P_0 + P = 2\int_{E_{v,h1}}^{\Delta E_v} \rho_v(E'')[1 - f_v(E'')]\, dE'', \tag{11a}$$

$$P_0 + P \approx N_0 + N = 2\int_{E_{c,1}}^{\Delta E_c} \rho_c(E') f_c(E')\, dE', \tag{11b}$$

where $\rho_c$ and $\rho_v$ are the densities of states (disregarding spin degeneracy) of the conduction and valence band, respectively, and the $\rho$ are related to the D of Eq. (1) according to the proper equivalence of Eq. (5). For lasing, injection levels are usually high compared with the doping levels, i.e., $N, P \gg N_0, P_0$, and the equilibrium carrier densities $N_0$, $P_0$ can be neglected.

In a similar way, the spontaneous emission is calculated through the following equation [29]:

$$R_{sp}(E) = r_0(E)|M_b|^2 \sum_{n,j} \rho_{r,jn}(E)[f_c(1 - f_v)]_{E,jn}, \tag{12a}$$

where the spontaneous-emission prefactor is given by

$$r_0(E) = e^2 \bar{n}'_g E / \pi m_0^2 \hbar^2 c^3 \varepsilon_0; \tag{12b}$$

here, $\bar{n}'_g$ is the group index of the QW material (see Chapter 1). It should be noted that the form of the spectra $g(E)$ and $R_{sp}(E)$ are largely determined by the density of states functions $\rho_{r,jn}$ and the Fermi–Dirac functions $f_c, f_v$. The prefactors $g_0(E)$ and $r_0(E)$ are relatively slow functions of $E$.

To obtain the spectral gain maximum versus current density, the radiative QW component of the current in a MQW laser is calculated from the integral over the spontaneous emission spectrum of the QWs:

$$J_r = \nu_{QW} L_z e \int_{E_{g,h1}}^{E_{g,b}} R_{sp}(E)\, dE, \tag{13a}$$

$\nu_{QW}$ being the number of QWs. This expression implies uniform well width $L_z$ and uniform injection into the wells, i.e., flat quasi-Fermi levels across the MQW region. In describing MQW lasers with various $\nu_{QW}$, it is advantageous to introduce a current density normalized to a single QW, i.e.,

$$J_{r,1} = J_r/\nu_{QW} = L_z e \int_{E_{g,h1}}^{E_{g,b}} R_{sp}(E)\, dE. \tag{13b}$$

Notice that the integration is over the transitions involving QW states only, with the lower cutoff at the effective bandgap $E_{g,h1}$ of the QW, and the upper cutoff at the bandgap $E_{g,b}$ of the barrier. As pointed out before, contributions beyond $E_{g,b}$ are regarded to be undesirable leakage. The leakage and nonradiative parts of the current will be considered in Section 3. These parts include, additionally, recombination at each hetero-interface, bulk recombination at defects, spilling of carriers to the barrier and guiding layers, and the Auger recombination currents.

## 2.2. Accurate Model

In the "simplified" gain model just discussed, spectral broadening effects as well as effects resulting from the anisotropy of the QW have been ignored. Both the gain spectrum $g(E)$ as well as the spontaneous emission spectrum $R_{sp}(E)$ exhibit unrealistic sharp steps in this case as a direct consequence of the step-like density of states function of the QW. Spectral broadening results from a number of mechanisms, the most important of which is intraband carrier relaxation due to scattering, since the scattering times involved are known to be very short (about 0.1 ps in GaAs or InGaAs). These broadening mechanisms smooth out the spectral envelope and, in particular, reduce (and thereby energetically shift) the spectral maxima. In case of the gain spectrum, this peak reduction is counteracted, however, by a gain enhancement experienced by light polarized in the plane of the QW, i.e., for the TE polarized mode of the vertical QW waveguide structure. The TE mode exhibits the lowest lasing threshold except under unusual circumstances [31], and hence, TM polarization does not concern us here. Elaborate models have been developed to include these broadening and polarization effects, resulting in theoretical gain spectra that match experimental ones extremely well (e.g., [33, 34].) Details on spectral broadening are presented in Chapter 2.

As pointed out already, for determining laser threshold we are concerned only with the magnitude of the gain peak, not its position or other fine details of the complete gain spectrum, and, hence, a simplified broadening model suffices. We assume that each energy state at a particular $k$ value is homogeneously broadened by the intraband carrier relaxation process, forming a Lorentzian lineshape for each state. The linewidth is determined from an energy-independent carrier scattering time, and $k$ conservation is imposed for the optical transition between these energetically broadened $k$ states in the conduction and valence bands. Such a broadening of the energy states, in fact, can be interpreted as a violation of energy conservation due to the uncertainty principle. The TE polarization enhancement of the gain is accounted for in the usual way, by employing the properly modified transition matrix element [10].

### *2.2.1. Gain Broadening by Intraband Relaxation*

Strict $k$ and energy conservation predicts the disappearance of spontaneous emission at energies below the bandgap, a result inconsistent with experiments. In order to explain the low-energy tails of cxperimental emission spectra, two models have been proposed: (1) The band tail model [23] assumes the existence of a continuous distribution of localized energy states below the bandgap, giving rise to a violation of the $k$ selection rule. This produces an energy dependence for the transition matrix element. The argument used for this model is the screening effect caused by impurities or holes in the active region. As an extreme case, nonconservation of $k$ is postulated for the complete spectrum [12]. Additionally, excitonic effects may modify transitions below the bandgap. (2) The intraband carrier relaxation model [2] uses broadening of energy states to explain the observed transitions. A Lorentzian lineshape is used to describe a homogeneous broadening of the energy states. Homogeneous broadening results from the assumption that the carriers become indistinguishable due to the high scattering rate (scattering times $< 1$ps). The latter model also explains the gain broadening effect in quantum well lasers very well, therefore, it is commonly used in gain calculations [1, 10, 35]. As already mentioned, we adopt here the most simplified description of the intraband relaxation model for estimating the influence on the spectral gain maximum. The bulk matrix element remains energy-independent in this case.

The intraband relaxation model considers carrier–carrier, carrier–phonon, and carrier–impurity scattering. As pointed out by Yamada and Suematsu [2], the energy broadening due to scattering, $\delta E_{\text{in}}$, is much larger than the corresponding broadening due to radiative recombination, $\delta E_{\text{R}}$; i.e., making use of the uncertainty principle,

$$\delta E_{\text{in}} = \hbar/\tau_{\text{in}} \gg \delta E_{\text{R}} = \hbar/\tau_{\text{r}}, \tag{14}$$

where $\tau_{\text{in}}$ and $\tau_{\text{r}}$ are the intraband scattering (relaxation) time and the radiative lifetime, respectively, as a direct result of $\tau_{\text{in}} \ll \tau_{\text{r}}$. Thus, we may treat the scattering effect separately from the broadening effect of the finite radiative lifetime. In our calculation, we assume the $k$-selection rule to be valid, thus neglecting the broadening due to the finite radiative lifetime. It is, however, controversial whether the $k$-selection rule is still valid to govern the transitions in this case [12].

Without broadening, the number of electrons per unit energy and unit volume at an energy $E'$ above the conduction band edge of the bulk semiconductor (Fig. 2) is $N(E') = 2\rho_{\text{c}}(E')f_{\text{c}}(E')$, where $\rho_{\text{c}}(E')$ is the density of states (without spin degeneracy) and $f_c(E')$ the Fermi occupancy, cf. Eq. (11b). With relaxation broadening, the energy distribution of carriers is broadened. Thus, the effective number of electrons at energy $E'$ becomes

$$N(E') = 2\int_{-\infty}^{\infty} \rho_c(\hat{E}) f_c(\hat{E}) L_c(E' - \hat{E})\, d\hat{E}, \tag{15}$$

where

$$L_c(E' - \hat{E}) = (\delta E_c/2\pi)[(E' - \hat{E})^2 + (\delta E_c/2)^2]^{-1} \tag{16a}$$

is the normalized Lorentzian function (valid when $\delta E_c \ll E'$), i.e.,

$$\int_{-\infty}^{\infty} L_c(E' - \hat{E})\, d\hat{E} = 1. \tag{16b}$$

The energy broadening $\delta E_c$ is given by

$$\delta E_c = \hbar/\tau_c, \tag{16c}$$

where $\tau_c$ is the intraband scattering or relaxation time of an electron in the conduction band. For mathematical symmetry the integration limits for $\hat{E}$ have been chosen to be $-\infty$ and $+\infty$, but the unphysical values of $\hat{E}$ in Eq. (15) are of course eliminated by the density of states function $\rho$, since the actual lower limit is determined by the lowest $n = 1$ level in the QW, $E_{c1}$, and the upper limit, in our approach of counting QW states in $\rho$ only, by the top of the well $\Delta E_c$.

Similarly, we have, for holes,

$$P(E'') = 2\int_{-\infty}^{\infty} \rho_v(\hat{E})[1 - f_v(\hat{E})] L_v(E'' - \hat{E})\, d\hat{E}, \tag{17}$$

$$L_v(E'' - \hat{E}) = (\delta E_v/2\pi)[(E'' - \hat{E})^2 + (\delta E_v/2)^2]^{-1}, \tag{18a}$$

$$\delta E_v = \hbar/\tau_v, \tag{18b}$$

where the energy $E''$ is counted downwards from the valence band edge of the bulk semiconductor (Fig. 2), and $\tau_v$ is the relaxation time of a hole in the valence band.

The lineshape of a transition between two homogeneously broadened energy states remains Lorentzian, with a broadening time $\tau_{cv}$ given by

$$1/\tau_{cv} = 1/\tau_c + 1/\tau_v, \tag{19a}$$

if the broadening mechanisms are independent. Introducing the intraband relaxation time for the electron–hole dipole,

$$\tau_{in} = 2\tau_{cv}, \tag{19b}$$

the Lorentzian factor for a dipole of energy $E$ is given by ($\delta E_{\text{in}} \ll E$)

$$L(E - \hat{E}) = (\delta E_{\text{in}}/\pi)[(E - \hat{E})^2 + (\delta E_{\text{in}})^2]^{-1}, \tag{20a}$$

with

$$\delta E_{\text{in}} = \hbar/\tau_{\text{in}}. \tag{20b}$$

For the broadened distribution of carriers, we may convolve the unbroadened gain with the Lorentzian broadening factor to obtain the gain spectrum [36]:

$$g(E) = \int_{-\infty}^{\infty} g_{\text{u}}(\hat{E}) L(E - \hat{E})\, d\hat{E}. \tag{21a}$$

Inserting for $g_{\text{u}}(\hat{E})$ the explicit form for the unbroadened gain $g(E \to \hat{E})$ given by Eq. (6a), and limiting the integration to our practical boundaries for $\rho_{\text{r}}$, we obtain[3]

$$g(E) = g_0(E)|M_{\text{b}}|^2 \sum_{j,n} \int_{E_{\text{g},jn}}^{E_{\text{g,b}}} \rho_{\text{r},jn}(\hat{E})[f_{\text{c}} - f_{\text{v}}]_{\hat{E},jn} L(E - \hat{E})\, d\hat{E}, \tag{21b}$$

where the transition matrix element $M_{\text{b}}$ is assumed to be energy-independent and the prefactor $g_0(E)$ has been taken out of the integral, since its energy dependence is relatively slow. The integration in Eq. (21) is subject to the conditions of $k$ and energy conservation for the unbroadened dipole, cf. Eqs. (8) and (9),

$$k_{\text{c}}(E'_{jn}) = k_{\text{v}}(E''_{jn}) \qquad (k\text{-selection rule}), \tag{22a}$$

$$\hat{E} = E_{\text{g},jn} + E'_{jn} + E''_{jn} \qquad (\text{energy conservation}). \tag{22b}$$

Thus, once $\hat{E}$ is fixed, $E'_{jn}$ and $E''_{jn}$ are both determined. Similarly, the broadened spontaneous emission spectrum follows from

$$R_{\text{sp}}(E) = \int_{-\infty}^{\infty} R_{\text{sp,u}}(\hat{E}) L(E - \hat{E})\, d\hat{E}, \tag{23a}$$

and, using for $R_{\text{sp,u}}(\hat{E})$ the explicit form of the unbroadened spectrum $R_{\text{sp}}(E \to \hat{E})$ of Eq. (12a) and considering the practical integration limits, we obtain

$$R_{\text{sp}}(E) = r_0(E)|M_{\text{b}}|^2 \sum_{j,n} \int_{E_{\text{g},jn}}^{E_{\text{g,b}}} \rho_{\text{r},jn}(\hat{E})[f_{\text{c}}(1 - f_{\text{v}})]_{\hat{E},jn} L(E - \hat{E})\, d\hat{E}, \tag{23b}$$

[3] In our numerical calculations, the upper integration boundary was actually extended slightly above $E_{\text{g,b}}$ (by at least $3\,\delta E_{\text{in}}$) with the reduced QW density of states $\rho_{\text{r}}$ extrapolated beyond $E_{\text{g,b}}$ in order to eliminate unphysical effects in $g(E)$ near $E_{\text{g,b}}$ caused by the abrupt cutoff for the density of states $\rho_{\text{r}}$ occurring, in effect, at the upper integration boundary.

where the prefactor $r_0(E)$, as $g_0(E)$ in Eq. (21b), has been taken out of the integral because of its relatively weak energy dependence. Broadening of the emission spectrum should not influence the total radiative current obtained by integrating over the whole spectrum, Eq. (13), because the $k$-selection rule has been retained in the model described. Only a normalized Lorentzian lineshape function was introduced to the spontaneous emission spectrum in a convolution process, and, hence, the area under the spectral envelope should remain the same. Therefore, Eq. (13) with the unbroadened spectral function of Eq. (12) is still valid [30].

For $\delta E_{\text{in}} \to 0$ ($\tau_{\text{in}} \to \infty$), the Lorentzian factor $L(E - \hat{E})$ approaches the Dirac delta function $\delta(E - \hat{E})$, and one can easily verify that the previous results for the unbroadened spectra are obtained by using the properties of the delta function in the integrals over $\hat{E}$. Our simplified description of broadening breaks down, however, for large $\delta E_{\text{in}}$, where the convolution procedure becomes inadequate as a result of the Fermi–Dirac distribution factor being applied to the unbroadened density of states. Obviously, when the full-width-half-maximum $2\,\delta E_{\text{in}}$ of the Lorentzian broadening function (20a) approaches the spectral width of the gain and/or spontaneous emission, such a convolution leads to erroneous results.

Another way to evaluate the broadening effect is to integrate with respect to the wavevector $k$ instead of the transition energy $\hat{E}$ [35]. In this way, the nonparabolicity of the conduction band could be taken into account with relative ease.

### *2.2.2. Discussion of the Broadening Model*

Our approach to the spectral broadening effect by intraband scattering has been simplified in order to describe the major phenomenon of smoothing out the sharp peaks in the spectra caused by the step-like density of states functions in QWs. Such an approach appears to be appropriate for estimating the gain peak, which determines lasing threshold.

If carrier–carrier scattering is the dominant mechanism for the relaxation broadening, then one should expect different scattering times for carriers at different energies with different densities of states. Therefore, an energy- and density-dependent broadening time should be used [33, 37]. On the other hand, Yamada *et al.* [38] estimated the relaxation time by fitting theoretical results on gain spectra to experimental data. They found that the relaxation time remains unchanged as the injection current levels vary, but increases toward lower temperature. This would imply that the broadening effect does not depend much on the carrier density, but rather on the number of phonons, which points to carrier–phonon scattering as the dominant mechanism. The relaxation time should be energy-independent in this case.

Contrary to this latter conclusion, a more accurate analysis by Asada [36] involving a detailed description of all possible scattering mechanisms leads (1) to an energy-dependent scattering time $\tau_{in}(\hat{E})$ and (2) to a deviation of the lineshape from the Lorentzian, a striking reduction of its low-energy tail in particular, a phenomenon already pointed out by Yamanishi and Lee [39]. The results would certainly affect some details of the spectrum, particularly farther away from the gain peak toward lower energies. For this reason, non-Lorentzian lineshapes have been used by Chinn *et al.* [21] to improve the gain calculations. However, the effect on the gain peak itself is judged to be minimal with the choice of a properly averaged scattering parameter $\tau_{in}$ [10].

On the other hand, we did not adopt a further simplification in evaluating the integral in Eq. (21b): It was suggested by Zee [40] that the term $f_c - f_v$ varies slowly enough with energy as compared with the relaxation broadening and can be taken out of the integral. A careful evaluation of the rate of change of the Fermi–Dirac function as compared with the Lorentzian factor does not justify such a procedure for typical relaxation times observed in GaAs. The full-width-half-maximum

$$2\,\delta E_{in} = 2\hbar/\tau_{in}$$

of the Lorentzian lineshape is about 26 meV for $\tau_{in} = 0.1$ ps. At room temperature, the Fermi–Dirac function

$$f = [1 + \exp\{(E' - F)/k_B T\}]^{-1}$$

exhibits a substantial change from 0.73 to 0.27 when $E'$ varies by 52 meV (or only about $4\,\delta E_{in}$ for $\tau_{in} = 0.1$ ps) from a value centered at the Fermi level $F$. Thus, such a simplifying step of integration is not justified in the vicinity of the quasi-Fermi levels.

Our theoretical results show three major effects on the gain spectrum due to intraband relaxation: (a) a reduction in maximum gain with increasing broadening; (b) an increase in the width of the spectrum due to a softening of the abrupt edge; and (c) a shift of the peak gain toward shorter wavelength with increasing density of injected carriers. These effects are consistent with previous published results [3, 21].

Figure 3 shows the gain spectrum, Eq. (21), with the relaxation time $\tau_{in}$ as a parameter, at a fixed carrier density injected into QW states of $N = P = 3.04 \times 10^{18}\ \text{cm}^{-3}$, Eq. (11). The spectrum is plotted for a wide wavelength range, including the transition to absorption at short wave-

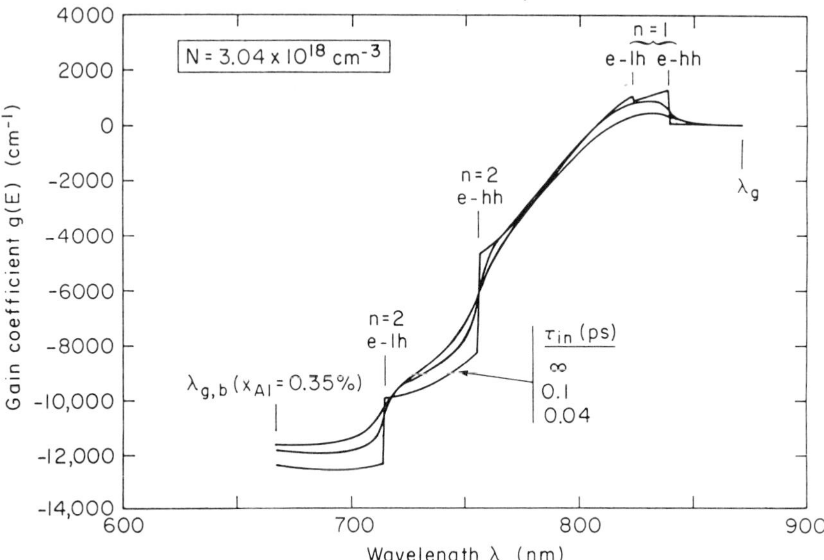


**Fig. 3.** Complete QW gain spectra for different broadening parameters $\tau_{in}$ at a moderate QW injection level $N = 3.04 \times 10^{18}\ \mathrm{cm}^{-3}$.

lengths[4]. Near the region of interest, i.e., close to the QW band edge at the $n = 1$ subband transitions, the broadening effect decreases the gain peak and generates long wavelength tail below the band edge. Figure 4 presents the details of the gain and spontaneous emission spectra, Eqs. (21) and (23), around their respective maxima at the $n = 1$ subband transition. Again, $\tau_{in}$ is used as a parameter. Curve discontinuities are smoothed out by the broadening effect and, thus, the maximum of gain or emission can occur at any energy between or around the e–lh and e–hh transitions. Note that, for the injection level chosen, the maximum gain corresponds to the e–hh transition, whereas the spontaneous emission exhibits its maximum near the energetically higher e–lh position (only at higher injection levels does the maximum gain shift to the e–lh position; see Fig. 6 following). For very low

[4] For completeness we have included the results for short wavelengths down to the transition wavelength $\lambda_{g,b}$ corresponding to the barrier bandgap $E_{g,b}$; it should be noted, however, that the mathematical results from the convolution model of broadening become increasingly inaccurate when approaching wavelengths that are close to $\lambda_{g,b}$, the somewhat arbitrary upper limit of the convolution integral, Eq. (21b). This inadequacy is a consequence of neglecting the bulk-like energy states above the edge of the well and the fact that by convolving the gain spectrum with a normalized Lorentzian lineshape, the area between the integration limits under the spectrum is forced to remain unchanged in the process, as pointed out before. See also footnote 3 on p. 123.

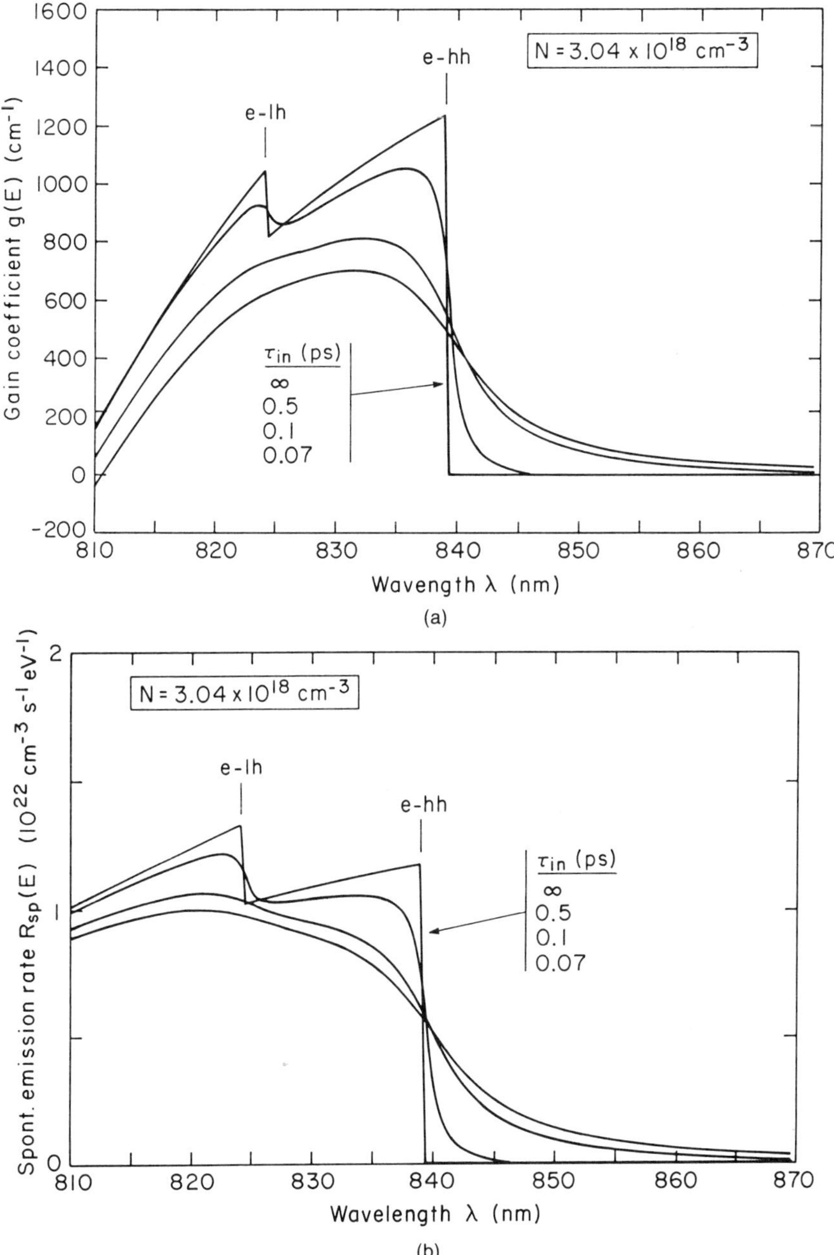


**Fig. 4.** QW gain spectra (a) and spontaneous emission spectra (b) near the gain maximum for several broadening parameters $\tau_{in}$ at a moderate QW injection level $N = 3.04 \times 10^{18}\ cm^{-3}$.

injection carrier density, when absorption prevails, the spontaneous emission maximum stays also at the e–hh transition. This is demonstrated in Fig. 5, where the injection level has been reduced to $N = 2 \times 10^{17}\ \mathrm{cm}^{-3}$. Since there is no gain peak in this case, we have defined the value of the (negative) gain coefficient at the e–hh subband edge as a meaningful "maximum gain" for the purpose of extrapolating gain versus injection level into the range below transparency ($g = 0$).

The influence of different injected QW carrier densities $N$ (from $1 \times 10^{18}\ \mathrm{cm}^{-3}$ to $6 \times 10^{18}\ \mathrm{cm}^{-3}$) on the gain spectrum is depicted in Fig. 6 for a fixed relaxation time of $\tau_{\mathrm{in}} = 0.1$ ps, an approximate value generally accepted for GaAs [3, 21]. Notice that at low carrier densities the e–hh transition determines the peak gain, whereas the e–lh transition becomes observable and eventually even takes over the peak gain at sufficiently high injection carrier levels.

Figure 7 shows a plot of gain-peak wavelength of the $n = 1$ electron–hole QW subband transition against $N$ for $\tau_{\mathrm{in}} = 0.1$ ps, comparing it with the unbroadened case ($\tau_{\mathrm{in}} = \infty$). When gain broadening is neglected, the gain peak occurs right at either the e–hh or the e–lh subband edge transition. This would lead to a discontinuous jump of the emission wavelength at a critical injection level, contary to experiments. The wavelength shift is strongly softened, however, when broadening is taken into account. Thus, broadening effects are important when considering lasing wavelength, which is determined by the spectral position of the gain peak. For improved accuracy, bandgap renormalization effects caused by the elevated carrier injection levels encountered in lasing should be taken into account [21]. These effects would shift the gain peak to a somewhat longer wavelength. It is interesting to note that at very high injection levels, when the $n = 2$ gain peak becomes dominant (not shown in the figure), a more dramatic shift to the $n = 2$ QW transition may occur, which cannot be smoothed out by the broadening effect (see also Section 2.3). Such a shift has actually been observed experimentally [41].

### *2.2.3. TE Polarization Enhancement of the Matrix Element*

The planar symmetry of the electronic wavefunctions in a QW structure results in a polarization dependence of the stimulated optical transitions. This leads to a modification of the transition matrix element, resulting in a difference between the dipole transitions e–lh and e–hh (see Chapter 1). Additionally, the QW matrix element becomes dependent on the transition energy $E = \hbar\omega$ [10]. It can be related to the bulk matrix element $M_{\mathrm{b}}$ of

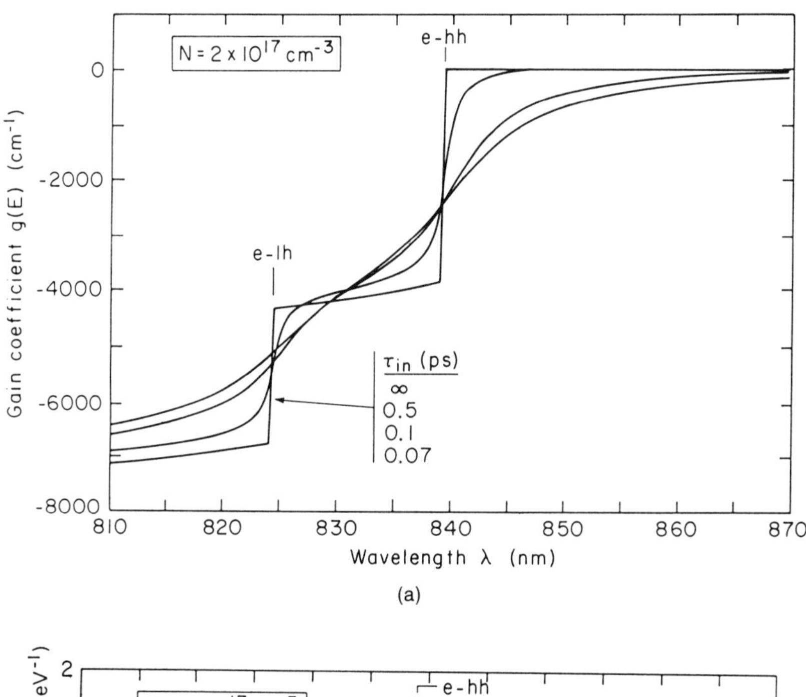


(a)

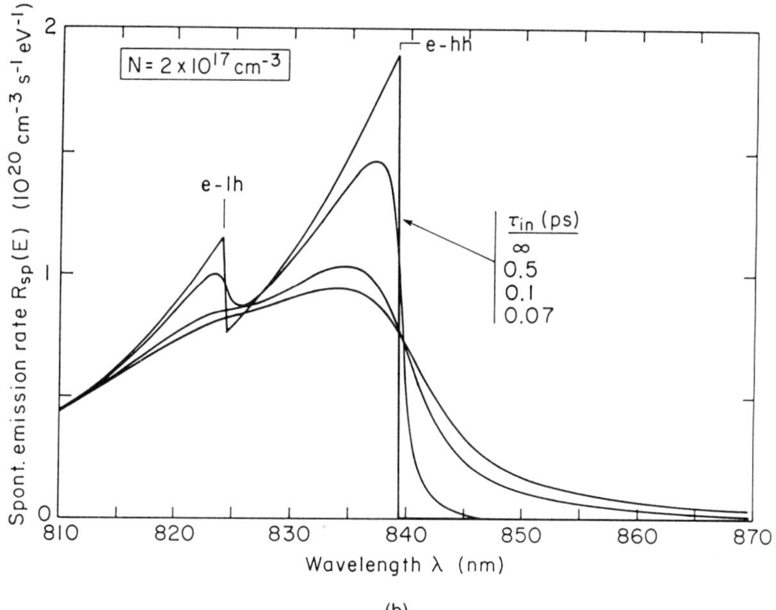


(b)

**Fig. 5.** QW gain spectra (a) and spontaneous emission spectra (b) for several broadening parameters $\tau_{in}$ at a QW injection level below transparency, $N = 2 \times 10^{17}\ \mathrm{cm}^{-3}$.

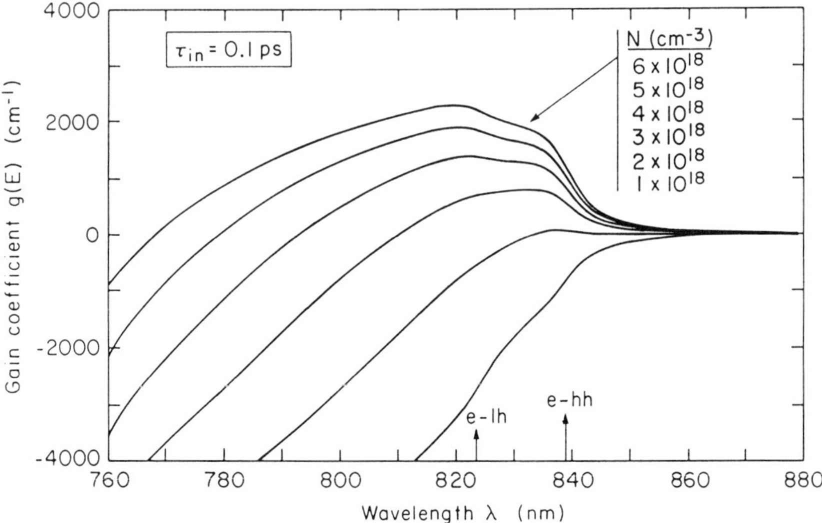


**Fig. 6.** Broadened QW gain spectra for $\tau_{in} = 0.1$ ps at several QW injection levels $N$.

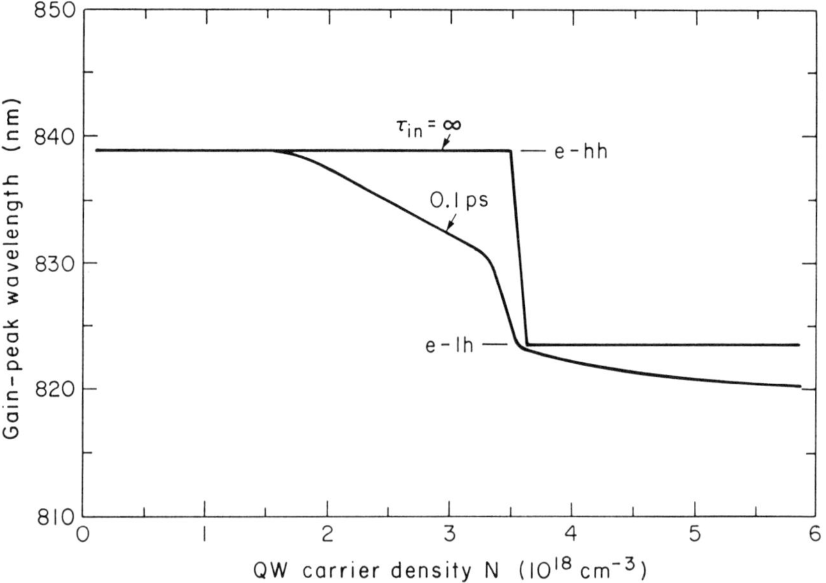


**Fig. 7.** Gain-peak wavelength of the $n = 1$ QW transition versus QW injection level $N$ for a broadening parameter $\tau_{in} = 0.1$ ps compared with the unbroadened case ($\tau_{in} = \infty$).

Eq. (7) by formulating, for each QW subband pair e$n$/$j$h$n$,

$$|M_{jn}(E)|^2 = \mu_{jn}(E)|M_b|^2 = (2/3)\mu_{jn}|M|^2, \tag{24a}$$

where $\mu_{jn}(E)$ is a polarization-dependent anisotropy factor. The feedback condition for lasing usually selects TE over TM polarization, even when the gain is polarization-independent, as in a DH structure. The anisotropy factor in the QW provides an enhancement of the oscillator strength for TE polarization at photon energies near the gain peak as opposed to TM polarization, where the oscillator strength is diminished. Thus, for a QW structure, stability of lasing in the TE mode is improved further, and TM polarization need not be considered. The TE polarization anisotropy factors for the e–hh and e–lh transitions are, respectively, [10]

$$\mu_{jn}(E) = \begin{cases} (3/4)[1 + (E_{c,n} + E_{v,hn})/(E - E_g)], & \text{for } j = \text{h}, \\ (1/4)[5 - 3(E_{c,n} + E_{v,ln})/(E - E_g)], & \text{for } j = \text{l}, \end{cases} \tag{24b}$$

with $E \geq E_{g,jn}$. Notice that the strongest enhancement is observed for transitions close to the hh subband gap $E_{g,hn}$, where $\mu_{hn} = 3/2$, i.e., an enhancement of 50%; at very large photon energies, $\mu_{hn}$ decreases toward an asymptotic value of 3/4. On the other hand, the e–lh transitions are diminished close to the lh subband gap $E_{g,ln}$, where $\mu_{ln} = 1/2$ with $\mu_{ln}$ approaching 5/4 at very large photon energies. These e–lh transitions are less frequent, however, because of the reduced density of states in the lh subband, and, thus, provide only a small contribution.

With the modified matrix element (24), the broadened gain spectrum of Eq. (21) eventually becomes

$$g(E) = g_0(E)|M_b|^2 \sum_{j,n} \int_{E_{g,jn}}^{E_{g,b}} \mu_{jn}(\hat{E})\rho_{r,jn}(\hat{E})[f_c - f_v]_{\hat{E},jn} L(E - \hat{E})\, d\hat{E}. \tag{25}$$

The momentum anisotropy factor $\mu_{jn}(\hat{E})$, which depends both on energy and subband index, has to remain under the sum and integral signs. At energies around the spectral gain peak, the diminished e–lh transitions contribute relatively little to the summation in Eq. (25) and, hence, the gain peak is enhanced for TE polarization. An overall diminishing effect is only observed at the high-energy tail of the gain spectrum.

No modification in the expression for the spontaneous emission spectrum of QWs is necessary, and Eq. (23b) is still valid [30]. The anisotropy of the structure is unimportant in that case, since both TE and TM polarizations are randomly emitted. The radiative QW current is still given by Eqs. (12) and (13).

## 2.3. Results from the Two Models

A gain spectrum calculated from the accurate model, Eq. (25), is shown in Fig. 8 and compared with spectra calculated for the same injection conditions from the simplified model, Eq. (6), and the relation including broadening only, Eq. (21). It is obvious that the reduction of the spectral gain peak due to broadening is largely compensated for by the TE enhancement of the gain. This compensation is maintained over a wide range of injection levels, as seen from Fig. 9, in which the gain peak $g$ corresponding to the $n = 1$ electron–hole QW subband transition is plotted against the carrier density $N$ injected into the QW states, Eq. (11), both according to the simplified Eq. (6) and accurate Eq. (25). Thus, the simplified model can be used for rough estimates of the magnitude of the peak gain and, consequently, the threshold current for lasing. The peak position, on the other hand, is strongly influenced by broadening and TE enhancement and needs to be determined from the accurate model including bandgap renormalization effects, as pointed out in Section 2.2.2.

An important feature of Fig. 9 is the rapid increase of gain at low injection levels and the pronounced saturation toward high injection levels. This

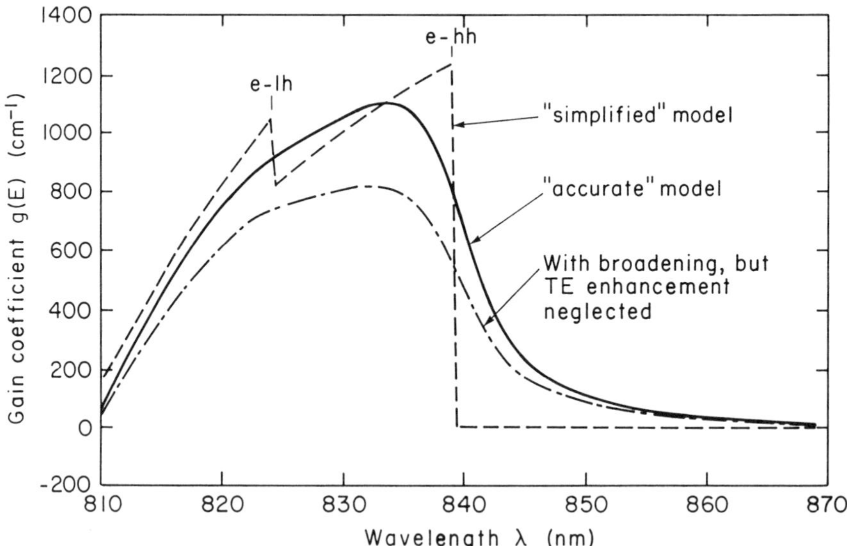


**Fig. 8.** The QW gain spectrum at $N = 3.04 \times 10^{18}$ cm$^{-3}$ for the simplified model without broadening and TE enhancement (Section 2.1) and for the more accurate model taking broadening wih $\tau_{in} = 0.1$ ps and TE enhancement into account (Section 2.2). The broadened spectrum without TE enhancement is also shown for comparison.

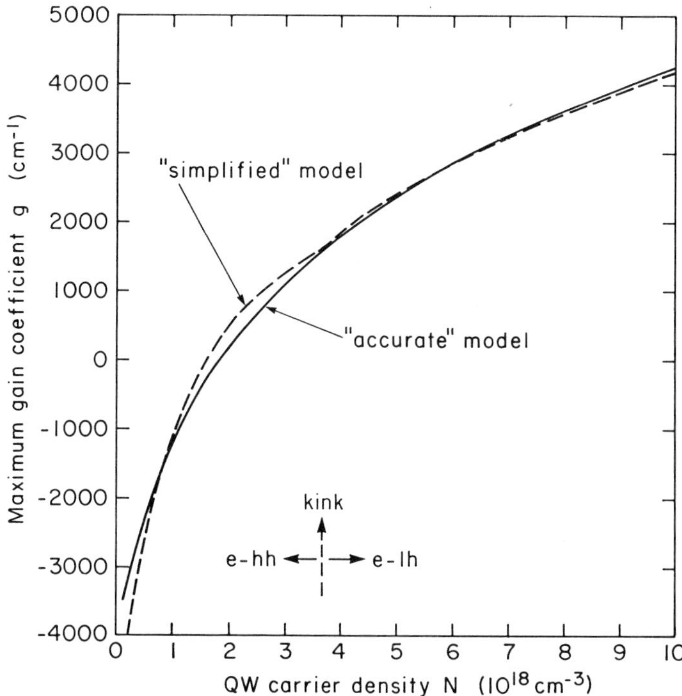


**Fig. 9.** The relationship between the spectral peak gain $g$ of the $n = 1$ QW subband transition and the QW injection carrier density $N$ for the simplified and the more accurate model.

behavior results from band filling as a direct consequence of the step-like density of states function in the QW [1] and, as we shall see in Section 4, needs crucial attention in the design of MQW lasers. Obviously, one should attempt to tailor the MQW structure in such a way as to allow for lasing to occur near the initial steep part of the gain curve at low injection carrier densities. It is in this gain region, as a result of the large differential gain factor $\beta = dg/dN$, that QW lasers perform far superior than conventional DH lasers. It not only provides a low threshold current with a low temperature sensitivity (high characteristic temperature $T_0$: [11, 32, 41]), but also a high modulation bandwidth [42–44]. For details on the subjects of ultralow threshold current and highspeed modulation, the reader should consult Chapters 4 and 5, respectively. The large $\beta$ factor in QW lasers also leads to a reduced linewidth enhancement factor, $\alpha = -(4\pi/\lambda\beta)(d\bar{n}'/dN)$ [45], with the benefits of a smaller lasing linewidth: lower intensity noise, and lower chirp during modulation as compared with DH lasers. However, in this case the corresponding change in the QW refractive index, $d\bar{n}'/dN$, partially compensates the favorable effect of $\beta$, resulting in an optimum injection range somewhat above the steepest gain increase at low injection [46–48]. It is noteworthy that an additional enhancement of the $\beta$ factor with a further reduction of the $\alpha$ factor can be achieved in QW lasers by proper control of the Fermi–Dirac term $f_c - f_v$ in the gain expression, Eq. (25), using p-type modulation-doping of the barrier material [17, 49].

A minor feature of the gain curve is a slight kink near $N \approx 3.7 \times 10^{18}\ \text{cm}^{-3}$, which is more pronounced in the simplified model. This kink can also be seen in Fig. 11 (following) and corresponds to the switch from the e–hh to the e–lh transition of the $n = 1$ subbands, as demonstrated in Fig. 7. At very high injection levels, when the quasi-Fermi level moves beyond the $n = 2$ QW electron level (see Fig. 10a), the $n = 2$ transition is expected to become dominant. This would produce an additional kink in the gain curve if the $n = 2$ gain peak were taken into account. Since our analysis is aimed at avoiding excessively high injection levels in lasing operation, the relationship for the $n = 2$ gain peak has not been included in Figs. 9 and 11. It should be pointed out, however, that a shift from the $n = 1$ to the $n = 2$ QW transition has been observed experimentally when investigating the threshold of short cavity SQW lasers as a function of temperature [41], a fact we have already mentioned when discussing the spectral position of the gain peak (Section 2.2.2).

The position of the electron quasi-Fermi level $F_c$ relative to the conduction band edge, based on Eq. (11), and the normalized radiative QW current density $J_{r,1}$, following from Eq. (13), are plotted against the injected QW carrier density $N$ in Fig. 10. Figure 10a is useful for estimating the injection levels at which the occupancy of higher electron energy levels, such as the $n = 2$ quantum level $E_2$ or the bulk-like levels above $\Gamma_{(b)}$, would start to play a significant role and, hence, should be avoided in a preferred laser design. From Figs. 9 and 10b the relationship between gain and current density is obtained by eliminating $N$, as shown in Fig. 11a. Such a relationship is useful for determining the threshold current of a particular MQW laser structure, as will be discussed in Section 4. Before we can deal comprehensively with this subject, however, we have to consider the nonideal contributions to the total device current. Anticipating the results for the contribution from nonradiative Auger recombination (GaAs, see Section 3.3) the relation of Fig. 11a has been modified to include this contribution in Fig. 11b, demonstrating the expected degradation of the gain/current curve. In this figure, the characteristic linear gain saturation parameters $g_0$ and $J_0$, introduced by McIlroy *et al.* [8], are also indicated. These are very useful

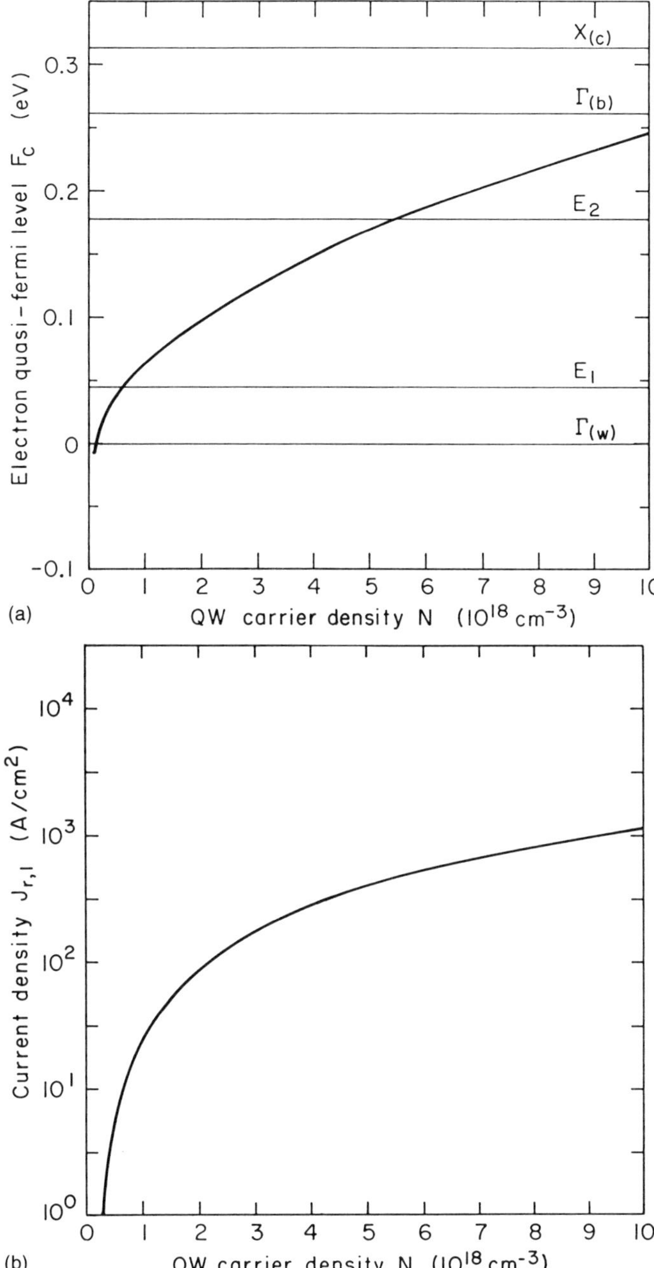


**Fig. 10.** (a) Position of electron quasi-Fermi level $F_c$ and (b) radiative QW current density ($\log J_{r,1}$) versus QW injection carrier density $N$. For the definition of the characteristic energy levels indicated in (a), see Fig. 1.

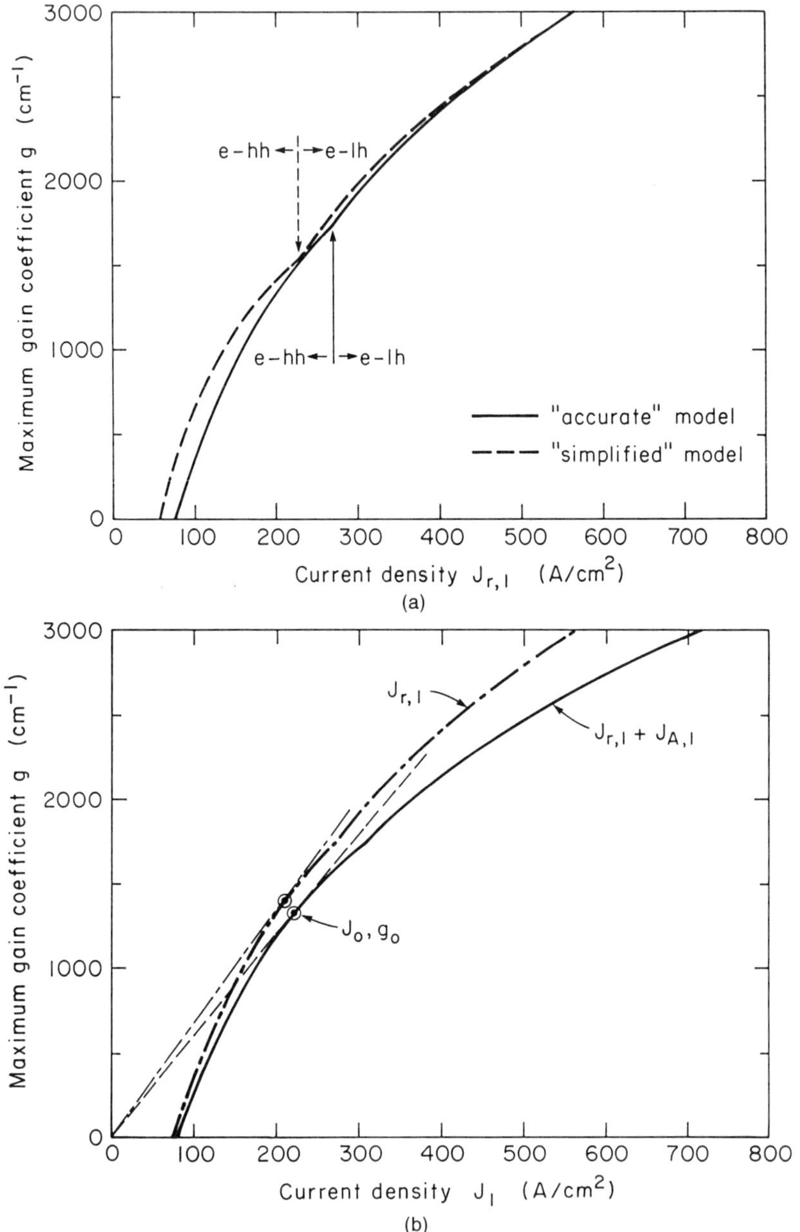


**Fig. 11.** (a) Spectral peak gain $g$ of the $n = 1$ QW subband transition versus radiative QW current density $J_{r,1}$ for the simplified and the more accurate model; (b) degradation of the gain curve (accurate model) due to Auger recombination (GaAs, $C_A = 5 \times 10^{-30}$ cm$^6$/s). $J_0$ and $g_0$ are McIlroy's [8] linear gain saturation parameters.

parameters for describing the quality of a particular gain curve and provide, e.g., a measure for such a degradation.

## 3. NONRADIATIVE TRANSITIONS AND CARRIER LEAKAGE

### 3.1. General Considerations

So far, we have confined our analysis of MQW lasers to the radiative transitions involving the QW energy states only. The "radiative" current resulting from these transitions, Eq. (13), is an idealization, however, and does not describe the total current in the MQW structure. Unavoidable nonradiative and leakage current components are known to contribute to the total current, leading to possibly substantial deviations from the ideal conditions. A number of mechanisms for such nonideal contributions have been considered in the literature (see, e.g., [4–6]): (1) nonradiative transitions inside the QWs, which includes bulk and interface recombination via defect states in the bandgap, as well as Auger recombination; (2) carrier "leakage" over the top of the QWs into the continuum of the bulk-like states, including spillage sideways to the barrier layers, and, eventually, to the cladding layers, possibly involving recombination from the higher lying indirect conduction band valleys ($L$ and $X$).

The leakage mechanisms under group 2 require the occupancy of higher energy levels in the overall MQW structure, a condition that becomes stronger as the quasi-Fermi levels move farther into the bands with increased injection (see Fig. 10a). On the other hand, we concluded in Section 2 that, in order to fully take advantage of the high QW differential gain coefficient, the design of laser devices should aim at the lowest injection levels possible. This approach should also minimize the occupancy of higher energy levels, including the $L$ and $X$ valleys above the top of the QWs, or spillover into the barrier and cladding layers.

The severity of the higher level occupancy can be estimated from the energy steps involved [5]. In GaAs, the $L$ level is at 0.284 eV above the $\Gamma$ level, and the $X$ level is at 0.467 eV above (Fig. 1). Carrier spillage out of the QW is controlled by the potential height of the barriers as, e.g., in the GaAs/AlGaAs system determined by the Al concentration. We chose relatively high values for our representative MQW structure of Fig. 1: 35% Al concentration in the AlGaAs barrier layers and 55% in the AlGaAs cladding. Values quoted in the literature are often lower (an Al concentration of 20% in the AlGaAs barriers is not uncommon, e.g., [10, 22]), making carrier leakage more likely. With 35% Al, the AlGaAs alloy provides a well (w)–barrier (b) potential step of 0.262 eV for the $\Gamma_{(w)}$–$\Gamma_{(b)}$ conduction band edge separation, still slightly lower than the position of the $L$ level above the GaAs well; with 55% Al, however, the alloy becomes indirect, leading to well (w)–cladding (c) potential steps of 0.313 eV for the $\Gamma_{(w)}$–$X_{(c)}$ conduction band edge separation, 0.362 eV for the $\Gamma_{(w)}$–$L_{(c)}$ valley separation, and 0.422 eV for the $\Gamma_{(w)}$–$\Gamma_{(c)}$ valley separation. These values are based on the bandgaps and band offsets given in Table I. The corresponding barriers for holes in the guide and cladding are 0.175 and 0.275 eV, respectively. Even though these barriers are lower than the corresponding electron barriers, the spillage of holes out of the QWs is far less severe. This is a consequence of the large density of states effective mass in the valence band, which keeps the hole quasi-Fermi level close to the valence band edge.

Recombinations from $L$ and $X$ valleys in the QWs are indirect and so is most of the recombination in the 55% AlGaAs cladding layers. This means that the carriers involved exhibit substantially longer lifetimes (at least an order of magnitude) than those recombining radiatively from the central $\Gamma$ valley in the QW and, thus, contributions to the total current are expected to be negligible in most cases, despite the large density of states available in the $L$ and $X$ valleys. It should be pointed out, however, that lower cladding barriers, particularly when direct recombination becomes dominant in the cladding layers, can result in substantial leakage because of the relatively large spillover volume of the cladding layers in combination with the carrier sinks located at the ohmic contacts to the claddings (cf. [21].)

From the various possibilities of the group 2 leakage paths involving higher lying levels in the composite MQW structure, the immediate leakage out of the top of the well into the $\Gamma$-like bulk continuum, particularly spillover to the 35% AlGaAs guiding layers, remains as the most likely one to generate an appreciable influence. It exhibits the lowest available energy levels and, in addition, still provides direct recombination with a carrier lifetime comparable with that for radiative recombination from the QW states. We chose to include this mechanism in our analysis as a representative of the high-energy leakage recombinations in order to provide some quantitative insight into its contribution to the total current in a MQW structure. This current component comprises the whole waveguide region and, therefore, to first order, is nearly independent of the number of QWs.

In addition, we will analyze the group 1 nonradiative paths within the QWs themselves. These group 1 contributions to the total current are proportional to the number of QWs and, hence, can be directly compared with the radiative QW current in its normalized form for a single QW, Eq. (13b). In estimating these contributions, we neglect any carriers occupying energies beyond the top of the QWs, as we did for estimating the radiative current. Auger recombination, being a three particle process, becomes relatively more important at high injection levels similar to the group 2 leakage recombination. On the other hand, nonradiative recombination in the QW via defects can be significant even at low injection levels, depending on the number of nonradiative recombination centers that may be generated during the epitaxial growth process. Experimental evidence suggests that these centers are concentrated near the hetero-interfaces [51, 52]. Therefore, we will describe the dominant nonradiative process by interface recombination. This interface recombination is of particular concern in MQW lasers because of their multiple interfaces adjacent to the active recombination volume of the QWs.

### 3.2. Interface Recombination

Interface recombination is modeled with a phenomenological interface recombination velocity $v_s$ for the carriers injected into the QWs (density $N$), assuming identical interfaces in all of the structure. Hence, we simply have

$$J_s = 2\nu_{QW} eNv_s, \tag{26a}$$

$$J_{s,1} = J_s/\nu_{QW} = 2eNv_s, \tag{26b}$$

where the QW carrier density $N$ is obtained from Eq. (11). Figure 12 plots $J_{s,1}$ against $N$ with $v_s$ as a parameter. For comparison, the radiative current density $J_{r,1}$ of Fig. 10b is also shown. The figure clearly demonstrates the severity of interface recombination, particularly at low $N$ values. Even at the relatively small interface recombination velocities between 100 and 200 cm/s, $J_{s,1}$ is of the same order of magnitude as $J_{r,1}$ near injection levels corresponding to the transparency condition $g = 0$ ($N_{tr} \approx 1.8 \times 10^{18}$ cm$^{-3}$; see Fig. 9). At higher injection levels the situation becomes more favorable, because the radiative current increases more rapidly due to its bimolecular nature (roughly proportional to $N^2$). It needs to be stressed that interface recombination is the only nonideal contribution to the current that affects primarily the region of lower injection carrier density, which is the preferred operating region of QW lasers. Therefore, technological attention to the quality of the interfaces during epitaxial growth is of utmost importance.

### 3.3. Auger Recombination

In the Auger effect, the energy released by a recombining electron is absorbed by another electron (or hole), which then dissipates the energy by the

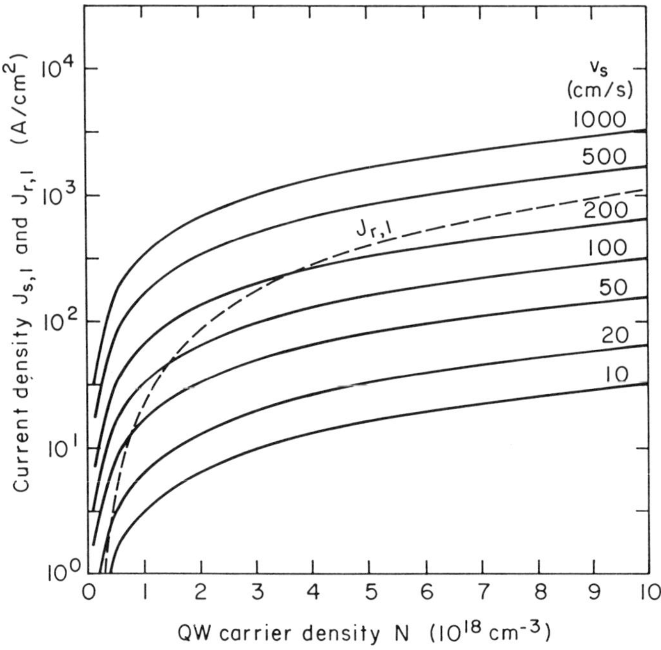


**Fig. 12.** Interface recombination current density (log $J_{s,1}$) for different recombination velocities $v_s$ versus QW injection carrier density $N$, compared with radiative QW current density (log $J_{r,1}$).

emission of phonons. Thus, the transition does not generate any photons. Since the process depends on a three carrier interaction, it becomes, in the relatively high injection regime relevant for lasing ($N = P$), proportional to $N^3$. The Auger current density is then simply given by [4]

$$J_A = v_{QW} L_z e C_A N^3, \tag{27a}$$

$$J_{A,1} = J_A / v_{QW} = L_z e C_A N^3, \tag{27b}$$

where $N$ is again obtained from Eq. (11), and $C_A$ is the Auger coefficient. For the GaAs/AlGaAs QWs we use the room temperature GaAs bulk value, $C_A = 5 \times 10^{-30}$ cm$^6$/s, as quoted by Reisinger *et al.* [4]. Auger recombination may be somewhat reduced in QWs [21], an effect we disregard in our estimate because of its uncertainty.

In Fig. 13, $J_{A,1}$ with $C_A = 5 \times 10^{-30}$ cm$^6$/s is compared with the normalized radiative current density, $J_{r,1}$, again as a function of the injection level $N$. It is apparent that Auger recombination is largely negligible in GaAs except for very high injection densities. Its contribution is below 20% for

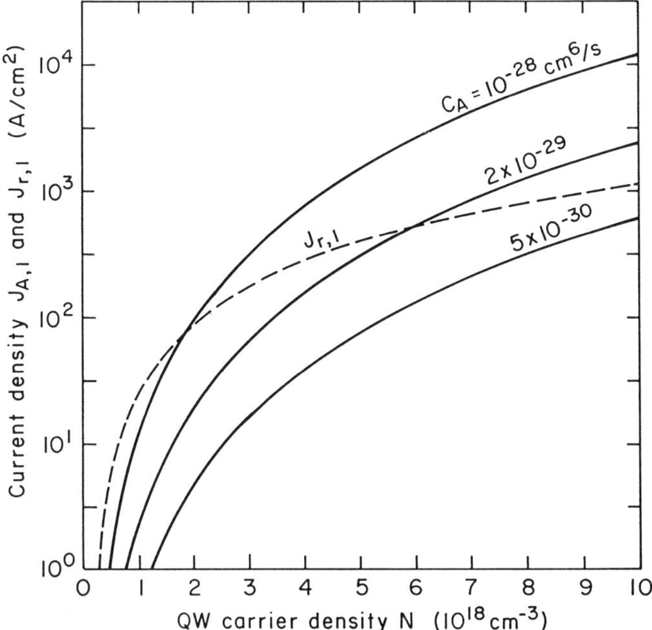


**Fig. 13.** Auger recombination current density ($\log J_{A,1}$) versus QW injection carrier density $N$, compared with radiative QW current density ($\log J_{r,1}$), for several $C_A$ values: $C_A = 5 \times 10^{-30}$ cm$^6$/s, representative for GaAs; $C_A = 2 \times 10^{-29}$ cm$^6$/s, representative for InGaAsP ($\lambda_g = 1.3$ $\mu$m); and $C_A = 10^{-28}$ cm$^6$/s, representative for InGaAs ($\lambda_g = 1.65$ $\mu$m).

$N < 5 \times 10^{18}$ cm$^{-3}$ and falls below 10% for $N < 3 \times 10^{18}$ cm$^{-3}$. For illustrative purposes, we also show two curves with larger $C_A$ values: $C_A = 2 \times 10^{-29}$ cm$^6$/s is representative of InGaAsP ($\lambda_g = 1.3$ $\mu$m); and $C_A = 10^{-28}$ cm$^6$/s corresponds to InGaAs ($\lambda_g = 1.65$ $\mu$m) lattice matched to InP [9, p. 122]. In the InGaAs case, Auger recombination becomes dominant ($J_{A,1} \approx J_{r,1}$) already at the low injection levels near transparency ($N_{tr} \approx 1.8 \times 10^{18}$ cm$^{-3}$). The situation is actually slightly more unfavorable than the figure suggests, since the radiative current in InGaAs/InP QWs is somewhat lower than in GaAs/AlGaAs QWs, the case used for the radiative current density in our figure. In case of 1.3-$\mu$m InGaAsP, the Auger contribution starts at 20% of $J_{r,1}$ near transparency. Thus, with long wavelength MQW lasers in the range 1.3–1.5 $\mu$m, the emerging dominance of Auger recombination requires even more careful attention in the design of the laser structures to minimize injection carrier density at threshold (see Chapter 7 for additional considerations).

### 3.4. Leakage Recombination in the Guiding Region

The carrier leakage out of the QWs within the guiding region can be estimated from the potential steps formed at the QW hetero-interfaces and the positions of the quasi-Fermi levels [5]. Since the carrier diffusion lengths are generally large compared with the typical thickness of the overall guiding region in MQW structures, the quasi-Fermi levels can be treated approximately as flat throughout this region. Consistent with the description of the gain and spontaneous emission calculation in Section 2, we assume flat well and barrier potentials and impose charge neutrality only for the carriers trapped in the QWs. This means that minor space charge effects resulting from small charge imbalances near the well–barrier interfaces are neglected. These imbalances would form self-consistently from the actual condition of integral charge neutrality over all available energies of the entire guiding region in conjunction with the imposition of flat quasi-Fermi levels (cf. [16]). As a consequence of our simplification, strict charge neutrality is violated in the barrier layers (which is of course inconsistent with the assumption of completely flat barrier potentials). However, since we anticipate the carrier leakage to be a small contribution to the total current, this simplification is well justified.

As a further simplification, we introduce a phenomenological lifetime $\tau$ for the electrons injected into the guiding region above the barrier potential and disregard the holes in order to describe the spill leakage. Strictly speaking, this lifetime would depend on the injection level, a dependence we ignore in our estimate. We then may write, for the leakage current density,

$$J_{\mathrm{lk}} = e[v_{\mathrm{QW}} L_z N_{\mathrm{w}}/\tau_{\mathrm{w}} + (d_{\mathrm{g}} - v_{\mathrm{QW}} L_z) N_{\mathrm{b}}/\tau_{\mathrm{b}}], \tag{28}$$

which is a sum of two contributions, one from the higher energy bulk-like states in the QW material itself (w) and the other from the barriers (b). The leakage carrier concentrations $N_{\mathrm{w}}$ and $N_{\mathrm{b}}$ are calculated by summing over the electrons occupying the conduction band states above the top of the QWs in the well and the barrier material, respectively. As pointed out before, we may ignore the indirect valleys in this summation, because we are only interested in the faster direct recombination. Additionally, we limit our consideration to moderate injection levels in which the quasi-Fermi level stays sufficiently below the top of the QWs to justify Boltzmann statistics (cf. Fig. 10a). Hence, within the parabolic band approximation, we have

$$N_{\mathrm{w}} = N_{\mathrm{tc,w}} \exp[(F_{\mathrm{c}} - \Delta E_{\mathrm{c}})/k_{\mathrm{B}} T], \tag{29a}$$

$$N_{\mathrm{b}} = N_{\mathrm{c,b}} \exp[(F_{\mathrm{c}} - \Delta E_{\mathrm{c}})/k_{\mathrm{B}} T], \tag{29b}$$

where $N_{\mathrm{tc,w}}$ is the well effective density of states in a truncated parabolic

Γ conduction band, i.e., counting only energies above the top, $\Delta E_c$, of the QWs; and $N_{c,b}$ is the regular effective density of states in the Γ conduction band of the barrier, i.e.,

$$N_{c,b} = 2(m_{c,b}k_B T/2\pi\hbar^2)^{3/2}. \tag{30}$$

The truncation increases $N_{tc,w}$ somewhat against the regular effective density of states $N_{c,w}$ in the untruncated conduction band of the well material. On the other hand, the effective electron mass $m_{c,w}$ in the well (GaAs) is smaller than that in the barrier ($Al_{0.35}Ga_{0.65}As$); see Table I. Thus, the recombination rates $N_w/\tau_w$ and $N_b/\tau_b$ are expected to be of the same order of magnitude, meaning that the dependence of $J_{lk}$ on $\nu_{QW}$ is only weak. To simplify the expression even further, we assume $N_w/\tau_w = N_b/\tau_b$, which is equivalent of neglecting this weak dependence on $\nu_{QW}$ completely; thus,

$$J_{lk} = ed_g N_b/\tau_b. \tag{31}$$

For our representative structure of Fig. 1, this current density is plotted in Fig. 14 against the QW carrier density $N$, Eq. (11), with $\tau_b$ as parameter,

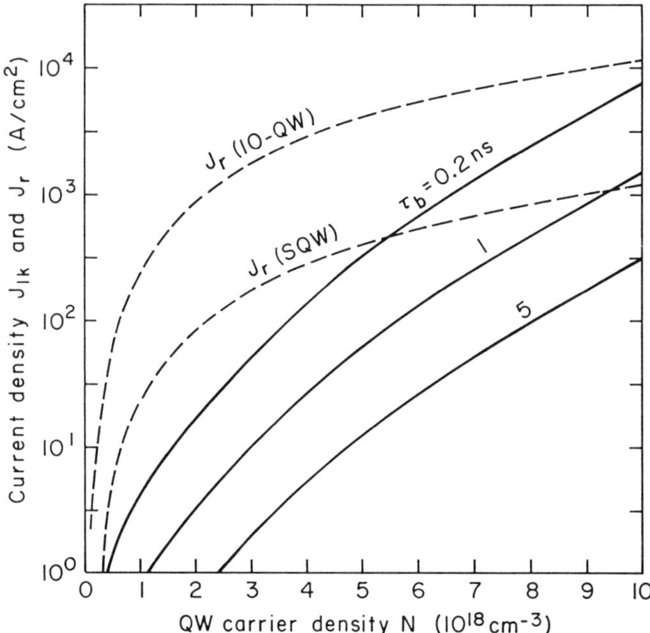


**Fig. 14.** QW leakage current density (log $J_{lk}$) versus QW injection carrier density $N$ for several carrier lifetimes $\tau_b$ in the bulk-like energy states of the waveguide region, compared with radiative QW current density (log $J_r$) for a SQW and a 10-QW structure.

and is compared with the radiative QW current densities of a SQW and a 10-QW structure. Using a realistic carrier lifetime of about 1 ns, the leakage component behaves very similarly to the Auger component in a SQW structure, being negligible except at a high injection. There, its contribution reaches values somewhat larger than the Auger current when $N > 6.2 \times 10^{18}\ \text{cm}^{-3}$. By increasing the number of QWs, the situation obviously becomes more favorable. It is worth noting that, according to Eq. (29), the carrier leakage out of the QW is very temperature sensitive, being more pronounced at higher temperature. This may lead to a substantial contribution to the temperature dependence of threshold current in a QW laser [41, 53].

## 4. THRESHOLD CURRENT RELATIONS

### 4.1. Criterion for Threshold

The threshold gain $g_{\text{th}}$ is obtained from the optical mode losses via the balance equation (e.g., [21])

$$G_{\text{th}} = \Gamma g_{\text{th}} = [\alpha_{\text{i}} + \ln(1/R)/L]. \tag{32}$$

Here, the mode gain $G$ is expressed in terms of the local gain coefficient $g$ and the gain confinement factor $\Gamma$. The mode loss is modeled by a combination of mirror loss,

$$\alpha_{\text{M}} = \ln(1/R)/L, \tag{33a}$$

with a reflectivity parameter defined by the two mirror reflectivities $R_1$ and $R_2$,

$$R = \sqrt{R_1 R_2}, \tag{33b}$$

and intrinsic loss $\alpha_{\text{i}}$. The latter is caused by the free carrier loss $\alpha_{\text{fc}}$ in the active QW layers, the scattering loss $\alpha_{\text{sc}}$ out of the optical waveguide, and some loss $\alpha_{\text{pg}}$ in the passive portions of the guide and in the cladding:

$$\alpha_{\text{i}} = \alpha_{\text{sc}} + \Gamma\alpha_{\text{fc}} + (1 - \Gamma)\alpha_{\text{pg}}. \tag{33c}$$

For simplicity, we assume a constant $\alpha_{\text{i}} = 5\ \text{cm}^{-1}$ [4], consistent with published experimental data for QW lasers (e.g., [54, 55]). This implies that the main contribution would stem from $\alpha_{\text{sc}}$, and the dependence on carrier density and gain confinement $\Gamma$, which sometimes is taken into account (e.g., [34]), is neglected [8, 56]. Free carrier absorption, however, can become an important effect for the external differential quantum efficiency, particularly at high injection levels when carriers spill over to the barrier layers of the guiding region, thus contributing to $\alpha_{pg}$ [57]. In many cases the influence of $\alpha_i$ is marginal, since most of the loss stems from $\alpha_M$, which, for a typical laser length of $L = 400$ $\mu$m and the natural mirror reflectivities at the GaAs/air interface of $R_1 = R_2 = 0.3$, is about 30 cm$^{-1}$.

The mirror loss $\alpha_M$ and the gain confinement factor $\Gamma$ are the most important parameters determining the threshold gain $g_{th}$ in QW lasers. Both an increasing $\alpha_M$ and a decreasing $\Gamma$ push $g_{th}$ toward higher values and eventually into the saturating regime [8]. The mirror loss $\alpha_M$, according to Eq. (33a), is controlled either by the cavity length $L$ or by the reflectivity parameter $R$, whereas the gain confinement factor $\Gamma$ is governed by the number of QWs. Because of the inherently small size of QWs, $\Gamma$ is obviously small and can be increased only by adding more QWs to the guiding region. This means that particularly in SQW lasers the local gain may be substantially higher than the mode gain, an effect unfavorable for remaining in the desirable lower injection carrier density regime (Fig. 9).

Once $g_{th}$ is known, the required injected carrier density follows immediately from the relationship of Fig. 9, and the total current density is obtained by adding all current contributions,

$$J_{th} = J_{r,th} + J_{s,th} + J_{A,th} + J_{lk,th}, \tag{34}$$

using the relations of Figs. 10b and 12–14. In our threshold calculation, we assume a fixed $R_1 = R_2 = R = 0.3$ and analyze the role of the number of QWs, $\nu_{QW}$, and of the cavity length $L$. Notice though that other values of the reflectivity parameter $R$ can be described by simply renormalizing the cavity length $L$ using the invariance of the expression of Eq. (33a).

## 4.2. The Role of Quantum Well Number and Cavity Length

We will now investigate the effects of $\alpha_M$ and $\Gamma$ on the threshold current using the cavity length $L$ and the number of QWs, $\nu_{QW}$, as a variable parameter, respectively. The relation between $\alpha_M$ and $L$ is straightforward; see Eq. (33a). The confinement factor $\Gamma$ depends on the waveguide parameters, but primarily on the number of QWs. A good approximation for a separate confinement MQW laser is a linear relationship between $\Gamma$ and $\nu_{QW}$ [1, 8], which, for our representative structure of Fig. 1 with $L_z = 7.5$ nm and $d_g = 250$ nm, reads

$$\Gamma = 0.023\ \nu_{QW}. \tag{35}$$

Figure 15 is a representation of the threshold current density $J_{th}$ as a function of the injected QW carrier density at threshold $N_{th}$ in a grid of the parameters

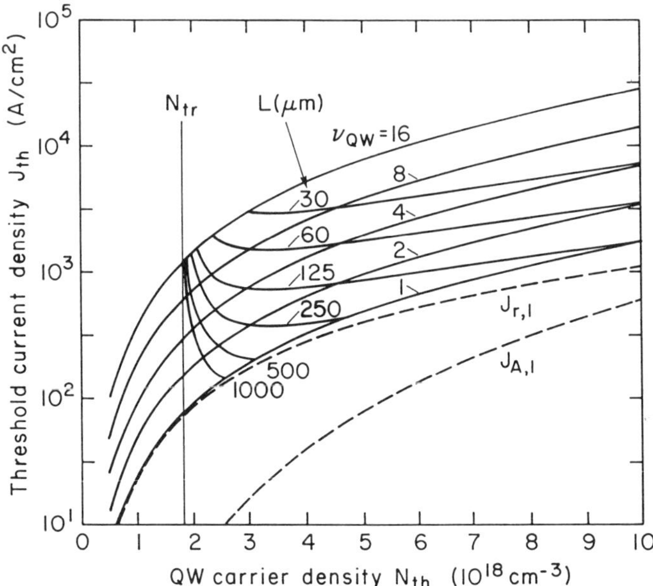


**Fig. 15.** Threshold current density $J_{th}$ (gain from accurate model, $C_A = 5 \times 10^{-30}$ cm$^6$/s, $v_s = 0$, $\tau_b \to \infty$) versus injection carrier density $N_{th}$ in the two parameter space of $\nu_{QW}$ and $L$. For comparison, $J_{r,1}$ and $J_{A,1}$ are also indicated.

$\nu_{QW}$ and $L$, for simplicity specified for the idealized case of $v_s = 0$, $\tau_b \to \infty$. In order to illustrate the role of Auger recombination, the radiative and Auger QW currents of the SQW case are included separately in the figure. The maximum number of QWs that can be accommodated with the chosen waveguide parameters is 16, the highest number shown. One observes the expected general trend of increasing $N_{th}$ with decreasing $L$ and/or $\nu_{QW}$. This trend becomes particularly severe toward the lower end of $L$ and/or $\nu_{QW}$ as a consequence of the QW gain saturation. Note that this is also the region with a larger contribution of the Auger current (and the carrier leakage out of the QWs; cf. Fig. 14). The series of $J_{th}$ curves for constant $L$ exhibit a shallow minimum near $N_{th} \approx 3.5 \times 10^{18}$ cm$^{-3}$. Below about $N_{th} = 2.5 \times 10^{18}$ cm$^{-3}$, $J_{th}$ starts to increase rapidly, approaching a vertical asymptote that corresponds to the transparency carrier density of $N_{tr} \approx 1.8 \times 10^{18}$ cm$^{-3}$. Also note that interface recombination would become relatively more severe at these lower $N_{th}$ values (cf. Fig. 12).

The intermediate $N_{th}$ range between about 2.5 and $3 \times 10^{18}$ cm$^{-3}$ can be considered as optimal for all practical purposes. For operation in this favorable injection regime, one can choose either long cavities or a large number of QWs as indicated by our analysis. To keep, e.g., $N_{th}$ below about

$3 \times 10^{18}\ \text{cm}^{-3}$ one needs at least 16 QWs for $L = 30\ \mu$m, 8 QWs for $L = 60\ \mu$m, 4 QWs for $L = 125\ \mu$m, and 2 QWs for $L = 250\ \mu$m. For longer cavities, the $J_{th}$ minimum disappears [7]. Obviously, less is to be gained by increasing $\nu_{QW}$ when the laser cavity is longer. On the other hand, even then an increase of $\nu_{QW}$ for lowering $N_{th}$ would still be desirable in such cases where carrier leakage out of the QWs and/or Auger recombination is already substantial at the lower $N$ values (cf. Figs. 13 and 14), but interface recombination is not a major technological problem (cf. Fig. 12). This conclusion forms the basis for the preference of MQW lasers over SQW ones, with relatively long cavities, in lower bandgap materials, such as the InGaAsP/InP system.

Besides injected carrier density $N_{th}$ and threshold current density $J_{th}$, absolute threshold current $I_{th}$ is also an important operational parameter of a semiconductor laser. The results for the behavior of $I_{th}$ versus $\nu_{QW}$ as derived from Fig. 15 are given in Fig. 16, where cavity length $L$ is used as

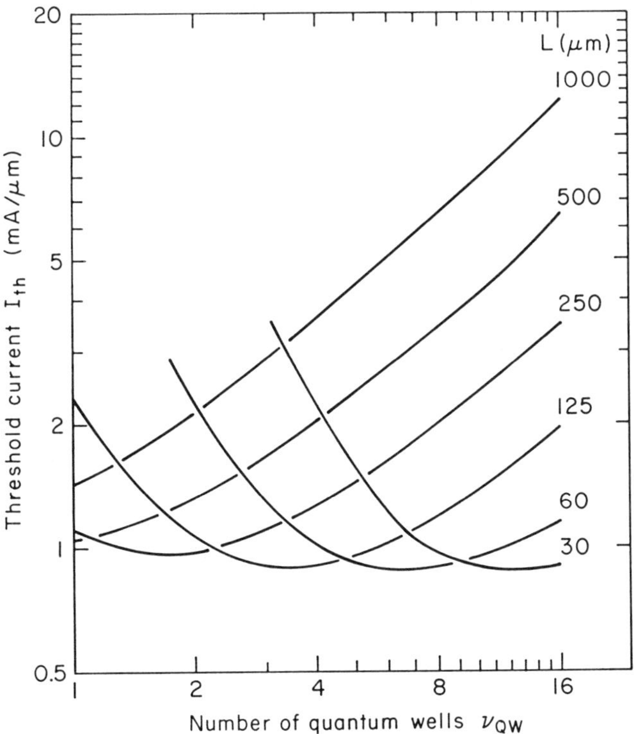


**Fig. 16.** Threshold current per unit cavity width, $I_{th}$ (gain from accurate model, $C_A = 5 \times 10^{-30}\ \text{cm}^6/\text{s}$, $v_s = 0$, $\tau_b \to \infty$), versus number of QWs, $\nu_{QW}$, for various cavity lengths $L$.

parameter. For convenience, $I_{th}$ has been normalized to 1 $\mu$m of cavity width. The curves with $L \leq 250\ \mu$m exhibit minima near 0.9 mA/$\mu$m, which tend to decrease slightly toward smaller $L$ (larger $v_{QW}$). These minima are largely determined by the reflectivity parameter $R$ through the relation [8]

$$I_{th}^{min} = I_0 G_{th} L = I_0[\ln(1/R) + \alpha_i L]. \tag{36}$$

Here, the threshold gain $G_{th}$ is expressed by Eq. (32), and the quantity $I_0$ is essentially a function of the QW dimensions only. This quantity follows from McIlroy *et al.*'s [8] saturation parameters $g_0$ and $J_0$ (see Fig. 11b) as

$$I_0 = J_0/G_0, \text{ with } G_0 = g_0\Gamma/v_{QW}. \tag{37}$$

For our representative laser structure of Fig. 1 with $\Gamma/v_{QW}$ given by Eq. (35), we have $I_0 \approx 0.73$ mA/$\mu$m (Fig. 11b, curve $J_{r,1} + J_{A,1}$). It should be noted that the $I_{th}$ minima correspond to the $J_{th}$ minima in the curves $L = \text{const}$ of Fig. 15 near $N_{th} \approx 3.5 \times 10^{18}$ cm$^{-3}$. Obviously, there is an optimum value of $v_{QW}$, which, at a particular $L$, leads to the lowest value of $I_{th}$ [7]; using the model of McIlroy *et al.* [8], this optimum QW number can be obtained from

$$v_{QW}^{(opt)} = \text{Int}(G_{th}/G_0). \tag{38a}$$

The quantity $G_0$, already introduced in Eq. (37), depends like $I_0$ on the QW dimensions only, yielding $G_0 \approx 30$ cm$^{-1}$ according to curve $J_{r,1} + J_{A,1}$ of Fig. 11b. The rising $I_{th}$ branches at lower $v_{QW}$ should be avoided, since there $N_{th}$ rapidly increases, whereas those at larger $v_{QW}$ could be tolerated, or even preferred, since they correspond to decreasing $N_{th}$, as pointed out before.

The alternate representation of $I_{th}$, with cavity length $L$ as the variable parameter, is presented in Figs. 17–19, which we chose for demonstrating the influence of interface recombination, Auger recombination, and carrier leakage, respectively. In these figures, a SQW, a 3-QW, and a 6-QW laser are compared with each other based on the simplified model of Section 2.1, which was judged accurate enough for analyzing the influence of these nonradiative and leakage mechanisms. Deviations from the accurate model of Section 2.2 are illustrated with Fig. 20, which, in the case of interface recombination, compares the results from the two models for a 3-QW laser. It is obvious that the simplified model shows very good qualitative ageement and quantitatively simulates somewhat larger $v_s$ values when interpreting threshold current data.

For Figs. 17–20, rather than using Eq. (35), the gain confinement $\Gamma$ was more accurately calculated by using a multilayer guiding model (e.g., [58]) with known refractive index relationships for the AlGaAs alloys [18, 59] and

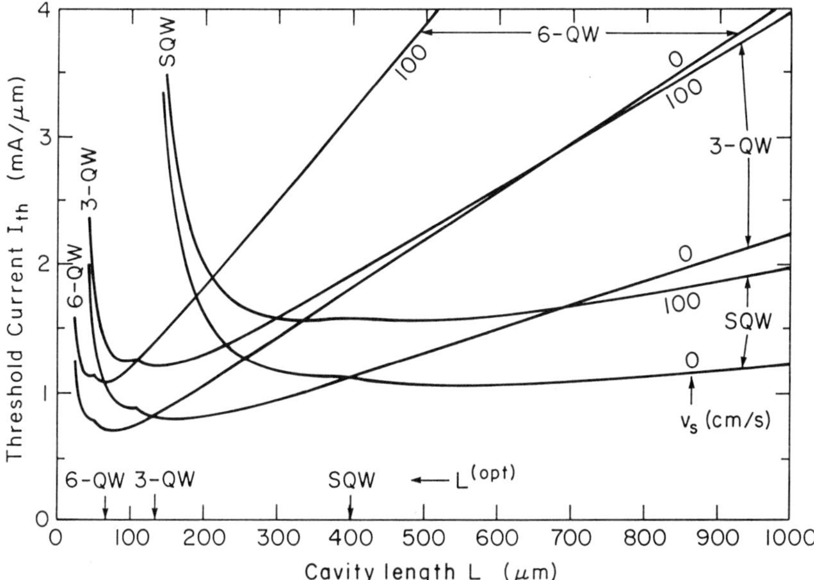


**Fig. 17.** Threshold current per unit cavity width, $I_{th}$ (gain from simplified model, $C_A = 5 \times 10^{-30}$ cm$^6$/s, $\tau_b = 1$ ns), versus cavity length $L$ for a SQW, a 3-QW, and a 6-QW laser, demonstrating the influence of interface recombination (parameter: $v_s$).

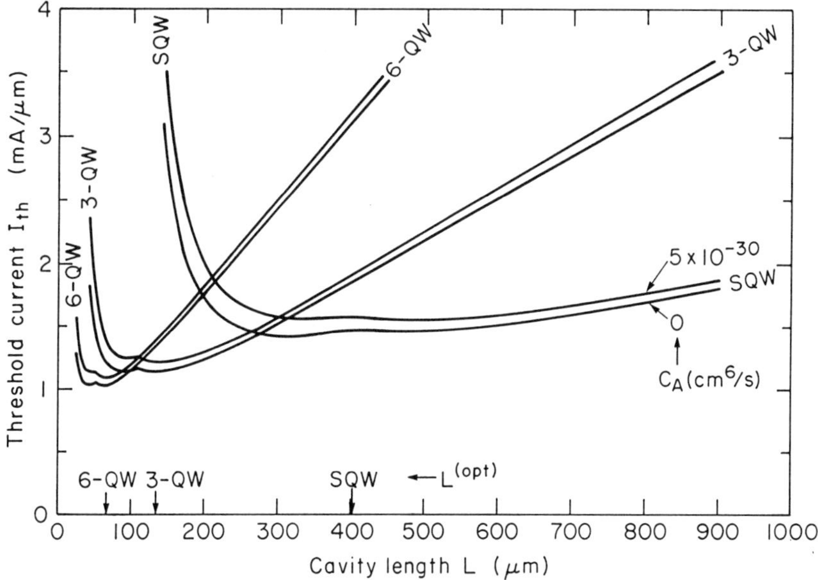


**Fig. 18.** Threshold current per unit cavity width, $I_{th}$ (gain from simplified model, $v_s = 100$ cm/s, $\tau_b = 1$ ns), versus cavity length $L$ for a SQW, a 3-QW, and a 6-QW laser, demonstrating the influence of Auger recombination (parameter: $C_A$).

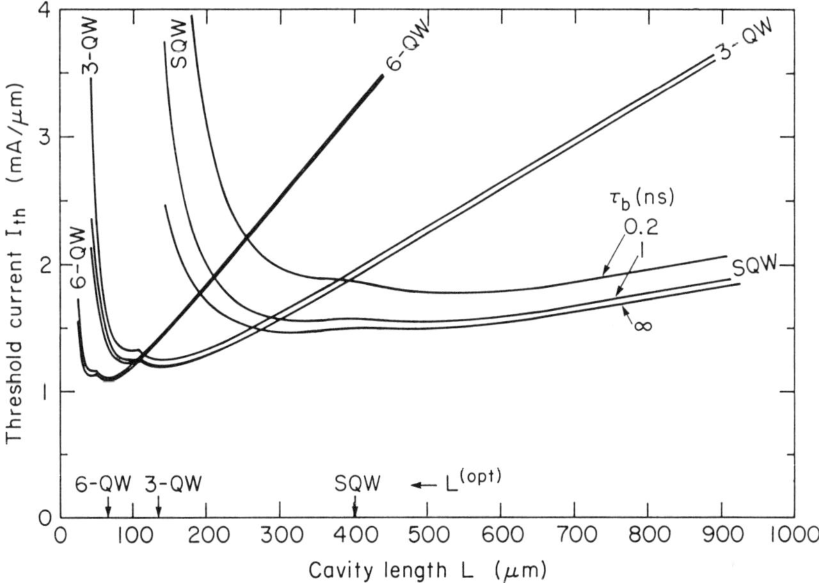


**Fig. 19.** Threshold current per unit cavity width, $I_{th}$ (gain from simplified model, $C_A = 5 \times 10^{-30}$ cm$^6$/s, $v_s = 100$ cm/s), versus cavity length $L$ for a SQW, a 3-QW, and a 6-QW laser, demonstrating the influence of carrier leakage out of the well (parameter: $\tau_b$).

neglecting quantization effects. In this procedure, the thickness of the guiding region ($d_g$) was not kept constant but optimized for each $\nu_{QW}$ to yield a maximum Γ. The calculated Γ is 0.023 for the SQW structure, 0.074 for the 3-QW structure, and 0.16 for the 6-QW structure, showing an increase slightly stronger than proportional to $\nu_{QW}$; the corresponding $d_g$ values are 260, 220, and 215 nm, respectively.

The figures show a striking increase of $I_{th}$ toward short cavity length, which is a direct consequence of the gain saturation in QWs at high injection [4–6, 8, 60, 61]. With increasing QW number, this seemingly anomalous increase shifts to smaller $L$ values. Consistent with our previous analysis, the position of the "anomalous" increase is affected by Auger recombination and, at lower $\nu_{QW}$, by carrier leakage, but only marginally by interface recombination. The latter, however, has a dominant effect at longer cavity lengths, including the flat threshold current minimum, where the contributions of the former two mechanisms become small. This demonstrates again the importance of minimizing interface recombination. To emphasize this point, Fig. 20 contains a full series of curves for a 3-QW laser with $v_s$ ranging up to 500 cm/s.

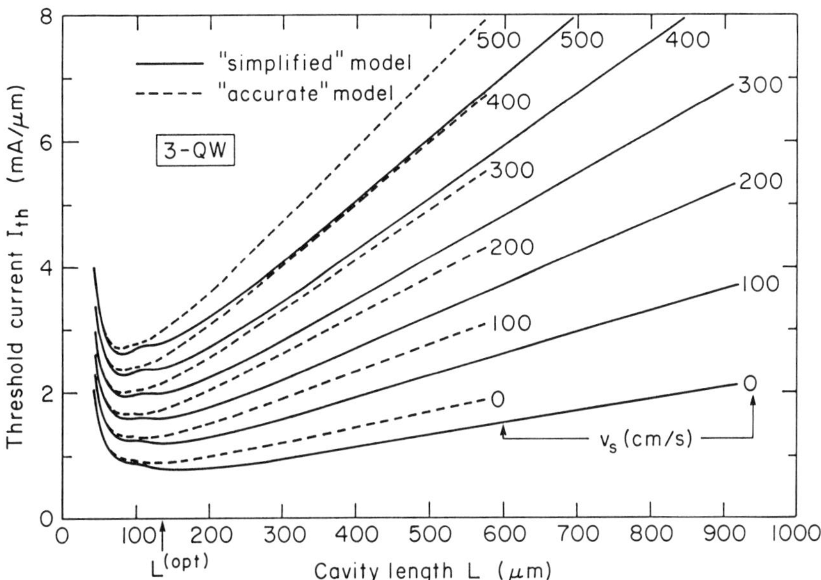


**Fig. 20.** Threshold current per unit cavity width, $I_{th}$ ($C_A = 5 \times 10^{-30}$ cm$^6$/s, $\tau_b = 1$ ns), versus cavity length $L$ for a 3-QW laser with interface recombination velocities up to 500 cm/s, calculated using both the simplified and the more accurate model for the gain maximum.

The $I_{th}$ versus $L$ curves are qualitatively similar to the $I_{th}$ versus $v_{QW}$ curves of Fig. 16. This similarity results from the fundamental equivalence of the two parameters $\alpha_M$ and $\Gamma$ in determining threshold, Eq. (32), and actually can be formulated also in a more quantitative way [8]. Thus, the optimum cavity length $L$ at the minimum $I_{th}$ can be estimated from

$$L^{(opt)} \approx \ln(1/R)/(v_{QW}G_0). \tag{38b}$$

The values for $L^{(opt)}$ obtained from this estimate, with $G_0 \approx 30$ cm$^{-1}$ (see the preceding; nonideal current contributions neglected except for Auger recombination), are indicated in Figs. 17–20 as well as in Figs. 22–24, to be discussed in Section 5. Due to the flatness of the current minimum, this remains a good estimate even when the nonideal contributions from $J_s$ and $J_{lk}$ are included, as is evident from inspecting the figures (note that the discrepancy in Figs. 17–20 is somewhat larger, sine they are based on the simplified model).

Figure 21 compares the influence of interface recombination in the 6-, 3- and SQW laser for two different cavity lengths, clearly showing the increased sensitivity to $v_s$ in MQW lasers. This can lead to the loss of their

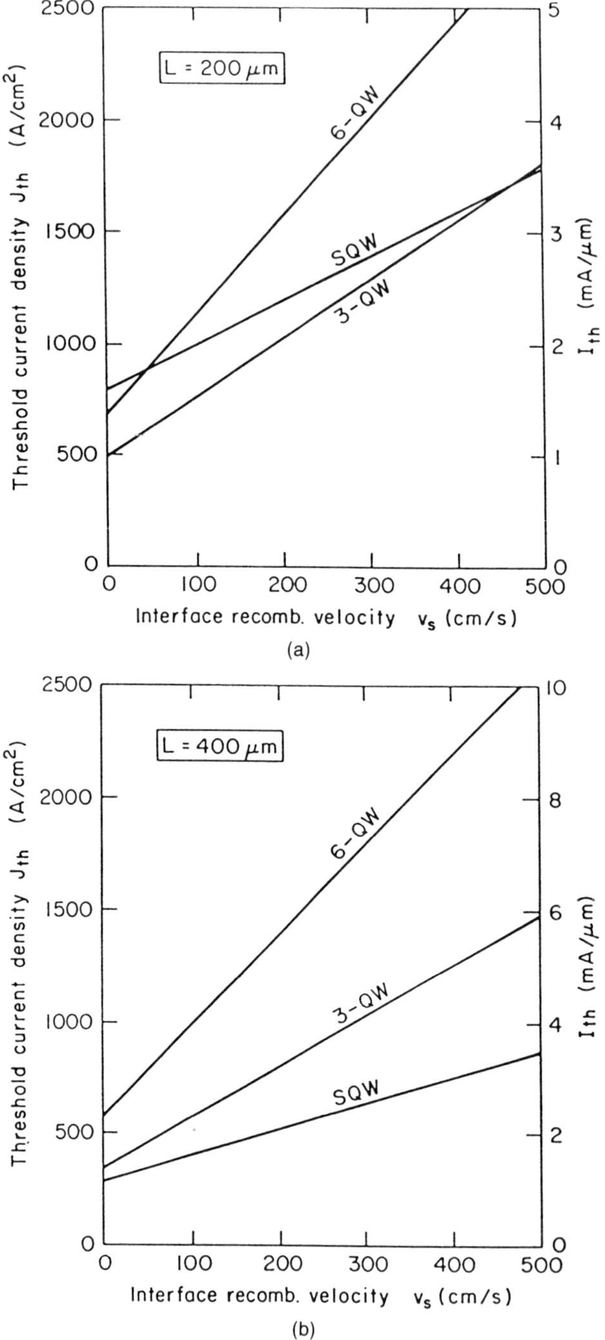


**Fig. 21.** The influence of the interface recombination velocity $v_s$ on threshold current density $J_{th}$ and/or threshold current per unit cavity width $I_{th}$ (gain calculated from the accurate model, $C_A = 5 \times 10^{-30}$ cm$^6$/s, $\tau_b = 1$ ns), for two cavity lengths: (a) $L = 200$ μm; (b) $L = 400$ μm

$I_{th}$ advantage against SQW lasers at higher $v_s$ values; e.g., with $L = 200\ \mu$m, the 6-QW laser loses its advantage for $v_s > 50$ cm/s, the 3-QW laser for $v_s > 420$ cm/s.

### 4.3. Multi versus Single Quantum Well Lasers

From the foregoing, certain general conclusions can be drawn regarding the disadvantages or advantages of MQW lasers as compared with SQW versions. The lowest threshold current density is clearly achievable in long SQW lasers [8]. Lowest carrier density, however, is provided in MQW lasers with the maximum $\Gamma$ possible. Operation in this regime ensures the highest differential gain factor $\beta = dg/dN$ with its resulting advantages in laser performance. Additionally, contributions to the total current from Auger recombination and carrier leakage out of the QW are minimized. On the other hand, the contributions from nonradiative recombination at crystal defects in the QW, mainly at the well–barrier interfaces, are relatively more harmful in MQW lasers than in SQW versions. Quality control in the growth of multilayer structures is therefore extremely important. Interface recombination in InGaAsP/InP long wavelength systems is known to be less severe than in the GaAs/AlGaAs system. However, the long wavelength stystems are plagued by severe Auger recombination. This can be overcome only by operating immediately above transparency in the lowest carrier density regime possible. As a result, MQW lasers with longer cavities are the preferred implementation for long wavelength systems [50, 62, 63]. Another important consideration for MQW structures is obviously the control over the uniformity of the QW thickness $L_z$ during the epitaxial growth.

In high speed optoelectronic integrated circuits and integrated optics, short lasers with low threshold current are desirable. Shortening the laser cavity increases modulation speed by reducing the photon lifetime. For a particular cavity length, an optimum QW number results, which increases for smaller lengths. Recent experiments by Wolf *et al.* [44] (also see [64]) revealed, e.g., a modulation bandwidth of up to 14 GHz and relaxation oscillation frequencies of up to 36 GHz in 150–200 $\mu$m long GaAs/AlGAs 3-QW lasers (cf. Fig. 20) that were designed to minimize the influence of parasitic reactances.

Theoretically, the lowest threshold current is achieved with the highest QW number that can be accommodated in the optical waveguide, and the cavity length is adjusted to the corresponding shortest optimum value. For the case of our representative separate confinement structure of Fig. 1 with $d_g = 250\ \mu$m, this leads, e.g., to $v_{QW} = 16$ with $L = 25\ \mu$m and $I_{th} = 0.88$ mA/$\mu$m based on Eqs. (36) and (37). However, the largest QW number also exhibits the highest threshold current density, which could be of concern for reliability reasons. Thus, from a practical point of view somewhat longer cavities with less QWs should be preferred, which not only reduces $J_{th}$ but also $N_{th}$, without incurring a large penalty in $I_{th}$. In the preceding example, by reducing $\nu_{QW}$ to 6 and increasing $L$ to 125 $\mu$m, one can reduce $J_{th}$ from about 3.5 kA/cm$^2$ to 830 A/cm$^2$, and $N_{th}$ from about $3.5 \times 10^{18}$ cm$^{-3}$ to $2.6 \times 10^{18}$ cm$^{-3}$ (cf. Figs. 15 and 16) with an increase in $I_{th}$ from 0.88 to 1.04 mA/$\mu$m, or a penalty of less than 20%. The threshold current can be reduced further with higher facet reflectivities [56, 65], which, however, calls for adding sophisticated coating technology to the device processing and also would reduce modulation speed by increasing the photon lifetime.

## 5. COMPARISON WITH EXPERIMENTS

Various authors have measured and discussed the cavity length dependence of threshold in SQW [60, 66] as well as MQW lasers [5, 6, 56, 67]; others have focused on the role of the number of QWs only [68]. The "anomalous" increase of threshold at short cavity length as originally reported by Zory *et al.* [60] for SQW lasers was confirmed as a characteristic feature and was also observed in MQW lasers. Additionally, the theoretical conclusions regarding the number of QWs were supported by several of these measurements. In this section, we compare our modeling results with the experiments of Shieh *et al.* [6] for illustrating the importance of interface recombination in GaAs/AlGaAs QW lasers.

### 5.1. The Experimental GaAs/AlGaAs MQW Structures

GaAs/AlGaAs QW lasers were prepared with a single well, three wells, and six wells. The QW structures were grown by molecular beam epitaxy (MBE) on a Si doped n-type substrate according to our representative separate confinement structure of Fig. 1 with the waveguide size $d_g$ optimized for maximum $\Gamma$. The bottom cladding and guiding layers were Si doped ($3 \times 10^{17}$ cm$^{-3}$); those on the top were Be doped (also $3 \times 10^{17}$ cm$^{-3}$); and the MQW region was undoped, yielding a slightly p-type background (mid $10^{14}$ cm$^{-3}$). The p–n junction, thus, was located at the lower boundary of the MQW region, and the acceptor concentration in the MQW region was low enough to ensure uniform injection into the QWs for lasing [14].

The QW lasers were processed by the shallow Schottky mesa stripe technology [69]. To minimize current spreading effects at the stripe edges, a large sripe width of 50 $\mu$m has been chosen. Thus, the measured threshold currents can be directly compared with the theoretical model, which is based on a uniform current distribution. The wafers were cleaved into bars with cavity length ($L$) ranging from 80 to 1000 $\mu$m. No facet coating was applied. The threshold current ($I_{th}$) was measured on the unmounted die under pulsed operation (1-$\mu$s pulses, 1-kHz repetition rate) to avoid heating. Near field analysis indicated that the devices lased quite uniformly over the width of the relatively wide stripe. The measured $I_{th}$ was averaged over five lasers for each QW structure investigated.

## 5.2. Interpretation Based on the Model Calculations

For comparison with the experiments, we calculated the threshold current using the accurate model. Auger recombination was included with the coefficient $C_A = 5 \times 10^{-30}$ cm$^6$/s, appropriate for GaAs. The leakage current in the guiding region was modeled with a constant carrier life time of $\tau_b = 1$ ns. As we have seen in Figs. 18 and 19, both the Auger and the carrier leakage contribution lead to a relatively small correction. Since we concluded that the most important nonradiative transition in the QWs is interface recombination, we used the interface recombination velocity ($v_s$) as a variable parameter in this comparison.

The calculated threshold current for the SQW laser is plotted against the cavity length for various interface recombination velocities $v_s$ in Fig. 22a. The experimental points are shown on the same graph (bars indicate the standard deviation from the five lasers in each group). Agreement is reasonably good if we assume $v_s \approx 100$ cm/s. This value is of similar order as measured by time-resolved photoluminescence for the AlGaAs/GaAs interface in MBE material [51].

The "anomalous" rise of the threshold current toward short laser lengths caused by gain saturation is very sensitive to the magnitude of the transition matrix element. Without the enhancement of the transition matrix for TE polarization in the QW as, e.g., implied in the calculations of Wilcox *et al.* [61], the calculated rise is shifted to considerably larger $L$, which is shown for $v_s = 100$ cm/s by the dashed line in Fig. 22b. The deviation from the experimental points at small cavity length is obvious. A significant deviation would remain even when assuming $v_s = 0$ and neglecting Auger recombination ($C_A = 0$). On the other hand, a further increase of the absolute square of the momentum matrix, $|M|^2$, by a factor of 1.27 (leading to $\xi = 3.38$), which was postulated by Yan *et al.* [30] as a consequence of the interaction between the conduction band valleys, would reduce the anomalous rise to

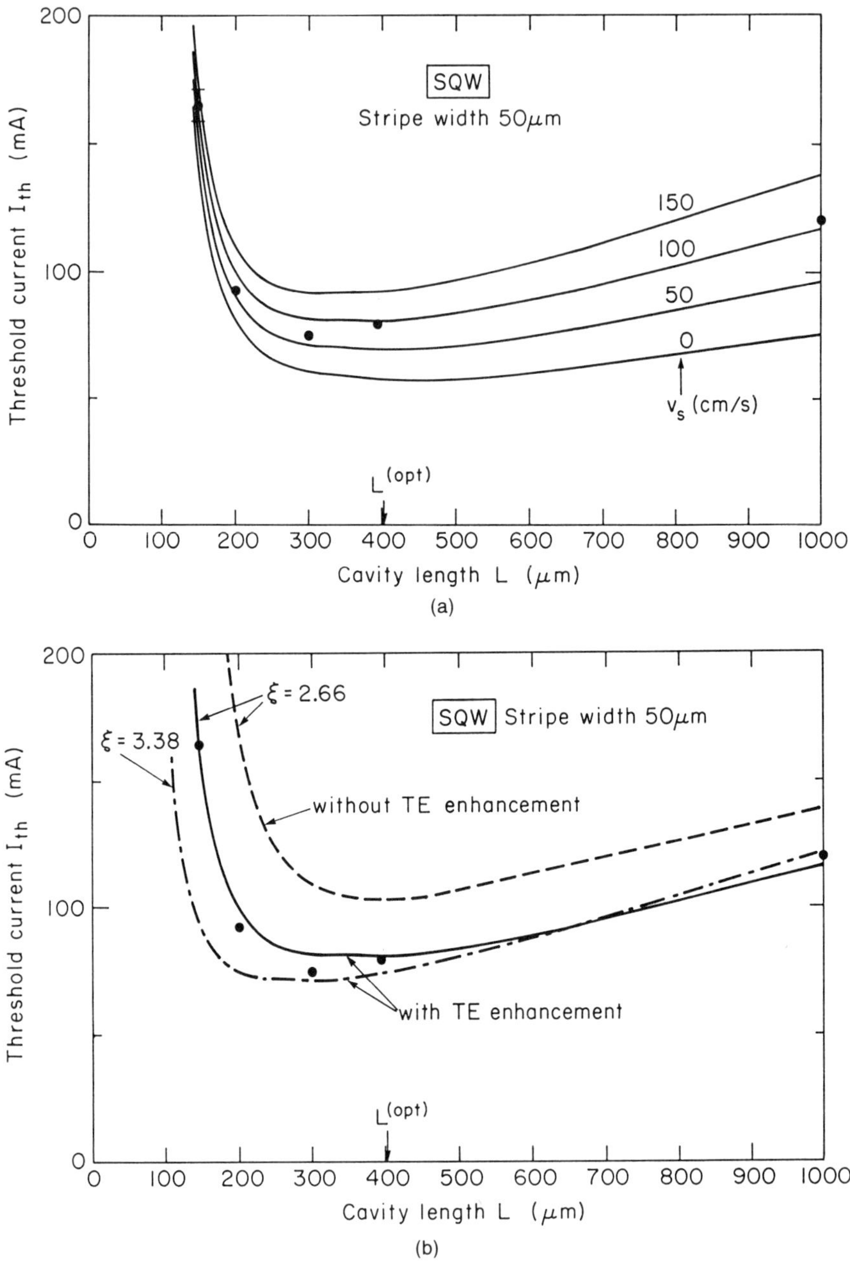


**Fig. 22.** Threshold current $I_{th}$ ($C_A = 5 \times 10^{-30}$ cm$^6$/s, $\tau_b = 1$ ns) of a 50 μm wide stripe laser with a SQW versus cavity length $L$, (a) calculated with several interface recombination velocities $v_s$ using the more accurate model and compared with experimental data; (b) calculated with $v_s = 100$ cm/s as in (a) and demonstrating the effect of neglecting TE enhancement ($\mu_{jn} = 1$; dashed curve) and of increasing the matrix element prefactor $\xi$ by 27% to 3.38 (dash–dotted curve).

below the experimental data points. We, therefore, tentatively conjecture that band-mixing effects in QWs [24, 25] compensate for this increase of $|M|^2$ in our case.

Similar calculations based on the accurate model were performed for the 3- and the 6-QW structure and compared with the experimental results, as shown in Figs. 23 and 24. The only difference between the SQW laser and the MQW lasers is their confinement factor $\Gamma$ and the small change in the size of the waveguide region $d_g$. Reasonable agreement is obtained. The inferred $v_s$ values range from about 30 cm/s to 270 cm/s; this large variation indicates difficulties in controlling this parameter in MBE growth [51]. As shown in the graphs, the anomalous threshold current increase of the MQW structures, in accordance with their larger $\Gamma$, occurs at a cavity length shorter than available in the experiments by facet cleaving. The optimum cavity length $L^{(opt)}$, estimated from Eq. (38), is marked on the $L$ axis in the figures, demonstrating a fairly good match as pointed out before.

The lowest threshold current has been obtained for the 6-QW laser with a cavity length of 80 $\mu$m, namely, 51 mA, or 1.02 mA/$\mu$m. As seen from Fig. 24, the effect of interface recombination on this value is small and, hence, it

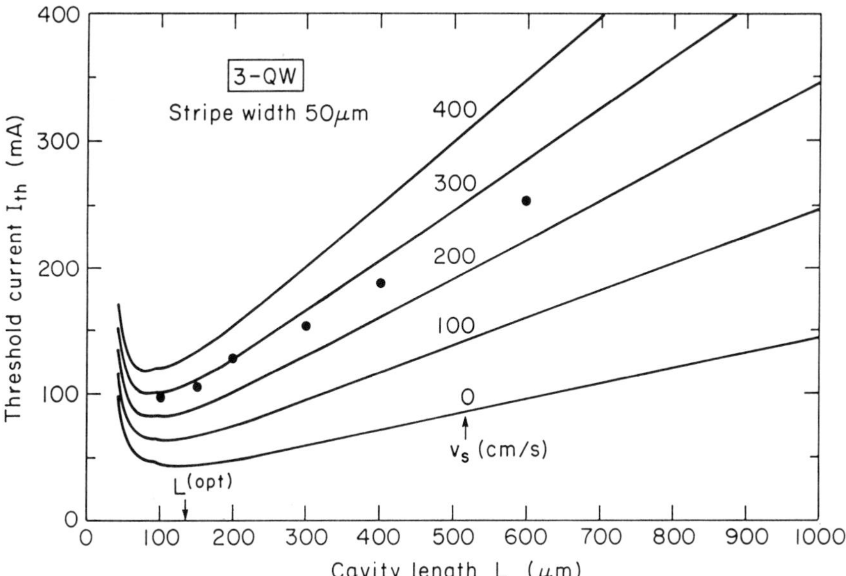


**Fig. 23.** Threshold current $I_{th}$ ($C_A = 5 \times 10^{-30}$ cm$^6$/s, $\tau_b = 1$ ns) of a 50 $\mu$m wide stripe laser with 3 QWs versus cavity length $L$, calculated with several interface recombination velocities $v_s$ using the more accurate model and compared with experimental data.

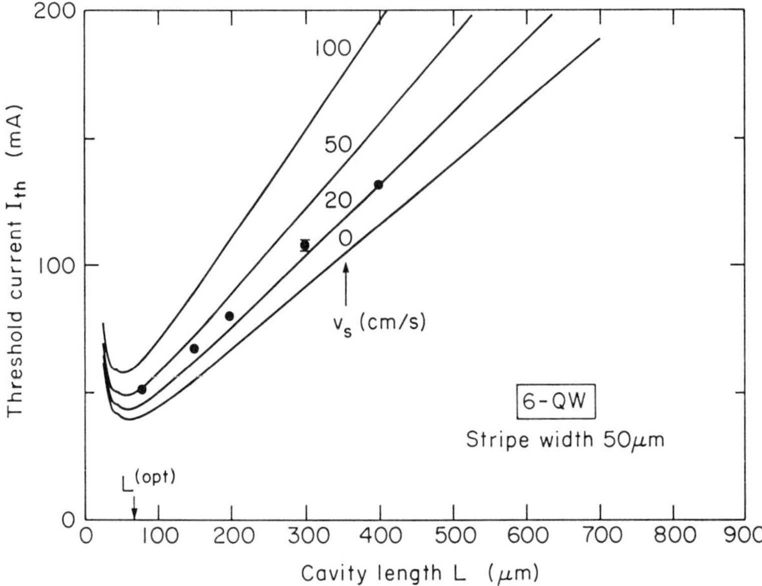


**Fig. 24.** Threshold current $I_{th}$ ($C_A = 5 \times 10^{-30}$ cm$^6$/s, $\tau_b = 1$ ns) of a 50 $\mu$m wide stripe laser with 6 QWs versus cavity length $L$, calculated with several interface recombination velocities $v_s$ using the more accurate model and compared with experimental data.

is close to the ideal value expected from Fig. 16. The theoretical model suggests that a further reduction of the cavity length for a 6-QW laser would not lead to an appreciable improvement. An additional threshold reduction can only be achieved with an increase in facet reflectivity [56]. Dry-etching for fabricating the laser facets, and facet anodization with simple self-aligned metallic facet coating is an effective way for fabricating low threshold current, short cavity MQW lasers [65].

## 6. CONCLUSIONS

The gain and current calculations for MQW lasers were reviewed. It was shown that TE polarization enhancement of gain in QWs counteracts the effect of gain broadening on the magnitude of the spectral gain maximum, which determines lasing threshold. At the broadening parameter typical for GaAs (intraband scattering time of 0.1 ps) and with commonly used QW sizes (around 7.5 nm), the compensation is nearly complete. This allows neglecting both effects for rough estimates of threshold current with considerable savings in computation time. Using such a "simplified" model, the resulting threshold current versus cavity length relations were shown to be almost identical to those determined from the "accurate" model except for an adjustment of interface recombination velocities to somewhat higher values.

A striking feature of the threshold current versus cavity length relations is a pronounced rise in threshold current toward shorter cavity length, which is caused by the strong gain saturation at high injection current densities, typical for QWs. This seemingly anomalous rise is particularly pronounced in a SQW laser because of its low gain confinement $\Gamma$. It is enhanced by Auger recombination and carrier loss out of the QW, but its location is relatively insensitive to the magnitude of interface recombination. On the other hand, the threshold current rise is shifted substantially toward shorter cavity lengths by increasing the number of QWs in the guiding region, which augments the fraction of the optical mode subject to gain (increased $\Gamma$) and thereby eliminates the gain saturation problem even at relatively short cavity lengths. Thus, e.g., the threshold of a 6-QW laser can be lowered by reducing the cavity length to values as short as 80 $\mu$m. Theoretically, the lowest threshold current is obtained with the maximum number of QWs that can be accommodated in the waveguide at an optimized short cavity length if interface recombination could be eliminated. Further reduction could be achieved by using high reflective facet coatings to reduce optical mode losses.

SQW lasers can be operated with the lowest threshold current densities when designed with long cavities. However, it is concluded that in general MQW lasers with the proper number of QWs are preferable to SQW lasers for fully utilizing the QW advantage of the high differential gain coefficient at low injection carrier densities just above the transparency level, leading to low temperature sensitivity of threshold (high $T_0$; [32, 41]) and high modulation bandwidth [43, 44, 64]. Optimization for lowest linewidth enhancement, however, requires operation at somewhat higher injection levels and, hence, the number of QWs is limited to a lower value for improved coherency (smaller linewidth, lower intensity noise) and reduced chirping during modulation [46–48]. A design with an increased number of QWs is particularly advantageous when considering lasers with short cavities for integrated optics. Such designs also ensure the lowest contributions from carrier leakage out of the well and from Auger recombination, the latter being of special concern in long wavelength InGaAsP/InP structures. For these reasons, MQW lasers are the preferred choice for the long wavelength communications lasers [50, 62, 63].

The calculations based on the accurate model that include broadening and TE gain enhancement in the QWs were compared with experimental data for cleaved-facet single-, three- and six-QW GaAs/AlGAs lasers using interface recombination velocity as an adjustable parameter [6]. The most significant nonradiative transition in GaAs/AlGaAs QW lasers is interface recombination, since its relative contribution to the total current increases at low injection. As a consequence, it also affects MQW lasers more strongly than SQW versions. For the experimental MBE samples tested, the variation in threshold between different QW growth runs was interpreted by a variation in interface recombination velocity $v_s$ estimated to range between 30 and 270 cm/s. This variation is indicative of the difficulty in controlling the quality of the interfaces during epitaxial growth. The lowest values inferred from our analysis are consistent with values measured by Wolford [52] for interfaces grown under well controlled conditions by organometallic chemical vapor deposition (MOCVD) where $v_s \leq 40$ cm/s.

Additional comparison of the experimental data with the accurate model showed that, when broadening effects are included in the calculation of the gain spectrum, the enhancement of the oscillator strength for TE polarization in QWs has to be taken into account to properly predict the location of the anomalous rise of the threshold current at short cavity lengths. On the other hand, an additional increase of the absolute square of the momentum matrix element by a factor of 1.27, as expected from the effects of higher conduction band valleys on the electron effective mass and proposed by Yan *et al.* [30], is inconsistent with the experimental location of the anomalous threshold current rise. It is conjectured that subband-mixing effects in QWs, which give rise to nonparabolic bands [24, 25], provide a full compensation for this proposed increase in the case of the investigated QW size of $L_z = 7.5$ nm.

## ACKNOWLEDGMENTS

Some of this work was performed when the authors were affiliated with Siemens Corporate Research, Inc., Princeton, New Jersey. We thank Kambiz Alavi, now at the University of Texas at Arlington, for the MBE growth, and Joe Mantz, now at the Boeing High Technology Center in Seattle, for his able assistance in device processing and characterization. George Valliath from the University of Rhode Island (now at the Motorola Corporate Manufacturing Research Center, Schaumburg, Illinois) kindly assisted in the computer simulations, and Barbara Ryall from the Oregon Graduate Institute provided the excellent art work. Thanks are also due to Nu Yu from the Oregon Graduate Institute (now at Hoechst Japan Ltd.) for stimulating discussions and to Chuck Zmudzinski from the University of Florida in Gainesville for very helpful comments leading to an improvement of the manuscript. Last, but not least, we thank the editor, Peter Zory, for his patient guidance and friendly encouragement during the preparation of this chapter.

## SYMBOLS

| | |
|---|---|
| $C_A$ | Auger recombination coefficient |
| $c$ | speed of light in free space |
| $D(E')$ | 2D density of states at energy $E'$, spin degeneracy *not* included ($D_c$ in conduction band; $D_v$ in valence band; $D_r$ reduced for electron–hole dipole) |
| $d_g$ | guiding region thickness |
| $E = \hbar\omega$ | electron–hole dipole transition energy = photon energy |
| $E_g$ | bulk bandgap of QW semiconductor |
| $E_{g,b}$ | bandgap of barrier semiconductor |
| $E_{g,jn}$ | bandgap between the $n$th subband pair involving the $j$ valence band |
| $E_n$ | quantized energy level in QW ($E_{c,n}$ in conduction band; $E_{v,jn}$ in $j$-hole valence band) |
| $\hat{E}$ | resonance energy of broadened electron–hole dipole |
| $E'$ | electron energy from bottom of conduction band in bulk semiconductor |
| $E''$ | hole energy from top of valence band in bulk semiconductor |
| $E'_{jn}$ | electron energy from bottom of conduction subband $n$ for transition to valence subband $jn$ (Fig. 2) |
| $E''_{jn}$ | hole energy from top of valence subband $jn$ (Fig. 2) |
| $e$ | electron charge |
| $F$ | quasi-Fermi level ($F_c$ for conduction band electrons measured from bulk band edge upwards; $F_v$ for valence band holes measured from bulk band edge downwards) |
| $f$ | Fermi–Dirac function for electrons ($f_c$ in conduction band; $f_v$ in valence band) |
| $G$ | maximum modal gain coefficient for $n = 1$ QW subband transition |

| | |
|---|---|
| $g$ | maximum local gain coefficient for $n = 1$ QW subband transition |
| $g_0$ | gain saturation parameter (Fig. 11b) |
| $g(E)$ | gain coefficient at photon energy $E = \hbar\omega$ |
| $\hbar$ | reduced Planck's constant ($h/2\pi$) |
| $I_{th}$ | threshold current (also threshold current per unit width) |
| $J$ | current density ($J_A$ from Auger recombination in QW; $J_{lk}$ from carrier leakage out of QW; $J_r$ from radiative recombination in QW; $J_s$ from QW interface recombination) |
| $J_0$ | current-density saturation parameter (Fig. 11b) |
| $J_{th}$ | threshold current density |
| $j$ | = h, l (index for heavy or light hole subband) |
| $k$ | electron momentum/$\hbar$, in QW restricted to plane of QW ($x$, $y$ plane) |
| $k_B$ | Boltzmann's constant |
| $L$ | cavity length |
| $L_{(i)}$ | indirect conduction band valleys at zone boundaries with $k$ along $\langle 111 \rangle$ crystal directions for the well ($i$ = w), the barrier ($i$ = b), and the cladding ($i$ = c) bulk materials |
| $L_z$ | size of QW |
| $M$ | base function momentum matrix element |
| $M_b$ | energy-independent transition matrix element averaged for the electron–hole dipole transition (spin degeneracy included) in bulk semiconductor |
| $M_{jn}$ | transition matrix element averaged for the electron–hole dipole transition of subband pair $jn$ (spin degeneracy included) in QW |
| $m_0$ | free electron mass |
| $m_c$ | electron effective mass in conduction band |
| $m_n$ | effective mass of $n$th QW subband |
| $m_r$ | reduced effective mass of electron–hole dipole |
| $m_v$ | hole effective mass in valence band |
| $N$ | well electron concentration injected into QW subband energy states |
| $N_b$ | electron concentration injected into the barrier |
| $N_{tr}$ | QW electron concentration at transparency |

| | |
|---|---|
| $N_w$ | well-region electron concentration injected into bulk-like energy states beyond the top of the QW |
| $n$ | quantum number of QW energy state |
| $\bar{n}_{eq}$ | equivalent refractive index for optical waveguide mode |
| $\bar{n}'$, $\bar{n}'_g$ | phase and group index, respectively, of refraction in QW |
| $R$ | reflectivity parameter ($R_1$, $R_2$ facet reflectivities) |
| $R_{sp}(E)$ | spontaneous emission rate at photon energy $E = \hbar\omega$ |
| $v_s$ | interface recombination velocity |
| $X_{(i)}$ | indirect conduction band valleys at zone boundaries with $k$ along $\langle 100 \rangle$ crystal directions for the well ($i$ = w), the barrier ($i$ = b), and the cladding ($i$ = c) bulk materials |
| $\alpha$ | laser linewidth enhancement factor |
| $\alpha_i$ | internal absorption coefficient of optical mode |
| $\beta$ | differential gain coefficient |
| $\Gamma$ | gain confinement factor |
| $\Gamma_{(i)}$ | direct conduction band valleys in zone center ($k = 0$) for the well ($i$ = w), the barrier ($i$ = b), and the cladding ($i$ = c) materials |
| $\delta E$ | energy broadening parameter due to intraband scattering ($\delta E_c$ in conduction band; $\delta E_{in}$ for electron–hole dipole; $\delta E_v$ in valence band) |
| $\Delta$ | spin–orbit splitting in valence band of QW semiconductor |
| $\Delta E_c$, $\Delta E_v$ | band offsets at conduction, valence band edge, respectively |
| $\Delta E_g$ | bulk bandgap difference between barrier and QW |
| $\varepsilon_0$ | permittivity of free space |
| $\lambda = 2\pi c/\omega$ | free-space wavelength |
| $\lambda_g$ | free-space wavelength corresponding to bandgap energy in QW |
| $\mu_{jn}$ | matrix-element anisotropy factor for $en$-$jhn$ subband transition in QW |
| $v_{QW}$ | number of quantum wells in MQW layer |
| $\xi$ | matrix element prefactor, Eq. (7) |
| $\rho$ | volume density of states, spin degeneracy *not* included ($\rho_c$ in conduction band; $\rho_v$ in valence band; $\rho_r$ reduced for electron–-hole dipole) |
| $\tau_b$ | carrier lifetime in barrier layers |
| $\tau_{in}$ | intraband scattering (relaxation) time for electron–hole dipole |

## REFERENCES


1. Arakawa, Y., and Yariv, A. (1986). *IEEE J. Quantum Electron.* **QE-22**, 1887.
2. Yamada, M., and Suematsu, Y. (1981). *J. Appl. Phys.* **52**, 2653.
3. Asada, M., Kameyama, A., and Suematsu, Y. (1984). *IEEE J. Quantum Electron.* **QE-20**, 745.
4. Reisinger, A. R., Zory, P. S., Jr., and Waters, R. G. (1987). *IEEE J. Quantum Electron.* **QE-23**, 993.
5. Nagarajan, R., Kamiya, T., and Kurobe, A. (1989). *IEEE J. Quantum Electron.* **QE-25**, 1161.
6. Shieh, C., Engelmann, R., Mantz, J., and Alavi, K. (1989). *Appl. Phys. Lett.* **54**, 1089.
7. Arakawa, Y., and Yariv, A. (1985). *IEEE J. Quantum Electron.* **QE-21**, 1666.
8. McIlroy, P. W. A., Kurobe, A., and Uematsu, Y. (1985). *IEEE J. Quantum Electron.* **QE-21**, 1958.
9. Agrawal, G. P., and Dutta, N. K. (1986). *Long-wavelength semiconductor lasers.* Van Nostrand Reinhold Company, New York.
10. Yamada, N., Ogita, S., Yamagishi, M., and Tabata, K. (1985). *IEEE J. Quantum Electron.* **QE-21**, 640.
11. Dutta, N. K. (1982). *J. Appl. Phys.* **53**, 7211.
12. Landsberg. P. T., Abrahams, M. S., and Osinski, M. (1985). *IEEE J. Quantum Electron* **QE-21**, 24.
13. Saint-Cricq, B., Lozes-Dupuy, F., and Vassilieff, G. (1986). *IEEE J. Quantum Electron.* **QE-22**, 625.
14. Dutta, N. K. (1983). *IEEE J. Quantum Electron.* **QE-19**, 794.
15. Kroemer, H., and Okamoto, H. (1984). *Japan. J. Appl. Phys.* **23**, 970.
16. Casey, H. C., Jr. (1978). *J. Appl. Phys.* **49**, 3684.
17. Uomi, K. (1990). *Japan. J. Appl. Phys.* **29**, 81.
18. Casey, H. C., Jr., and Panish, M. B. (1978). *Heterostructure lasers.* Academic Press, Orlando, Florida.
19. Duggan, G. (1985). *J. Vac. Sci. Technol.* **B3**, 1224.
20. Kolbas, R. M. (1979). Ph.D. Thesis, University of Illinois, Urbana, Illinois (unpublished).
21. Chinn, S. R., Zory, P. S., and Reisinger, A. R. (1988). *IEEE J. Quantum Electron.* **QE-24**, 2191.
22. Kasemset, D., Hong, C. S., Patel, N. B., and Dapkus, P. D. (1983). *IEEE J. Quantum Electron.* **QE-19**, 1025.
23. Stern, F. (1971). *Phys. Rev. B* **3**, 2636.
24. Colak, S., Eppenga, R., and Schuurmans, M. F. H. (1987). *IEEE J. Quantum Electron.* **QE-23**, 960.
25. Ahn, D., and Chuang, S.-L. (1990). *IEEE J. Quantum Electron.* **QE-26**, 13.
26. Dingle, R. (1975). In *Festkörperprobleme XV, Advances in solid state physics,* (Queisser, M. J., ed.) p. 21. Pergamon-Vieweg, Braunschweig.
27. Bastard, G. (1981). *Phys. Rev. B* **24**, 5693.

28. Alavi, K., Pearsall, T. P., Forrest, S. R., and Cho, A. Y. (1983). *Electron. Lett.* **19**, 227.
29. Lasher, G., and Stern, F. (1964). *Phys. Rev.* **133**, A533.
30. Yan, R. H., Corzine, S. W., Coldren, L. A., and Suemune, I. (1990). *IEEE J. Quantum Electron.* **QE-26**, 213.
31. Zhu, L. D., Zheng, B. Z., Xu, Z. Y., and Feak, G. A. B. (1989). *IEEE J. Quantum Electron.* **25**, 1171.
32. Zhu, L. D., Zheng, B. Z., and Feak, G. A. B. (1989). *IEEE J. Quantum Electron.* **25**, 2007.
33. Zielinski, E., Schweizer, H., Hausser, S., Stuber, R., Pilkuhn, M. H., and Weimann, G. (1987). *IEEE J. Quantum Electron.* **QE-23**, 969.
34. Zielinski, E., Keppler, F., Hausser, S., Pilkuhn, M. H., Sauer, R., and Tsang, W. T. (1989). *IEEE J. Quantum Electron.* **QE-25**, 1407.
35. Asada, M., and Suematsu, Y. (1985). *IEEE J. Quantum Electron.* **QE-21**, 434.
36. Asada, M. (1989). *IEEE J. Quantum Electron.* **QE-25**, 2019.
37. Martin, R. W., and Störmer, H. L. (1977). *Solid State Comm.* **22**, 523.
38. Yamada, M., Ishiguro, H., and Nagao, H. (1980). *Japan. J. Appl. Phys.* **19**, 135.
39. Yamanishi, M., and Lee, Y. (1987). *IEEE J. Quantum Electron.* **QE-23**, 367.
40. Zee, B. (1978). *IEEE J. Quantum Electron.* **QE-14**, 727.
41. Zory, P. S., Reisinger, A. R., Waters, R. G., Mawst, L. J., Zmudzinski, C. A., Emanuel, M. A., Givens, M. E., and Coleman, J. J. (1986). *Appl. Phys. Lett.* **49**, 16.
42. Lau, K. Y., Bar-Chaim, N., Harder, C., and Yariv, A. (1983). *Appl. Phys. Lett.* **43**, 1.
43. Yuasa, T., Yamada, T., Asakawa, K., Ishii, M., and Uchida, M. (1987). *Appl. Phys. Lett.* **50**, 1122.
44. Wolf, H. D., Lang, H., and Korte, L. (1989). *Electron. Lett.* **25**, 1246.
45. Burt, M. G. (1984). *Electron. Lett.* **20**, 27.
46. Hausser, S., Idler, W., Zielinski, E., Pilkuhn, M. H., Weimann, G., and Schlapp, W. (1989). *IEEE J. Quantum Electron.* **QE-25**, 1469.
47. Dutta, N. K., Temkin, H., Tanbun-Ek, T., and Logan, R. (1990). *Appl. Phys. Lett.* **57**, 1390.
48. Takahashi, T., Schulman, J. N., and Arakawa, Y. (1991). *Appl. Phys. Lett.* **58**, 881.
49. Uomi, K., Mishima, T., and Chinone, N. (1990). *Japan. J. Appl. Phys.* **29**, 88.
50. Uomi, K., Sasaki, S., Tsuchiya, T., Nakano, H., and Chinone, N. (1990). *IEEE Photon. Technol. Lett.* **2**, 229.
51. Sermage, B., Pereira, M. F., Alexandre, F., Beerens, J., Azoulay, R., and Kobayashi, N. (1988). In *GaAs and related compounds 1987*, *Inst. of Phys. Conf. Series* **91**, p. 605. Institute of Physics, Bristol.
52. Wolford, D. J. (1991). Private communication.
53. Blood, P., Fletcher, E. D., Woodbridge, K., Heasman, K. C., and Adams, A. R. (1989). *IEEE J. Quantum Electron.* **QE-25**, 1459.
54. Hagen, S. H., van't Blik, H. F. J., Boermans, M. J. B., and Eppenga, R. (1988). *Appl. Phys. Lett.* **52**, 2015.
55. Kesler, M. P., and Harder, C. (1990). *IEEE Photon. Technol. Lett.* **2**, 464.

56. Kurobe, A., Fukuyama, H., Naritsuka, S., Sugiyama, N., Kokobun, Y., and Nakamura, M. (1988). *IEEE J. Quantum Electron.* **QE-24**, 635.
57. Wilcox, J. Z., Ou, S. S., Yang, J. J., and Jansen, M. (1989). *CLEO 1989, Technical Digest Series* **11**, paper **ThM1**, 330.
58. Kressel, H., and Butler, J. K. (1977). *Semiconductor laser and heterojunction LEDs*, Chapter 5. Academic Press, New York.
59. Afromowitz, M. A. (1974). *Solid State Commun.* **15**, 59.
60. Zory, P. S., Reisinger, A. R., Mawst, L. J., Costrini, G., Zmudzinski, C. A., Emanuel, M. A., Givens, M. E., and Coleman, J. J. (1986). *Electron. Lett.* **22**, 475.
61. Wilcox, J. Z., Peterson, G. L., Ou, S., Yang, J. J., Jansen, M., and Schechter, D. (1988). *Electron. Lett.* **24**, 1218.
62. Zah, C. E., Bhat, R., Menocal, S. G., Favire, F., Andreadakis, N. C., Koza, M. A., Caneau, C., Schwarz, S. A., Lo, Y., and Lee, T. P. (1990). *IEEE Photon. Technol. Lett.* **2**, 231.
63. Tanbun-Ek, T., Logan, R. A., Temkin, H., Olsson, N. A., Sergent, A. M., and Wecht, K. W. (1990). *Photon. Technol. Lett.* **2**, 453.
64. Lang, H., Wolf, H. D., Korte, L., Hedrich, H., Hoyler, C., and Thanner, C. (1991). *IEE Proc., Part J* **138**, 171.
65. Shieh, C., Mantz, J., Alavi, K., and Engelmann, R. (1990). *Photon. Technol. Lett.* **2**, 159.
66. Wagner, D. K., Waters, R. G., Tihanyi, P. L., Hill, D. S., Rosa, A. J., Jr., Vollmer, H. J., and Leopold, M. M. (1988). *IEEE J. Quantum Electron.* **QE-24**, 1258.
67. Yuasa, T., Yamada, T., Asakawa, K., and Ishii, M. (1988). *J. Appl. Phys.* **63**, 1321.
68. Blood, P., Fletcher, E. D., and Woodbridge, K. (1985). *Appl. Phys. Lett.* **47**, 193.
69. Amann, M. C. (1979). *Electron. Lett.* **15**, 441.

# Chapter 4

# ULTRALOW THRESHOLD QUANTUM WELL LASERS

**Kam Y. Lau**

*Department of EECS*
*University of California at Berkeley*
*Berkeley, California*



## 1. INTRODUCTION

Quantum well lasers have threshold currents that are an order of magnitude below that of conventional laser diodes fabricated from bulk materials. Few other characteristics of quantum well lasers can boast of performance improvements by such a wide margin. And yet much of this improvement has its origin *not* in the quantum confinement of electrons and holes, but from a simple scaling of the physical size of the active region. As one reduces





ISBN 0-12-781890-1

the thickness of the active region from a conventional 0.1 $\mu$m to 100 Å, one obtains a reduction in the active volume by roughly a factor of ten and an improvement in lasing threshold by approximately the same factor. Further reducing the active region (quantum well) thickness brings about little further reduction in threshold, and at thicknesses below 50 Å the threshold actually increases due to quantum confinement. These effects will be discussed in detail later.

If further improvements in threshold current cannot be brought about by an indefinite shrinkage of the thickness of the quantum well, a logical alternative is to shrink its lateral dimensions, eventually becoming a *quantum wire* and a *quantum dot* when both lateral dimensions are reduced to sizes comparable with the electronic radius. A further order of magnitude reduction in threshold currents result from this lateral shrinkage, but again this reduction has relatively little to do with quantization of the carriers.

One can then conceptually separate the effects due to quantization of the electrons and holes from those due to simple physical scaling, even though this distinction is somewhat artificial, since as the physical dimension is reduced quantization will naturally take place. *Simple physical scaling* refers to effects that can be predicted simply by geometrical considerations without even invoking quantization. The latter is often referred to in the existing literature as *quantum size effects*. To avoid confusion, we shall simply refer to these as *quantum effects*, and reserve the term physical scaling for those effects arising primarily from the small physical size of the device.

To summarize what is presently understood on the subject, physical scaling (along with reduction of the valence band effective mass) accounts for most of the superior threshold characteristics of quantum-confined lasers. This is not to say that quantum effects need not be considered in quantum-confined lasers, since they account for most of the expected improvements in high speed modulation as well as a host of other dynamic effects observed in quantum-confined lasers, which will be discussed in Chapter 5.

This chapter deals with the low threshold nature of quantum-confined lasers. It is not an extensive bibliography of the work done to date in this field, nor is it an exposition of the latest record achieved. (These tend to become outdated very quickly.) Fundamentally, what has been asserted thus far constitutes a major part of what is presently understood about the low threshold behavior of quantum-confined lasers. The remainder of this chapter will provide a firm basis for the preceding assertions. The organization is such that the conceptual base of the subjects are first presented, followed by a sampling of some recent representative experimental results that illustrate the concepts.

## 2. OPTICAL GAIN

The study of the threshold current of a semiconductor laser is largely a study of the optical gain characteristics of the lasing medium. Since detailed gain calculations of quantum wells including many subtle effects have been presented in earlier chapters, the following treatment makes no attempt at repeating them but will be done in a simplified fashion that leads directly to the conclusion that physical scaling is a major effect in determining the threshold current. It will be seen later on that this conclusion depends only weakly on the exact band structure, and hence for simplicity the parabolic band approximation will be used throughout. We consider the state of a forward-biased junction as specified by the quasi-Fermi levels of the electrons and holes. The electron and hole densities are given respectively by

$$N = \int \rho_c(E_c) f_c(E_c, R_{fc})\, dE_c, \tag{1}$$

$$P = \int \rho_v(E_v) f_v(E_v, E_{fv})\, dE_v, \tag{2}$$

where

$$f_c(E_c, E_{fc}) = \frac{1}{\exp(E_c - E_{fc}) + 1} \tag{3}$$

$$f_v(E_v, E_{fv}) = \frac{1}{\exp(E_v - E_{fv}) + 1} \tag{4}$$

are the Fermi distributions, with $E_c$ measuring positive into the conduction band from the conduction band edge, and $E_v$ measuring positive into the valence band from the valence band edge. The quantities $\rho_{c,v}$ are the density of state functions, which in a conventional (three dimensional) material are proportional to the square root of the energy, and in an ideal quantum well takes the form of a "staircase" [1]. In general, though, due to the interactions between the different valence bands and the nonparabolicity of the bands, the density of states functions assume forms more complicated than the simple staircase model. The optical absorption coefficient, with $k$-selection rule enforced, is

$$\alpha(E) = \xi \rho_r (1 - f_c - f_v), \tag{5}$$

where

$$E = \hbar\omega - E_g = E_c + E_v \tag{6}$$

is the reduced photon energy, $\xi$ is a material constant, and $E_g$ is the *equivalent* bandgap of the quantum well material, which is the sum of the lowest quantization energies of the electrons and that of the holes, and the bandgap of the parabolic band. The quantity $\rho_r(E)$ is the reduced density of states, which in the case of a two dimensional quantum well (staircase approximation) is

$$\frac{1}{\rho_r} = \sum_i \left(\frac{1}{\rho_c} + \frac{1}{\rho_{ui}}\right). \tag{7}$$

The summation is over all heavy- and light-hole bands. For a one dimensional quantum wire and a zero dimensional quantum dot, the expressions for the density of states have been given elsewhere [2]. The choice of whether to enforce the $k$-selection rule depends on the material under consideration: For heavily doped compounds, the transition is dominated by band tail states where the $k$-selection rule does not apply; on the other hand, in lightly doped quantum well materials the $k$-selection should apply, as experimental data seem to indicate. The gain coefficient $g(E)$ is simply the negative of the absorption coefficient. For the material to experience gain at some photon energies, $g(E) > 0$ for at least some $E$, and the following condition can be obtained from Eq. (5):

$$\frac{1}{\exp(E_c - E_{fc})/kT + 1} + \frac{1}{\exp(E_v - E_{fv})/kT + 1} > 1, \tag{8}$$

which can then be further reduced to

$$E_{fc} + E_{fv} > E_c + E_v. \tag{9}$$

This is the well-known Bernard–Duraffourg condition for optical transparency in a direct gap semiconductor, independent of the band structure. The electron density (and hence the injection current) needed to attain this point can be found by evaluating $E_{fc}$ and $E_{fv}$ subject to the condition Eq. (9) and simultaneously to the charge neutrality condition, $N = P$.

## 3. THE TRANSPARENCY CONDITION

We shall argue that the electron density for transparency for a quantum well laser is roughly the same as that of bulk material. Once this is established, it is trivial that since the quantum well has a volume that is one order of magnitude smaller than that of a typical bulk laser, the injection current needed to sustain transparency is an order of magnitude lower. The *transparency current density* represents a fundamental limit to achieving the lowest lasing threshold for semiconductor lasers in general. The current needed for lasing is composed of two parts: the first part being the current needed for maintaining the electron density at the optical transparency level, and beyond that a second part to attain the necessary gain to overcome all the losses in the laser cavity. It can be argued (and can actually be demonstrated experimentally) that a laser cavity can be designed such that the losses are minimal, but this can only reduce the second part of the threshold current while the first part, that responsible for optical transparency, is unaffected. The key to building an ultralow threshold laser is thus to design a laser cavity with a very low loss, with a material that has the lowest transparency *current* density. A single quantum well structure is one that possesses both of these qualities and, when combined with high reflectivity coatings to minimize mirror loss, results in some of the lowest lasing threshold currents achieved to date.

To put it simply, since the Bernard–Duraffourg condition states that transparency occurs when the separation in quasi-Fermi levels equals the gap between the lowest and highest available states for the electrons and holes, respectively, the relative locations of the Fermi levels with respect to these states are not too different between bulk and quantum well, wire, or dot materials. Since the relative location of the Fermi level with respect to the edge of a band determines the electron density, it follows that the transparency electron density is not too different in all of these materials either.

Quantitatively, the Bernard–Duraffourg condition implies that

$$E_{\mathrm{fc}} + E_{\mathrm{fv}} = \varepsilon_{\mathrm{c1}} + \varepsilon_{\mathrm{v1}}, \tag{10}$$

where $E_{\mathrm{fc,fv}}$ are the quasi-Fermi levels, measured from the band edges and positive *into* the bands as before, and $\varepsilon_{\mathrm{c1,v1}}$ are the first quantized energies in the conduction and valence bands, respectively. The electron and hole densities are given by

$$N, P = \int \rho_{\mathrm{c,v}} f_{\mathrm{c,v}}(E, E_{\mathrm{fc,v}})\, dE, \tag{11}$$

where $f_{\mathrm{c,v}}$ are the Fermi distributions as given in Eqs. (3) and (4), and $\rho_{\mathrm{c,v}}$ are the conduction and valence band density of states functions, which in the simple single-band approximation is given by

$$\rho_{\mathrm{c,v}} = \frac{m_{\mathrm{c,v}}}{\pi\hbar^2 L_z} \sum_i H(E - \varepsilon_{\mathrm{c}i,\mathrm{v}i}) \tag{12}$$

where $H(x)$ is the Heaviside function, $\varepsilon_{ci,vi}$ are the $i$th quantized energies of the conduction and valence bands, given by

$$E_{ci,ui} = \frac{i^2\pi^2\hbar^2}{2m_{c,vi}L_z^2} \tag{13}$$

and $L_z$ is the width of the quantum well. Substituting Eq. (12) into (11), one obtains

$$N = \frac{m_c}{\pi\hbar^2 L_z} \sum_i E_{fc} - E_{ci} + kT \ln(1 + \exp[(E_{ci} - E_{fc})/kT)]. \tag{14}$$

A corresponding formula exists for holes. The condition for charge neutrality requires that $N = P$ (for an undoped quantum well), and puts a relation between $E_{fc}$ and $E_{fv}$. This together with the transparency condition Eq. (10) uniquely determine the Fermi levels and hence the electron density at transparency. The computed transparency electron sheet density is shown in Fig. 1 [3]. For comparison, the result for an infinite well is also shown. The transparency density decreases linearly with well thickness (thus implying a constant transparency *electron density*) until approximately 100 Å, and from then on stays approximately constant between 100 Å and 50 Å—at the latter point, the electrons are no longer confined in the quantum well.

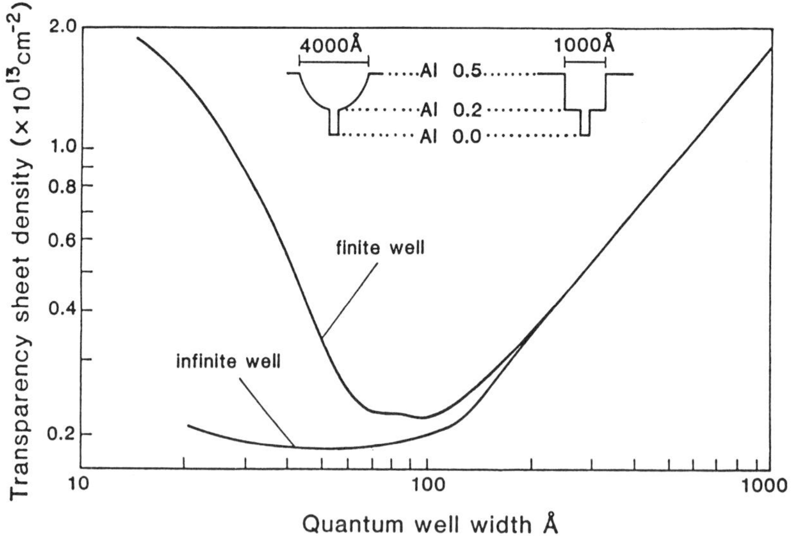


**Fig. 1.** Computed threshold current density as a function of well thickness for a single quantum well laser.

Experimentally, a nearly constant threshold current dependence of a very long laser device (which approaches the transparency current) on well width between 40 and 100 Å was observed [4].

When the calculation is carried out in full, one will discover that not only is the transparency current density lower, but the optical gain is considerably higher in a quantum well than in conventional bulk materials. However, since its small physical dimension results in a very weak confinement of the optical mode, there is *no* substantial overall advantage in a single quantum well as far as optical modal gain is concerned. The problem associated with a small optical confinement in quantum wells can be circumvented by using multiple quantum wells in the active region of a laser. A calculated result for the optical modal gain is shown in Fig. 2 [2]. The cases for multi-quantum wells are simply scaled from the results of a single quantum well by assuming that the injected electrons are equally divided among all the wells, which are

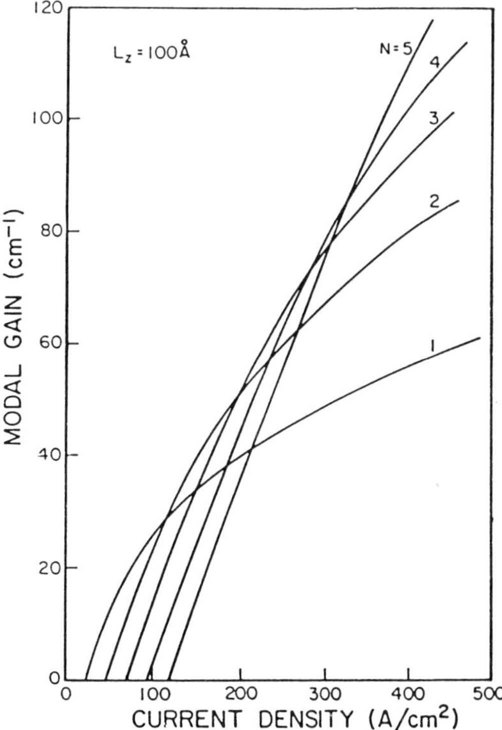


**Fig. 2.** Calculated gain versus current characteristic for single and multiple quantum wells (from [2]; © 1986 *IEEE*).

uncoupled electronically, and that the optical confinement factor is multiplied by the number of wells. Neglecting the fact that these assumptions are not quite true in reality due to coupling between wells and the nonuniform injection of carriers as well as interfacial effects, multiple quantum wells are therefore expected to have a higher transparency current density but a higher optical gain. If one makes no attempt to absolutely minimize the loss of the laser, then a combination of a relatively small current for transparency (compared with bulk materials) and a higher optical gain enable multi–quantum well buried heterostructure lasers to be built with threshold current as low as 2.5 mA [5]. If one minimizes the cavity loss by applying highly reflective coatings, then a single quantum well should be used, which cannot supply as much optical gain as multiple quantum wells but has the absolutely lowest transparency density, which results in a threshold current as low as 0.5 mA [3].

While it appears that lowering the thickness of a single quantum well to much below 100 Å will not aid in lowering the transparency current, it has been proposed that further reduction can result from reducing the effective mass of the heavy-hole band [6, 7]. This can be accomplished using strained layer structures, as discussed in Section 5.

## 4. THRESHOLD CALCULATIONS

Given the preceding understanding of the quantum well gain characteristics, it is elementary to calculate the threshold current from the gain curves of Fig. 2. In principle, one should take advantage of the low transparency current density of a single quantum well by minimizing the active area (so that the total current needed to attain transparency is scaled down proportionally) while at the same time reducing the cavity loss to minimize the gain required for lasing. The former implies a narrow active region and a short cavity length. However, a short cavity length implies a large gain per unit length required to overcome mirror loss, so that to minimize the required gain the mirror loss itself has to be minimized. The result is a very short cavity laser, with a very narrow active region and a high reflectivity coating applied to the facets. However, a high reflectivity coating usually leads to a low differential quantum efficiency, since the latter is basically the ratio of the mirror loss to the total cavity loss. To maintain a reasonable differential quantum efficiency, one needs to simultaneously reduce the internal loss of the waveguide. Luckily, passive waveguides that incorporate a single layer of quantum well as an active region do have very low losses (a structure known as *graded index separate confinement* [5], due to the low optical confinement factor, which on one hand reduces the gain of the optical mode but, on the other hand, reduces the loss associated with the active region as well. Losses in the neighborhood of 3–6 $cm^{-1}$ have been measured. The result is that even with facet reflectivities as high as 90% the differential quantum efficiency of a single quantum well laser can still be maintained at close to 80% of that of a conventional laser. All these factors combined makes it possible to realize an ultralow threshold laser with good quantum efficiency.

Using the gain characteristics of Fig. 2 and equating gain to cavity loss in the usual manner, one can obtain the threshold electron density for various device configurations. Furthermore, using the bimolecular recombination relation to relate the electron density to current density, one arrives at the threshold current relationships. These results are shown in Fig. 3, which

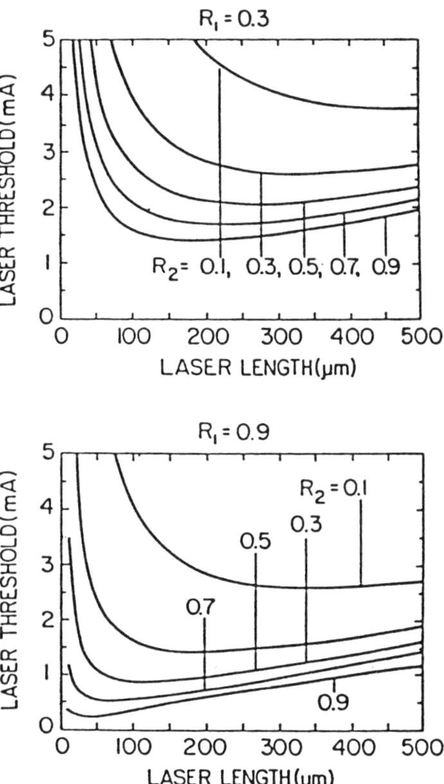


**Fig. 3.** Calculated threshold current for a buried heterostructure GRIN SCH SQW laser with various cavity lengths and facet reflectivities.

quantitatively shows the results of the preceding discussions. One should note that without mirror coatings the lowest threshold is obtained with a relatively long cavity length (around 300 $\mu$m). This is about the same as that for conventional lasers. However, for conventional lasers, applying high reflectivity coatings does *not* help the situation, since the high transparency electron density prevents any substantial reduction in threshold by the coatings, which furthermore has the undesirable consequence of reducing the quantum efficiency.

## 5. STRAINED LAYER QUANTUM WELLS

It can be shown that [6, 7] the absolute lowest transparency electron density will result when the density of states functions of the electrons and holes are both small and identical. Unfortunately, this is not the case in III-V material, in which the effective mass of the hole is an order of magnitude larger than that of the electron. Consider the case where the electron and hole density of states functions are identical in functional form and where the hole density of states is $D$ times that of the electron density of states: $\rho_v(E) = D\rho_c(E)$. Then charge neutrality gives the following:

$$\int \rho_c \, dE\left(\frac{1}{\exp(E - E_{fc})/kT + 1} - \frac{D}{\exp(E - E_{fv})/kT + 1}\right) = 0. \quad (15)$$

If $D = 1$, then, from symmetry, at the point of optical transparency we must have $E_{fc} + E_{fv} = 0$. If $D > 1$, then the integral contribution of the second term in the parenthesis in Eq. (15) must be reduced by requiring that $E_{fv} < 0$, which forces $E_{fc} > 0$, thus resulting in a higher electron density.

Strained layer superlattices can be used to reduce the effective mass of holes, thereby reducing the transparency electron density and subsequently the lasing threshold. Detailed discussions of strained layer materials and band structures can be found in Chapters 1 and 7. Here, we again concentrate on a simplified treatment that serves to illustrate their roles in low threshold lasers. In conventional bulk materials, the degeneracy of heavy- and light-hole bands at zone center implies domination of the heavy-hole band in hole occupation. It was known that, by applying strain to the bulk crystal, the degeneracy can be broken so that under a biaxial compressive stress the light-hole band is lifted above the heavy hole band in the $k$-vector directions parallel to the applied strain (Fig. 4) [8, 9]. Since the heavy hole is five times heavier than the light hole, the resultant effective hole mass can be reduced by a factor of five. The strain can be built in by growing lattice-mismatched

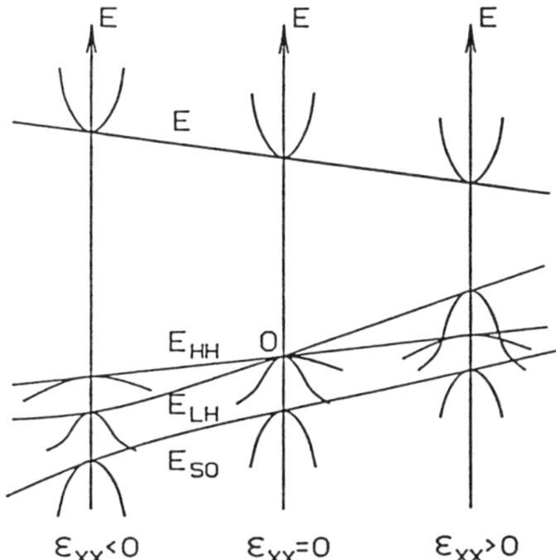


**Fig. 4.** Schematic representation of the band structure of a III-V compound under uniaxial strain and stress.

$In_xGa_{1-x}As$ on GaAs substrate. Such lattice-mismatched epitaxial layers cannot be arbitrarily thick for it to be defect free, the maximum thickness being only about 100 Å for $x = 0.2$. Thus, all strained layer structures are by default quantum well structures, and it appears that the same argument as before, that of lifting the light-hole band above the heavy-hole band by strain, can be applied to predict that strained quantum wells have an effective hole mass five times lower than that of unstrained quantum wells. This argument is largely incorrect, because in an unstrained quantum well the lowest hole quantum state corresponds to the band that is heavy in the direction perpendicular to the well, and this same band happens to be light in the plane of the well. Hence, the effective mass of the lowest quantum state in a conventional quantum well is always that of a light hole, even *without* strain. This can be seen very clearly in band structure calculations of quantum wells; the valence subband structure of a GaAs–$Ga_{0.3}Al_{0.7}As$ quantum well in the (100) orientation is shown in Fig. 5 [10]. The question then becomes, what is the advantage of applying strain to a quantum well, since the effective hole mass is already so light? In fact, Shubnikov–de Hass measurements show that the effective hole mass of a strained quantum well is reduced only by 20% below that of an unstrained quantum well [11], which is a minimal reduction in terms of influencing the threshold of a laser. The answer comes from a detailed examination of the band diagram Fig. 5, where it is seen that the highest hole band, the HH1 subband, which has a hole mass of 0.2, deviates from the parabolic characteristic beyond approximately 10 meV of the band edge, becoming substantially heavier as energy

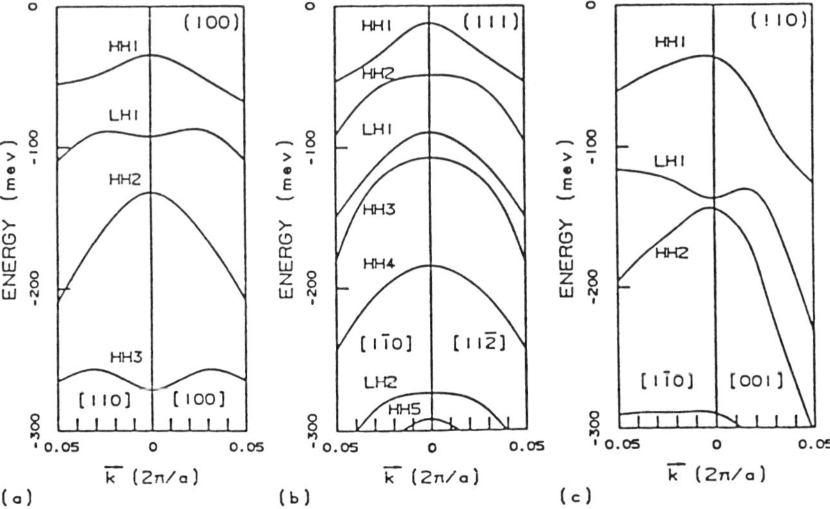


**Fig. 5.** Detailed structure of the hole bands of a single quantum well with various crystal orientations.

increases. This is very undesirable for laser operation, since under the heavy carrier concentration of $>10^{18}\ \mathrm{cm}^{-3}$ in common laser operation the hole energies extend well beyond 10 meV into the band and effectively sees a very heavy mass. One can easily convert the band structure into a density of states function, and the nonparabolicity is reflected very clearly in Fig. 6, which shows a very low density at the band edge but increasing rapidly beyond. The nonparabolicity comes from the interaction between the subbands, and an applied strain can separate the bands further, thus reducing the subband interaction, resulting in the parabolicity being maintained even deep into the band. (In addition, the narrower bandgap InGaAs leads to a lower hole mass to start with.) This is illustrated in Fig. 7 [12], where an almost flat density of states is maintained until the second hole band is encountered.

It is clear from the discussions in the preceding sections that, if one compares two structures with an approximately flat density of states function (as in a 2D strained quantum well, Fig. 7), then a low density of states results in a low transparency electron density. What is not so clear is how a low density of states results in a higher gain; after all, a low density of states means that at any particular photon transition energy there are fewer states available, and hence the peak gain should be reduced. It will be shown that this is not quite the case, that the peak gain attainable is approximately independent of the density of states, except at very high carrier densities.

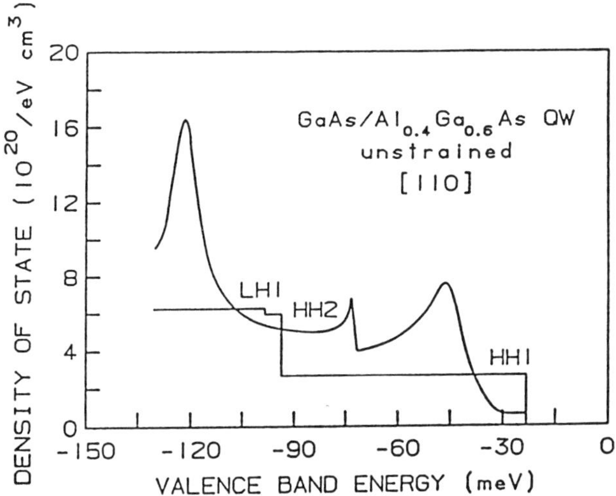


**Fig. 6.** Computed density of states function of a single quantum well. The staircase function is the ideal single quantum well density of states without taking into account the crystal orientation of the quantum well and band-to-band interaction (from [12]).

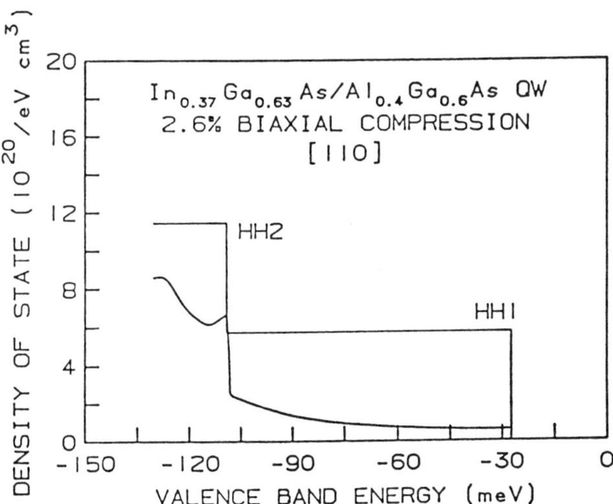


**Fig. 7.** Computed density of states of a strained quantum well with 2.6% biaxial strain (from [12]).

More important is the fact that, in the structure with a low density of states, the gain reaches its peak value at a lower electron density. There is thus every reason to believe that the strained quantum well laser should exhibit a lower threshold current than an unstrained one under all reasonable conditions of operation, a fact repeatedly demonstrated in recent experiments.

To see how the density of states affect the gain characteristic, we use Eq. (5) for the optical gain:

$$g(E) = \xi\rho_r(f_c + f_v - 1), \tag{16}$$

where

$$\zeta = \frac{\hbar\pi q^2|M|^2}{\varepsilon_0 m_0^2 c\bar{\mu}E} \tag{17}$$

is a material-dependent parameter, $E = \hbar\omega - E_g$ is the reduced photon energy, and $f_{c,v}$ are the Fermi factors as defined in Eqs. (3) and (4). As before, the energies are measured positive into the conduction band from the (quantized) conduction band edge and positive into the valence from the valence band edge. For a 2D quantum well whose density of states is independent of energy, the reduced density of states $\rho_r$ is the harmonic mean of the electron and hole density of states:

$$\frac{1}{\rho_r} = \frac{1}{\rho_c} + \frac{1}{\rho_v} = L_z\hbar^2\pi\left(\frac{1}{m_c} + \frac{1}{m_v}\right). \tag{18}$$

For $k$-selection transition, the electron and hole energies are related by

$$E_{c,v} = E\frac{m_r}{m_{c,v}} = \frac{\rho_r}{\rho_{c,v}}. \tag{19}$$

The quasi-Fermi levels are related to the electron and hole densities by Eqs. (1) and (2), which can be evaluated for a constant density of state (as in the case of strained layer quantum for energies below 100 meV, Fig. 7) to give

$$N, P = \rho_{c,v}\{E_{fc,fv} + kT\ln[1 + \exp(-E_{fc,fv}/kT)]\}. \tag{20}$$

Let the hole density of states be $D$ times that of the electron density of states. Then, equating $N$ and $P$ under the charge neutrality condition yields the following relation between $E_{fc}$ and $E_{fv}$:

$$e^{E_{fc}/kT} + 1 = (e^{E_{fv}/kT} + 1)^D \tag{21}$$

The gain spectrum of the strained quantum well is thus

$$g(E) = \xi\rho_r\left(\frac{1}{1 + e^{(E_c - E_{fc})/kT}} + \frac{1}{1 + e^{(E_v - E_{fv})/kT}} - 1\right). \tag{22}$$

It can be easily seen that the maximum of this gain curve occurs at $E = 0$, i.e., $E_c = E_v = 0$. Thus, the peak gain $g_p$ is given by

$$g_p = \xi\rho_r\left(\frac{1}{1 + e^{-E_{fc}/kT}} + \frac{1}{1 + e^{-E_{fv}/kT}} - 1\right), \tag{23}$$

which, using Eq. (21), can be converted to a relation that contains only a single variable, the electron quasi-Fermi level:

$$g_p = \xi\rho_r\left(\frac{1}{(e^{E_{fc}/kT} + 1)^D} + \frac{1}{e^{E_{jc}/kT} + 1} - 1\right). \tag{24}$$

It has been shown that the carrier lifetime in a quantum well structure is almost inversely proportional to the carrier density $N$, with the experimentally obtained dependence being $\tau_s = (2.5 \times 10^9\ \text{s} - \text{cm}^{-3})/N$ [13], from which the injection carrier density can be found for a given $N$. Using Eqs. (20) and (24), one can plot the maximum gain versus the injected current density, as shown in Fig. 8, for various values of $D$. It will be assumed throughout this calculation that the electron mass is unchanged, while $D$ is varied by varying the hole mass. The current density where $g_m$ crosses zero is the transparency level, and it decreases with decreasing $D$, as expected. The maximum achievable gain is also reduced for small $D$, but this won't be felt until one reaches injection current densities of approximately 2.5 kA/cm$^2$ or higher.

## 6. LOWER DIMENSION QUANTUM STRUCTURES: QUANTUM WIRES AND DOTS; AND VERTICAL EMITTING LASERS

The combination of the topics of lower dimension quantum-confined lasers and vertical emitting lasers seems odd at first glance, but as far as obtaining a low threshold current is concerned, they have many conceptual similarities. As discussed in the very beginning of this chapter, if reducing the thickness of a quantum well beyond a certain point does not result in further reduction of threshold current, then we try shrinking the lateral dimensions, which eventually results in quantum wires and quantum dots. Simple physical scaling considerations yield threshold currents in the microampere range for quantum wire and dot lasers. The basic principles are the same as in quantum

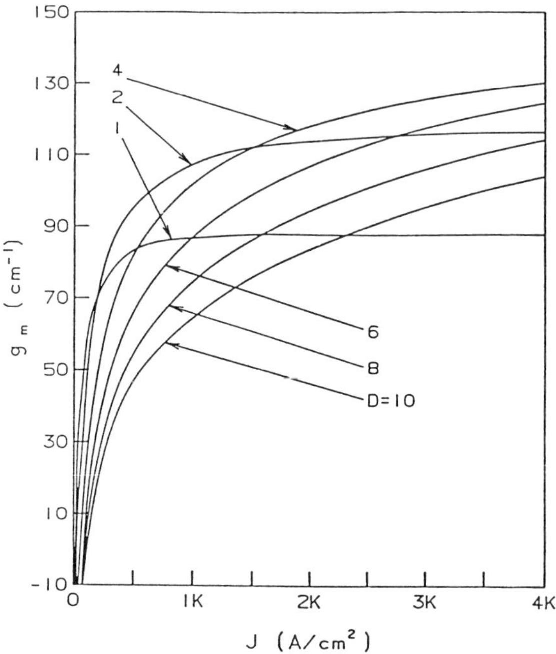


**Fig. 8.** Computed mode gain versus current density for a single quantum well with the electronic density of states equal to that of a regular single quantum well, while the hole density of states is $D$ times that of the electronic density of states.

well lasers: As one shrinks the physical size of the active region, the transparency current density scales down proportionally but so does the optical confinement factor and hence the modal gain. A schematic diagram of a quantum wire laser is shown in Fig. 9, where the wires are placed perpendicular to the laser cavity to take advantage of a larger gain for light polarized in a direction parallel to it. It is obvious that the active volume is much reduced by this arrangement, but the gain experienced by a photon traveling one round trip inside the cavity is also proportionally reduced. The concern is whether one can make an optical cavity with such a low loss that the drastic reduction in modal gain becomes an irrelevant matter. There are thus two issues to consider for lower dimension quantum-confined lasers as far as lasing threshold is concerned: first, the technology needed for the fabrication of nanostructures (for a review of fabrication technologies, see the review paper [14] and Chapter 10, this book); second, the ability to take advantage of the small physical dimensions through the implementation of ultralow loss laser cavities [15–17].

These two issues lead naturally to vertical emitting lasers, since they are

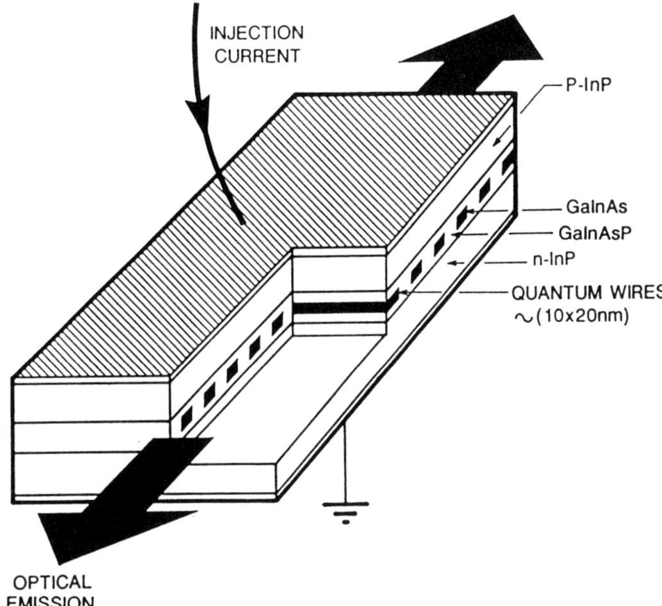


**Fig. 9.** Schematic diagram of a quantum wire laser (from [14]; © *IEEE* 1991).

exactly the two issues that confront vertical emitting lasers. In a vertical emitting cavity structure (Fig. 10) [18, 19], the optical beam traverses a small number of quantum wells, from which the optical gain is derived, and hence the optical confinement is extremely small in the longitudinal direction. Looking sideways at the cross section, one can see many similarities between

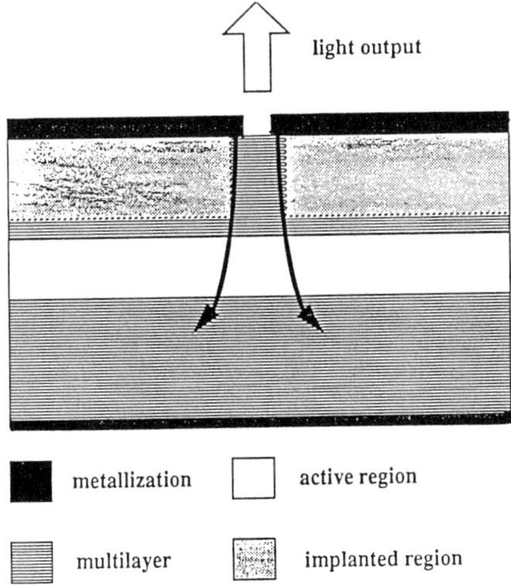


**Fig. 10.** Schematic diagram of a vertical emitting laser (from [20]).

the quantum wire laser in Fig. 9 and the vertical emitting laser in Fig. 10. In both cases, a very low loss cavity with superior high reflectivity coatings on both mirrors are needed. Both of these structures have now achieved sub-milliampere lasing threshold [16, 18, 19]. This is still very far from the fundamental limit of microampere threshold currents, and many technological hurdles need to be removed before that can be achieved. The fundamental principles involved in these efforts, though, are those outlined in this chapter.

## 7. EXPERIMENTAL LOW THRESHOLD LASERS

As mentioned in the beginning of this chapter, the experimental results discussed here are by no means exhaustive but are representative and informative, and they serve to illustrate some of the simple principles outlined in the preceding. For the most recent experimental results on low threshold quantum well and strained layer quantum well lasers, see Chapter 8 and 9 of this book.

A vivid illustration of the connection between transparency current density, cavity loss, and differential quantum efficiency, discussed in Section 4, is shown in the following experiment.

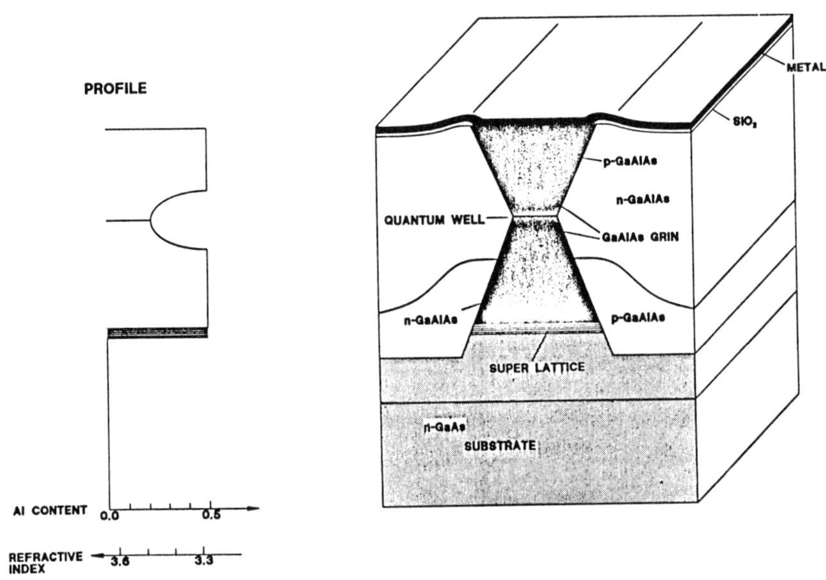


**Fig. 11.** Schematic diagram of the laser used in the sub-milliampere threshold experiment.

One starts out with single quantum well lasers in a graded index separate confinement structure (GRIN SCH SQW), fabricated into a buried heterostructure geometry for lateral carrier and optical confinement with a lateral dimension of approximately 1.0 $\mu$m (Fig. 11). The quantum well thickness is 100 Å, and the graded index structure is 0.4 $\mu$m thick, placed symmetrically about the quantum well. The net optical gain of the laser at various bias currents are derived from the spectral modulation depth of the laser emission when biased below lasing threshold [21, 22]. The intrinsic optical gain was obtained by adding to the measured net gain the internal loss and mirror loss of the device. The latter two parameters were obtained independently by measurement of the differential quantum efficiencies of the lasers under different reflectivity coatings. Starting with an uncoated device with mirror reflectivities of 0.3 and with almost identical power emission from both facets, a high reflectivity coating is applied to one facet. The ratio of optical emission from the front and back facets gives an accurate value for the actual reflectivity of the coated facet, which can be used, together with the total differential quantum efficiency (from both front and back facets) of the devices as compared with that before coating, to obtain the internal loss of the device. The process can be repeated by applying additional layers of coatings on both facets. Measured values for the internal loss lie in the range of 3–9 cm$^{-1}$. This loss includes the free carrier loss, which should be very small (1–2 cm$^{-1}$) in a GRIN SQW structure due to a low optical confinement, together with the lateral waveguide scattering loss in the buried optical guide. The intrinsic optical gain is then obtained by adding the internal loss to the net gain as computed from the spectral modulation data.

The experimental intrinsic optical gain in the three devices, with cavity lengths of 120, 200, and 250 $\mu$m, are shown in Fig. 12a. The injection current at which the intrinsic optical gain goes to zero represents a basic limit for the lowest threshold current attainable in that device. The black circles are data points obtained from spectral modulation measurements, while the blank circles are the actual threshold currents of the devices under different high reflectivity coatings. When the injection currents of the three lasers are scaled with the length of the devices, the data in Fig. 12a can be replotted in Fig. 12b to give the normalized gain curve for this type of laser. Data for the three independent devices matches extremely well, as indicated by the lack of substantial scattering of the data points in Fig. 12b. It shows that, for a 100 $\mu$m long device with a typical active region width of 1 $\mu$m, the lowest threshold that can be obtained is 0.3 mA. This corresponds to a transparency injection current density of 300 A cm$^{-}$. It should be noted that this value is measured in a realistic device that includes leakage currents

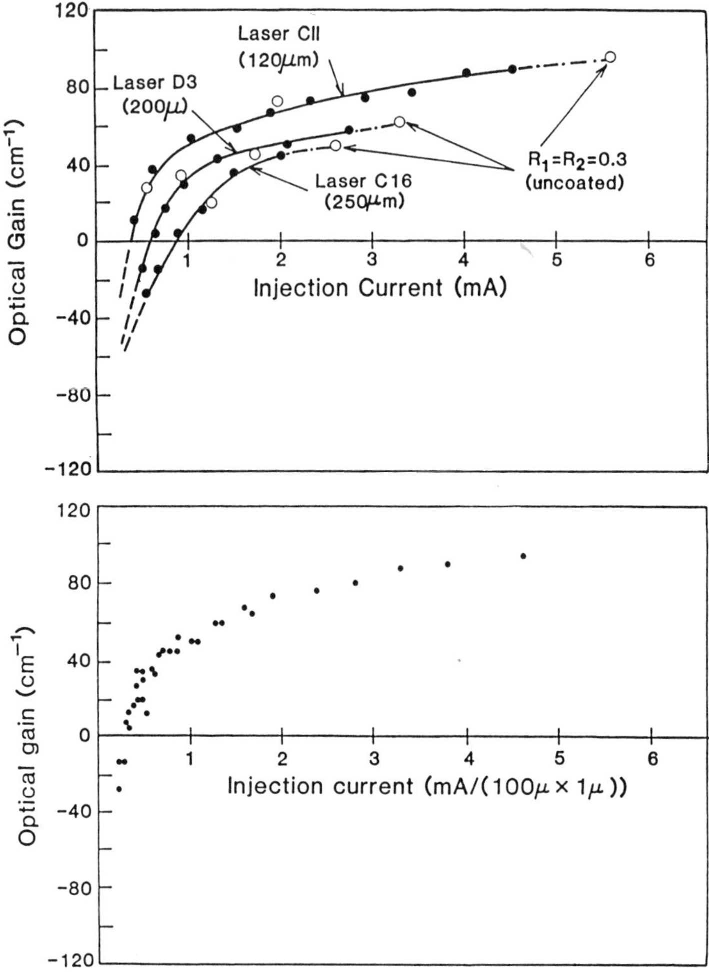


**Fig. 12.** (a) Measured intrinsic optical gain of SQW BH lasers versus injection current; (b) intrinsic optical gain versus injection current density for the three lasers in (a).

and interface recombination currents at the numerous carrier-confining heterojunctions. A shorter or narrower device will lead to an even lower threshold, since it scales with the area of the device. Lasing thresholds in the microampere range will be possible by using a very small lateral dimension—i.e., a quantum wire.

The measured light–current characteristics from both facets of a 120 $\mu$m long SQW BH laser is shown in Fig. 13 [3]. The lasing threshold was a

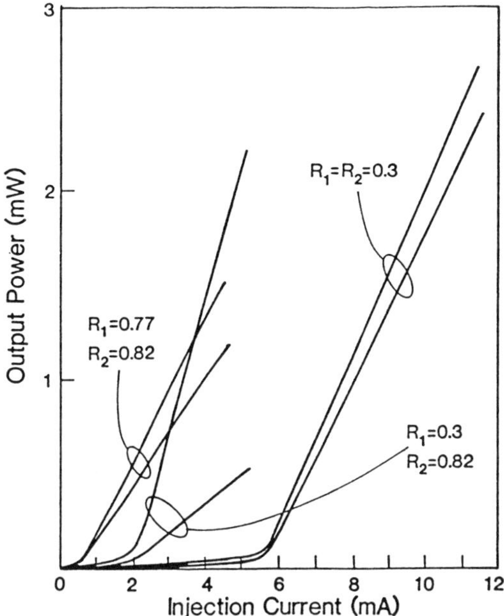


**Fig. 13.** Light versus current characteristics of a 120 μm long laser with different facet reflectivities.

relatively high 5.5 mA without coating, and it is reduced by 10-fold to 0.55 mA after both facets are coated to approximately 80%. The threshold data for different coatings for this device are represented as blank circles in the curve labelled by "laser c11" in Fig. 12a. One should also note that the differential quantum efficiency of the final coated device is only slightly degraded from that of the uncoated device.

While the experiment just described concerned only one particular type of device, the results illustrate the main principles involved in low threshold quantum well lasers, which should be applicable to any device geometry. On narrowing the lateral dimension of the waveguide, one can project from the preceding results the threshold performance of a quantum wire laser. On scaling the cavity length and mirror reflectivities, one can project the performance of vertical emitting lasers. Implementation of these structures is not a trivial matter though, and it requires both superb materials and fabrication technology. These are still subjects of intense interest today, and any results quoted here will likely be superceded by better results when this article sees print. Nevertheless, one should note the progress offered by "bandstructure engineering," in particular strained layer quantum well devices and those grown on misoriented substrates, which have produced the lowest threshold current density to date [23–28]. These subjects are discussed in considerable detail in Chapters 1, 7, and 8. Complementary to these new materials are new structures aimed at obtaining the lowest possible threshold, those of the quantum wire [14] and vertical emitting [18, 19] geometry. It is expected from these results that lasers with thresholds in the tens of microamperes if not in the microampere range will become a reality in the not too distant future.

## 8. FAST DIGITAL SWITCHING

Although the main topic of modulation of quantum well lasers will be discussed in the following chapter, the aspect of fast digital switching is discussed here since it pertains more to the low threshold characteristics and physical scaling effects than the intrinsic quantum effects. The latter pertains to small signal modulation and other dynamic effects such as mode-locking and gain-lever, which will be described in the following chapter. Among the practical aspects of low threshold lasers, power consumption is perhaps the one that is given the most attention. This issue becomes very important in integrated optoelectronic circuits, since power dissipation is one major limiting factor in the practical realization of these circuits. On the other hand, one might ask why, other than for academic purposes, should one continue to pursue microampere threshold lasers given the fact that hundred-microampere threshold lasers are now already feasible. In other words, hundred-microampere threshold lasers are already of sufficiently low threshold as far as power dissipation is concerned that it may no longer be a dominating factor, and further lowering of the threshold is perhaps of little practical value. This assertion may (or may not) be true (it obviously depends on the design of the circuit) when considering quiscent power dissipation, but is certainly *not* true when considering high speed on–off digital switching. It will be seen from the discussion following that if one desires fast on–off digital switching of these lasers (in the Gbit/s range), then the power requirement for the driver circuit may be excessive even with hundred-microampere lasers, and further lowering of the threshold is certainly desirable.

In conventional digital transmitter design, the laser must be biased slightly above threshold, and that bias point must be stabilized using active monitoring and feedback. Failing to do so, the bias point might drift upward, which reduces the optical extinction ratio and results in added system penalty; or it might drift downwards below threshold, which results in the following: (1)

a substantial delay exists between the onset of lasing and the current pulse; (2) severe relaxation oscillation distorts the optical pulse; and (3) pattern effect occurs as a result of charge left over from the previous pulse(s).

Other than the drawbacks (1)–(3), which are indeed severe, there are many reasons why it is desirable to operate the laser at zero bias current. First, it will not be necessary to provide optical monitoring and feedback, and this is a big advantage in integrated optoelectronic circuit design. Secondly, the extinction ratio is always 100%. It turns out that a low threshold quantum well laser can operate under the zero-current bias mode *without* the preceding three drawbacks. This constitutes a major reason why low threshold quantum well lasers are considered so important in the field of integrated optoelectronics. (The other factor is obviously power dissipation.)

First, consider the issue of switch-on time delay. The time delay in the onset of lasing is determined mainly by the time required by the injected carrier to reach the threshold density. Assuming a constant carrier lifetime, the charging time is

$$t_{\mathrm{d}} = \tau_{\mathrm{s}} \ln(I/I - I_{\mathrm{th}}) \rightarrow \tau_{\mathrm{s}} \frac{I_{\mathrm{th}}}{I} \tag{25}$$

for large $I$, where $I$ is the pump current, $I_{\mathrm{th}}$ is the cw threshold current, and $\tau_{\mathrm{s}}$ is the carrier lifetime. For example, if one requires, for multigigabit digital transmission, a delay time of no more than 50 ps and taking $\tau_{\mathrm{s}} = 2$ ns, a laser with a cw threshold of 25 mA will require a drive pulse of amplitude equal to 1 A! A sub-milliampere laser can achieve this time delay with a pulse amplitude of less than 40 mA, compatible with logic-type drive levels.

The light–current characteristic of a BH GRIN SCH SQW laser with a cavity length of 250 $\mu$m and a threshold of 0.95 mA is shown in Fig. 14. The response of this laser to a direct pulse current drive *without* bias is shown in Fig. 15. The drive pulse amplitude is controlled by a broadband continuously variable RF attenuator, while the pulse shape is maintained in all the measurements. In Fig. 15 the drive pulse is shown in the top picture, and the bottom picture is a superposition of the laser responses to five different pulse amplitudes. Taking into account that diode current turns on at 1.4 V, the five drive levels correspond to currents of 7.4, 10.8, 16.8, 23, and 30 mA, respectively. Switch-on delays are very evident at the lower drive levels.

To accurately measure the switch-on delay down to $<50$ ps time scale, it is assumed that the laser turns *off* without delay at the termination of the current pulse. This assumption is well supported by the observed data described in the following. The switch-on delay is then given by the difference in pulse width between the input and optical pulse. Since the drive pulse has

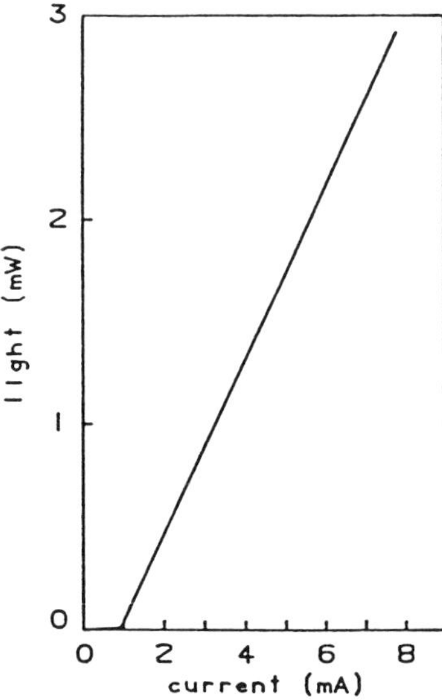


**Fig. 14.** Cw light versus current characteristic of the BH GRIN SQW laser used in the switching experiment.

a finite risetime of approximately 100 ps and it takes 1.4V to turn on the diode current, the actual pulse width of the drive current is slightly different for different drive levels and must be taken into account in the measurement. The result is shown in Fig. 16, which demonstrates that this laser has a switch-on delay of between 20–50 ps with a modest 30 mA pulse drive. The laser was driven with a current amplitude of up to 150 mA without damage, at which point the input and laser output pulse widths as measured on the oscilloscope were identical. This provides strong support for the preceding assumption that the laser output turns off simultaneously with the current pulse, an assumption that is the basis for the preceding switch-on delay measurements.

Due to the high carrier concentration involved in a quantum well structure, bimolecular recombination should be taken into account to give a more accurate prediction of the time delay. The time delay $t_d$ can be calculated from the carrier equation $\dot{n} = J/ed - Bn^2$, which gives

$$t_d = \sqrt{\frac{ed}{JB}} \tanh^{-1} n_{th} \sqrt{\frac{Bed}{J}}, \tag{26}$$

where $n_{th}$ is the carrier density for the quantum well gain to attain threshold and $d$ is the quantum well thickness. Given device parameters, $n_{th}$ can be

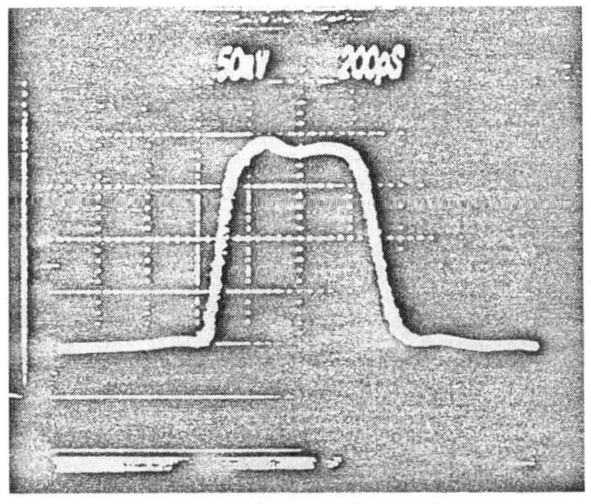


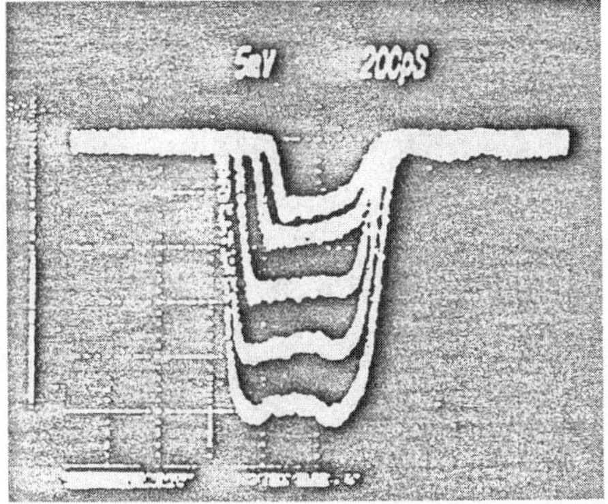


**Fig. 15.** Response of a BH GRIN SQW laser with 0.9 mA threshold to direct pulse drive *without* bias. Top trace: pulse input. Bottom traces: optical response for pulse current = 7.4, 10.8, 16.8, 23, and 30 mA. Hor.: 200 ps/div.

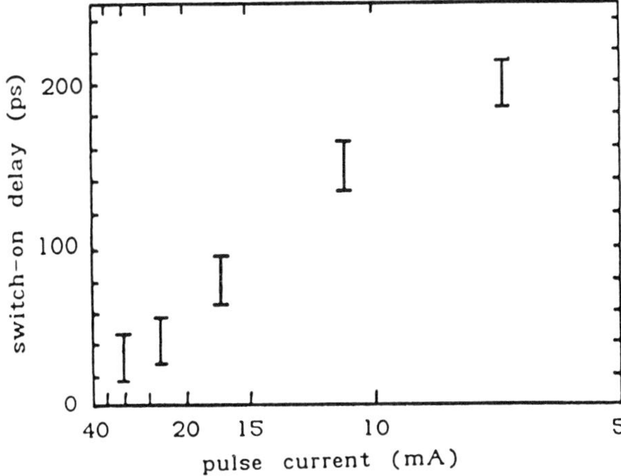


**Fig. 16.** Measured switch-on delay versus pulse current amplitude.

calculated using previously computed quantum well gains [5]. Assuming an internal loss of 15 $cm^{-1}$, the theoretical switch-on delay versus cavity length of a BH GRIN SCH SQW laser with 100 Å well width is plotted in Fig. 17 for various facet reflectivities, assuming a drive current pulse of 20 mA. By using very short cavities with high reflectivity coatings, it is possible to obtain delays of less than 20 ps.

A striking feature of the laser responses in Fig. 15 is the total absence of relaxation oscillation despite being turned on from zero bias. This is of course a very desirable feature in digital transmissions. This absence of ringing cannot be a result of parasitics, since it would have slowed down the rise and fall times in the laser output. Nor can it be explained by the spontaneous emission factor [29] alone without adopting unreasonable values (>0.1). It was not recognized at the time, but in retrospect this was perhaps the first indication that quantum well lasers have an *abnormally* high level of gain compression, a factor that contributes to overdamping of the modulation response and is responsible for the limitations in the modulation bandwidth of injection lasers [20, 30].

Next, consider the issue of pattern effect. In situations where the laser is modulated by a real digital pulse stream, the starting condition for each pulse is affected by the charge left over from the previous pulse, if one existed [31, 32]. The carrier density typically stays at slightly below the threshold

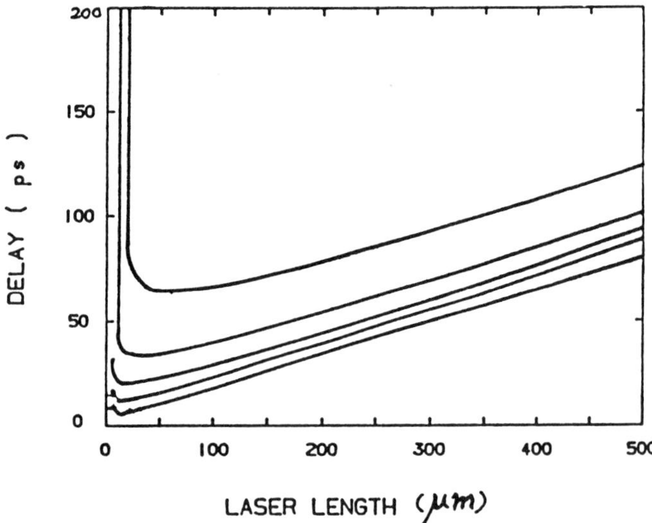


**Fig. 17.** Calculated switch-on delay versus cavity length, for facet reflectivities $R_1 = 0.9$ and $R_2 = 0.1, 0.3, 0.5, 0.7$, and 0.9, assuming a drive current pulse of 20 mA.

level following an optical pulse, and it decays with a time constant equal to the carrier lifetime. The second of two consecutive optical pulses will have a higher amplitude—the pattern effect—and the ratio can be estimated, using simple considerations, to be

$$\frac{I_{\text{pulse}}}{I_{\text{pulse}} - I_{\text{threshold}}}. \tag{27}$$

This ratio will exceed unity by only a few percent at most for a sub-milliampere threshold laser pumped by current pulses of 20–40 mA. This has been confirmed experimentally by driving the laser with two consecutive current pulses separated by as little as 200 ps (limited by the risetime of the pulser).

What was demonstrated in the preceding was a laser diode that can be modulated directly with logic-level drives without the need for current prebias. The laser exhibits truly *on–off* switching with switch-on delays of $<50$ ps, enabling multigigahertz digital modulation rates. The common relaxation oscillation resulting from turning on a laser from below threshold is not observed, and pattern effect does not exist. All of these performance factors, except the absence of relaxation oscillation, can be traced to the low threshold current and hence the physical scaling effect. The heavy damping, on the other hand, has its origin in quantum effects, and it will be discussed in the following chapter (Chapter 5).

## REFERENCES


1. Agrawal, G. P., and Dutta, N. K. (1986). *Long wavelength semiconductor lasers* Van Nostrand Reinhold.
2. Arakawa, Y., and Yariv, A. (1986). *IEEE J. Quantum Electron.* **QE-22**, 1887.
3. Lau, K. Y., Bar-Chaim, N., Derry, P. L., and Yariv, A. (1987). *Appl. Phys. Lett.* **51**, 69.
4. Hayakawa, T., Suyama, T., Kakahashi, T., Kondo, M., Yamamoto, S., and Hijikata, T. (1987). *IEDM 1987*, Washington, DC.
5. Tsang, W. T. (1981). *Appl. Phys. Lett.* **39**, 786.
6. Adams, M. J. (1986). *Electron. Lett.* **22**, 250.
7. Yablonovitch, E., and Kane, E. O. (1986) *IEEE J. Lightwave Tech.* **LT-4**, 504.
8. Obsborn, G. C. (1986). *IEEE J. Quantum Electron.* **QE-22**, 1677.
9. Schirber, J. E., Fritz, I. J., and Dawson, L. R. (1985). *Appl. Phys. Lett.* **46**, 187.
10. Houng, M. P., Chang, Y. C., and Wang, W. I. (1988). *J. Appl. Phys.* **64**, 4609.
11. Iye, Y., Mendez, E. E., Wang, W. I., and Esaki, L. (1986). *Phys. Rev. B* **33**, 5854.

12. SUEMUNE, I., COLDREN, L. A., YAMANISKI, M., and KAN, Y., (1988). *Appl. Phys. Lett.* **53**, 1378.
13. ARAKAWA, Y., SAKAKI, H., NISHIOKA, M., YOSHINO, J., and KAMIYA, T. (1985). *Appl. Phys. Lett.* **46**, 519.
14. KAPON, E. (1991). *Proceedings of IEEE.*
15. KAPON, E., SIMHONY, S., BHAT, R., and HWANG, D. M. (1989). *Appl. Phys. Lett.* **26**, 2715.
16. KAPON, E., SIMPHONY, S., HWANG, D. M., COLAS, E., and STOFFEL, N. G. (1990). *12th international semiconductor laser conference*, Davos, paper **F-2**.
17. SIMPHONY, S., KAPON, E., COLAS, E., BHAT, R., STOFFEL, N. G., and HWANG, D. M. (1990). *Photon. Technol. Lett.* **2**, 305.
18. JEWELL, J. L., SCHERER, A., MCCALL, S. L., LEE, Y. H., WALKER, S., HARBISON, J. P., and FLOREZ, L. T. (1989). *Electron. Lett.* **25**, 1123.
19. GEELS, R. S., and COLDREN, L. A. (1990). *Appl. Phys. Lett.* **57**, 1605.
20. BOWERS, J. E., and POLLACK, M. A. (1988). *Optical fiber telecommunications*, Chapter 13 (MILLER, S., and KAMINOW, I., eds.) Academic Press.
21. DUTTA, N. K. (1983). *IEEE J. Quantum Electron.* **QE-19**, 794.
22. HAKKI, B. W., and PAOLI, T. L. (1975). *J. Appl. Phys.* **46**, 1299.
23. CHEN, H. Z., GHAFFARI, A., MORKOC, H., and YARIV, A. (1987). *Electron. Lett.* **23**, 1334.
24. ENG, L. E., CHEN, T. R., SANDERS, S., ZHUANG, Y. H., ZHAO, B., YARIV, A., and MORKOC, H. (1989). *Appl. Phys. Lett.* **55**, 1378.
25. BEERNINK, K. J., YORK, P. Y., and COLEMAN, J. J. (1989). *Appl. Phys. Lett.* **25**, 2585.
26. CHOI, H. K., and WANG, C. A., (1990). *Appl. Phys. Lett.* **57**, 321.
27. TEMKIN, H., DUTTA, N. K., TANBUN-EK, T., LOGAN, R. A., and SERGENT, A. M. (1990). *Appl. Phys. Lett.* **57**, 1610.
28. TANBUN-EK, T., LOGAN, R. A., TEMKIN, H., BERTHOLD, K., LEVI, A. F. J., and CHU, S. N. G. (1989). *Appl. Phys. Lett.* **55**, 2283.
29. BOERS, P. M., and VLAADINGERBROEK, M. T., (1975). *Electron. Lett.* **11**, 206.
30. OLSHANSKY, R., FYE, D. M., MANNING, J., and SU, C. B., (1985). *Electron. Lett.* **21**, 893.
31. LEE, T. P., and DEROSIER, R. M. (1974). *Proc. IEEE* **62**, 1176.
32. DANIELSEN, M. (1976). *IEEE J. Quantum Electron.* **QE-12**, 657.
33. DERRY, P. L., YARIV, A., LAU, K. Y., BAR-CHAIM, N., LEE, K., and ROSENBERG, J. (1987). *Appl. Phys. Lett.* **50**, 1773.
34. LAU, K. Y., WANG, W., XIN, S., MITTLESTEIN, M., and BAR-CHAIM, N. (1989). *Appl. Phys. Lett.* **55**, 1173.

# *Chapter 5*

# DYNAMICS OF QUANTUM WELL LASERS

**Kam Y. Lau**

*Department of EECS*
*University of California at Berkeley*
*Berkeley, California*



## 1. INTRODUCTION

Apart from the obvious advantage of having superior DC characteristics such as an ultralow threshold current, which is the subject of discussion in the last chapter, the other major attraction of quantum well lasers is their predicted advantages in dynamic characteristics. These include a very high modulation speed, a low frequency chirp ($\alpha$-parameter) and a narrow linewidth. While the low frequency chirp and the narrow linewidth are





ISBN 0-12-781890-1

obviously related, both of them result from a small $\alpha$-parameter, which in turn is largely a consequence of a high differential gain—the derivative of the optical gain with respect to carrier density—which is also responsible for the predicted high modulation speed. The high differential gain arises primarily from a modification of the density of states functions and thus can be regarded as a direct consequence of the quantum confinement of the carriers. A factor of 3–4 improvement in modulation bandwidth was predicted for quantum well lasers as compared with bulk lasers, and a further improvement of a factor of 2–3 was predicted for strained quantum well lasers, thus putting the predicted modulation bandwidth of the latter at around 90 GHz. To date, these superior performances have not been realized in practice, and the modulation bandwidth of the best quantum well and strained quantum well lasers are roughly on par with or slightly better than the best bulk lasers, which have a modulation bandwidth in the upper teens and low twenties of gigahertz. It has recently been realized that the differential gain of quantum well lasers might not be as high as previously predicted due to a finite intraband carrier scattering time, which broadens the transition [1]. However, it is a general consensus that, even taking into account a finite scattering time, quantum well lasers should still have an enhanced differential gain over bulk lasers. A consequence of a small but finite intraband scattering time is a small amount of inhomogeneous broadening in the semiconductor gain saturation characteristic, which leads to *gain compression*: a reduction in the optical gain *beyond* that expected from carrier depletion at a high optical power. It is well known that gain compression leads to an upper limit in the modulation bandwidth that is considerably lower than that predicted by simple theory, and this may constitute a fundamental limit to the highest modulation bandwidth possible for semiconductor lasers. Recently, it was realized that the capture of carriers from the separate confinement region into the quantum well, previously thought to be too fast (picosecond range) to have any significance in the direct modulation response of quantum well lasers, can contribute to an enhanced gain compression. Furthermore, diffusion of carriers in the separate confinement region also constitutes a limitation in the modulation bandwidth. Much of the work related to these transport issues is still ongoing at the time this chapter is being written. Nevertheless, a discussion of recent results will be given in this chapter.

Aside from the potential for high modulation bandwidth due to a high differential gain, there is ample evidence that the gain versus carrier density characteristic of quantum well lasers shows a highly saturated behavior. By taking advantage of this characteristic, a number of new dynamic device functions emerge, which include very high frequency mode-locking and gain-lever. These functions have their origin in the unique density of states function of the quantum wells and thus can be considered a manifestation of quantum confinement of the carriers.

## 2. OPTICAL GAIN AND DIRECT MODULATION SPEED

The *differential* optical gain (derivative of gain versus carrier density) has a direct influence on the modulation speed of the laser. To see this, one starts with the basic description of laser dynamics, which involves a pair of rate equations governing the photon and carrier densities inside the laser medium [2–7, 65, 66]:

$$\frac{dN}{dt} = \frac{J}{ed} - \frac{N}{\tau_s} - vg(N)S, \tag{1}$$

$$\frac{dS}{dt} = \Gamma vg(N)S - \frac{S}{\tau_p} + \beta R_{sp}, \tag{2}$$

where $N$ is the carrier density, $S$ is the photon density in a mode of the laser cavity, $\Gamma$ is the optical confinement factor, $J$ is the pump current density, $d$ the thickness of the active region, $\tau_s$ is the recombination lifetime (radiative and nonradiative) of the carriers, $\tau_p$ is the photon lifetime, $g(N)$ is the optical gain as a function of the carrier density, expressed in $cm^{-1}$, $v$ is the group velocity, $\beta$ is the fraction of spontaneous emission entering the lasing mode, $R_{sp}$ is the spontaneous emission rate, and $e$ is the electronic charge. It should be noted that the optical gain $g$ here implicitly represents the gain at the *peak* of the gain spectrum, the quantity $g_p$ in the last chapter. The implicit assumption is that the laser will lase in a longitudinal mode that is always at or near the peak of the gain spectrum, a good assumption for Fabry–Perot lasers with standard cavity lengths ($>100\ \mu m$) but not necessarily so for DFB or very short cavity lasers. Also, since most of the gain calculations and measurements express the result in inverse distance ($cm^{-1}$) and yet in a dynamic situation one almost always deals with gain in terms of inverse time ($s^{-1}$), we designate a new symbol $G$ for the latter. They are simply related by $G(s^{-1}) \equiv vg(cm^{-1})$. These two quantities will be used interchangeably in this chapter. Since the electron density never fluctuates too far from its steady state value, the gain $g(N)$ can be expressed as a linear function of electron density as

$$G(N) = \frac{G_0 + G'n}{1 + \varepsilon S} \tag{3}$$

where $G_0$ is the average optical gain, $G'$ is the differential gain, $n$ is the small signal deviation of the electron density from the steady state value, and $\varepsilon$ is the gain compression parameter, whose origin will be discussed later. The relative amount of gain compression, $\varepsilon S$, is very small compared with 1 even at very high optical power, so that its effect on the DC lasing characteristics is only secondary. Its effect on the dynamic characteristic is, however, profound. To put it simply, this is due to the fact that the dynamics of the laser depends not so much on the absolute value of the gain but on the difference between the gain and the cavity loss. This difference is no more than a few percent of the gain itself, and thus even a very small gain compression translates into a substantial effect. Quantitatively, a small signal analysis of the rate equations gives the following modulation response function for the photon density:

$$s(\omega) = \Gamma G' S_0 (j/ed)/f(\omega), \tag{4a}$$

$$f(\omega) = (i\omega + \beta R_{sp}/S_0 + \varepsilon S_0/\tau_p)(i\omega + 1/\tau_s + G'S_0) + \frac{G'S_0}{\tau_p(1 + \varepsilon S_0)}, \tag{4b}$$

where $s(\omega)$ is the small signal photon density, $S_0$ is the average photon density, $j$ is the small signal modulation current. Equation (4) represents a conjugate pole pair (second order low pass filter) type of response, and under the assumption of a small $\varepsilon$ the resonance frequency is approximately given by

$$\omega_r = \sqrt{\frac{G'S_0}{\tau_p}} \equiv \sqrt{\frac{vg'S_0}{\tau_p}}. \tag{5}$$

The poles of this response function are shown in Fig. 1. As $\varepsilon$ increases, the poles move in the direction shown, eventually reaching the real axis, and split. At this point, the response makes the transition from being critically damped to overdamped, thus reducing the $-3$ dB modulation bandwidth [5, 7–9]. The reduction is more severe at high optical power because of a stronger compression effect, thus negating the advantage of a higher modulation bandwidth at high optical power as Eq. (5) implies. The effect of gain compression on the modulation response and the $-3$ dB modulation bandwidth as a function of $S_0$ are shown in Figs. 2a and b for different compression parameters.

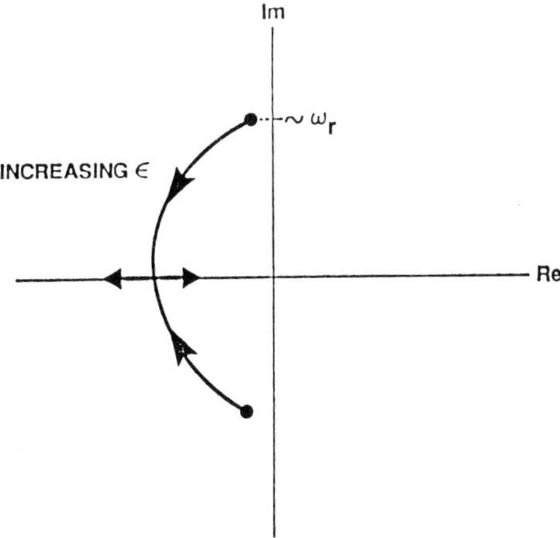


**Fig. 1.** Location of the poles of the modulation response function as gain compression is increased.

## 3. QW AND STRAINED LAYER QW LASERS

We shall revisit the problem of gain compression and transport effects later, but for the meantime consider the potential advantages offered by quantum well lasers due to a high differential gain. A high differential gain should lead to a high modulation bandwidth as the simple formula Eq. (5) implies.

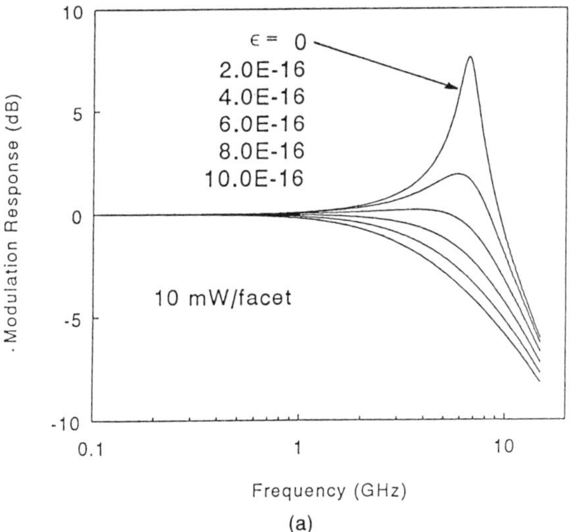


**Fig. 2.** Dependence of the resonance frequency on output optical power for various values of gain compression.

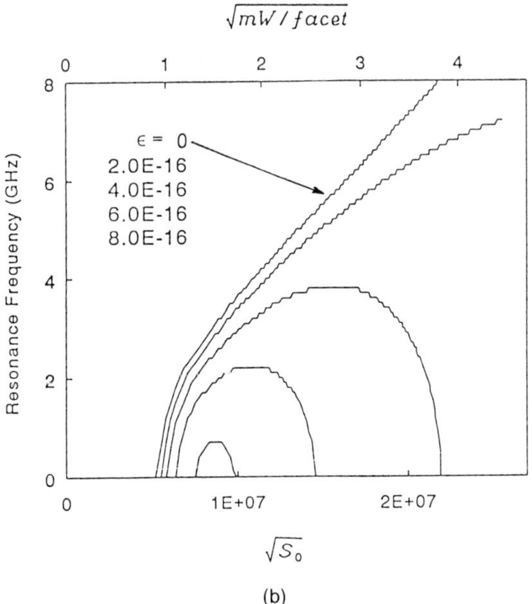


(b)

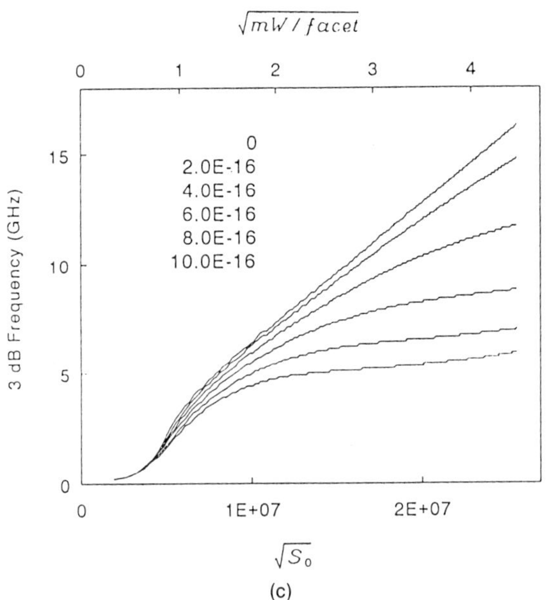


(c)

**Fig. 2.** (*Continued*).

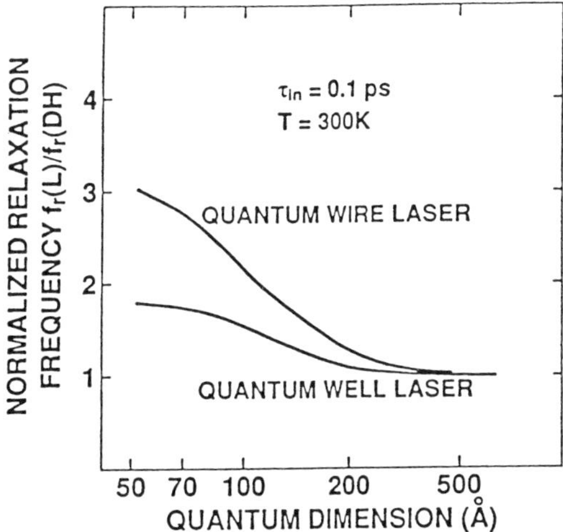


**Fig. 3.** Dependence of the relaxation oscillation frequency on well dimension for quantum well and quantum wire (from [10, 11, 12]).

This is the basis on which predictions of high modulation bandwidths in quantum well lasers are made. Sample results of these predictions are shown in Fig. 3 [10–12] and Fig. 4 [13]. Without taking into account gain compression, these results tell only half the story. Modification of the results in [10–12] after accounting for gain compression is shown in Figs. 5 and 6 [14]. Figure 5 shows how the differential gain is reduced as the optical

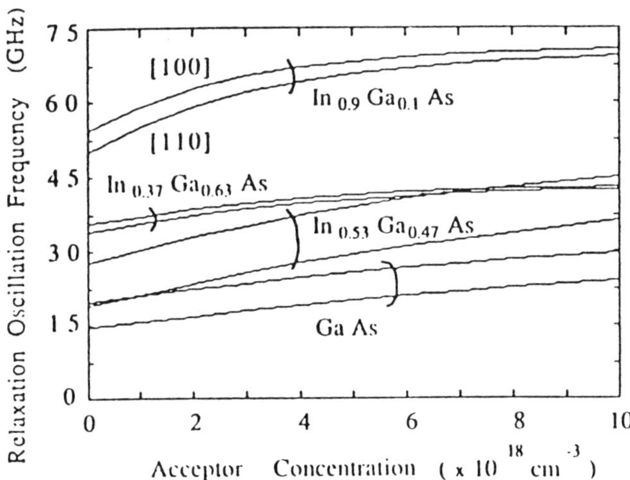


**Fig. 4.** Relaxation oscillation frequency of quantum well and strained quantum well laser diodes (from [13])

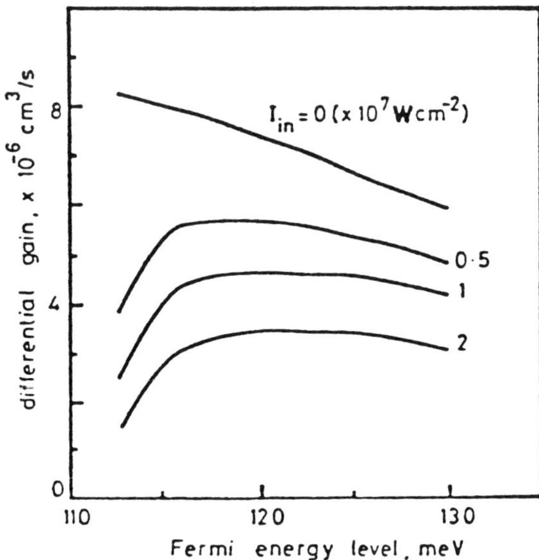


**Fig. 5.** Calculated differential gain of a quantum well as a function of the Fermi level of the electrons (a measure of the electron density) for various optical intensities (from [14]).

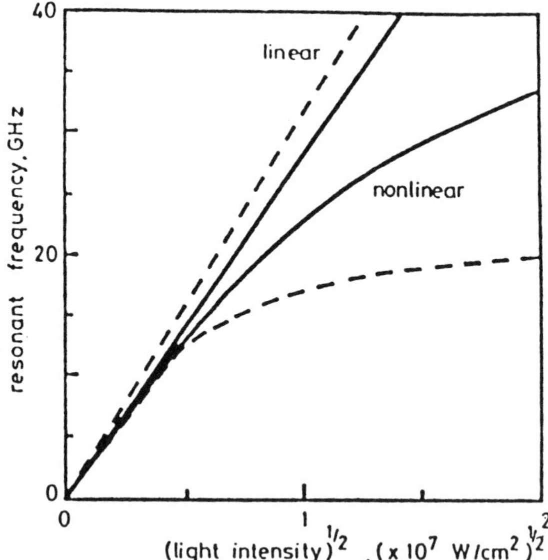


**Fig. 6.** Relaxation oscillation frequency of quantum well lasers with gain compression taken into account (from [14]). Solid lines for 2 wells, dashed lines for 10 wells.

intensity is increased, and Fig. 6 shows the consequent reduction of the relaxation oscillation frequency.

Results such as Fig. 4 induce high hope for strained layer quantum well lasers as high speed optical sources. This is based again on the large $G'$ predicted for these lasers due to a reduced valence band density of states as a result of a reduced valence band effective mass [15–18]. It should be recognized that the maximum optical gain available from transitions between two quantized states (such as the first quantized states of the electrons and holes) is finite, and to first order does not depend strongly on the density of states. It follows that, if one considers transitions involving the first quantized states only (optical gain below 100 cm$^{-1}$) [19], a high differential gain cannot be maintained over an arbitrarily wide range of electron density [20], since at high electron densities the gain must necessarily saturate (as a function of electron density). The advantage of a strained layer quantum well laser as compared with a unstrained quantum well laser is that the former can attain a higher differential gain at a nominal range of optical gain encountered by devices of usual design. To quantify these advantages, we use the expression for relaxation oscillation frequency (without taking into account gain compression) given by Eq. (5), where the photon lifetime is related to the optical gain by

$$\frac{1}{\tau_p} = \Gamma v g \equiv \Gamma G. \tag{6}$$

The relaxation oscillation frequency is

$$f_r = \frac{1}{2\pi} v \sqrt{\Gamma g g' S_0}. \tag{7}$$

The important quantity is thus the product $gg'$. A closed form solution can be obtained for strained layer quantum well [21]:

$$g' = \frac{dg}{dN} = \frac{dg}{dE_{fc}} \frac{dE_{fc}}{dN}, \tag{8}$$

where

$$\frac{dg}{dE_{fc}} = \frac{\xi \rho_r e^{E_{fc}/kT}}{kT} \left( \frac{D}{(e^{E_{fc}/kT})^{D+1}} + \frac{1}{(e^{E_{fc}/kT} + 1)^2} \right), \tag{9a}$$

$$\frac{dN}{dE_{fc}} = \frac{\rho_c}{kT} \left( 1 - \frac{1}{e^{E_{fc}/kT}} \right), \tag{9b}$$

where $\xi$ is a material dependent parameter, $f_{c,v}$ are the Fermi factors, and $\rho_r$ is the reduced density of states. To proceed further, one needs to evaluate the optical gain $g$ (which, as explained in Section 2, is really the peak in the optical gain spectrum) as a function of carrier density. For a constant density of states that is approximately independent of energy (as in the case of strained layer quantum well for energies below 100 meV [13], the quasi-Fermi levels are related to the electron and hole densities by

$$N, P = \rho_{c,v}\{E_{fc,fv} + kT\ln[1 + \exp(-E_{fc,fv}/kT)]\}. \tag{10}$$

Let the hole density of states be $D$ times that of the electron density of states. Then, charge neutrality yields the following relation between $E_{fc}$ and $E_{fv}$:

$$(e^{E_{fc}/kT} + 1)^{1/D} = e^{E_{fv}/kT} + 1. \tag{11}$$

The maximum of the gain curve occurs at band edge transition, and it is given by

$$g = \xi\rho_r\left(\frac{1}{(e^{E_{fc}/kT} + 1)^{1/D}} + \frac{1}{e^{E_{fc}/kT} + 1} - 1\right). \tag{12}$$

It has been shown that the carrier lifetime in a quantum well structure is almost inversely proportional to the carrier density $N$, with the experimentally obtained dependence being $\tau = (2.5 \times 10^9 \text{ s-cm}^{-3})/n$ [22, 23], from which the injection carrier density can be found for a given $N$. Using Eqs. (10) and (12), one can plot the mode gain, given by $g_m \equiv \Gamma g$, versus the injection current density, as shown in Fig. 7, for various values of $D$. It is assumed that the electron mass is unchanged, and different values of $D$ reflect different hole masses. One of the principal effects of inducing strain is to lower the value of $D$. The current density where $g_m$ crosses zero is the transparency level, and it decreases with decreasing $D$, as expected. The maximum achievable gain is also slightly reduced for small $D$. Using the preceding results, we have

$$g' = \frac{\xi}{\rho_c + \rho_v}\left(\frac{\rho_c}{(e^{E_{fc}/kT} + 1)^{1/D}} + \frac{\rho_v}{e^{E_{fc}/kT} + 1}\right). \tag{13}$$

Figure 8 shows plots of $\Gamma g'$ as a function of $g_m \equiv \Gamma g$.

Furthermore, one can use Eq. (7) and the results in Fig. 8 to obtain a plot of $f_r$ versus $g_m$, which is shown in Fig. 9, for different values of $D$, at the same photon density $P_0$. One striking feature of this plot is that the relaxation oscillation frequency of a regular quantum well laser ($D = 10$) is quite independent of $g_m$, while that of a strained layer ($D = 2$ or 1) shows a clear maximum. Thus, by fitting actual modulation data to theory, one can actually obtain an estimation of the value $D$, the ratio of the hole to the

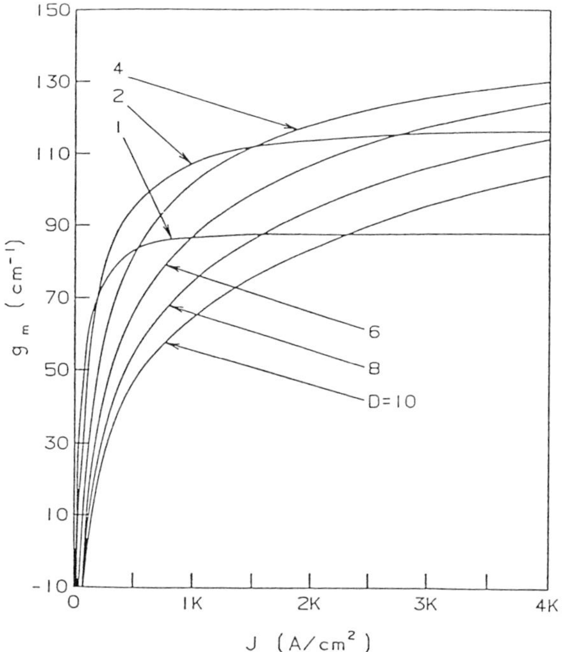


**Fig. 7.** Computed mode gain versus current density for a single quantum well with the electron density of states equal to that of a regular single quantum well, while the hole density of states is $D$ times that of the electron density of states.

electron densities of states, and hence the amount of strain in the quantum well. The plot also shows that the optimum value of $D$ is around 2, at which point the relaxation oscillation frequency can be enhanced by a factor of approximately 1.8 by applying strain. Notice also that, in practical cases of InGaAs over GaAs strain structures, the bandgap is lowered, and hence lasing wavelength is increased by up to 25% over GaAs. Therefore, if one were to compare the modulation speed at the same optical power, the enhancement is even higher.

The general behavior can be observed by using devices of various lengths, and the data points are superimposed on the theoretical plots of Fig. 9 for both single quantum well lasers and strained layer single quantum well lasers, operated at an identical (linear) optical power density of 3 mW/$\mu$m. One striking feature one observes is that the relaxation oscillation frequency of standard single quantum well lasers is relatively indpendent of gain, while that for strained layer quantum well shows an obvious peak at an optical gain of around 60 $cm^{-1}$. This observation is consistent with the theoretical

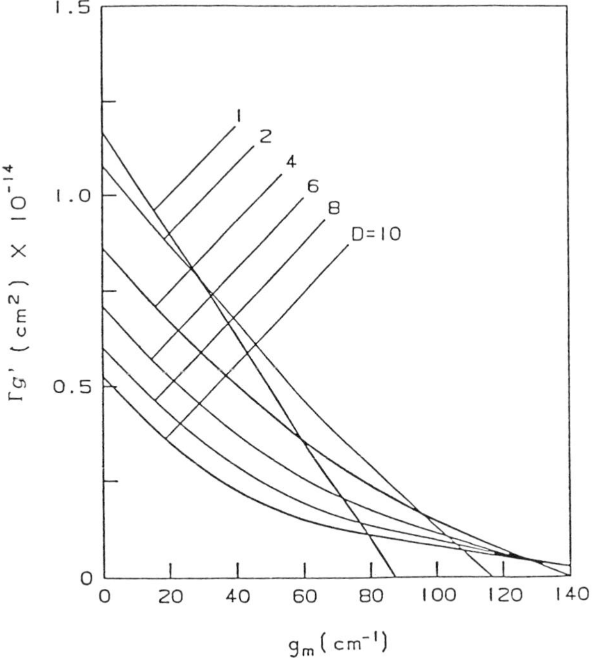


**Fig. 8.** The differential gain versus the peak gain for various values of *D*.

expectations based on a reduced hole mass for the strained layer structure. In fact, the peak of the observed data matches the case for $D \cong 2$. The improvement is about a factor of two, and can be sustained only over a narrow range of optical gain, so that to take advantage of this improvement it is necessary to pay some attention to the device design. These conclusions are corroborated by recent detailed measurements of the optical gain [24] and modulation responses [25] of strained quantum well lasers. Other results of modulation response measurements of quantum well lasers can be found in [26, 27].

Since quantum confinement in general leads to a higher differential gain, it is logical to expect higher order quantum-confined structures (quantum wires and dots) to exhibit even higher differential gains than quantum wells. Theoretical calculations showed that this is indeed the case [28, 29]. Avoiding the difficulty of actually fabricating these structures, their dynamic properties have been studied through a series of experiments in which quantum well lasers are placed in a high magnitude field, which produces quantization of carriers moving perpendicularly to the field

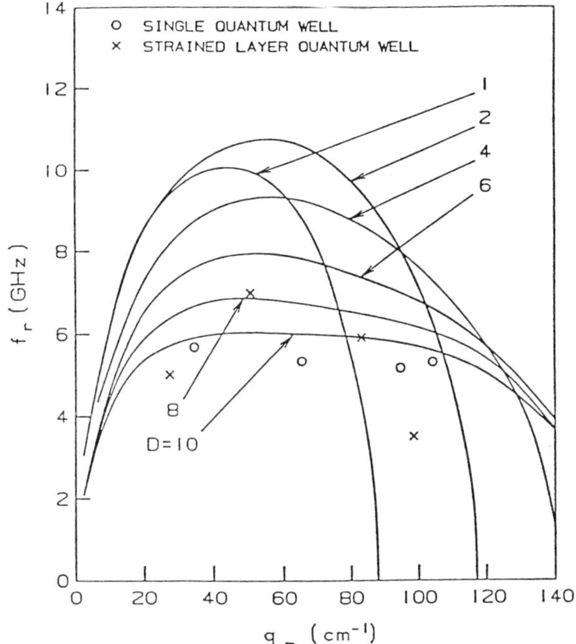


**Fig. 9.** Relaxation oscillation frequency versus peak gain for various values of D.

[10–12, 22, 23, 30]. The problem of fabrication is indeed a severe one, since even a small degree of physical inhomogeneity can lead to the "smearing" of quantum-confined effects to the degree that they become irrelevant [31, 32]. (Note that even if quantum-confined effects are irrelevant, so that little enhancement in the modulation bandwidth results, these lower dimension lasers still possess advantages in terms of a low lasing threshold due to their small physical size alone; see the previous chapter). Setting aside the question of fabrication tolerance, there still remains the question of gain compression and carrier transport in these lower dimension structures. To the extent that these issues are not completely settled in 2D quantum well lasers, it may be premature to judge these issues for lower dimension structures.

## 4. CARRIER TRANSPORT IN QW LASERS

As we saw in the last section, quantum well lasers, in particular strained quantum well lasers, are expected to possess very large modulation bandwidths, owing to a large differential gain [33]. On the other hand, if these large differential gains are inevitably accompanied by a large gain compression, then the predicted advantages could largely be negated [34]. Whether this is true or not depends on the physical origin of gain compression. As mentioned before, a number of such mechanisms have been proposed for the origin of gain compression, including spectral hole burning and carrier heating. Both of these effects are intrinsically present in any semiconductor lasers and will definitely contribute to gain compression, but the question is whether they yield numerical values and trends consistent with experimental observations. To date, as of this writing, it appears that these mechanisms generally give gain compression values too small compared with experiments, and the dependence on various laser parameters, such as device length and threshold density, are not consistent with observations either. These imply that some other physical mechanisms are present in quantum well lasers that dominate over the preceding mechanisms. A new model (the "reservoir" model) recently proposed by Rideout *et al.* [35] considered the consequence of a small but finite capture and escape time of the carriers between the separate confinement (SCH) region and the quantum well(s). These time constants are in the picosecond time scale and have in the past been considered too small to be of any consequence in the direct modulation response of injection lasers up to 100 GHz. What Rideout *et al.* proposed, however, was that these small time constants contribute significantly, in principle, to the gain compression parameter. Furthermore, the model predicts a linear scaling relationship between the gain compression coefficient ($\varepsilon$) and differential gain ($dg/dn$), with the proportionality constant depending on these small transport times. If this model were correct there would exist an ultimate limit in the modulation bandwidth of quantum well lasers despite a high differential gain in these lasers.

While the carrier capture and escape time constants used in the original reservoir model were described only phenomenologically, Nagarajan *et al.* [36] attributed carrier diffusion across the SCH region to be a major contributor to the capture time, and classical thermionic emission as the major physical mechanism for carrier escape from the quantum well, as supported by experimental results from quantum well lasers with various SCH widths. Furthermore, when one considers the rate equations involved in the transport in more detail, it is found that the effect of transport is more than just an increased gain compression—it has other consequences in terms of modifying the relaxation oscillation frequency and the shape of the response function [36, 37]. We shall consider some of these issues in this section.

The reservoir model is illustrated schematically in Fig. 10. The confined

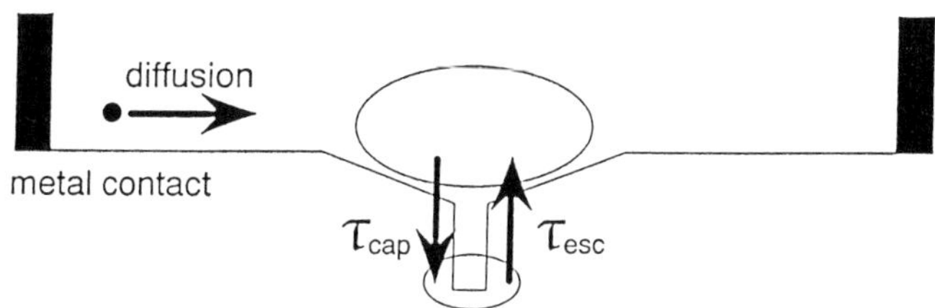


**Fig. 10.** Schematic illustration of the "reservoir" model.

electrons (in the quantum well) and the unconfined electrons in the SCH region are represented by distinct reservoirs. The exchange of carriers between these two reservoirs is described by time constants $\tau_{cap}$ and $\tau_{esc}$, for the capture and escape processes, respectively. The capture and escape processes are intrinsically described by phonon-assisted quantum transitions, but for lasers with a very wide SCH region, the capture time has been attributed primarily to classical carrier diffusion in the SCH region [36]. Rate equations can then be written that describe exchange of carriers between the reservoirs of unconfined and confined carriers, and between the latter and the lasing photons. The small signal rate equations are [35]

$$\frac{ds}{dt} = \Gamma G' S_0 n_a, \tag{14}$$

$$\frac{dn_2}{dt} = \left(\frac{n_3}{\tau_{cap}} - \frac{n_2}{\tau_{esc}}\right) - G' S_0 n_a - \frac{s}{\Gamma \tau_p} - \frac{n_2}{\tau_2}, \tag{15}$$

$$\frac{dn_3}{dt} = -\left(\frac{n_3}{\tau_{cap}} - \frac{n_2}{\tau_{esc}}\right) + j_{pump} - \frac{n_3}{\tau_3}, \tag{16}$$

where $s$, $n_2$, and $n_3$ are the (small signal) photon density, carrier density in the confined quantized states, and carrier density in the unconfined states in the SCH regions (reservoir), respectively, $\Gamma$ is the confinement factor, $G' = v\, dg/dn_a \equiv vg'$, $S_0$ is the photon density at the bias point, $\tau_2$ and $\tau_3$ are the recombination times, $\tau_p$ is the photon lifetime, and $j_{pump}$ is the pumping current density. As in [35], no gain compressison coefficient is explicitly incorporated into the preceding equations. The effective gain compression coefficient $\varepsilon_{cap}$ deduced from these equations is purely due to the carrier capture effect. The total $\varepsilon$ is then the sum of the effective $\varepsilon_{cap}$ and $\varepsilon_{other}$, where $\varepsilon_{other}$ is due to other mechanisms, such as spectral hole burning [38], carrier heating [39], and standing wave effect [40, 41].

The response function $s/j_{\text{pump}}$ in this model was considered by Rideout *et al.* [35]. In the following, this response function will be studied using series expansion. The recombination terms in Eqs. (15) and (16) are neglected for simplicity as the recombination times ($\tau_2$ and $\tau_3$) are typically a few nanoseconds and, therefore, larger than the time scale of interest. From Eq. (16), we obtain an expression for net capture current as follows:

$$\frac{n_3}{\tau_{\text{cap}}} - \frac{n_2}{\tau_{\text{esc}}} = \frac{n_2}{\tau_{\text{esc}}}\left(\frac{-i\omega\tau_{\text{cap}}}{1 + i\omega\tau_{\text{cap}}}\right) + \frac{j_{\text{pump}}}{1 + i\omega\tau_{\text{cap}}}. \tag{17}$$

Following a standard analysis of the rate equations, (14) and (15), we obtain the modulation response:

$$\frac{s}{j_{\text{pump}}} = \frac{\Gamma G' S_0}{-\omega^2(1 + R + i\omega\tau_{\text{cap}}) + (1 + i\omega\tau_{\text{cap}})(i\omega G' S_0 + \omega_{\text{r}}^2)} \equiv \frac{\Gamma G' S_0}{F(\omega)}, \tag{18}$$

where $\omega_{\text{r}} = \sqrt{G' S_0/\tau_{\text{p}}}$ (the conventional relaxation oscillation frequency) and $R = \tau_{\text{cap}}/\tau_{\text{esc}}$. In the limit of extremely small capture time ($\omega\tau_{\text{cap}} \to 0$), $F(\omega) \to F^{(0)}(\omega)$:

$$F^{(0)}(\omega) = -\omega^2(1 + R) + i\omega G' S_0 + \omega_{\text{r}}^2. \tag{19}$$

This expression differs from the conventional modulation response [35] in that the inertia of the relaxation oscillation is increased by the factor $1 + R$, resulting in a lowering of the relaxation oscillation frequency and the damping rate by the same factor [36, 37]. Note that the modification of the resonant oscillation inertia depends only on the ratio $R = \tau_{\text{cap}}/\tau_{\text{esc}}$ and not on the absolute values of $\tau_{\text{cap}}$ or $\tau_{\text{esc}}$.

In the limit of small but finite $\tau_{\text{cap}}$ ($\omega\tau_{\text{cap}} \ll 1$), the denominator of the modulation response can be factored as

$$F(\omega) \approx F^{(1)}(\omega) = \left(1 - \frac{\omega}{\omega_3}\right)\left\{-\omega^2(1 + R) + i\omega G' S_0\left[1 + \tau_{\text{cap}}\frac{R}{1 + R}\left(\frac{1}{\tau_{\text{p}}}\right)\right] + \tilde{\omega}_{\text{r}}^2\right\}, \tag{20}$$

where $\tilde{\omega}_{\text{r}}^2 = \omega_{\text{r}}^2[1 - \tau_{\text{cap}}R/(1 + R)^2 G' S_0]$ and $1/\omega_3 \cong -i\tau_{\text{cap}}/(1 + R)$. A corresponding $F(\omega)$ can be obtained using a differential gain constant $G'_0(1 - \varepsilon S)$ in the conventional rate equations [42]:

$$F(\omega) = -\omega^2 + i\omega G'_0 S_0\left[1 + \frac{\varepsilon}{G'_0}\left(\frac{1}{\tau_{\text{p}}}\right)\right] + \omega_{\text{r}}^2. \tag{21}$$

By comparing Eqs. (20) and (21) and neglecting the pole at $\omega = \omega_2$ in

(20), it is observed that the carrier capture effect is equivalent to a gain compression effect. However, one has to be very careful in making the comparison, because the factor $1 + R$ in the $\omega^2$ term in Eq. (20) reduces the damping rate (the coefficient of $i\omega$ term) as well as the resonance frequency. The effective $\varepsilon_{cap}$ due to the carrier capture time effect can then be expressed as

$$\varepsilon_{cap} = \tau_{cap} \frac{R}{1+R} G_0' = \tau_{cap} \frac{R}{1+R} v(g')_{eff}, \tag{22}$$

where $(g')_{eff} = g'/(1 + R)$ is the effective differential gain. Note that our expression for $\varepsilon_{cap}$ differs from the result in [35] by an additional factor of $1 + R$. The quantity $(g')_{eff}$ can be deduced experimentally from the observed resonant frequency $\omega_r^{ex}$ as $(g')_{eff} = (\omega_r^{ex2}/S_0)(\tau_p/v)$.

The linear dependence between $\varepsilon$ and $g'$ has been experimentally observed recently [37]. The laser used was a tensile strained 1.55-$\mu$m quantum well laser with the following structure: a 200 Å thick $In_{0.41}Ga_{0.59}As$ single quantum well was sandwiched by SCH layers as shown in the inset of Fig. 11. Lateral confinement was provided by a 1.5 $\mu$m wide waveguide buried with semi-insulating InP. The lasers were then cleaved into different cavity lengths with a one-sided HR coating applied on some of the devices. The setup for measuring the modulation response of the lasers was an optical modulation method similar to that reported previously [43]. The laser used

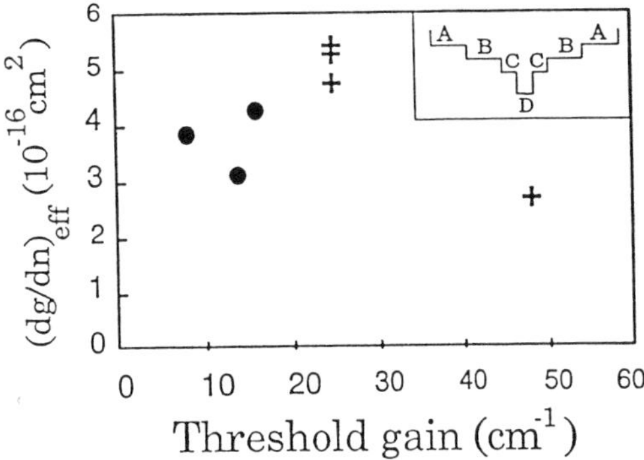


**Fig. 11.** Plot of $(dg/dn)_{eff}$ versus threshold gain using data in Table I. Solid circles indicate lasing in the first quantized state; crosses indicate lasing in the second quantized state. Inset: the separate confinement heterostructure of the InGaAs tensile-strained single QW laser where $\lambda_g$ and $d$ for the regions $A$, $B$, and $C$ are, respectively, 1.0 $\mu$m and 145 nm, 1.1 $\mu$m and 145 nm, and 1.2 $\mu$m and 10 nm, and region $D$ is the $In_{0.41}Ga_{0.59}As$ quantum well.

for modulation injection was a 1.3-$\mu$m diode laser with the 3 dB bandwidth of 10 GHz, at approximately 10 mW output power. The measurement results are summarized in Table I. For both sets of HR coated and uncoated samples, the internal quantum efficiency and internal loss are determined to be around 47% and 2.62 $cm^{-1}$, respectively. The differential gain and the gain compression coefficient ($\varepsilon$) were determined from the measured slopes of relaxation oscillation frequency squared versus power and the measured $K$ factor [44]. The resulting values are also listed in Table I.

One notices from Table I that samples 1–3 lase in the first quantized state, while samples 4–7 lase in the second quantized state; as is evident from the lasing wavelengths, which are separated by 56 meV (corresponding to a 200-Å well) between the two sets of samples. Figure 11 plots $g'$ versus threshold gain. The solid circles indicate lasing in the first quantized state, while the crosses denote lasing in the second quantized state. The transition takes place at a threshold gain of around 20 $cm^{-1}$. Note the abrupt increase in $g'$ at the transition to second quantized state lasing and the subsequent decrease in $g'$ at very high gain. This variation of $g'$ with threshold gain is consistent with a standard model of quantum well gain [19]. The linear dependence of the gain compression coefficient $\varepsilon$ on $dg/dn$ can be visualized by plotting the ratio $\varepsilon/(v_g\, dg/dn)$, which is the $K$ factor less the part involving photon lifetime (i.e., $K' = K - 4\pi^2\tau_p$), versus threshold gain, as shown in Fig. 12. The factor $v$ is the group velocity of light. These experimental results show that $K'$ is basically constant to within $\pm 15\%$, despite the variation in $g'$ by a factor of two among the samples.

**Table I.**

Summary of Measurement Results

| Device number | Cavity length ($\mu$m) | Estimated threshold gain ($cm^{-1}$) | Lasing wavelength ($\mu$m) |
|---|---|---|---|
| 1 | 1000 + HR | 8.0 | 1.54 |
| 2 | 1000 | 13.6 | 1.53 |
| 3 | 500 + HR | 15.7 | 1.52 |
| 4 | 500 | 24.7 | 1.475 |
| 5 | 500 | 24.7 | 1.472 |
| 6 | 500 | 24.7 | 1.471 |
| 7 | 240 | 47.9 | 1.454 |

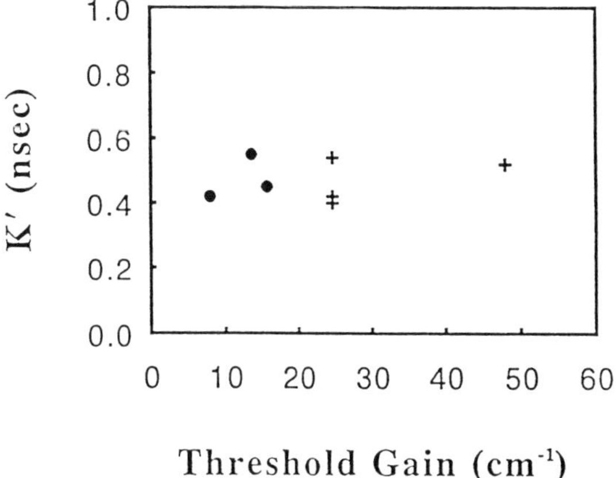


**Fig. 12.** Plot of $K'$ $(= K - 4\pi^2\tau_p)$ versus threshold gain using data in Table I.

As discussed earlier, the total $\varepsilon = \varepsilon_{cap} + \varepsilon_{other}$ is the sum of contributions from carrier capture effect, $\varepsilon_{cap}$, and other mechanisms, $\varepsilon_{other}$. According to Eq. (22), the proportionality constant between $\varepsilon_{cap}$ and $(g')_{eff}$, defined as $B$, is $v_g\tau_{cap}R/(1 + R)$. In Fig. 13, $\varepsilon$ is plotted against $(g')_{eff}$ using the experimental results in Table I. The data are linearly fitted using a least square regression method. The $y$ intercept ($\varepsilon_{other}$) is $0.63 \times 10^{-17}$ cm$^3$ and the slope of the line ($B$) is 0.087 cm. This data implies that $\tau_{cap} = 13$ ps, assuming that $v_g = 10^{10}$ cm s$^{-1}$ and $R = 2$ [35]. This value of $\tau_{cap}$ is comparable with values

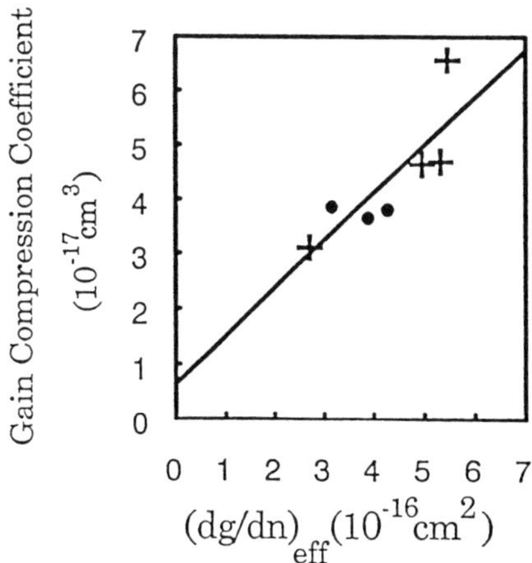


**Fig. 13.** Plot of the $\varepsilon$ versus $(dg/dn)_{eff}$ using data in Table I. The straight line is a least square linear regression fit of the data.

measured independently in other experiments [45]. The value of $\varepsilon_{\text{other}}$ derived from our data is also comparable with the value derived from other mechanisms [38–41]. Dependency of $\varepsilon$ on $g'$ can also be obtained in the spectral hole burning theory for quantum well lasers. A recent calculation [46] illustrated this fact. However, the predicted value of the proportionality constant between $\varepsilon$ and $g'$ in [46] based on spectral hole burning is one order of magnitude smaller than that observed experimentally. This implies that while spectral hole burning produces gain compression characteristics similar to that of the transport mechanism, the former may not be a dominant effect. As of this writing, intense effort is still underway to confirm or clarify these issues.

## 5. QUANTUM CAPTURE AND ESCAPE IN QW LASERS

As illustrated in the last section, carrier capture and escape processes play important roles in limiting the modulation bandwidth of quantum well lasers [35, 36]. Rideout *et al.* [35] expressed the effect using an equivalent gain compression coefficient in the limit of short capture time. Nagarajan *et al.* [36] considered theoretically the carrier diffusion in the separate confinement region and neglected the quantum capture effect. They showed experimentally that the relatively long capture time, which causes the reduction of modulation bandwidth, is predominantly due to carrier diffusion. However, regardless of the physical origins of the capture time, both formulations [35, 36] are mathematically equivalent, with the result that the *capture* time constant ($\tau_{\text{cap}}$) and the *ratio* of the *capture* to *escape* time constant ($R \equiv \tau_{\text{cap}}/\tau_{\text{esc}}$) are two important parameters that directly affect the modulation response. In particular, a large ratio $R$ increases both the damping and inertia [35–37] of the relaxation oscillation and is thus detrimental to high speed modulation. While the diffusion process is well understood, the intrinsic quantum capture and escape rates have been dealt with only through phenomenologically equivalent time constants. In order to have a clearer understanding of the relative importance of diffusion and the intrinsic quantum capture processes, it is necessary to consider the physical nature of the latter. It is well accepted that the dominant processes of transferring carrier into and out of III-V quantum well (QW) are via longitudinal optical (LO) phonon emission and absorption [33, 47]. Based on these phonon-assisted processes, considerable effort has been undertaken (both theoretical and experimental) in establishing the rate of capture into the quantum well, while little attention was paid to the relative magnitude of capture and escape rates [48]. Here, simple analytical expressions for the net carrier capture current are derived [49], which are then used to compute the capture and escape time constants ($\tau_{cap}$ and $\tau_{esc}$) in a unified way and, therefore, are able to calculate the ratio $R$ under different operating conditions and for different QW structures [49].

In quantum well structures, electronic states are classified into confined states $|c\rangle = |E_z^c, \mathbf{k}_{\parallel}^c\rangle$ and unconfined states $|u\rangle = |\mathbf{k}^u\rangle$ where $E_z^c$ is the longitudinal energy of the quantized state in quantum well, $\mathbf{k}^u = (k_z^u, \mathbf{k}_{\parallel}^u)$ is the wavevector of unconfined state, $\mathbf{k}_{\parallel}^c$ and $\mathbf{k}_{\parallel}^u$ are the wavevectors parallel to the QW layers, and $k_z^u$ is the wavevector perpendicular to the layers. Assuming transitions between $|c\rangle$ and $|u\rangle$ are LO phonon-assisted, an electron in $|u\rangle$ is captured into $|c\rangle$ by either emitting or absorbing a LO phonon as shown in Fig. 14. An unconfined electron of energy $E^u$ is scattered into the confined states of constant energy $E^c = E^u \pm E_{ph}$ on Ring-C. We assume a constant $E_{ph}$ ($\sim 36$ meV).

Let $w_1(\mathbf{k}^u)$ be the total departure rate for the electron from an unconfined state $|\mathbf{k}^u\rangle$ to Ring-C by emitting one LO phonon:

$$w_1(\mathbf{k}^u) = \int_{\text{Ring-C}} r(u \to c)\rho(\mathbf{k}_{\parallel}^c)\, d\mathbf{k}_{\parallel}^c, \tag{23}$$

where $r(u \to c)$ is the scattering rate from $|\mathbf{k}^u\rangle$ to $|E_z^c, \mathbf{k}_{\parallel}^c\rangle$ with emission of one LO phonon, and $\rho(\mathbf{k}_{\parallel}^c)$ is the density of states. The total capture current

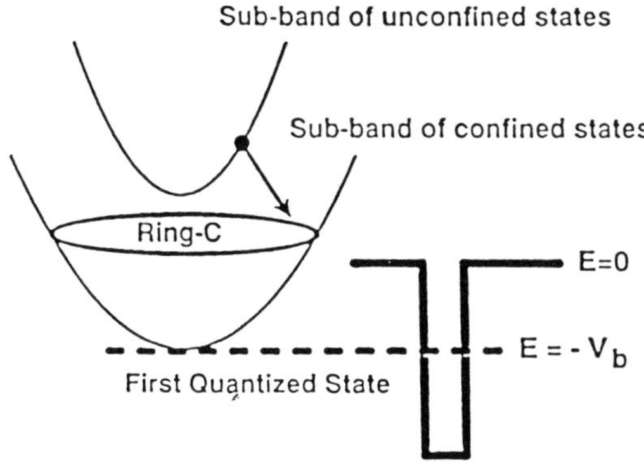


**Fig. 14.** Carrier capture process in QW structures. The parabolic subband associated with each longitudinal energy ($E_z^c$ or $E_z^u$) is due to the transverse energy of the electron $(\hbar \mathbf{k}_{//}^{(c,u)})^2/2m^*$.

density due to LO phonon emission may then be expressed as

$$J_{\text{cap}}^{\text{em}} = 2(n_{\text{ph}} + 1) \int_{-\infty}^{\infty} \frac{dk_z^{\text{u}}}{2\pi} \int_{-\infty}^{\infty} \int_{-\infty}^{\infty} \frac{d^2\mathbf{k}_{\|}^{\text{u}}}{4\pi^2} w_1(\mathbf{k}^{\text{u}}) f^{\text{u}}(1 - f^{\text{c}}), \tag{24}$$

where $f^{\text{c}}$ and $f^{\text{u}}$ are the Fermi–Dirac distributions for the confined and unconfined states, respectively, and $n_{\text{ph}}$ is the phonon occupation number. If we assume that the Fermi level is near or below the band edge of the barrier, so that electrons are localized only in the states near $|\mathbf{k}^{\text{u}}| = 0$, we may neglect the $\mathbf{k}^{\text{u}}$ dependence in $w_1$. We then integrate over $\mathbf{k}_{\|}^{\text{u}}$ in Eq. (24):

$$J_{\text{cap}}^{\text{em}} = n_0(n_{\text{ph}} + 1)n_{\text{ph}}^{-} w_1 \int_0^{\infty} dk_z^{\text{u}} \operatorname{Log}\left\{\frac{1 + \exp[(E_{\text{f}}^{\text{c}} - E_z^{\text{u}} + E_{\text{ph}})/kT]}{1 + \exp[(E_{\text{f}}^{\text{u}} - E_z^{\text{u}})/kT]}\right\}, \tag{25}$$

where $E_{\text{f}}^{\text{c}}$ and $E_{\text{f}}^{\text{u}}$ are the Fermi levels for the confined and unconfined states, respectively, $n_0 = m^*kT/\pi^2\hbar^2$, $m^*$ is the effective mass (the difference of $m^*$ in different layers is neglected), and

$$n_{\text{ph}}^{-} = 1/\{\exp[E_{\text{ph}}/kT] \exp[-(E_{\text{f}}^{\text{u}} - E_{\text{f}}^{\text{c}})/kT] - 1\}.$$

The top of the barrier is taken to be the zero energy reference and, therefore, the longitudinal energy of unconfined state $E_z^{\text{u}}$ equals $(\hbar k_z^{\text{u}})^2/2m^*$. The effective barrier height $V_{\text{b}}$ in Fig. 14 is the energy difference between the top of the barrier and the first quantized state of the QW.

The reverse of the capture process is an escape process in which an electron on Ring-C jumps to $|\mathbf{k}^{\text{u}}\rangle$ by absorbing a LO phonon. Because of the reversal symmetry of the scattering rate, $w_1(\mathbf{k}^{\text{u}})$ is also the total arrival rate to $|\mathbf{k}^{\text{u}}\rangle$ from the confined states on Ring-C. Similarly, we let $w_2(\mathbf{k}^{\text{u}})$ be the total departure rate for the case of capture processes via phonon absorption. Again, because of the reversal symmetry of the scattering rate, $w_2(\mathbf{k}^{\text{u}})$ is also the total arrival rate for escape processes via phonon emission. Using the same approach for $J_{\text{cap}}^{\text{em}}$, we obtain the following expressions for $J_{\text{esc}}^{\text{ab}}$, $J_{\text{cap}}^{\text{ab}}$, and $J_{\text{esc}}^{\text{em}}$:

$$J_{\text{esc}}^{\text{ab}} = n_0(n_{\text{ph}}^{+} + 1)n_{\text{ph}} w_1 \int_0^{\infty} dk_z^{\text{u}} \operatorname{Log}\left\{\frac{1 + \exp[(E_{\text{f}}^{\text{c}} - E_z^{\text{u}} + E_{\text{ph}})/kT]}{1 + \exp[(E_{\text{f}}^{\text{u}} - E_z^{\text{u}})/kT]}\right\}, \tag{26}$$

$$J_{\text{esc}}^{\text{em}} = n_0(n_{\text{ph}} + 1)n_{\text{ph}}^{+} w_2 \int_0^{\infty} dk_z^{\text{u}} \operatorname{Log}\left\{\frac{1 + \exp[(E_{\text{f}}^{\text{u}} - E_z^{\text{u}})/kT]}{1 + \exp[(E_{\text{f}}^{\text{c}} - E_z^{\text{u}} - E_{\text{ph}})/kT]}\right\}, \tag{27}$$

$$J_{\text{cap}}^{\text{ab}} = n_0(n_{\text{ph}}^{-} + 1)n_{\text{ph}} w_2 \int_0^{\infty} dk_z^{\text{u}} \operatorname{Log}\left\{\frac{1 + \exp[(E_{\text{f}}^{\text{u}} - E_z^{\text{u}})/kT]}{1 + \exp[(E_{\text{f}}^{\text{c}} - E_z^{\text{u}} - E_{\text{ph}})/kT]}\right\}, \tag{28}$$

where $n_{\text{ph}}^{+} = 1/\{\exp[E_{\text{ph}}/kT] \exp[(E_{\text{f}}^{\text{u}} - E_{\text{f}}^{\text{c}})/kT] - 1\}$.

The total capture current ($J_{cap} = J_{cap}^{em} + J_{cap}^{ab}$), escape current ($J_{esc} = J_{esc}^{em} + J_{esc}^{ab}$), and net current ($J_{net} = J_{cap} - J_{esc}$) in a typical GaAs quantum well laser structure are calculated from Eqs. (25)–(28) and plotted against $\Delta E_f \equiv E_f^u - E_f^c$ in Fig. 15. Assuming $V_b \gg E_{ph}$, the Ring-C in Fig. 14 for processes via phonon emission is very close to the Ring-C for processes via phonon absorption. Therefore, we take $w_1 \cong w_2 = w_0 = 0.2\ \text{ps}^{-1}$ [33, 47]. In a typical quantum well laser, the net current density $J_{net}$, which determines $\Delta E_f$, is in the range of 100 A/cm$^2$ (near threshold operation) to 2000 A/cm$^2$ (high power operation). We have assumed only one quantized state exists in the QW. Existence of multiple quantized states will complicate the model due to the intersubband transitions and is currently under investigation. We consider the electron capture process only, assuming that the quantum capture of electrons is slower or that the capture process is ambipolar [45, 49]. However, the same formalism can be applied in the study of hole capture.

Using Eqs. (25)–(28), we calculate the capture time constant ($\tau_{cap}$) and the ratio of the capture to escape time constant ($R \equiv \tau_{cap}/\tau_{esc}$)—the two determining parameters of the carrier capture effect on the modulation response—for different effective barrier heights ($V_b$) and carrier densities of confined states of quantum well ($N_c$) in the range $J_{net} = 0$–$2000$ A/cm$^2$. The effective barrier height ($V_b$) is determined by the bandgap offset of QW and the quantum well width. The carrier density of confined states ($N_c$) is actually

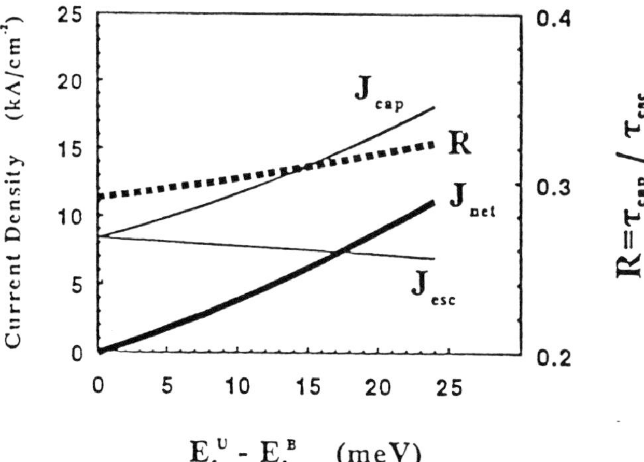


**Fig. 15.** Various current components in a QW structure, with $V_b = 0.15$ eV and $N_c = 4 \times 10^{-12}\ \text{cm}^{-2}$ at $T = 300$ K, and the ratio $R$ are plotted against $\Delta E_f$.

the threshold carrier density, which is determined by the photon lifetime and number of QWs. The two time constants ($\tau_{cap}$ and $\tau_{esc}$) are defined as $\tau_{cap} \equiv \partial J_{net}/\partial N_u$ and $\tau_{esc} \equiv \partial J_{net}/\partial N_c$, in the small signal rate equations of quantum well lasers [37], where $N_u$ is the carrier density of unconfined states. Note that $\tau_{cap}$ is determined by both the capture and escape currents, since $J_{net} = J_{cap} - J_{esc}$. We assume $w_0$ to be constant, independent of $V_b$, in this calculation. Recent calculations [48] indicated that $w_0$ oscillates with well thickness. However, such dependence has not been observed experimentally [47]. Nevertheless, our result for $R$ is independent of $w_0$, since this factor is cancelled out. From our calculations, we find that when $J_{net}$ is changed from 0 to 2000 A/cm$^2$, $\tau_{cap}$ and $R$ are almost constant (within a few percent), even though $J_{cap}$ is much larger than $J_{esc}$ when $J_{net}$ is large. This is illustrated in Fig. 15 by plotting the ratio $R$ versus $\Delta E_f$. The average values of $\tau_{cap}$ and $R$ over the range $J_{net} = 0$–2000 A/cm$^2$ for lasers with different $V_b$ and $N_c$ are listed in Table II.

Simple analytic expressions for $\tau_{cap}$ and $R$ can be obtained based on the fact that when $R$ is comparable with or larger than unity, $(\Delta E_f)_{2000}$ is very

**Table II.**

Calculated Values of $\tau_{cap}$ and $R$ for Different Structural Parameters. The Value of $\Delta E_f$ when $J_{net} \approx 2000$ A/cm$^2$ is denoted by $(\Delta E_f)_{2000}$

| $N_c$ ($10^{12}$ cm$^{-2}$) | $R$ | $\tau_{cap}$ (ps) | $(\Delta E_f)_{2000}$ (meV) |
|---|---|---|---|
| | $V_b = 0.2$ eV | | |
| 2 | 0.003 | 0.30 | 85 |
| 3 | 0.013 | 0.31 | 45 |
| 4 | 0.054 | 0.33 | 18 |
| 5 | 0.19 | 0.36 | 7 |
| 6 | 0.48 | 0.40 | 4 |
| | $V_b = 0.1$ eV | | |
| 2 | 0.15 | 0.36 | 10 |
| 3 | 0.43 | 0.40 | 5 |
| 4 | 0.79 | 0.42 | 3 |
| 5 | 1.12 | 0.43 | 2 |
| 6 | 1.39 | 0.43 | 1 |

small compared with $kT$. We then expand the expression for $J_{net}$ to first order in $\Delta E_f/kT$:

$$J_{net} = \left(\frac{\Delta E_f}{kT}\right) J_0 \int_0^\infty dk_z^u \, \mathrm{Log}\left\{\frac{1 + \exp[(E_f^c - E_z^u + E_{ph})/kT]}{1 + \exp[(E_f^c - E_z^u - E_{ph})/kT]}\right\}, \tag{29}$$

where $J_0 = n_0 n_{ph}(n_{ph} + 1)w_0$. The rate constants $1/\tau_{cap}$ and $1/\tau_{esc}$ can now be expressed as

$$\begin{aligned} \frac{1}{\tau_{cap}} &= \frac{1}{dN_u/dE_f^u} J_0 \frac{1}{kT} I_{log}, \\ \frac{1}{\tau_{esc}} &= \frac{1}{dN_c/dE_f^c} J_0 \frac{1}{kT}\left[I_{log} - \frac{E_f^u - E_f^c}{kT} I_f\right] \\ &\approx \frac{1}{dN_c/dE_f^c} J_0 \frac{1}{kT} I_{log} \qquad \text{as } \Delta E_f \ll kT, \end{aligned} \tag{30}$$

where $I_{log}$ is the integral in Eq. (29), $I_f = \int_0^\infty dk_z^u\,(f^+ - f^-)$, and

$$f^{\pm} = \frac{1}{\{\exp[(\pm E_{ph} - E_f^c + E_z^u)/kT] + 1\}}.$$

The ratio $R \equiv \tau_{cap}/\tau_{esc}$ then becomes

$$R = \left(\frac{dN_u}{dE_f^u}\right) \Big/ \left(\frac{dN_c}{dE_f^c}\right). \tag{31}$$

This factor is a function of carrier densities and densities of states of the confined and unconfined states, which depends on the QW structure. The dependence of $R$ on the quantum well thickness is shown in Fig. 16 for a quantum well laser operating at different threshold carrier densities ($N_c$). Note that the calculation for very wide quantum wells, which probably have a second quantized state, is for reference only, since our model is valid only for wells with one quantized state. This is also the case for data in Table I for large $V_b$.

As discussed earlier, the gain compression coefficient of the laser increases with $R$ and $\tau_{cap}$, and the relaxation oscillation frequency is reduced by a factor $1 + R$. The modulation response of quantum well lasers is thus degraded appreciably when $R \approx 1$. One notes from Table II that $\tau_{cap}$ is relatively constant with respect to the effective barrier height $V_b$ and the carrier density of confined states $N_c$ (which is the threshold carrier density of the laser), while the ratio $R$ varies strongly with $V_b$ and $N_c$. Furthermore, $R$ is rather independent of the net injection current, as is evident from Fig.

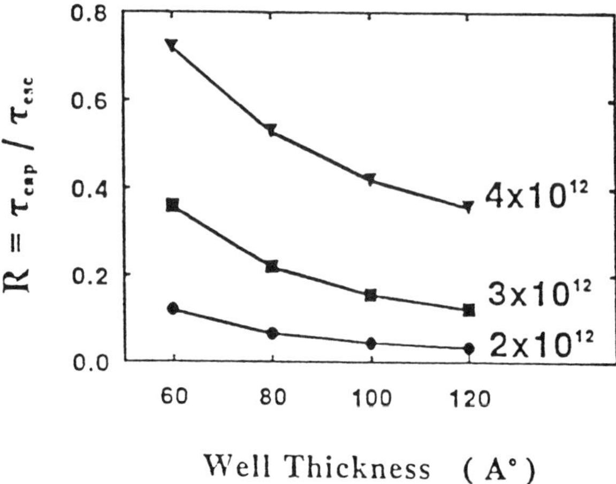


**Fig. 16.** The ratio $R$ versus well thickness of $Al_{0.3}Ga_{0.7}As/GaAs$ QW for $N_c = 2, 3,$ and $4 \times 10^{12}$ $cm^{-2}$.

15. The implications are as follows: (1) A multi–quantum well laser has a smaller $R$ than a single quantum well laser due to the lower threshold *carrier* density of the former; (2) Lasers with narrow or shallow QWs exhibit higher values of $R$ due to a smaller effective barrier height $V_b$; (3) These effects are relatively independent of the bias current (and hence optical power). To completely understand the structural dependence of the modulation response, the effect of carrier diffusion in the separate confinement region of the device has to be included. The preceding calculations consider the quantum capture effect only. They reveal dependences on device structure that are consistent with observed experimental results to date [36, 50, 51], thus suggesting that quantum capture can play a significant role relative to the carrier diffusion effect.

## 6. INTRINSIC EQUIVALENT CIRCUIT OF QW LASERS

Intrinsic equivalent circuit of semiconductor lasers is useful not only for designing electronic circuitry with which the laser operates, but also for furthering the understanding of laser dynamics and noise properties [52–54]. The intrinsic equivalent circuit of a bulk semiconductor laser was developed a decade ago [52, 53]. Similar approaches were applied to quantum well lasers [54]. However, with the recent understanding of the importance of transport in quantum well lasers, as described in the last two sections, it becomes necessary to revise the well-accepted equivalent circuit models to include these effects [55]. It is assumed here that the SCH region in the laser structure is sufficiently narrow so that carrier diffusion is not a significant factor [36]. Using this circuit model, certain qualitative features of the modulation dynamics and their dependence on the capture time constants can be visualized [55].

Let $n_3$ be the small signal carrier density in the SCH region, and let $n_2$ be the density of carriers in the confined states of QW. The small signal rate equation describing the transfer of carriers into and out from QW is [35]

$$\frac{dn_3}{dt} = \frac{i_{\text{inj}}}{qAw} - \frac{n_3}{\tau_3} - \frac{i_{\text{net}}}{qAw}, \tag{32}$$

where $q$ is the electronic charge, $A$ is the junction area, $w$ is the quantum well width, and $\tau_3$ is the spontaneous lifetime. It is assumed that carriers in the SCH region are fed directly by an external injection current $i_{\text{inj}}$. The net current flowing into the QW (denoted by $i_{\text{net}}$) is the difference between the quantum capture and escape currents, $n_3(qAw)/\tau_{\text{cap}} - n_2(qAw)/\tau_{\text{esc}}$, where $\tau_{\text{cap}}$, $\tau_{\text{esc}}$ are the capture and escape time constants, respectively [35].

The external voltage across the device (neglecting parasitics) is given by the difference of the quasi-Fermi levels in the SCH region ($V_3 = E_{\text{f}}^{\text{c}} - E_{\text{f}}^{\text{v}}$), which is related to the density of unconfined carriers. The small signal relationship can be written as $n_3(qAw_{\text{s}}) = c_3 v_3$, where $w_{\text{s}}$ is the width of the SCH region, $v_3$ is the small signal of $V_3$, and $c_3$ is the small signal inversion capacitance associated with the unconfined carriers [52, 53]. Similarly, the change of voltage across the quantum well is given by the difference of the quasi-Fermi levels in the quantum well, i.e., $V_2 = E_{\text{f}}^{\text{c}}(\text{QW}) - E_{\text{f}}^{\text{v}}(\text{QW})$, which is related to the density of confined carriers: $n_2(qAw) = c_2 v_2$, where $v_2$ is the small signal of $V_2$, and $c_2$ is the inversion capacitance associated with the confined carriers in QW [54]. The expressions for $c_2$ and $c_3$ are given in [53, 54]. Using these relations between carrier densities and voltages, we can rewrite Eq. (1) as

$$i_{\text{inj}} = c_3 \frac{dv_3}{dt} + \frac{v_3}{r_3} + \left(\frac{v_3}{r_{\text{cap}}} - \frac{v_2}{r_{\text{esc}}}\right), \tag{33}$$

where $r_3 = \tau_3/c_3$, $r_{\text{cap}} = \tau_{\text{cap}}/c_3$, and $r_{\text{esc}} = \tau_{\text{esc}}/c_2$. Note that $v_3$ is the actual AC voltage on the device. The net current ($i_{\text{net}} = v_3/r_{\text{cap}} - v_2/r_{\text{esc}}$) flowing into the QW can be expressed as the current flowing between two nodes in a circuit by making the following transition:

$$i_{\text{net}} = \frac{v_3 - v_2(\tau_{\text{cap}}/\tau_{\text{esc}})}{\tau_{\text{cap}}}. \tag{34}$$

The node voltage $v_2(r_{cap}/r_{esc})$ is denoted as $\tilde{v}_2$, and the ratio $r_{cap}/r_{esc}$ as $B$.

The equivalent circuit can now be constructed by using the small signal rate equations for the confined carriers in QW and the photons and taking $i_{net}$ as the effective injection current:

$$\frac{dn_2}{dt} = \frac{i_{net}}{qAw} - \frac{n_2}{\tau_2} - vg'S_0 n_2 - vg_0(1 - \varepsilon S_0)s, \tag{35}$$

$$\frac{ds}{dt} = \Gamma vg'S_0 n_2 + \Gamma vg_0(1 - \varepsilon S_0)s + \Gamma\beta\frac{n_2}{\tau_2} - \frac{s}{\tau_p}, \tag{36}$$

where $\tau_2$ is the spontaneous lifetime, $v_g$ is the group velocity of light, $g_0$ is the optical gain at the bias point, $g'$ is the differential gain, $\varepsilon$ is the gain compression coefficient, $S_0$ is the photon density at the bias point, $\Gamma$ is the optical confinement factor, and $\beta$ is spontaneous emission coupling factor. An equivalent circuit of the quantum well alone (without regard for the SCH region) can then be constructed using the standard procedure, resulting in an RLC resonance circuit (Fig. 17a) whose impedance is given by

$$\frac{1}{Z_{QW}(f)} = \frac{i_{net}}{v_2} = j2\pi f c_2 + \frac{1}{r_2} + \frac{1}{j2\pi f L + r_s}. \tag{37}$$

Combining Eq. (37) with Eqs. (33) and (34), the overall equivalent circuit

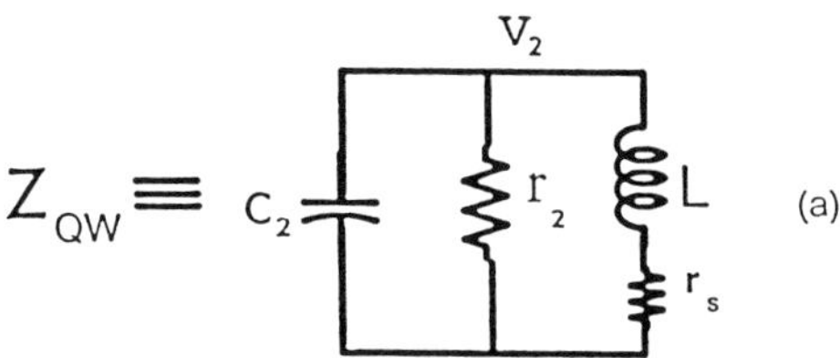


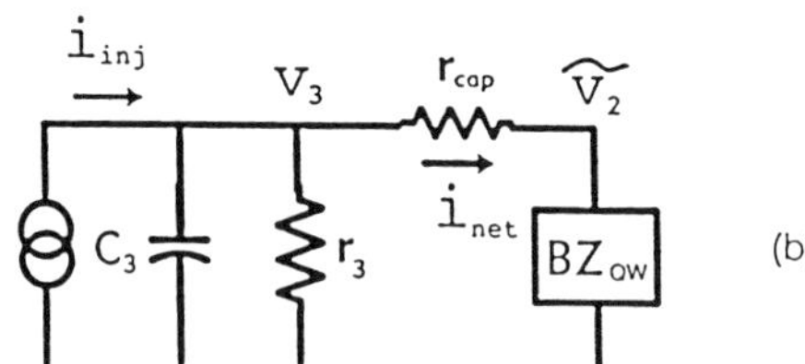


**Fig. 17.** (a) Equivalent circuit of photon–carrier interaction in QW; (b) complete intrinsic equivalent circuit of quantum well laser.

of the device is shown in Fig. 17b. The total impedance $Z_{\text{int}}(f)$ is given by

$$\frac{1}{Z_{\text{int}}(f)} = j2\pi f c_3 + \frac{1}{r_3} + \frac{1}{r_{\text{cap}} + BZ_{\text{QW}}(f)}. \tag{38}$$

Note that the AC optical output is proportional to the current flowing through the inductor, $i_L = v_g g_0(1 - \varepsilon S_0)s(qAw)$.

We now consider several qualitative features of the equivalent circuit. First, when the capture time is infinitely long ($r_{\text{cap}} \to \infty$), the equivalent circuit bcomes a simple RC circuit (Fig. 18a) with a time constant $r_3 c_3 = \tau_3 =$ the spontaneous lifetime, which is on the order of a few nanoseconds. This represents the situation in which the QW is effectively nonexistent, and the

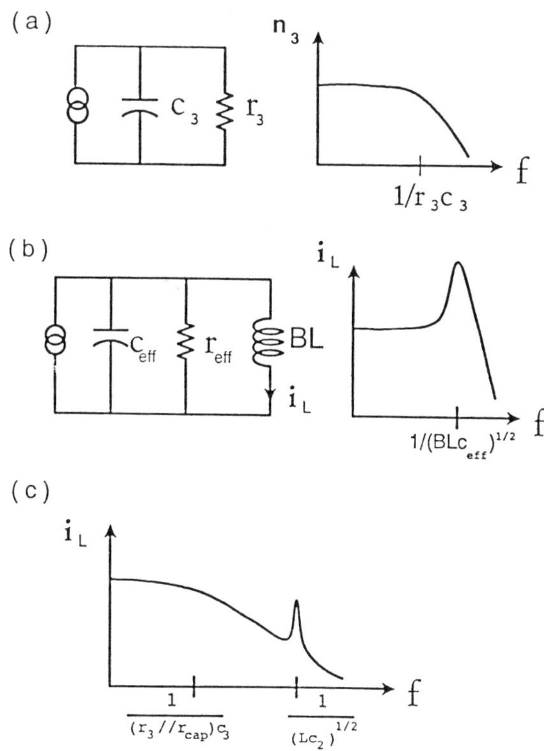


**Fig. 18.** Equivalent circuit and schematic modulation response in the cases: (a) long capture time; (b) short capture time with finite $B$; and (c) intermediate case and high output power. ($n_3$ is the carrier density in SCH region and $i_L$ is the inductor current, which is proportional to the laser output.)

inversion capacitance $c_3$ is responsible for limiting the modulation response of an LED (emitting light in the SCH region) to well below 1 GHz. In another extreme, where the carrier capture is infinitely fast ($r_{\text{cap}} \to 0$, $r_{\text{esc}} \to 0$) but $B$ remains nonzero, the equivalent circuit reduces to a RLC circuit as shown in Fig. 18b. In this case, the inversion capacitance ($c_3$) *does not* produce a low frequency roll-off as in the previous case, but is added to the inversion capacitance for confined carriers ($c_2$) and thus lowers the relaxation oscillation frequency ($\sim 1/\sqrt{(BL)C_{\text{eff}}}$), where $C_{\text{eff}} = c_3 + Bc_2$. It is clear from Fig. 18b that, even in the event of infinitely fast carrier capture, the relaxation oscillation frequency of a quantum well laser is affected by the inversion capacitance for unconfined carriers ($c_3$) and the capture/escape ratio $B = r_{\text{cap}}/r_{\text{esc}}$.

For the intermediate case where the capture time is finite, a simple circuit analysis shows that, at high bias power (in which case $L \to 0$), the inductor current $i_L$ (which is proportional to the optical response) can be expressed as:

$$i_L = \frac{i_{\text{inj}}}{r_{\text{cap}}\left[j\omega c_3 + \dfrac{1}{r_3//r_{\text{cap}}}\right]\left(-\omega^2 Lc_2 + j\omega\dfrac{L}{r_2} + 1\right)}, \tag{39}$$

where // denotes a parallel combination of the circuit elements. There are thus two spectral features in the modulation response: (1) a RC roll-off at angular frequency

$$\omega = \frac{1}{c_3(r_3//r_{\text{cap}})} = \frac{1}{\tau_3} + \frac{1}{\tau_{\text{cap}}}; \tag{40}$$

and (2) a resonance at $\omega_r = 1/\sqrt{Lc_2}$. This is shown schematically in Fig. 18c. Note from Eq. (40) that the RC roll off is *not* simply due to the spontaneous lifetime $\tau_3$ but is a parallel combination of the spontaneous lifetime and capture time. Since the general estimates of capture time are on the order of 1–40 ps [45, 49], the RC roll-off is dominated by $\tau_{\text{cap}}$ at frequency $1/(2\pi\tau_{\text{cap}})$, which may or may not be higher than the relaxation oscillation frequency. A similar RC roll-off effect can also result from carrier diffusion across a wide SCH region [56].

The computed total impedance and the modulation response are plotted in Figs. 19a–c for various choices of parameters as described in the figure caption. Comparing the impedance plot in Figs. 19a and b, we notice a drastic reduction of the relaxation oscillation peak when $B$ is reduced from 0.6 to 0.006 by increasing $\tau_{\text{esc}}$. This is due to the reduction of the effect of $v_2$ by a factor of $B$ ($\tilde{v}_2 = Bv_2$) in the complete equivalent circuit. The

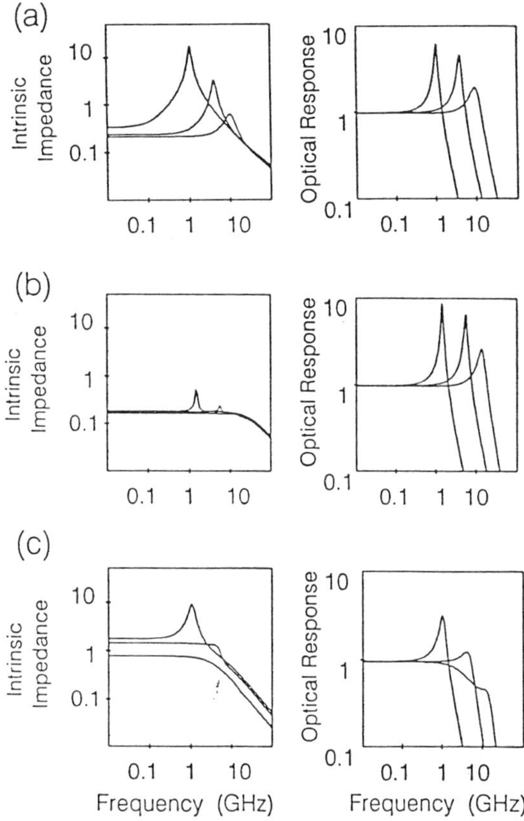


**Fig. 19.** Intrinsic impedance (left) and modulation response (right) in cases: (a) $\tau_{cap} = 5$ ps, $\tau_{esc} = 5$ ps, and $B = 0.6$; (b) $\tau_{cap} = 5$ ps, $\tau_{esc} = 500$ ps, and $B = 0.006$; (c) $\tau_{cap} = 50$ ps, $\tau_{esc} = 50$ ps, and $B = 0.5$. The parameters used for calculation are $\tau_2 = \tau_3 = 1$ ns, $\tau_p = 2$ ps, $A = 200\ \mu\text{m} \times 3\ \mu\text{m}$, $\beta = 10^{-4}$, $\varepsilon = 10^{-17}\ \text{cm}^3$, $\eta' = 6 \times 10^{-16}\ \text{cm}^2$, and $\Gamma = 0.03$. In each case, three curves are plotted, corresponding to three different current levels, $I = 1.2I_{th}$, $4I_{th}$, and $20I_{th}$, where $i_{th}$ is the threshold current.

physical meaning is that when $\tau_{esc}$ is long, the effect of relaxation oscillation inside the QW cannot be fed back to the SCH region. This result suggests that the ratio $B$ can be estimated from the impedance measurement with parasitic contributions being eliminated. A long $\tau_{esc}$ also leads to less damping in QW lasers [35], as can be seen in the modulation response plotted in Figs. 19a and b. Figure 19c illustrates the case of long capture time, which shows a RC roll-off and resonance peak as mentioned previously.

## 7. EFFECTIVE CAPTURE AND ESCAPE TIMES

The effect of transport in quantum well lasers has been described in the last few sections in terms of a "reservoir" model, Fig. 10, in which carriers inside and outside the quantum well are assumed to exchange at the rate described by the capture and escape times. These exchanges between the 2-D and the 3-D electrons involve phonons which supply the necessary momentum difference between the 2-D and 3-D electrons in the direction perpendicular to the well. These processes are briefly described in Section 5. However, it was also mentioned that in the event of a very wide separate confinement (SCH) barrier region, the time taken by the carriers to physically diffuse across this region, before reaching the quantum well, is not negligible and in some cases may even be dominant. This is the basis on which the effective capture time is sometimes taken to be the diffusion time instead of the intrinsic capture time as reported in [36]. In order to clarify the effects of diffusion and quantum capture, we derive in this section effective capture and escape times accounting for *both* carrier diffusion and quantum capture. These effective time constants are then the ones that should be used in the "reservoir" model for analyzing the modulation response of quantum well lasers.

It should be noted that the problem of carrier transport within the SCH region is not a simple one, considering that electrons and holes are injected separately from opposite ends of this region, and that dielectric relaxation and the resultant temporal modification of the band structure makes a simple ambipolar diffusion model appear overly simplistic. However, at the time of this writing, detailed carrier transport modelling and resultant physically observable effects on the direct modulation response of QW lasers, if any, have yet to be demonstrated. On the other hand, experimenta evidences such as those shown in [36] indicate that a simple ambipolar diffusion model might suffice to explain some of the gross (and key) features of the modulation response of certain QW lasers. It is with these issues in mind that the following analysis, referred to here as the *semiclassical capture* (SCC) model [57], is carried out.

The semi-classical description of the transport effect on modulation response involves solving the (classical) diffusion problem in the SCH region along with quantum capture at the quantum well [58] as described in Section 5. Interpretation of these results can be done most conveniently by reducing them to an equivalent "reservoir" rate equations model with effective capture and escape times. The role played by quantum capture will then be apparent.

We first consider one-dimensional transport in a quantum well structure. Carriers in the SCH region is described by the diffusion equation:

$Dd^2n_b(x)/dx^2 - i\omega n_b(x) = 0$ where $n_b(x)$ is the carrier distribution in the barrier layers. If the width of the SCH region ($L_b$) is much smaller than the diffusion length ($L_d \equiv \sqrt{D/\omega}$) (as most often is the case) then $n_b(x)$ is approximately constant. In this limit, we can derive a relation between the pump current $j_{pump}$, the quantum capture current $j_{Q2}$, and the active carriers $n_{Q2}$:

$$j_{pump} = j_{Q2}\left[1 + i\omega\left(\tau_{cap}^{Q}\frac{L_{sch}}{L_q} + \frac{L_b^2}{2D}\right)\right] + n_{Q2}i\omega L_{sch}\frac{\tau_{cap}^{Q}}{\tau_{esc}^{Q}} \tag{41}$$

where $L_q$ is the total width of the QWs, $L_{sch} = 2L_b + L_q$ is the full width of the SCH region, $\tau_{cap}^{Q}$ and $\tau_{esc}^{Q}$ are the quantum capture and escape times, respectively, which can be calculated from first principles as in Section 5 [58]. On the other hand, if one simply lumps the carriers in the SCH region into a single "reservoir" and consider the exchange of carriers between this reservoir and the quantum well through some effective capture and escape time constants ($\tau_{cap}^{E}$, $\tau_{esc}^{E}$, then one can obtain a corresponding result:

$$j_{pump} = j_{Q2}[1 + i\omega\tau_{cap}^{E}] + n_{Q2}i\omega L_q\frac{\tau_{cap}^{E}}{\tau_{esc}^{E}} \tag{42}$$

Comparing Eqns. 41 and 42 we obtain the expressions for the effective capture time and the ratio $\tau_{cap}^{E}/\tau_{esc}^{E}$:

$$\tau_{cap}^{E} = \frac{L_b^2}{2D} + \tau_{cap}^{Q}\frac{L_{sch}}{L_q}, \tag{43a}$$

$$R = \frac{\tau_{cap}^{E}}{\tau_{esc}^{E}} = \frac{\tau_{cap}^{Q}}{\tau_{esc}^{Q}}\left(\frac{L_{sch}}{L_q}\right). \tag{43b}$$

Therefore, the simple results of the reservoir model can accurately describe the dynamic transport situation, *provided that* the correct effective capture and escape times are used, as given in Eqn. 43a and b. For example, we plot in Fig. 20 the modulation responses calculated from the SCC model and the equivalent reservoir model based on the relations in Eqn. 43. Excellent agreement is obtained even for $L_b$ as large as $L_d$.

One important implication of the result in Eqn. 43 is that the effective capture time is the sum of the diffusion time AND the intrinsic quantum capture time up-scale by a factor $L_{sch}/L_q > 1$. For typical values of $\tau_{cap}^{Q} \sim 0.5$ ps and $L_{sch}/L_q \sim 10$, the maximum modulation bandwidth due to quantum capture alone is $1/(2\pi\tau_{cap}^{E})$ is 30 GHz.

To the extent that the above interpretation is sufficiently general that applies equally well to 2-D or 3-D transport [59], we infer the following

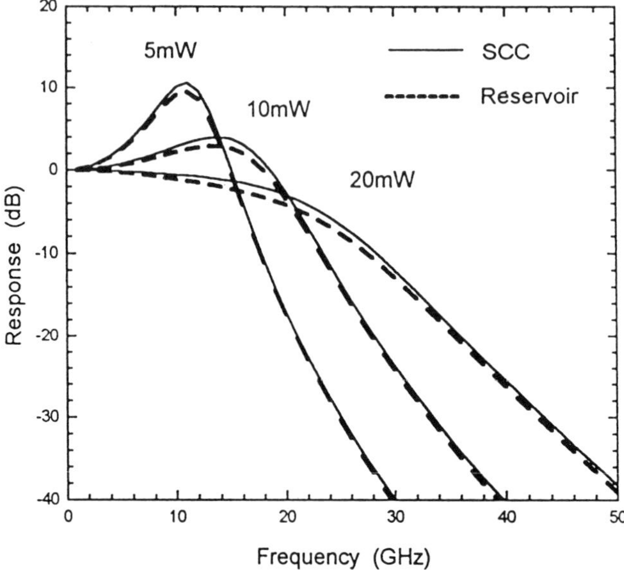


**Fig. 20.** Responses calculated from the SCC model and the equivalent reservoir model. Excellent fit is obtained using the derived expression for the effective capture and escape times.

effective capture times for quantum wire and quantum dot lasers such as those shown in Fig. 21:

$$\tau^{E}_{cap,\,wire} = \tau^{Q}_{cap,\,wire}(A_{sch}/A_{wire}) \tag{44a}$$

$$\tau^{E}_{cap,\,dot} = \tau^{Q}_{cap,\,dot}(V_{sch}/V_{dot}) \tag{44b}$$

where $A_{sch}$, $A_{wire}$ are the cross sectional areas of the confinement region and the quantum wires respectively, $V_{sch}$ and $V_{dot}$ are the corresponding volumes in a quantum dot laser. In other words, the effect of quantum capture on high speed modulation can be described in terms of the "packing density" of quantum wires and dots from high speed modulation consideration. These results point to the importance of maintaining a high packing density in these lower dimensional devices, not just from the point of view of attaining a reasonable optical gain, but in from the point of view of high speed modulation as well.

In Fig. 22, the maximum bandwidth $(1/2\pi\tau^{E}_{cap})$ for different quantum well, wire and dot laser structures are plotted versus the packing density $\rho$.

**Buried Quantum Well and Wire**

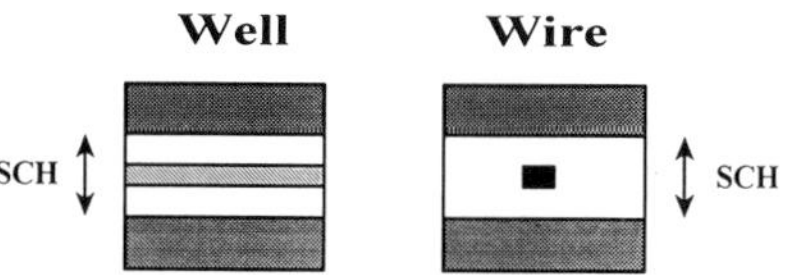


**Channelled-Well Quantum Wire**

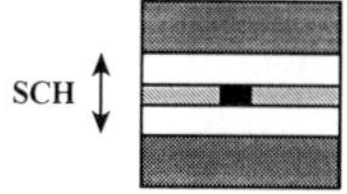


**Fig. 21.** Buried quantum well and wire lasers and channelled well-quantum wire lasers.

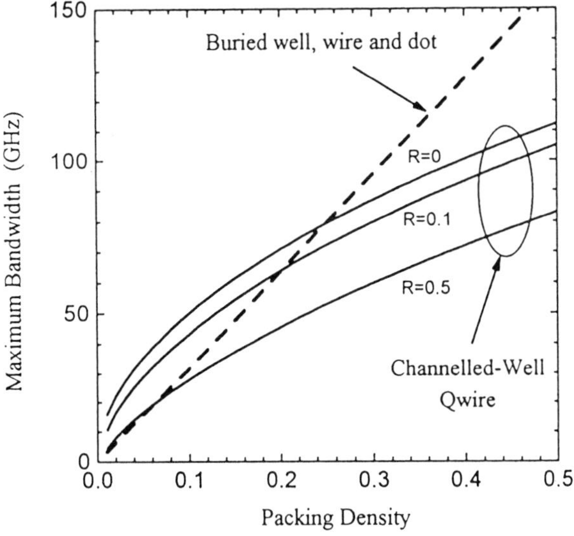


**Fig. 22.** The maximum bandwidth versus packing density of the various quantum well, wire and dot laser structures. The quantum capture time is taken to be 0.5 ps.

## 8. THE QW GAIN-LEVER EFFECT

Gain-lever is an effect that utilizes the highly sublinear nature of the quantum well gain characteristic to accomplish very high modulation efficiencies in intensity modulation (IM) and frequency modulation (FM), as well as broad wavelength tunability [60–63]. To appreciate the gain-lever effect, consider a single quantum well laser that has a typical gain characteristic as shown in Fig. 21 and that is inhomogeneously pumped in a tandem-contact geometry. The principle of operation is as follows: When the device is above lasing threshold, the sum of the optical gain of the sections is clamped at a constant value (equal to the cavity loss). If one section increases in optical gain by, say, injection of extra electrons, then the other section must automatically reduce its gain by the same amount by ejaculating the extra electrons. If each section is pumped by a constant current source, then the extra electrons cannot be ejaculated electrically but will do so optically through the emission of photons. According to Fig. 21, when the sections are biased as shown, then a small modulation applied to section "a" will result in a large modulation in the optical output by the "levering" effect.

Quantitatively, the modulation performance of this device is described by the rate equations

$$\dot{S} = S\left[\Gamma G_{\rm a}(1 - h) + \Gamma G_{\rm b}h - \frac{1}{\tau_{\rm p}}\right], \tag{45a}$$

$$\dot{N}_{\rm a} = \frac{J_{\rm a}}{ed} - BN_{\rm a}^2 - G_{\rm a}S, \tag{45b}$$

$$\dot{N}_{\rm b} = \frac{J_{\rm b}}{ed} - BN_{\rm b}^2 - G_{\rm b}S, \tag{45c}$$

where $S$, $N_{\rm a,b}$, $J_{\rm a,b}$, $G_{\rm a,b}$ are the photon density, carrier densities, injection currents, and optical gain in the respective sections; $\Gamma$ is the optical confinement factor; $h$ is the fractional length of the gain section (Fig. 23); and a bimolecular recombination is used that is shown experimentally to be appropriate for quantum well structures [22, 23]. A small signal analysis yields the following expression for the photon density modulation amplitude $s$:

$$\frac{s}{j_{\rm a}} = \frac{\Gamma G'_{\rm a0} S_0(1 - h)[(i\omega) + \gamma_{\rm b}]/ed}{(i\omega)^3 + (\gamma_{\rm a} + \gamma_{\rm b})(i\omega)^2 + A_1(i\omega) + A_2}, \tag{46a}$$

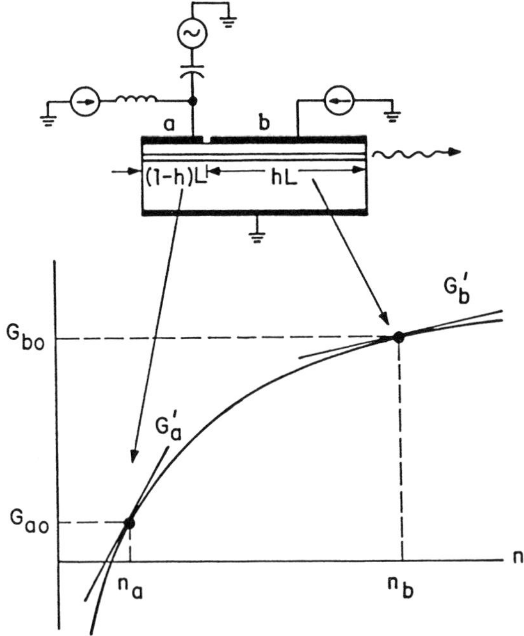


**Fig. 23.** Schematic diagram of the tandem-contact single quantum well laser, with the gain versus carrier density curve shown.

where

$$A_1 = G_{a0} G'_{a0}(1 - h) + G_{b0} G'_{b0} h + \gamma_a \gamma_b, \tag{46b}$$

$$A_2 = \Gamma S_0[G'_{a0} G'_{a0} \gamma_b(1 - h) + G'_{ao} G'_{a0} \gamma_a h], \tag{46c}$$

and $G_{a0/b0}$, $G'_{a0/b0}$ are the gain and differential gain of the two sections, $P_0$ is the cw photon density, $j_a$ is the amplitude of the modulation current density into section a, and

$$\gamma_{a,b} = \frac{1}{\tau_{a,b}} + G'_{a0,b0} S_0, \qquad \text{where } \frac{1}{\tau_{a,b}} = 2BN_{a0,b0} \tag{47}$$

are the inverse lifetimes. A further relation exists between the $G_{a0,b0}$:

$$G_{a0}(1 - h) + G_{bo} h = G_0 = \frac{1}{\Gamma \tau_p}. \tag{48}$$

If the device is pumped uniformly, then $G_{a0}/G_0 = G_{b0}/G_0 = 1$. Using measured values [64] of $G_{a0/b0}$ and $G'_{a0/b0}$, the modulation response is plotted in Fig. 24a for various values of $G_{a0}/G_0$. One observes an increase in the modulation efficiency, while the relaxation oscillation frequency $f_r$ remains largely unchanged. The constancy of $f_r$ can be explained as follows: Consider

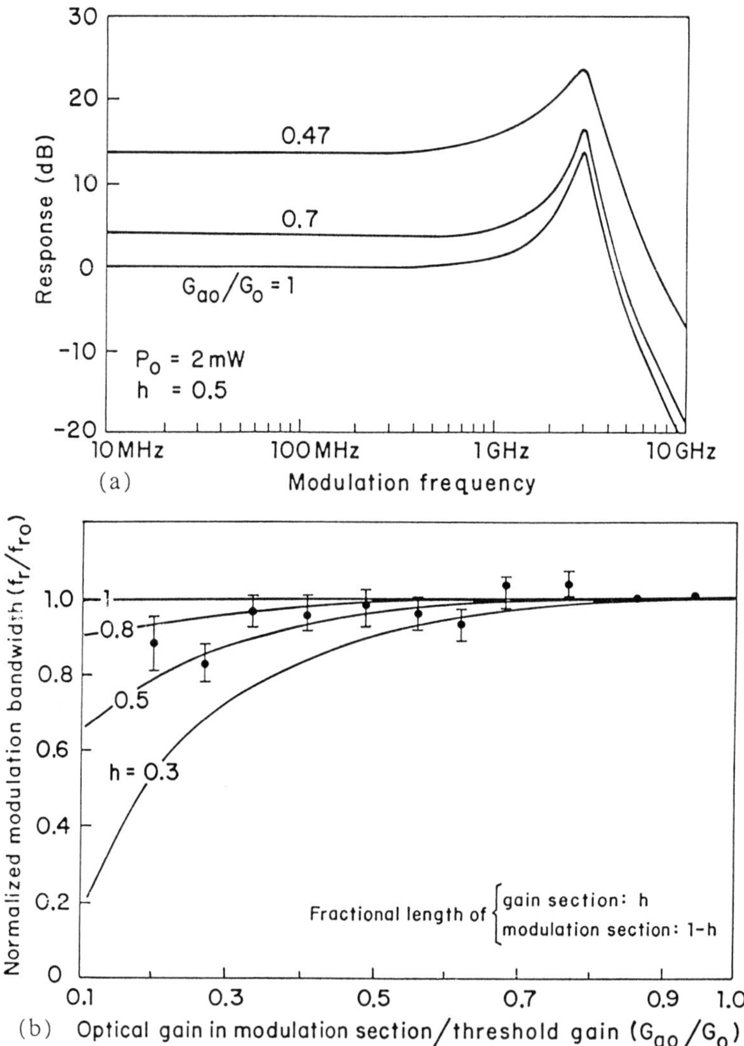


**Fig. 24.** (a) Modulation response with different bias on the modulation section. (b) Relaxation oscillation frequency as a function of modulation section gain, for various $h$.

modulation frequencies $\omega \gg 1/\gamma_a, 1/\gamma_b$; Eq. (46) can be expressed approximately as $s/j_a \sim 1/[(i\omega)^2 + \omega_r^2]$, where

$$\omega_r^2 = \Gamma S_0[G_{a0}G'_{a0}(1-h) + G_{b0}G'_{b0}h] \tag{49}$$

is the resonance frequency ($=2\pi f_r$). Note that, for a linear gain function, $G'_{a0} = G'_{b0} \equiv G'$ and $\omega_r^2 = G'S_0/\tau_p$, which is the standard formula for relaxation oscillation frequency [3, 4, 65, 66]. Because of the near-parabolic shape of the gain characteristics of single quantum well [10–12, 64], $G_{a0}G'_{a0} \cong G_{b0}G'_{b0}$ except for very large and very small $G_{a0,b0}$. As a result, $\omega_r$ is approximately constant, according to Eq. (49). Using the actual measured gain curve [64], we plot, in Fig. 24b, $\omega_r$ versus $G_{a0}/G_0$ for various values of $h$. One observes that if (1) the gain section occupies a larger fraction of the cavity ($h > 0.5$) and (2) $G_{a0}/G_0 > 0.2$, then the relaxation oscillation frequency remains virtually unchanged from that of a uniformly pumped device.

The relative modulation efficiency at frequencies below relaxation oscillation can be obtained from Eq. (46):

$$\eta = \frac{s/i_a}{s/i_a|_{h=0}} = \frac{\gamma_b}{(1-h)\gamma_b W + (h\gamma_a/k)Q}, \tag{50}$$

where $i_a$ is the modulation current amplitude, $W = G_{a0}/G_0$, $Q = G_{b0}/G_0$, and $k = G'_{a0}/G'_{b0}$. The quantities $W$ and $Q$ are related via Eq. (48): $Q = [1-(1-h)W]/h$. This modulation efficiency $\eta$ is normalized to that of a uniformly pumped device: When $h = 0$, $W = 1$, and $Q = 0$, then we obtain $\eta = 1$. To gain further insight into Eq. (50), consider cases near $h = 1$. Equation (50) is simplified to

$$\eta = \frac{\gamma_b}{\gamma_a}k. \tag{51}$$

At low optical power, the spontaneous lifetime dominates and $\gamma_b \cong \gamma_a$. The enhancement in modulation efficiency is thus approximately equal to the ratio of the slopes of the differential gain. At high optical power, the stimulated lifetime dominates and $\gamma_b/\gamma_a \cong 1/k$. The stimulated lifetime of the modulation section is short because of a high differential gain. Because of a shorter stimulated lifetime in the modulation section, a larger modulation current is needed to produce the same amount of modulation in the electron density, thus reducing the effectiveness of the levering effect. In fact, at a sufficiently high optical power, the levering effect is completely neutralized, and the modulation efficiency is no different from that of a uniformly pumped device. This implies a limited modulation bandwidth for gain-lever lasers, except for those with a very short photon lifetime (cavity length).

The gain-lever effect has been experimentally demonstrated first by optical modulation [61] and then by electrical modulation [60]. An example of the latter is shown in Fig. 25. The laser bandwidth is between 3 and 6 GHz at output power levels up to 4 mW for this particular device under uniform injection, as shown by curves labelled "uniform" in Fig. 25a for a 400-$\mu$m laser and (b) for a 220-$\mu$m device. The two sections are then biased separately, with the bias current into the modulation section reduced and that into the gain section increased correspondingly to keep the output power constant. The relaxation oscillation frequency changes slightly, and it is shown in the data points plotted in Fig. 24b for a device with $h = 0.45$. Microwave modulation is applied to the modulation section whose gain is estimated to be 0.2 that of threshold gain. The modulation responses are plotted in Figs. 25a and b as shown by curves labelled "tandem." The improvement for lower $f_r$ and for shorter devices is evident. The largest improvement observed was about 23 dB, with the 220-$\mu$m device at $f_r \cong 3$ GHz.

From an applications point of view, improvement in the modulation efficiency of a laser is not meaningful unless the intensity noise can be maintained at or close to the prior level. This is indeed the case, because the gain-lever effect depends on selective excitation of one of the sections with the appropriate characteristic (that with a larger differential gain), while intrinsic quantum noise excites the laser cavity more or less uniformly.

A straightforward extension of the analyses carried out for the uniformly pumped device [53, 67] can be used to model the noise characteristic of the gain-lever laser. Basically, we start with the rate equations and include the noise through the Langevin noise terms in the photon rate equation:

$$\dot{S} = S[\Gamma G_a(1 - h) + \Gamma G_b h - 1/\tau_p] + \Delta(t). \tag{52}$$

The electron rate equations remain the same as in Eq. (45). $\Delta$ is the Langevin noise source due to spontaneous emission given by its spectral density function:

$$\langle|\Delta(\omega)|^2\rangle = R_{sp}S_0, \tag{53}$$

where $P_0$ is the steady state photon density and $R_{sp}$ is the spontaneous emission rate. Upon linearizing and solving these equations with the small signal analysis, we obtain the following result for the relative intensity noise (RIN) in the inhomogeneously pumped tandem-contact single quantum well laser:

$$\mathrm{RIN}(\omega) = \frac{R_{sp}}{S_0} \frac{(\omega^2 + \gamma_a^2)(\omega^2 + \gamma_b^2)}{\omega^2[\omega^2[\omega^2 - (\omega_R^2 + \gamma_a\gamma_b)]^2 + [\omega^2(\gamma_a + \gamma_b) - \omega_R^2\Xi]^2} \tag{54}$$

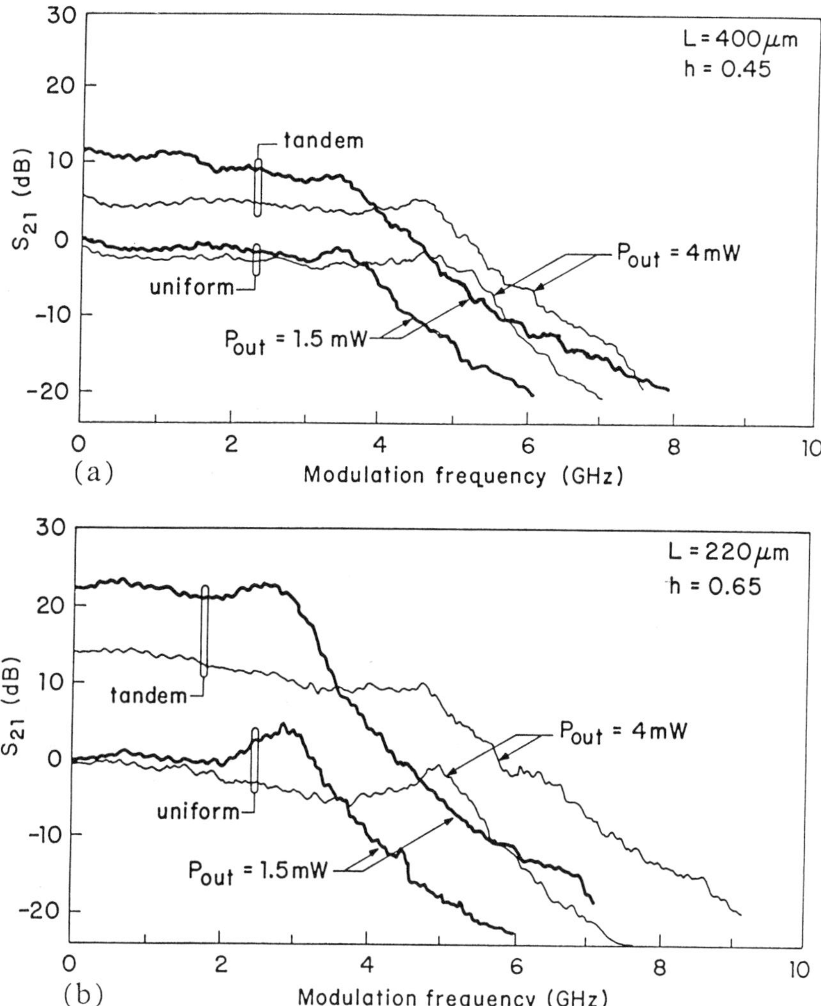


**Fig. 25.** Measured modulation response at output powers of 1.5 and 4 mW. Thin curves: uniformly pumped; thick curves: tandem pumped with $G_{a0}/G_0 \cong 0.2$. (a) A 400-$\mu$m cavity with $h = 0.45$; (b) a 220-$\mu$m cavity with $h = 0.65$.

where

$$\gamma_i = 1/\tau_i + G_i' S_0, \qquad i = \text{a, b, h}, \tag{55a}$$

$$1/\tau_i = 2BN_i^2, \tag{55b}$$

are electron lifetimes in the modulation section, gain section, and homogeneously pumped laser, respectively; $\omega_R$ is the relaxation resonance frequency, shown to be almost independent of the enhancement factor for a wide range of parameters [60];

$$\Xi = h\gamma_a + (1 - h)\gamma_b \tag{56}$$

is the measure of the effective asymmetry of the two sections, and

$$r = \gamma_a/\gamma_b. \tag{57}$$

The RIN is plotted in Fig. 26 for several values of the $r$ parameter, which correspond roughly to the modulation efficiency enhancements of 0, 10, 20 dB. These plots should be compared with the experimental results shown in Fig. 27. The pictures show the spectrum of the photodetector output with tandem-contact SQW laser biased at the constant optical output power $P_{out} = 1.5$ mW but at different currents in the gain and modulation sections. The top picture is the RIN of the uniformly pumped SQW laser. The middle has gain current $I_b = 15$ mA, control current $I_a = 2.5$ mA, and the modulation efficiency increase $\eta = 9$ dB. The bottom has gain current $I_b = 19$ mA, control current $I_a = 2$ mA, and the modulation efficiency increase $\eta = 19$ dB. The laser used in this experiment is similar in structure to the ultralow threshold SQW laser, which has been described in more detail

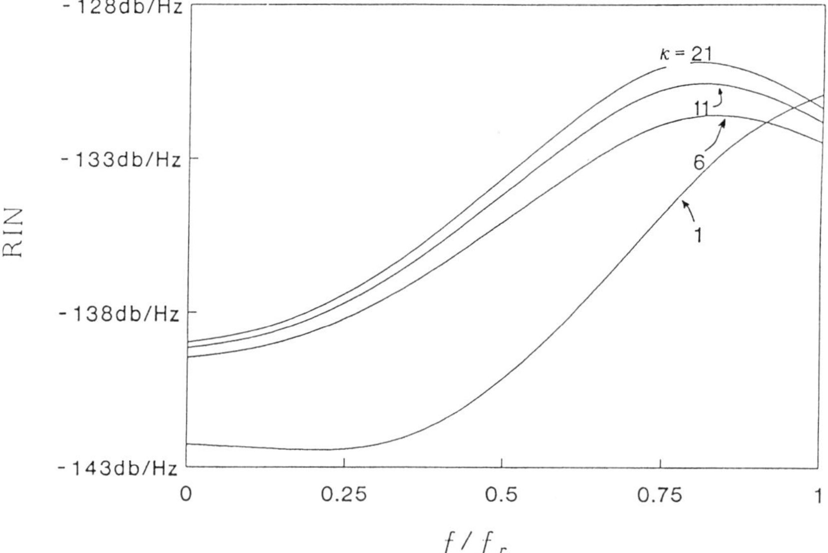


**Fig. 26.** Theoretical spectral plot of the tandem-contact SQW laser RIN for different $g$ values.

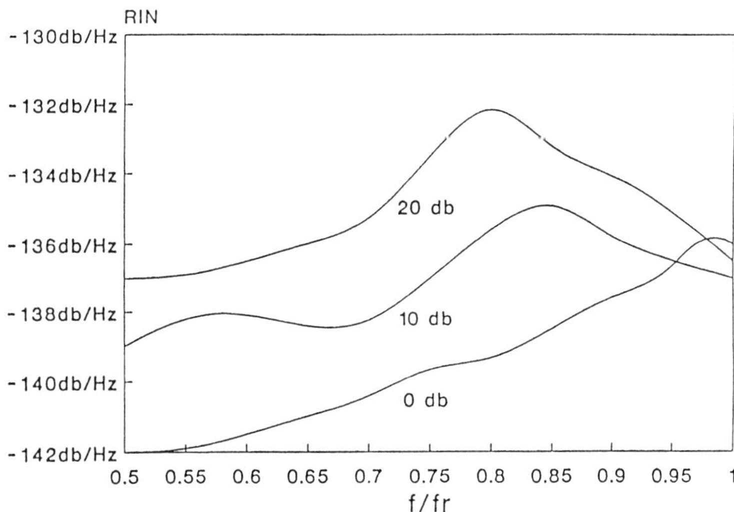


**Fig. 27.** Experimental results: photodiode spectral output at constant $P_{out} =$ 1.5 mW with horizontal scale 2–4 GHz. Top: $I_a$- $= I_b = 10$ mA, $\eta = 0$ dB; middle: $I_a = 2.5$ mA, $I_b = 15$ mA, $\eta = 9$ dB; bottom: $I_a = 2$ mA, $I_b = 19$ mA, $\eta = 19$ dB.

in [60, 64]. The horizontal scale of the plots in Fig. 27 is 2–4 GHz, which corresponds to the upper half of the theoretical plot. Theory predicts that, with the increasing efficiency, the noise peak moves slightly to the lower frequency, and the DC value of the RIN increases but by an amount substantially less than the increase in efficiency itself. Both the shift and the increase saturate for large $g$. These are exactly the effects observed experimentally.

One can also seek the device geometry that results in the lowest RIN. For example, the increase in the intensity noise at low frequency (relative to that of a uniformly injected device at an identical optical power) is shown in Fig. 28 as a function of the fractional length of the gain section and the $g$ value, which is proportional to the modulation efficiency. Quite simply, to minimize the RIN, the ratio of the length of the gain to control section should be maximized, a geometry that also maximizes the efficiency enhancement, as seen from the preceding discussions.

In addition to an enhancement in the intensity modulation efficiency, gain-lever can be used to enhance the frequency modulation (FM) efficiency and the wavelength tuning range of the laser *without* a corresponding increase in the FM noise, i.e., without increasing the linewidth. The principle

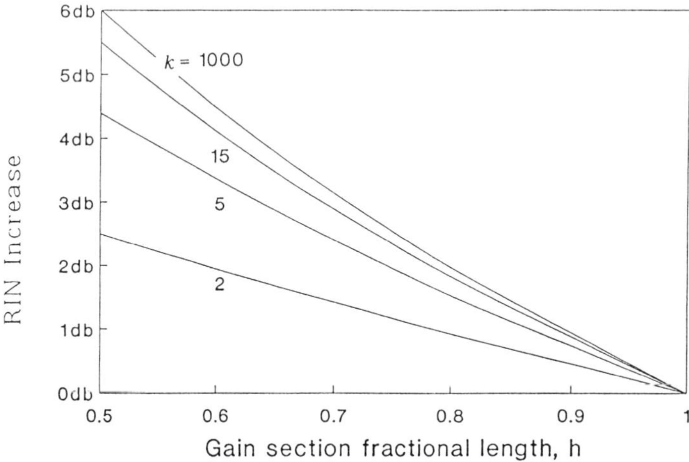


**Fig. 28.** Relative increases of the DC RIN versus gain section fractional length with $g$, the ratio of control and gain inverse electron lifetimes, as a parameter.

behind the FM enhancement is the lack of carrier clamping in the gain-lever laser (although gain clamping still exists). The large swing in the carrier density resulting from a small variation in the injection current causes a large change in the refractive index of the cavity, resulting in a large FM. The FM noise is not significantly enhanced for the same reason that the RIN is not significantly enhanced, as discussed before.

The FM enhancement can be observed using the tandem-contact GaAs SQW laser in [60, 68]. The FM characteristics of the laser was measured using a standard frequency discriminator arrangement consisting of a low-$Q$ etalon. The result is plotted in Fig. 29 for a gain-levered and a uniformly pumped arrangement, at an identical optical power of 2 mW/facet. The threshold of the uniformly pumped devices was 10 mA, while under gain-levered operation the modulation and gain section were biased at 3 and 20 mA, respectively. The physical lengths of the sections are 120 and 250 $\mu$m, respectively. The low frequency roll-off in the uniformly pumped case is due to thermal effects. In the "midband" range, the FM efficiency of the gain-levered laser is enhanced by a factor of almost 100, from about 0.2 to 20 GHz/mA. The uniformly pumped device has a resonance at around 3 GHz, while the gain-levered device rolls off at around 2 GHz. It is believed that, in the latter case, the single-pole roll-off is partly offset by the rising slope of the relaxation oscillation, resulting in a fairly flat FM response up to 2 GHz.

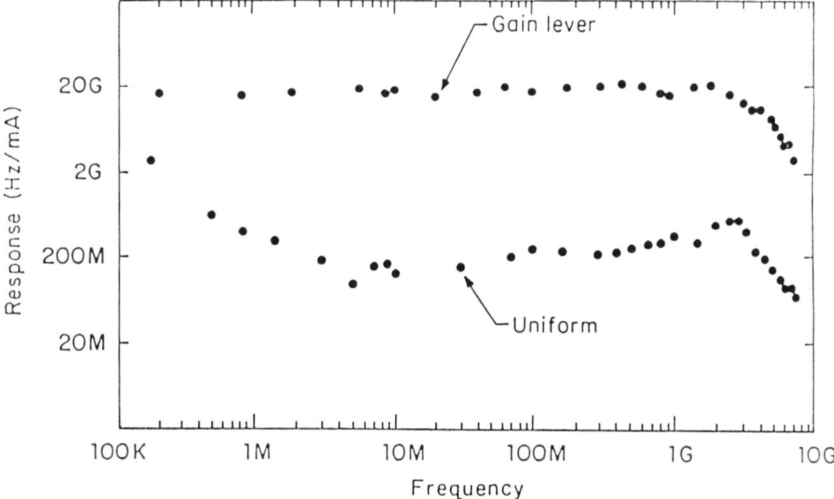


**Fig. 29.** Measured FM modulation characteristic of a uniformly pumped and gain-levered SQW laser.

A brief comparison of the properties of the quantum well gain-lever laser with other existing wavelength tunable structures—single section DFB and three-section DBR [69] lasers—is shown in Table III. A conventional two-section laser [70] is a limiting case of the gain-levered SQW device where the factor $g$ is small and the stimulated lifetime is somewhat longer. A study of Table III shows that while each of the structures has its attributes and shortcomings, the gain-levered SQW laser has a good combination of properties in terms of linewidth, tuning efficiency, range, and speed. (The range issue is rather complex, depending on whether continuous tuning is needed or not. See [69] for a detailed description.)

The preceding discussion introduced the essence of gain-lever effect. It has a high IM as well as FM efficiency. From an applications point of view, it is often undesirable for a FM laser to have a large residual IM, since the latter introduces added penalty in the FM demodulation process. The IM/FM ratio of a gain-lever laser is already an order of magnitude below that of a uniformly pumped laser, but a variation of the gain-lever laser, the so-called "inverted" gain-lever laser [71], exhibits an IM/FM ratio that is another order of magnitude below that of the "normal" gain lever. The inverted gain lever is one in which the role of the gain and modulation sections are inverted, so that the IM efficiency is not enhanced, while the FM is. For illustration, the IM and FM frequency responses of a normal

**Table III.**

Comparison of Characteristics of Various FM and Tunable Laser Structures

| | Linewidth | Tuning efficiency | Tuning range[1] | FM speed |
|---|---|---|---|---|
| Single section (QW or bulk) | $(1 + \alpha^2) \times$ (Schalow–Towns) | low, $<1$ GHz/mA (gain clamping) | small (gain clamping) | multi-GHz (relaxation oscillation) |
| 3-section DBR | a few times of single section (loss in passive section | medium, $\sim$3GHz/mA (free carrier injection | large (no restriction on free carrier density | $1/\tau_a < 1$ GHz (spontaneous lifetime) |
| 2-section gain-levered SQW | same as single section (whole cavity under inversion) | high $>10$ GHz/mA (gain-lever) | large (gain-lever) | $1/\tau_{stim} > \sim 1$ GHz stimulated lifetime) |

[1] not counting thermal effects

and an inverted gain-lever laser are measured, and the IM responses normalized to the FM responses are shown in Fig. 30, expressed in percentage IM per GHz of FM. Thus, even though gain-levered lasers have very high IM and FM efficiencies, its use for a pure FM application is not as desirable as that of an inverted gain-levered laser. It can be shown that the advantage of an inverted gain-lever laser is not limited to a small IM/FM ratio; the optical power and modulation bandwidth show improvement as well. The inverted gain-levered laser has overall characteristics that approach an ideal FM laser.

Since the large IM in the modulated output of a gain-levered laser is accompanied by a very large FM [62], it is natural to consider further enhancing the IM efficiency by interferometrically converting the large FM into IM. Since the IM and FM are correlated in a semiconductor laser, the converted FM can either add to or subtract from the existing IM. By the same token, the interferometrically converted phase noise can either enhance or suppress the intensity noise [72]. Thus, whether or not the noise figure can be improved at the interferometer output depends on the IM–FM

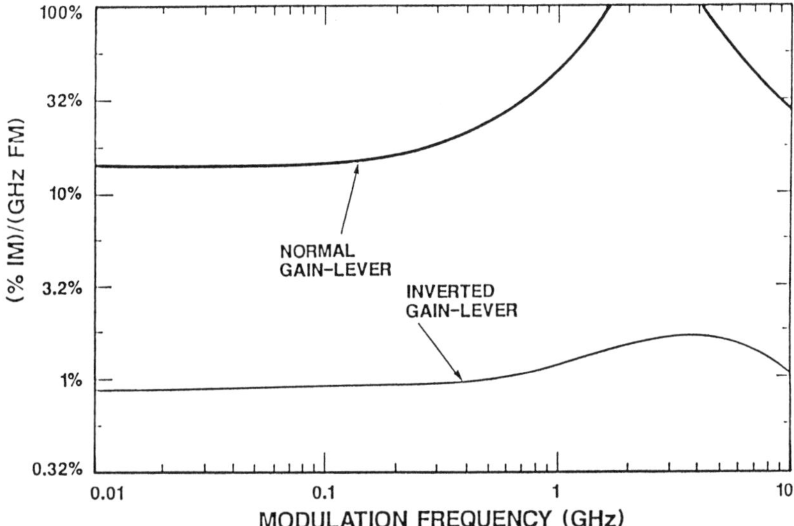


**Fig. 30.** Experimental IM/FM ratio, expressed in percentage IM per GHz in FM, for normal and inverted gain lever, at 1.0 and 3 mW output power per facet, respectively. For normal gain lever, $i_a = 3$ mA, $i_b = 23$ mA; for inverted gain lever, $i_a = 11.5$ mA, $i_b = 11$ mA.

correlation characteristics of the modulation and noise at the laser output.

The correlation characteristics between IM and FM in the modulation and noise are shown geometrically in Fig. 31. The ellipse represents the strongly but incompletely correlated IM and FM noise, nearly identical for all three cases. The slopes of the lines representing IM–FM modulation qualitatively demonstrate the relative strength and phases of IM and FM in each of the three cases.

Geometrically, in performing a pure intensity detection, one projects the noise ellipse and modulation lines in Fig. 31 onto the intensity axis, whereby the corresponding S/N ratio can be obtained. For a pure frequency detection, one rotates the observation plane by 90° onto the frequency axis. An interferometric device produces a linear combination of IM and FM and corresponds to rotating the observation plane to an intermediate angle (Fig. 31). It is easy to see that, in the case of an inverted gain-lever, an optimal situation can be found in which the signal can be maximized, while the intensity noise is simultaneously minimized. These results have been analyzed in detail and experimentally confirmed [73].

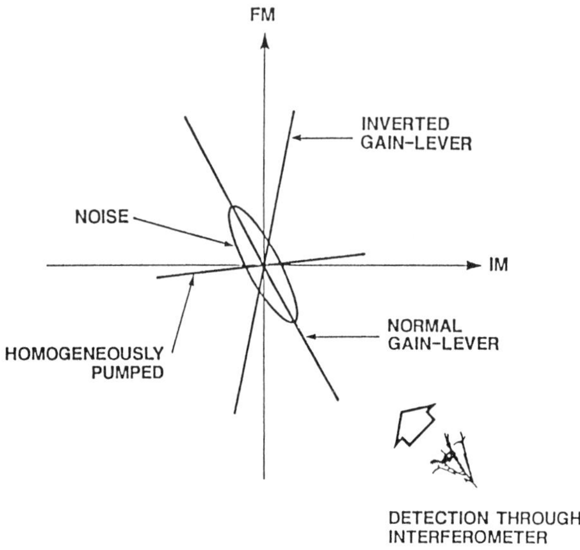


**Fig. 31.** IM–FM correlation characteristics of modulation and noise in a uniformly pumped, two-section gain-lever and inverted gain-lever lasers.

## 9. VERY HIGH FREQUENCY MODE-LOCKING

The connection between quantum well lasers and mode-locking, in particular passive mode-locking at very high frequencies (50 GHz and above), arises from the same characteristic that gives rise to gain-lever, namely the sublinear nature of the gain characteristic. To establish that connection, an overview of the subject of high frequency mode-locking is first provided.

Since the first demonstration of mode-locking of semiconductor lasers [74, 75], there has been a continuing effort to attempt mode-locking at ever higher frequencies. This arises in part from the potential for applications in communications and microwave systems and in part by the desire to avoid external cavity arrangements, which are needed for mode-locking at lower frequencies ($<20$ GHz). Despite early experimental demonstrations to the contrary [3, 4, 65, 66, 76], it was not intuitively obvious that mode-locking can take place at frequencies much above the intrinsic 3 dB cutoff in the direct modulation of these lasers—in the 10 GHz range. Later [63, 64], it was theoretically shown that, contrary to intuition, both active and passive mode-locking can indeed take place at frequencies much higher than the modulation cutoff—up to and beyond 100 GHz. The highest frequency at which mode-locking can be achieved, in fact, depends on the "degree of curvature" of the sublinear gain characteristics of the laser. In fact, for a passively mode-locked laser consisting of two sections, one reverse-biased and one forward-biased, the highest mode-locking frequency can be related to the ratio of the differential gain of these two sections, as shown in Fig. 32. Early experimental evidence [64] has given way to recent demonstrations that showed that mode-locking not only works at these high frequencies but works extraordinarily well, with 100% modulated pulses in the low-picosecond to subpicosecond range [77–80].

Mode-locking at frequencies above a few tens of gigahertz is most conveniently done passively with an intracavity saturable absorber, since it avoids direct modulation, which is severely hampered by device parasitics. Obviously, the absorber, which is implemented by reverse biasing certain section(s) of the laser, must, along with the gain medium, satisfy certain criteria for mode-locking to take place—hence defining the *parameter ranges*. It was long recognized that [81] the presence of a saturable absorber in a laser cavity can also cause "self-pulsation" (referred to as "relaxation

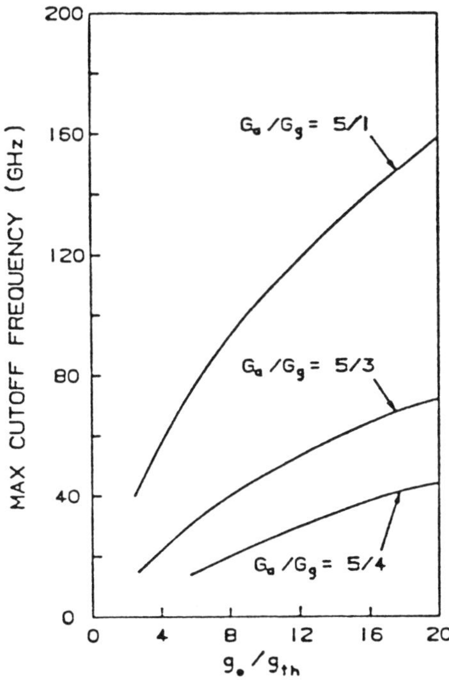


**Fig. 32.** Maximum cutoff frequency for passive mode-locking as a function of $g_0$, the steady state gain in the gain section, for various values of $G_a/G_g$.

oscillation" in [81]), which can hinder or even inhibit mode-locking. Indeed, recent experiments showed that self-pulsation rather than mode-locking takes place for a large range of bias conditions [79]. It is obviously desirable to understand and maximize the *nonoverlapping* region between the parameter ranges of self-pulsation and mode-locking, to make the process more reliably reproduced from device to device and in different material systems.

We experimentally map out the parameter ranges for mode-locking (at 65 GHz) and self-pulsation (a few GHz) using a GaAs multiple quantum well laser with graded index separate confinement. The structure has well and barrier widths of 100 and 75 Å, respectively, and an Al content of 0.3 in the barriers. Two separate top-contact sections define the gain and absorber regions with lengths of 400 and 120 $\mu$m, respectively, and the gap between them is 50 $\mu$m. DC forward/reverse biases were applied to the gain/absorber contacts and the mode-locked/self-pulsation output was observed on a high speed streak camera.

We represent the bias conditions by the gain current ($i_g$) and the absorber reverse bias voltage ($V_b$) and map the ranges in the $i_g$–$V_b$ plane. The result is shown in Fig. 33a. The threshold line defines lasing threshold. In principle, the mode-locking regime should extend almost all the way to the threshold [63, 81], but as seen in Fig. 33a there is hardly a region of mode-locking that is not affected by relaxation oscillation. The region labelled "strongly self-pulsing" is where self-pulsation completely quenches mode-locking, and in the "weakly self-pulsing" region, self-pulsation with $<100\%$ modulation depth coexists with mode-locking where they overlap.

The experiment was repeated with an identical laser coated with high reflectivity coating of 70% and 90% on the front and back facets, respectively. The result is shown in Fig. 33b. The striking contrast with Fig. 33a is that a mode-locked region is now clearly defined that is free of self-pulsation.

An example of a laser operating in the strongly self-pulsing regime is shown in Fig. 34a, while that of a mode-locked regime is shown in Fig. 34b. In the latter case, the measured pulse width is approximately 5 ps. Much of the jitter observed in Fig. 34b is due to the streak camera itself, a conclusion drawn from a microwave spectral analysis of the directly detected output from the laser, which shows an oscillation linewidth of 1.5 MHz centered at 65 GHz. Details of these results will be reported elsewhere. Figure 34c shows an example where self-pulsation simultaneously occurs with mode-locking, resulting in incomplete rendering of both.

To understand why high reflection coatings help to minimize the simultaneous occurrence of mode-locking and self-pulsation, we employ the

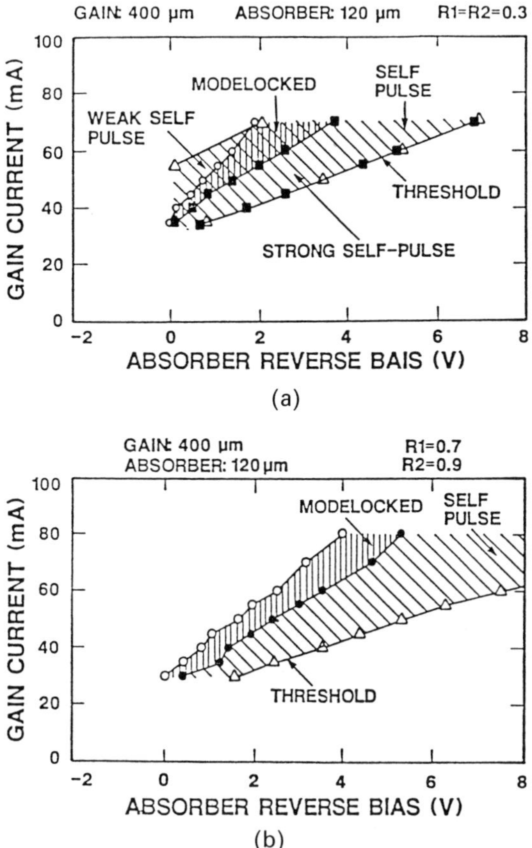


**Fig. 33.** Experimentally observed parameter ranges of 65-GHz mode-locking and low-GHz self-pulsation of an uncoated (a) and high-reflectivity coated (b) quantum well laser.

phasor diagram analysis originally used to predict the feasibility of ultrahigh frequency mode-locking [63]. In that approach, the photon intensity oscillation is represented by $s\exp(i\omega t)$ where $s$ is the real amplitude of the modulation. When considering mode-locking, $\omega = \Omega =$ longitudinal mode spacing ($>50$ GHz). When considering self-pulsation, $\omega = \omega_{SP} =$ self-pulsing frequency in the 1–5 GHz range for typical lasers. The optical modulation drives the gain and absorber media to produce gain and loss modulation, whose difference is the net gain modulation given by

$$g_{net} = \left\{ \frac{-g'_g g_0}{i\omega + 1/T_g} - \frac{-g'_a a_0}{i\omega + 1/T_a} \right\} s e^{i\omega t} \equiv \hat{g}_{net} e^{i\omega t}. \tag{58}$$

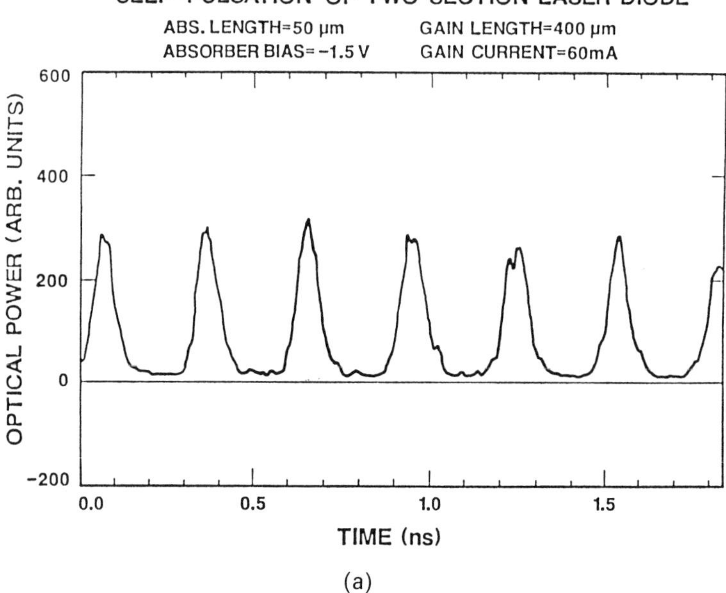


(a)

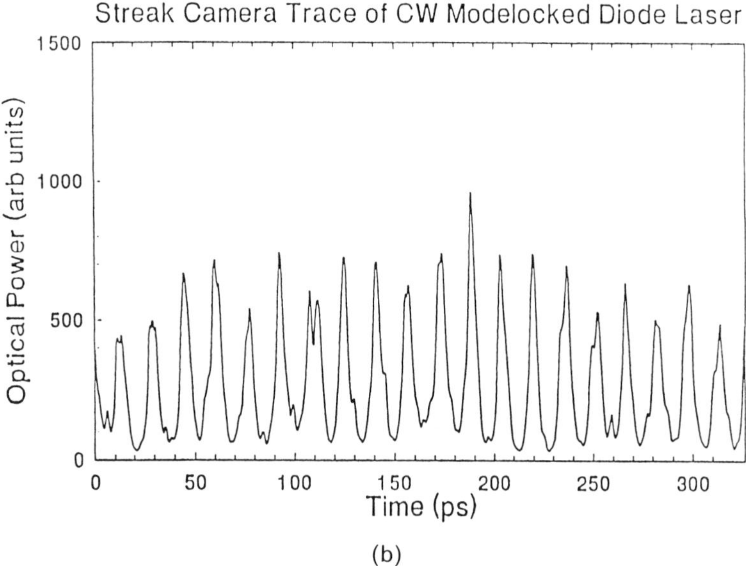


(b)

**Fig. 34.** Time domain trace of (a) self-pulsation, (b) mode-locking, and (c) simultaneous occurrence of both. Note the difference in time scales of these plots.

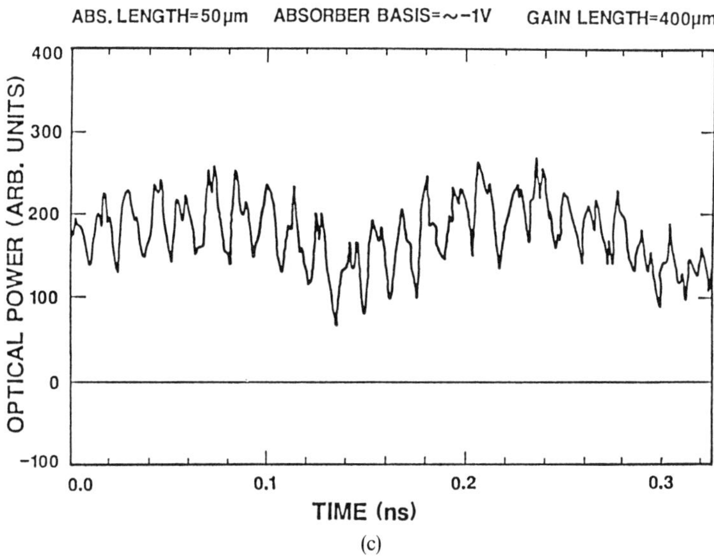


(c)

**Fig. 34.** (*Continued*).

$\hat{g}_{\text{net}}$ is a phasor quantity, represented as a vector in the complex plane, which is responsible for driving the optical modulation. In Eq. (58), $g'_{\text{g/a}}$ and $1/T_{\text{g/a}} = 1/\tau_{\text{g/a}} + g'_{\text{g/a}}S_0$ are the differential gain and effective lifetimes for the gain and absorber, respectively, $\tau_{\text{g/a}}$ and $S_0$ being the spontaneous lifetime and average photon density, respectively; $g_0$ and $a_0$ are the steady state gain and absorption, respectively. The first quantity in the braces in Eq. (58) is the gain modulation, and the second quantity the loss modulation.

For self-consistency, $\hat{g}_{\text{net}}$ must lie within certain quadrants of the complex plane. For mode-locking, it was shown that [63] $\hat{g}_{\text{net}}$ must lie in the right half-plane. For self-pulsation, the (relatively) low frequency net gain modulation drives the average photon density in the laser cavity via the small signal photon rate equation:

$$\dot{s} = \hat{g}_{\text{net}} S_0. \tag{59}$$

If self-pulsation were to take place, $s$ increases with time and, therefore, from Eq. (59), $\hat{g}_{\text{net}}$ must also lie in the right half-plane. These are diagrammatically illustrated in Fig. 35. Thus, the conditions for self-pulsation

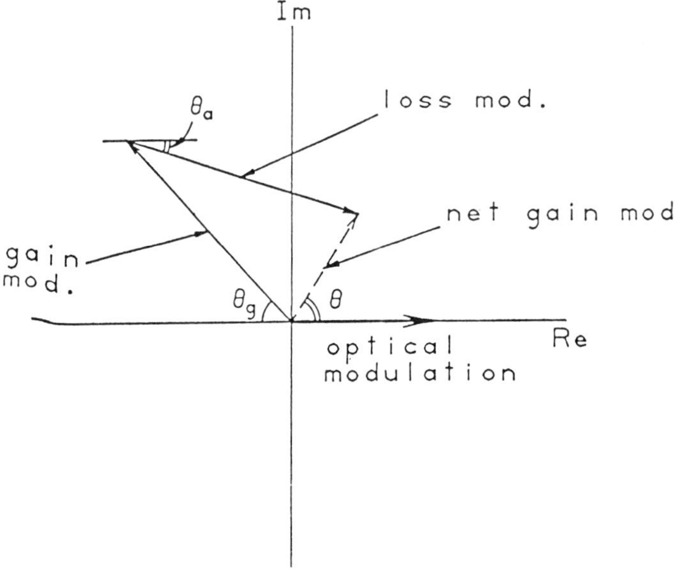


**Fig. 35.** Phasor diagram representation of regimes of mode-locking and self-pulsation. $\theta_a$ and $\theta_g$ are the phase lag of the loss and gain modulation, respectively, behind the optical modulation.

and mode-locking are actually very similar, given respectively by

$$\mathrm{Re}(\hat{g}_{\mathrm{net}}^{\mathrm{sp}}) > 0 \qquad \text{and} \qquad \mathrm{Re}(\hat{g}_{\mathrm{net}}^{\mathrm{ML}}) > 0, \tag{60}$$

where $\hat{g}_{\mathrm{net}}^{\mathrm{SP}}$ and $\hat{g}_{\mathrm{net}}^{\mathrm{ML}}$ are $\hat{g}_{\mathrm{net}}$ evaluated at $\omega = \omega_{\mathrm{SP}}$ and $\Omega$, respectively, and Re( ) denotes the real part. As noted in [63], in order for $\hat{g}_{\mathrm{net}}$ to be in the right half-plane it is preferable that the gain modulation lags in phase behind the loss modulation (Fig. 35). The single-pole roll-off nature of the gain and loss modulation (Eq. (58)) dictates that if gain lags behind loss at $\omega = \Omega$, then the same will happen at $\omega = \omega_{\mathrm{SP}}$. This is very unfortunate, since it implies self-pulsation will very likely accompany mode-locking.

Can one find a situation where mode-locking can take place without self-pulsation? Consider the situation where the various laser parameters satisfy the following criteria:

(*i*) $$g_g' S_0 \gg 1/\tau_g; \qquad g_a' S_0 \gg 1/\tau_a \tag{61a}$$

(*ii*) $$\omega_{\mathrm{SP}} \ll g_a' S_{a'} g_a' S_0 \ll \Omega \tag{61b}$$

These two conditions basically state that the stimulated lifetimes are short—in the sub-nanosecond range—which of course is much longer than $1/\Omega$, the cavity round-trip time. Both (*i*) and (*ii*) can be satisfied at high photon densities. Under these conditions,

$$\hat{g}_{\text{net}}^{\text{SP}} \sim (-g_0 + a_0)sS_0 \tag{62a}$$

and

$$\text{Re}(\hat{g}_{\text{net}}^{\text{ML}}) \sim (-g_0'^2 g_0 + g_a'^2 a_0)\frac{sS_0}{\Omega^2} \tag{62b}$$

But the average gain and loss must satisfy the threshold condition:

$$g_0 - a_0 = g_{\text{th}} > 0 \tag{63}$$

where $g_{\text{th}}$ is the threshold gain. Thus if conditions given by Eqs. (61a, b) are satisfied, then from Eqs. (60) and (62a), self-pulsation will *not* take place. On the other hand, from Eqs. (60), (62b) and (63), mode-locking will take place if

$$\left(1 - \frac{g_{\text{th}}}{g_0}\right) > \left(\frac{g_g'}{g_a'}\right)^2 \tag{64}$$

It is obvious that this last condition can be satisfied more easily if the threshold gain $g_{\text{th}}$ is small. Equations (61a, b) and (64) together serve as a set of prescriptions for achieving mode-locking without self-pulsation, and can be summarized as having a large (intra-cavity) photon density and a low threshold. These are readily achieved with high reflectivity coatings, as the experimental results just described have shown.

The above results can be physically interpreted as follows. Both self-pulsation and mode-locking represent dynamically unstable situations, whose occurrence requires that the loss saturates "easier" than the gain as the photon density increases. However, the unsaturated gain is always higher than the unsaturated loss in a laser above threshold, the difference being determined by the passive cavity $Q$. If the gain and loss saturates instantaneously with rising photon density (as in self-pulsation, where the frequency is relatively low), then it is inevitable that the reduction in gain is always larger than the reduction in loss. If, however, the saturation is not instantaneous (as in high frequency mode-locking), in that the response time of the gain is very long, then over a brief instant the absorber may be reduced by a large amount, while the gain has not yet responded to the rising photon density. This situation is particularly favored in a high-$Q$ cavity, since in this case the unsaturated loss is only slightly smaller than the unsaturated gain, and hence the reduction in loss due to increasing photon density is large compared with the corresponding reduction in gain.

## REFERENCES


1. Sakaki, H. (1980). *Japan. J. Appl. Phys.* **19**, L735.
2. Ikegami, T., and Suematsu, Y. (1967). *Proc. IEEE* **55**, 122.
3. Lau, K. Y., and Yariv, A. (1985). *Appl. Phys. Lett.* **46**, 1117.
4. Lau, K. Y., and Yariv, A. (1985). *Appl. Phys. Lett.* **46**, 326.
5. Olshansky, R., Fye, D. M., Manning, J., and Su, C. B. (1985). *Electron. Lett.* **21**, 893.
6. Su, C. B., and Lanzisera, V. A. (1986). *IEEE J. Quantum Electron.* **QE-22**, 1568.
7. Bowers, J. E., and Pollack, M. A. (1988). *Optical fiber telecommunications* (Miller, S., and Kaminow, I., eds.), Chapter 13. Academic Press.
8. Eom, J., Su, C. B., LaCourse, J., and Lauer, R. B. (1990). *Appl. Phys. Lett.* **56**, 518.
9. Westbrook, L., Fletcher, N., Coolper, D., Stevenson, M. Spurdens, P. (1989). *Electron. Lett.* **25**, 1183.
10. Arakawa, Y., Vahala, K., Yariv, A., and Lau, K. (1986). *Appl. Phys. Lett.* **48**, 384.
11. Arakawa, Y., Vahala, K., and Yariv, A. (1986). *Surface Sci.* **174**, 155.
12. Arakawa, Y., and Yariv, A. (1986). *IEEE J. Quantum Electron.* **QE-22**, 1887.
13. Suemune, I., Coldren, L. A., Yamaniski, M., and Kan, Y. (1988). *Appl. Phys. Lett.* **53**, 1378.
14. Arakawa, Y., and Takahashi, T. (1989). *Electron. Lett.* **25**, 170.
15. Adams, M. J., (1986). *Electron. Lett.* **22**, 250.
16. Yablonovitch, E., and Kane, E. O. (1986). *IEEE J. Lightwave Tech.* **LT-4**, 504.
17. Houng, M. P., Chang, Y. C., and Wang, W. I. (1988). *J. Appl. Phys.* **64**, 4609.
18. Iye, Y., Mendez, E. E., Wang, W. I. and Esaki, L. (1986). *Phys. Rev. B* **33**, 5854.
19. Mittelstein, M., Arakawa, Y., Larsson, A., and Yariv, A. (1986). *Appl. Phys. Lett.* **49**, 1689.
20. Zmudzinski, C. A., Zory, P. S., Lim, G. G., Miller, L. M., Beernik, K. J., Cockerill, T. L., Coleman, J. J., Hong, C. S., and Figuerod, (1991). *IEEE Photon. Tech. Lett.* **3**, 1057.
21. Lau, K. Y., Wang, W., Xin, S., Mittlestein, M., and Bar-Chaim, N. (1989). *Appl. Phys. Lett.* **55**, 1173.
22. Arakawa, Y., Sakaki, H., Nishioka, M., Yoshino, J., and Kamiya, T. (1985). *Appl. Phys. Lett.* **46**, 519.
23. Arakawa, Y., Vahala, K., Yariv, A., and Lau, K. (1985). *Appl. Phys. Lett.* **47**, 1142.
24. Rideout, W., Yu, B., LaCourse, J., York, P. K., Beernink, K. J., Coleman, J. J. (1991). *Appl. Phys. Lett.* **56**, 706.
25. Fukushima, T., Wasserbauer, J. G., Bowers, J. E., Logan, R. A., Tanbun-Ek, T.,

and TEMKIN, H. (1990). *IEEE International Semiconductor Laser Conference*, Davos, Switzerland, paper **PD-18**.
26. UOMI, K., MISHIMA, T., and CHINONE, N. (1985). *Appl. Phys. Lett.* **51**, 78.
27. SAKAKIBARA, Y., YAKEMOTO, A., NAKAJIMA, Y., FUJIWARA, M., YOSHIDA, N., and KAKIMOTO, S. (1989). *Electron. Lett.* **25**, 1530.
28. ARAKAWA, Y., VAHALA, K., and YARIV, A. (1990). *Appl. Phys. Lett.* **45**, 950.
29. ASADA, M., MIYAMOTO, Y., and SUEMATSU, Y. (1986). *IEEE J. Quantum Electron.* **QE-22**, 1915.
30. VAHALA, K. J., ARAKAWA, Y., and YARIV, A. (1987). *Appl. Phys. Lett.* **50**, 365.
31. VAHALA, K. J. (1988). *IEEE J. Quantum Electron.*, **QE-24**, 523.
32. ZAREM, H., VAHALA, K., and YARIV, A. (1989). *IEEE J. Quantum Electron.* **QE-25**, 705.
33. LESTER, L. F., OFFSEY, S. D., RIDLEY, B. K., SCHAFF, W. J., FOREMAN, B. A., and EASTMAN, L. F. (1991). *Appl. Phys. Lett.* **59**, 1162.
34. SHARFIN, W. F., SCHLAFER, J., RIDEOUT, W., ELMAN, B., LAUER, R. B., LACOURSE, J., and CRAWFORD, F. D. (1991). *IEEE Photon. Tech. Lett.* **3**, 193.
35. RIDEOUT, W., SHARFIN, W. F., KOTELES, E. S., VASSELL, M. O., and ELMAN, B. (1991). *IEEE Photon. Tech. Lett.* **3**, 784.
36. NAGARAJAN, R., FUKUSHIMA, T., CORZINE, S. W., and BOWERS, J. E. (1991). *Appl. Phys. Lett.***59**, 1835.
37. WU, T. C., KAN, S. C., VASSILOVSKI, D., LAU, K. Y., ZAH, C. E., PATHAK, B., and LEE, T. P. (1992). *Appl. Phys. Lett.* (in press).
38. TAKAHASHI, T., and ARAKAWA, Y. (1991). *IEEE J. Quantum Electron.* **QE-27**, 1824.
39. KESLER, M. P., and IPPEN, E. P. (1987). *Appl. Phys. Lett.* **51**, 1765.
40. SU, C. B. (1988). *Electron. Lett.* **24**, 370.
41. SU, C. B. (1988). *Appl. Phys. Lett.* **53**, 950.
42. TUCKER, R. S. (1985). *J. Lightwave Tech.* **LT-3**, 1180.
43. LANGE, C. H., and SU, C. B. (1989). *Appl. Phys. Lett.* **55**, 1704.
44. BOWERS, J. E. (1987). *Solid State Electron.* **30**, 129.
45. MORIN, S., DEVEAUD, B., CLEROT, R., FUJIWARA, K., and MITSUNAGA, K. (1991). *IEEE J. Quantum Electron.* **QE-27**, 1669.
46. ZHAO, B., CHEN, T. R., and YARIV, A. (1992). *Appl. Phys. Lett.* **60**, 313.
47. BLOM, P. W. M., MOLS, R. F., HAVERKORT, J. E. M., LEYS, M. R., and WOLTER, J. H. (1990). *Superlatt. Microstruct.* **7**, 319.
48. BRUM, J. A., and BASTARD, G. (1986). *Phys. Rev. B* **34**, 2500.
49. EISENSTEIN, G., WIESENFELD, J. M., WEGENER, M., SUCHA, G., CHEMLA, D. S., WEISS, S., RAYBON, G., and KOREN, U. (1991). *Appl. Phys. Lett.* **58**, 158.
50. SHIMIZU, J., YAMADA, H., MURATA, S., TOMITA, A., KITAMURA, M., and SUZUKI, A. (1991). *IEEE Photon. Tech. Lett.* **3**, 773.

51. Uomi, K., Aoki, M., Tsuchiya, T., Suzuki, M., and Chinone, N. (1991). *IEEE Photon. Tech. Lett.* **3**, 493.
52. Katz, J., Margalit, S., Harder, Ch., Wilt, D., and Yariv, A. (1981). *IEEE J. Quantum Electron.* **QE-17**, 4.
53. Harder, Ch., Katz, J., Margalit, S., Shacham, J., and Yariv, A. (1982). *IEEE J. Quantum Electron.* **QE-18**, 333.
54. Harder, Ch., Van Zeghbroeck, B. J., Kesler, M. P., Meier, H. P., Vettiger, P., Webb, D. J., and Wolf, P. (1990). *IBM J. Res. Develop.* **34**, 568.
55. Kan, S. C., and Lau, K. Y. (1992). *IEEE Photon. Tech. Lett.*, in press.
56. Nagarajan *et al.* (1992).
57. Kan, S. C., Vassilovski D., Wu T. C., and Lau K. Y. (1992). *Appl. Phys. Lett.* **7**, 752.
58. Kan, S. C., Vassilovski D., Wu T. C., and Lau K. Y. (1992). *Photonics Tech. Letters* **4**, 429.
59. Kan, S. C., Vassilovski D., Wu T. C., and Lau K. Y. (1993). *Appl. Phys. Lett.*, Jan. 1993.
60. Moore, N., and Lau, K. Y. (1989). *Appl. Phys. Lett.* **55**, 936.
61. Vahala, K. J., Newkirk, M. A., and Chen, T. R. (1989). *Appl. Phys. Lett.* **54**, 2506.
62. Lau, K. Y. (1990). *Appl. Phys. Lett.* **57**, 2068.
63. Lau, K. Y. (1990). *IEEE J. Quantum Electron.* **QE-26**, 250.
64. Lau, K. Y., Derry, P. L., and Yariv, A. (1988). *Appl. Phys. Lett.* **52**, 88.
65. Lau, K. Y., and Yariv, A. (1985). *IEEE J. Quantum Electron.* **QE-21**, 121.
66. Lau, K. Y., and Yariv, A. (1985). *Semiconductors and semimetals*, Vol. 22B, Chapter 2. Academic Press.
67. McCumber, D. E. (1966). *Phys. Rev.* **141**, 306.
68. Gajic, D., and Lau, K. Y. (1990). *Appl. Phys. Lett.*, Nov.
69. Koch, T. L., and Koren, U. (1990). *IEEE J. Lightwave Tech.* **LT-8**, 274.
70. Kuznetsov, M. A., Willner, A. E., and Kaminow, I. P. (1989). *Appl. Phys. Lett.* **55**, 1826.
71. Lau, K. Y. (1991). *IEEE Photon. Tech. Lett.* (inverted lever).
72. Vahala, K. J., and Newkirk, M. A. (1990). *Appl. Phys. Lett.* **57**, 974.
73. Lau, K. Y. (1991). *Appl. Phys. Lett.* **58**, 1715.
74. Ho, P. T., Glasser, L. A., Ippen, E. P., and Haus, H. A. (1978). *Appl. Phys. Lett.* **33**, 241.
75. Ippen, E. P., Eilenberger, D. J., and Dixon, R. W. (1980). *Appl. Phys. Lett.* **37**, 267.
76. Akiba, S., Williams, G. E., and Haus, H. A. (1981). *Electron. Lett.* **17**, 527–528.
77. Tucker, R. S., Koren, U., Raybon, G., Burrus, C. A., Miller, B. I., Kock, T. L., and Eisenstein, G. (1989). *Electron. Lett.* **25**, 621.

78. Bowers, J. E., Morton, P. A., Mar, A., and Corzine, S. W. (1989). *IEEE J. Quantum Electron.* **QE-25**, 1425.
79. Sanders, S., Eng, L., Paslaski, J., and Yariv, A. (1990). *Appl. Phys. Lett.* **56**, 310.
80. Wu, M. C., and Chen, Y. K. (1990). *Appl. Phys. Lett.* **57**, 759.
81. Haus, H. A. (1976). *IEEE J. Quantum Electron.* **QE-12**, 169.

# Chapter 6

# SINGLE QUANTUM WELL InGaAsP AND AlGaAs LASERS: A STUDY OF SOME PECULIARITIES

**D. Z. Garbuzov and V. B. Khalfin**

*Ioffe Physical-Technical Institute*
*Academy of Sciences of Russia*
*St. Petersburg, Russia*






ISBN 0-12-781890-1

# 1. INTRODUCTION

Scores of publications have appeared in the last few years having to do with the unusual threshold characteristics of MBE- and MOCVD-grown separate confinement heterostructure, single quantum well (SCH SQW) AlGaAs/-GaAs heterolasers. Many of the studies in this area were initiated by two papers published by Zory *et al.* [1, 2]. Starting from the fundamental studies carried out by Tsang [3] and Hersee *et al.* [4], and prior to the preceding papers, the interpretation of experimental threshold characteristics of SCH SQW lasers was based primarily on a qualitative model by which all specific features of these lasers were attributed to the step-like density of states distribution. In addition, the gain versus current dependence was assumed linear, just as is in the case with conventional DH lasers with thick ($\sim 1000$ Å) active regions. Since 1986, most publications on AlGaAs/GaAs SCH SQW lasers no longer use this assumption, and the results of these works differ substantially from the older works both in experimental data obtained and in the models employed in interpretation. In connection with this, it appears that studies of other types of SCH SQW lasers grown by techniques other than those widely used at present may turn out useful in revealing the essential features of these lasers.

This chapter contains, besides the introduction and conclusion, four main sections with material arranged as follows. The second and fourth sections deal primarily with LPE-grown, InGaAsP/GaAs and InGaAsP/InP SCH SQW structures and the parameters of the corresponding laser diodes. The third section discusses the theoretical calculations of the gain versus current characteristics in quantum well structures, with emphasis placed on interpretation of the physical aspects of the problem under study. The fifth section presents a short analysis and comparison with calculations of measured spectral characteristics of MBE-grown AlGaAs/GaAs SCH SQW lasers, which evidence the dominant effect on these characteristics of the higher lying quantized subbands.

# 2. STRUCTURES STUDIED. LUMINESCENCE EFFICIENCY AND OBSERVATION OF SPECIFIC FEATURES IN THE LASER CHARACTERISTICS

## 2.1 LPE-Grown InGaAsP Structures

Most of the results discussed in Sections 2 and 4 relate to SQW structures grown by a modified LPE technique, by which thin QW layers of $A^{\mathrm{III}}B^{\mathrm{V}}$

solid solutions are obtained by deposition from melt onto a fast-moving substrate in a narrow growth slot [5]. The associated technology represents a continuation of the growth studies initiated by Prof. N. Holonyak Jr. at the University of Illinois in the late 1970s–early 1980s [6, 7].

This technique can be used to advantage in producing SQW structures in the In-Ga-As-P system, which does not have easily oxidizable aluminum-containing components. The additional difficulties arising when applying this technique to the growth structures in the traditional Al-Ga-As system have been overcome in an investigation [8, 9] to be discussed at the end of this section. Nevertheless, in the case of the Al-Ga-As system, the simplest and most reliable modification of LPE permitting one to obtain SCH SQW structures is apparently low temperature epitaxy [10–12].

The photoluminescence experiments, combined with direct structural studies [5, 13, 14] carried out in the last five to six years, show that the just mentioned modifications of LPE are capable of yielding SQW structures with active layer thickness down to 30–50 Å, the interface extending over a few lattice parameters (Fig. 1).

Naturally, MOCVD and MBE are superior to liquid phase techniques in preparing SQW structures, in the reproducibility of samples with a given thickness of the active region and the degree of its spatial homogeneity. We believe, however, that the LPE method has an essential asset of providing a high stability in obtaining large values of the efficiency of spontaneous radiative recombination in the active region of SQW structures.

The bulk of the results discussed in the following were obtained for two types of In-Ga-As-P structures grown by a modified LPE technique on GaAs or InP substrates. In structures of the first type, where all layers matched GaAs in the lattice constant, the wide bandgap layers represented the ternary compound $In_{0.49}Ga_{0.51}P$ ($E_g = 1.9$ eV), and the active region was made up of quaternary solid solutions with a composition close to GaAs and corresponding radiation with a wavelength of 0.86 or 0.78–0.81 $\mu$m (Fig. 2a). In the narrower-band structures matching the InP substrate, the composition of the quaternary compound, acting as material for the active region ($In_{0.7}Ga_{0.3}As_{0.65}P_{0.35}$), corresponded to the position of the edge luminescence maximum at $\lambda = 1.3$ $\mu$m (Fig. 2b). In addition to the active region and cladding layers, the SCH SQW structures studied included waveguide layers made of quaternary solid solutions with compositions intermediate between those of the active region and the cladding layers. Besides the SCH SQW devices, we used structures without waveguide layers in the photoluminescence studies (SQW structures). Similar structures, but with an active region in excess of 300 Å, that were likewise employed in the experiments will be called here SCH and DH structures.

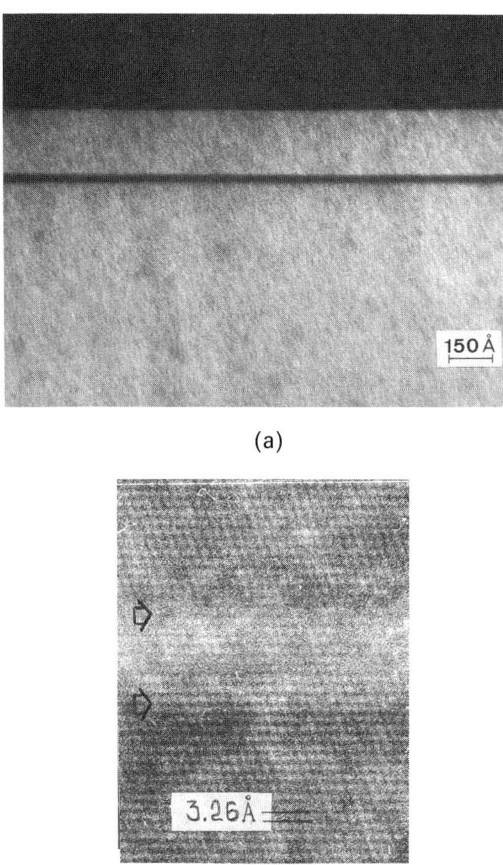


**Fig. 1.** (a) TEM dark-field image in reflection 002 of LPE InGaAsP/GaAs DH. The thickness of active region is about 30 Å. (b) TEM lattice image of the same DH as in Fig. 1a obtained with seven diffracted beams. Arrows mark interfaces.

## 2.2 Photoluminescence: Studies of Radiative Efficiency

A very essential point in any discussion of the characteristics of SCH SQW lasers is the internal efficiency of spontaneous radiative recombination of ($\eta_i$) in their active region. Photoluminescence studies [5] revealed that, in structures of both types with InP and GaAs substrates and wells down to $L_z \cong 100$ Å in thickness, more than 70% of nonequilibrium carriers at 300 K are trapped in the quantum well and recombine in it radiatively. Figure 3a presents measurements of the external quantum yield of luminescence ($\eta_e$)

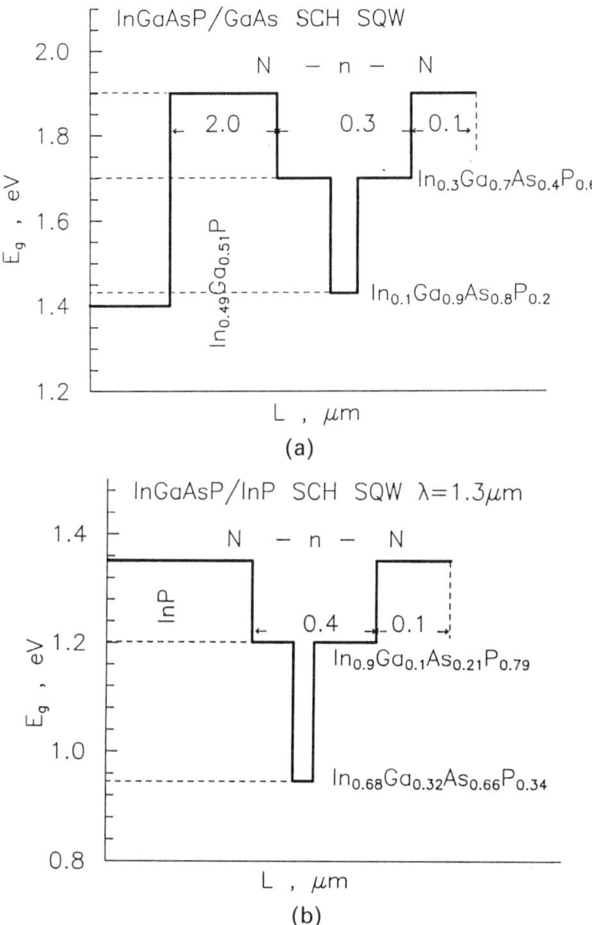


**Fig. 2.** Band schemes of N-n-N InGaAsP/GaAs (a) and InGaAsP/InP (b) SCH SQW structures.

that support this conclusion for InGaAsP/InP structures. The values of $\eta_e$ in the range of 1.2–1.6%, which are typical of SQW structures with $L_z \cong$ 100 Å, correspond to the values of $\eta_i$ approaching 100%.

A similar $\eta_e = f(L_z)$ dependence was obtained also for InGaAsP/GaAs structures with an active region composition corresponding to $\lambda \cong 0.78$ μm [5]. Subsequent experiments carried out on InGaAsP/GaAs structures with a more narrow-band active region ($\lambda = 0.86$ μm) yielded results shown in Fig. 3b. As seen from this figure, for structures without the waveguide layer the values of $\eta_e$ remain at the level of 1.3–1.5%, down to $L_z = 30$ Å. In

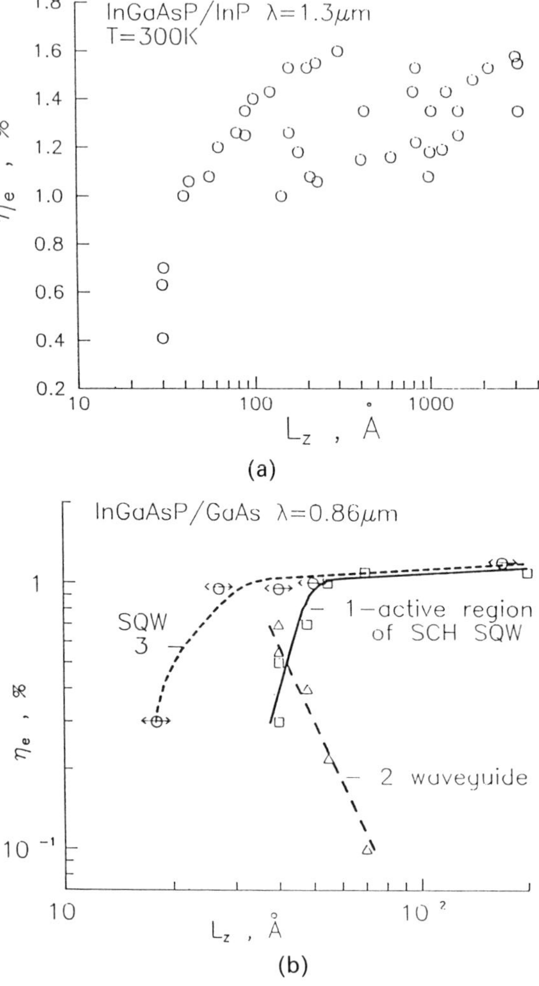


**Fig. 3.** Photoluminescence efficiency for heterostructures with different active region thickness. (a) InGaAsP/InP; active region radiation. (b) InGaAsP/GaAs; circles: active region radiation for SCH SQW structures; squares: the same for SQW structures; triangles: waveguide radiation for SCH SQW structures.

structures with waveguide layers, a fall off of $\eta_e$ becomes noticeable for $L_z \leq 100$ Å. However, the decrease in luminescence intensity from the active region is accompanied by an increase of radiation in the short wavelength band caused by recombination of nonequilibrium carriers in the waveguide layers, the total luminescence efficiency remaining close to its maximum possible value. Thus, the conclusion of the absence of nonradiative recombination losses in InGaAsP/GaAs SCH SQW structures can presently be extended to all structures with active region thickness in excess of 30 Å. To our knowledge, no published data on MBE- and MOCVD-grown AlGaAs/GaAs structures revealed dependencies similar to those presented in Fig. 3. In the course of implementation of the MBE and MOCVD techniques, we carried out measurements of the external efficiency of photoluminescence for a large number of isotopic AlGaAs/GaAs SCH SQW structures grown by these methods [10, 11]. It turned out that, in contrast to LPE, growth by MBE and MOCVD involves factors that affect strongly the radiative recombination efficiency and are difficult to control. It was established nevertheless that for $L_z \geq 100$ Å there are no fundamental reasons that could prohibit obtaining close to 100% internal efficiency of radiative transitions in the active region of MBE- and MOCVD-grown AlGaAs/GaAs SCH SQW structures. This conclusion is supported by curves 1 and 2 in Fig. 4a, relating to two MBE-grown AlGaAs/GaAs SCH SQW structures, with $L_z = 170$ and 100 Å. The band diagram for these structures is shown in Fig. 4b. External photoluminescence efficiencies in excess of 1% under pumping equivalent to 50–100 A/cm$^2$ (curves 1 and 2 in Fig. 4a) were obtained for several best samples selected from the AlGaAs/GaAs structures whose active region thickness was not less than 100 Å. For the best structure with $L_z = 50$ Å, the maximum value of $\eta_e$ was one-third (curve 3, Fig. 4a) that for the samples with $L_z = 100$ and 170 Å, the corresponding value of internal efficiency of radiative recombination not exceeding 40%. In accordance with Fig. 4a, as the temperature was lowered down to 100 K, the value of $\eta_e$ for the $L_z = 50$ Å sample increased nearly by a factor of three, approaching the limit of 1.5%, while the values of $\eta_e$ for structures with $L_z \geq 100$ Å remained practically constant (Fig. 4c). Approximately the same results were obtained on AlGaAs/GaAs SCH SQW structures grown by MOCVD [11].

Turning back to the photoluminescence data obtained on InGaAsP/GaAs and InGaAsP/InP (Fig. 3), it should be pointed out that while these results certainly characterize the quality of the SQW structures in question, nevertheless they cannot be used directly in an analysis of threshold characteristics of the corresponding lasers, since these data relate to lower excitation densities.

In connection with this, studies were carried out [15] of the luminescence efficiency versus excitation density dependence; the results are partially presented in Fig. 5. As seen from curve 1 in this figure, which relates to a InGaAsP/GaAs structure with $L_z = 150$ Å, the magnitude of $\eta_e$ remains constant up to photoexcitation densities equivalent to $10^3$ A/cm$^2$. Only at higher excitation levels is the luminescence efficiency seen to fall off smoothly. In contrast to this, in the case of InGaAsP/InP SQW structures with a similar active region thickness, an increase in optical excitation density leads to a much faster decrease of the luminescence efficiency (curve 2 in Fig. 5).

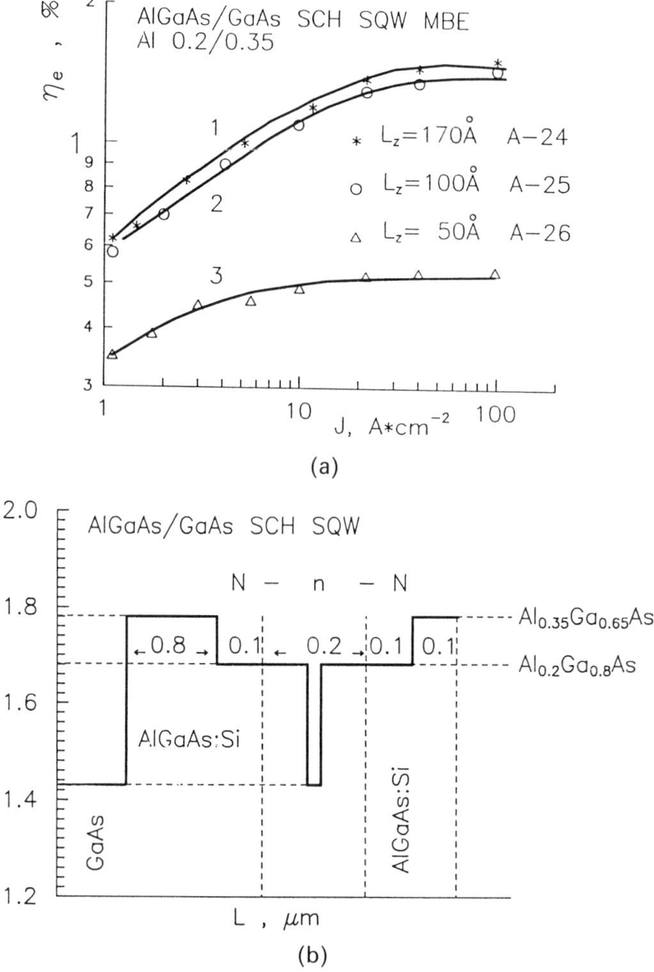


**Fig. 4.** (a) Photoluminescence efficiency as a function of pumping level for AlGaAs/-GaAs SCH SQW with different active region thickness. Optical pumping levels are expressed in units of equivalent current density. (b) Band diagram of the N-n-N AlGaAs/GaAs MBE-grown SCH SQW structures. (c) Temperature dependence of photoluminescence efficiency for two structures from part (a).

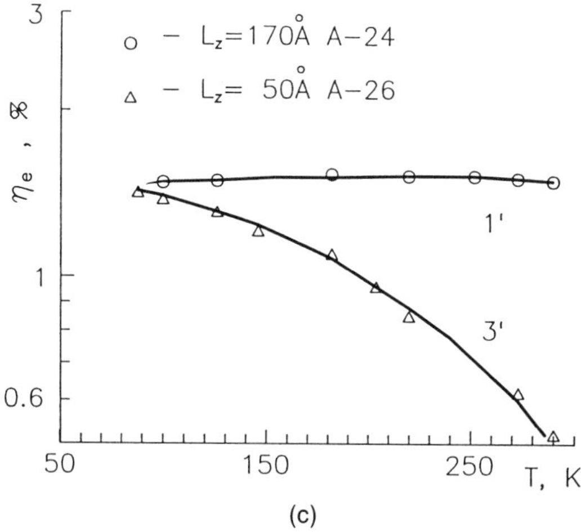


(c)

**Fig. 4.** (*Continued*).

A comparison of $\eta_e$ versus excitation density dependencies obtained for InGaAsP/InP structures with different thickness of the active layer shows that, both in structures with thick three dimensional active regions and in quantum well samples, the fall of luminescence with growing excitation

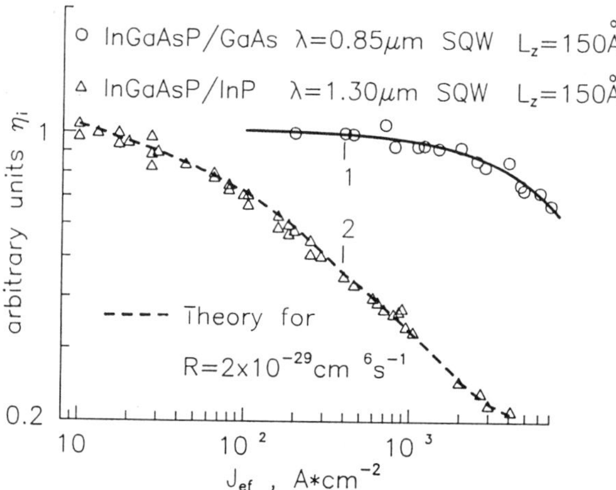


**Fig. 5.** Photoluminescence efficiency versus optical pumping density (in A × $cm^{-2}$ units) for InGaAsP/GaAs and InGaAsP/InP QWs with $L_z$ = 150 Å. Dashed curve: results of theoretical calculations with $R = 2 \times 10^{-29}$ $cm^6\ s^{-1}$. Solid line is a guide for eyes.

density follows the same pattern, and it results from an increasing effect of nonradiative Auger recombination. The results of the corresponding measurement [5] show that the recombination rate constant of this process practically does not change as one goes over from three to two dimensional conditions and is $2 \times 10^{-29}\ \mathrm{cm^6\ s^{-1}}$ at 300°C. The calculated dependence for $\eta_i$ constructed using this value of $R$ is shown by the dashed line in Fig. 5. The values of the radiative transition rates needed for this calculation can be found using expression (2) in Section 3.

## 2.3 Observation of Specific Features in the Threshold Characteristics of LPE-Grown, SCH SQW Lasers

The preceding results of photoluminescence studies of the InGaAsP/GaAs and InGaAsP/InP structures relate to undoped samples, all layers being n-type with electron concentration less than $10^{17}\ \mathrm{cm^{-3}}$. Figure 6 presents band diagrams of InGaAsP/GaAs and InGaAsP/InP p–n junction laser structures, studies of which will be discussed in this subsection and in Section 4. Just as in the case of the previously mentioned isotypic structures, the waveguide layers and the active region of these structures were left purposefully undoped, the p–n junction lying in the upper p-cladding 0.1–0.2 $\mu$m away from the waveguide layer boundary.

In the very first studies [16] of the parameters of injection lasers based on the InGaAsP/InP and InGaAsP/GaAs SCH structures in question, it was found that, in four cleaved facet samples with an active region a few hundreds of angströms thick, one can readily obtain threshold current densities ($J_{th}$) 5–6 times lower than those in similar samples of conventional DH lasers having an order of magnitude greater thickness of the active layer. It turned out, however, that, in normal laser samples with a Fabry–Perot resonator made of thin active region structures, the growth of output losses leads to a dramatic increase of threshold current densities ($J_{th}$); so that, in diodes with cavity length $L \leq 200\ \mu$m, the threshold densities become comparable with or even higher than those in DH lasers with the same cavity length [17, 18].

Particularly strong superlinear growth of $J_{th}$ with decreasing $L$ was observed to occur in InGaAsP/InP SCH lasers. The five- to six-fold threshold increase with $L$ decreasing from 1.5 to 0.15 mm, which is characteristic of such lasers, could not be accompanied by a change of the other parameters of these lasers as well. Indeed, it was found [19] that, in contrast to InGaAsP/InP DH lasers, the parameter $T_0$ characterizing the temperature dependence of the threshold current for InGaAsP/InP SCH lasers depends

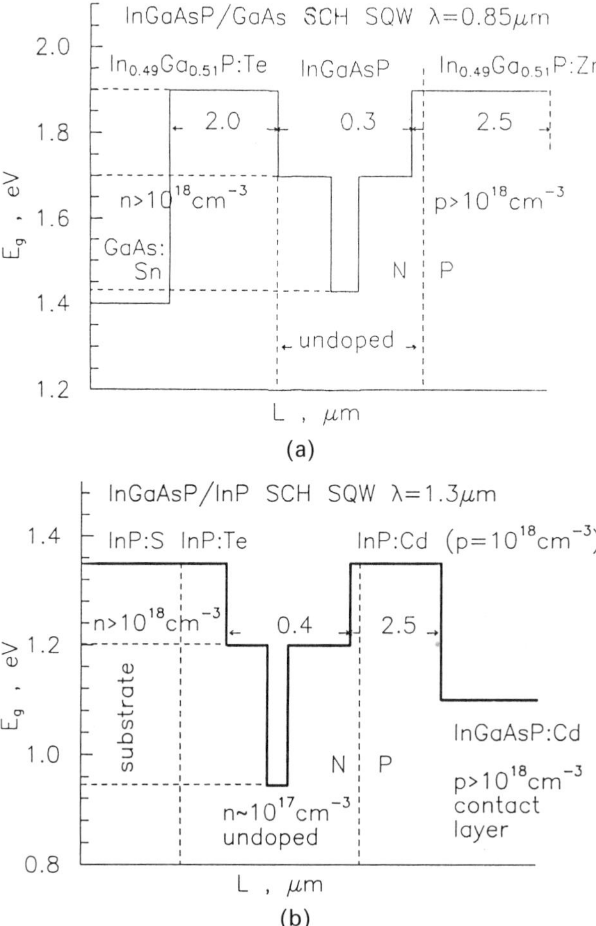


**Fig. 6.** Band diagrams for InGaAsP/GaAs (a) and InGaAsP/InP (b) laser structures.

strongly on cavity length (Fig. 7). To explain these features of the threshold characteristics of InGaAsP/InP SCH lasers, the just mentioned authors invoked the ideas of the effect of nonradiative Auger recombination on the threshold currents.

As for InGaAsP/GaAs SCH SQW lasers, no reasonable explanation for the specific $J_{th}$ $(1/L)$ dependencies has been found, and the corresponding experiments were continued with AlGaAs/GaAs lasers [8, 9] that were made, just as lasers in the In-Ga-As-P system, by LPE on a moving substrate. As seen from the data shown in Fig. 8, the minimum threshold current density for these AlGaAs/GaAs SCH SQW laser diodes with cavity 1.5 mm long

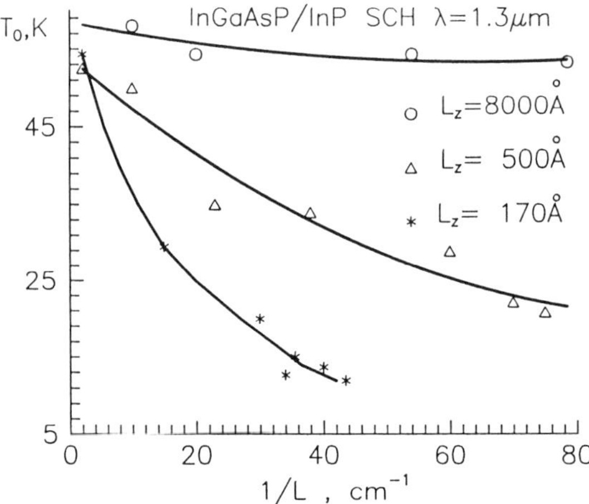


**Fig. 7.** Parameter $T_0$ as a function of reciprocal cavity length for InGaAsP/InP lasers fabricated from SCH structures with different thicknesses of active regions.

was about 260 A/cm$^2$. The rate of growth of $J_{th}$ increased with decreasing $L$ starting already from $L \leq 0.6$ mm, so that at $L = 250\ \mu$m the threshold current density reached 1.4 kA/cm$^2$. In Fig. 8, this dependence is compared with a similar dependence taken from the work of Zory *et al.* [2] mentioned in the introduction. While one cannot exclude a purely accidental coinci-

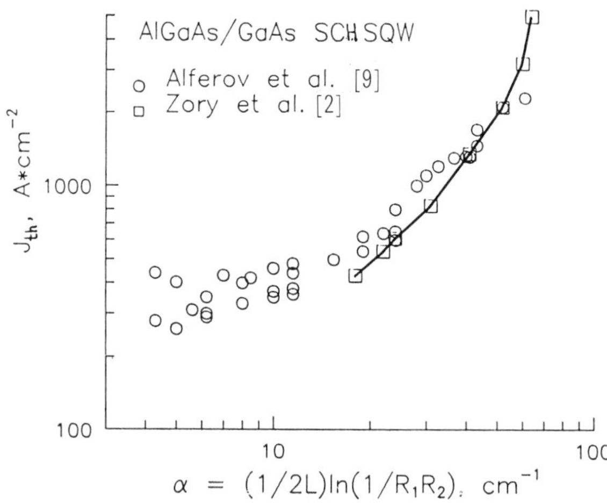


**Fig. 8.** Threshold current density as a function of output losses for AlGaAs/GaAs SCH SQW laser grown by MOCVD [2] and by modified version of LPE [9].

dence between the experimental data of these studies, there is no doubt that the $J_{th}(1/L)$ dependencies have a similar pattern.

The pattern of the threshold versus loss behavior observed for the SCH SQW lasers studied was totally unexpected for us in 1985 and disagreed with the earlier published data; this is why it initiated calculations, carried out in 1985–1986; their results and the model involved are discussed in the next section.

## 3. GAIN MODEL

### 3.1. Theoretical Model and Recombination Transition Rate

The problem of establishing the gain versus pump current dependence consists of two parts: (1) calculation of the gain spectrum in the active region for a given nonequilibrium carrier density; and (2) determination, for this carrier density, of the rate of recombination processes determining the corresponding current density. As follows from the preceding section, the calculation of the total rate of recombination transitions in narrow gap InGaAsP/InP ($\lambda = 1.3\ \mu$m) structure should include, besides the spontaneous radiative recombination, also the nonradiative Auger process. In the first approximation, the corresponding Auger component of the current density can be derived, just as in the 3D case, from the relation

$$J_A = qR_A NP^2 L_z, \tag{1}$$

where $N, P$ are the bulk carrier concentrations in the active region, and $R_A = 2 \times 10^{-29}\ \text{cm}^6\ \text{s}^{-1}$ for the active region compositions corresponding to $\lambda = 1.3\ \mu$m radiation.

In contrast to this, in the case of InGaAsP/GaAs-based lasers, all of the current may be assumed to be due to radiative recombination in the active region. The present subsection will deal primarily with a calculation of the radiative recombination rate. While the results of the corresponding calculations were published [20, 21], their procedure, just as that of the subsequent gain calculations for SCH lasers [22, 23], has not been described in detail.

A comparison with the parabolic model of Chinn *et al.* [24, 25] shows that our calculations may be considered as a somewhat simplified version of this model, the agreement between the two sets of data being not worse than 10–15%. The agreement between the Chinn *et al.* model and the more exact model using valence band mixing is discussed near the end of Chapter 1 in this book. In order for our analysis to be applicable to both 2D and 3D cases, we will formulate it in the most general form possible.

We start by formulating the assumptions underlying the calculations that bear on both spontaneous and induced radiative transitions:

1. The calculations were carried out in the single particle approximation neglecting all many body effects, including Coulomb interaction.

2. The free carrier wavefunction was represented within the effective mass model as a product of a slowly varying envelope with a Bloch function with zero quasimomentum $u(r)$.

3. With item 2 in mind for the direct gap $A^{\mathrm{III}}$–$B^{\mathrm{V}}$ compounds in question, the transition matrix element $|M_{\mathrm{T}}|^2$ between any states belonging to the conduction ($i$) and valence ($j$) bands is written as a product of the matrix element squared for the $u(r)$ functions, $|M_{\mathrm{B}}|^2$, with the square of the overlap integral $C_{ij}$ for the envelope wave functions (see Eq. (22), Chapter 1).

4. The representation of $M_{\mathrm{T}}$ in the preceding form corresponds to the case where the wavevector **k** conservation rules for radiative transitions are rigorously met. In the 3D case, the index $j$ can take on only two values, l and h, respectively, for the light- and heavy-hole subbands. In the 2D case, $i, j$ are the indexes of quantized electron and hole (light or heavy) subbands correspondingly. The transitions between subbands with the same index are taken into account only.

The weakening of the selection rules because of intraband scattering (see Chapter 2, this book) continues to be a subject of debate in many publications, including the already mentioned works [24, 25]. While this effect does not influence substantially the integral radiative transition rate, it may affect the behavior of the gain versus pump current density dependence, a point that will be dealt with later.

5. It can readily be shown that, using the quantities $|M_{\mathrm{B}}|^2$ and $C_{ij}$, the spontaneous recombination rate density can be described by expressions (2) and (2a), which are valid for both the 3D and 2D cases:

$$R_{\mathrm{sp}}(h\nu) = W \sum_{i,j} \rho_{ij} C_{ij} f_{\mathrm{c}i}(1 - f_{\mathrm{v}j}), \tag{2}$$

where

$$W = \frac{4\pi \bar{n}_{\mathrm{e}} E_{\mathrm{g}} e^2 |M_{\mathrm{B}}|^2}{\varepsilon_0 m_0^2 c^3 h^2 L_z}. \tag{2a}$$

A few more comments on the quantities entering (2) and the assumptions underlying their calculation are in order here.

1. The quantity $|M_{\mathrm{B}}|^2$ represents a direction-averaged momentum transition matrix element, which in Kane's approximation reduces to (no spin)

$$|M_{\mathrm{B}}|^2 \cong \frac{m_0^2 E_{\mathrm{g}}(E_{\mathrm{g}} + \Delta)}{6m_{\mathrm{c}}(E_{\mathrm{g}} + \frac{2}{3}\Delta)} \tag{3}$$

2. The coefficient $W$, which, as seen from Eqs. (2) and (3), depends only on four active region parameters (effective refractive index $\bar{n}$, $E_g$, $m_c$, and $\Delta$), is equal to the reciprocal lifetime for a pair of carriers that reside in states satisfying the selection rules. The quantity $1/W = \tau_m$ determines the minimum nonequilibrium carrier spontaneous lifetime that can be achieved experimentally in a given semiconductor, for instance, under very strong electron degeneracy, or high electron–hole plasma densities.

3. The quantity of $\rho_{ij}$ determines the density of the corresponding pairs of states; $f_{ci}$ and $f_{vj}$ are the electron Fermi functions characterizing the occupancy of these states. For the three dimensional case, $\rho_{ij}$ takes on only two values, $\rho_{cl}$ and $\rho_{ch}$, which correspond to the reduced density of states for transitions involving the light (l) and heavy (h) holes (spin included):

$$\rho_{cj} = \frac{8\sqrt{2}\,\pi}{h^3}(m_r^{cj})^{3/2}(h\nu - E_g)^{1/2}, \tag{4}$$

where

$$m_r^{cj-1} = m_c^{-1} + m_j^{-1}.$$

4. In the 2D case, the summation in (2) is performed over transitions between the quantized subbands. The reduced density of states $\rho_{ij}$ is calculated by the expression

$$\rho_{ij} = \frac{4\pi m_r^{ij}}{h^2} H(h\nu - E_{ij}), \tag{5}$$

where $H$ is Heavyside's function, and $E_{ij}$ is the energy gap between the bottoms of the $i$ and $j$ subbands.

5. In the calculations of the energy spectrum, densities of states, and carrier distributions in the quantum well, we neglected (i) light and heavy hole subband mixing, (ii) space charge effect, (iii) band nonparabolicity, and (iv) effects associated with different effective masses in the quantum well and the adjacent layers. The values of the effective mass in the subbands were assumed equal to those in the bulk ($m_c = 0.067, m_h = 0.45, m_l = 0.08$, for InGaAsP/GaAs, and $m_c = 0.058, m_h = 0.43, m_l = 0.07$ for InGaAsP/InP). When calculating the density of carriers in the well, it was assumed that $N_c = P_v$. The neglect of modulation doping is justified by the high values of $N_c$ and $P_v$ under the threshold conditions where $N_c, P_v \geq 10^{18}\ \text{cm}^{-3}$.

With these assumptions in mind, the calculations were performed in the following way in both the 3D and 2D cases. The position of the electron ($F_c$) and hole ($F_v$) Fermi-quasi levels were determined for fixed densities of nonequilibrium carriers. Substituting the quantities $F_c$ and $F_v$ into (2) and integrating in energy yielded the total radiative transition rate $R_{sp}$ and the quantity $R_{sp}e$, which is the driving current density. Next, the ratio $N_c/R_{sp}$, equal to the radiative lifetime of nonequilibrium carriers ($\tau_r$), was calculated for comparison with experimental data.

This approach and Eqs. (2)–(4) as applied to the calculation of $\tau_r$ in the three dimensional case were used in a number of studies (see references in [20, 26]) dealing with the lifetimes in the active region of double heterostructures in AlGaAs/GaAs, InGaAsP/InP, and InGaAsP/GaAs systems. The agreement between the absolute values of experimental and theoretical radiative lifetimes as well as between the temperature and concentration dependencies of $\tau_r$ argued for the validity of our model of radiative transitions. Similar results were obtained also by other researchers who investigated the radiative lifetimes in three dimensional heterostructures [27, 28].

Application of this approach to the two dimensional case leads to a conclusion that the free carrier spontaneous lifetime (in contrast to the case of excitons) should not decrease as a result of quantum well effects [21]. For instance, the minimum radiative lifetime that can be experimentally realized under strong degeneracy in both 3D and 2D conditions is obviously described by Eq. (2a), which for GaAs yields about 0.6 ns. From (2), (3), and (5), it follows also that under low excitation levels the quantity $\tau_r$ does not depend in the 2D case on well thickness provided the majority carrier density in these subwells is the same. The corresponding value of $\tau_r$ should be about the same as in a three dimensional material with the same position of the Fermi level for majority carriers.

Experimental studies performed under low excitation levels for AlGaAs/GaAs and InGaAsP/GaAs N-n-N quantum well heterostructures [29] are in agreement with theoretical calculations, both in the absolute values of the radiative lifetimes and in their weak dependence on $L_z$ in the range of $L_z$ from 50 to 600 Å. The linear temperature dependence of $\tau_r$ predicted by this theoretical model has also been confirmed by experimental data [30].

Summing up, one can say that the radiative transition model neglecting Coulomb interaction and including selection rules provides a good quantitative description of the radiative transition lifetimes for 300 K and $N$, $P \geq 10^{17}\ \mathrm{cm}^{-3}$ in both 3D and 2D cases. This justifies the application of this model also to calculation of the spectra and absolute values of gain, considered in the next section.

### 3.2 Calculation of Gain

By the quantity $g$ called *gain*, we will understand the parameter of the gain medium, which is related to the effective gain ($G$) for an electromagnetic mode propagating along a waveguide that includes a layer of this medium through the expression $G = \Gamma g$, where $\Gamma$ is optical confinement factor. In the 3D case, because of the cubic symmetry of the direct gap compounds in question, the gain in the active region does not depend on the polarization of the laser radiation. The gain spectrum in this 3D case can be described by the relation:

$$g(h\nu) \cong \frac{4\sqrt{2}\,\pi e^2 (h\nu - E_g)^{1/2}}{E_g \varepsilon_0 m_0^2 c h^2 \bar{n}} |M_B|^2 \sum_{j=\mathrm{h,l}} (m_r^{cj})^{3/2} (f_c - f_v), \tag{6}$$

where all the notation has the same meaning as in Eqs. (2)–(3) of the preceding section.

Spatial anisotropy in 2D conditions requires taking into account the polarization dependencies of the gain. In the gain calculations in question, the polarization effects were included in the same way as was done earlier by Yamanishi and Suemune [31]. The approach developed in this work is based on the assumption that the wavefunctions of light and heavy holes in a quantum well are constructed from the corresponding 3D wave functions. Then, just as in the 3D case, the orientation of the transition matrix element vector is determined by the direction of the quasimomentum of the recombining carriers. However, in the 2D case, when $g$ is calculated for a given $h\nu$, summation over quasimomenta is performed not over all directions, as is done for the 3D conditions, but only over the circles where the in-plane wavevector takes on fixed values corresponding to the given $h\nu$. Such a summation does not remove the polarization anisotrophy of gain, and polarization factors $\mu_{ij}$ appear in the expression for gain:

$$g(h\nu) \cong \frac{2\pi e^2 |M_B|^2}{E_g \varepsilon_0 m_0^2 c h \bar{n} L_z} \sum_{i,j} m_r^{ij} \mu_{ij} C_{ij} (f_c - f_v). \tag{7}$$

In the case of amplification of TE-polarized radiation, which is of practical interest, the $\mu_{ij}$ factors are $\frac{3}{4}(1 + \cos^2\theta_{ij})$ and $\frac{1}{4}(5 - 3\cos^2\theta_{ij})$, respectively, for the terms of the sum accounting for the transitions that involve heavy and light holes. The angle $\theta$ characterizes the slope of the general hole quasimomentum with respect to the plane of the structure:

$$\cos^2\theta_{ij} = \frac{E_i^c + E_j^v}{h\nu - E_g}. \tag{8}$$

In Eq. (8), $E_i^c$ and $E_j^v$ are the positions of the bottoms of the corresponding electron and hole quantized subbands measured from the band edges.

Equations (6)–(8) correspond to conditions where the selection rules in induced transitions are rigorously met and do not include the effect on the gain spectra of the energy level broadening associated with fast intraband relaxation processes. To take into account these processes, Yamada and Suematsu [32] introduced a parameter $\tau_{in}$ characterizing the decay time of the dipole formed by the recombining electron and hole, which interacts with the electromagnetic wave. By order of magnitude, $\tau_{in}$ should be equal to carrier quasimomentum relaxation time determined from mobility data. In the straightforward calculations considered here [22], just as in the studies of Yamada and Suematsu [32] and Yamada and Ishiguro [33], it was assumed that the result of the effect of intraband relaxation is equivalent to averaging in energy (with a Lorentzian weighting function of width $\hbar/\tau_{in}$) of the gain spectrum corresponding to the case where the momentum selection rules are rigorously satisfied. The corrected gain spectrum $g'(h\omega)$ was found using the relation

$$g'(h\nu) = \frac{1}{\pi}\int_{-\infty}^{+\infty} g(E')\frac{\hbar/\tau_{in}\,dE'}{(h\nu - E')^2 + (\hbar/\tau_{in})^2} \tag{9}$$

While this approach was shown subsequently [34–36] to contain substantial shortcomings, we will discuss in the next subsection some results obtained with the use of (9). For additional discussions about intraband relaxation, see Chapters 2 and 3, this book.

### 3.3 Results of Calculations for the Low Gain Region ($g \leq 100\ \text{cm}^{-1}$)

Using the theoretical framework developed in Sections 3.1 and 3.2, it is interesting to check to what extent the gain versus current dependencies derived from it agree with the well established experimental data for conventional DH lasers with three dimensional active regions. A convenient parameter for the three dimensional case is the volume current density $j$ (A cm$^{-2}$ $\mu$m$^{-1}$). Using $j$, one can construct a universal curve that depends only on fundamental parameters of the materials of the active region.

The calculation of the universal curve includes the following operations: (1) determination of the peak gain $g$ from Eq. (6) for a given nonequilibrium carrier density $N$ in the active layer of thickness $L_z$; (2) calculation and subtraction from $g$ of the absorption losses of free carriers in the active region ($\alpha_{in}$), so that $\bar{g} = g - \alpha_{in}$. For InGaAsP/InP ($\lambda = 1.3\ \mu$m) lasers, $\alpha_{in}$ includes absorption involving the split-off valence subband; (3) calculation by Eq. (2) (or by (1) and (2)) of the value of $j$ required to produce in the active region the nonequilibrium carrier concentration $N$. The results of the calculations of $\bar{g} = f(j)$ made by this program for the region $\bar{g} < 100\ \text{cm}^{-1}$ without the inclusion of relaxation processes ($\tau_{in} = \infty$) are shown by curves 1 and 2 in

Fig. 9 for GaAs and InGaAsP ($\lambda = 1.3$ $\mu$m) active regions, respectively. The same figure presents experimental data for AlGaAs/GaAs and InGaAsP/InP lasers reduced to the corresponding scales. The data for AlGaAs/GaAs DH lasers were taken from the well known work of Hakki and Paoli [38]. The saturation of the experimental values of $\bar{g}$ for $\bar{g} > 30\ \text{cm}^{-1}$ is due to the onset of lasing. Linear extrapolation in the region of $\bar{g} < 30\ \text{cm}^{-1}$ made in the form $\bar{g} = \beta[j - j_{tr}]$ yields the values of the parameters, $\beta \cong 50\ \text{kA}^{-1}$ $\mu$m cm and $j_{tr} \cong 4.5$ kA cm$^{-2}$ $\mu$m$^{-1}$, which are presently universally adopted for three dimensional GaAs active regions [39]. Considering the accuracy of the available experimental data and of the theoretical calculations, the agreement between the calculation and experiment for $j_{tr}$ and $\beta$ should be considered good. Introducing into the calculation $\tau_{in} = 10^{-12}$ increases $j_{tr}$, making the theory practically coincide with the experiment. However, using such a large value of $\tau_{in}$ could hardly be justified, whereas more realistic values, e.g., $\tau_{in} = 10^{-13}$ s, yield for $j_{tr}$ a value nearly twice that obtained experimentally.

The experimental $\bar{g} = f(j)$ dependence for InGaAsP/InP lasers ($\lambda = 1.3$ $\mu$m) was derived from the data obtained by Razeghi *et al.* [37] and relating to lasers with the lowest known values of the volume current density. As seen

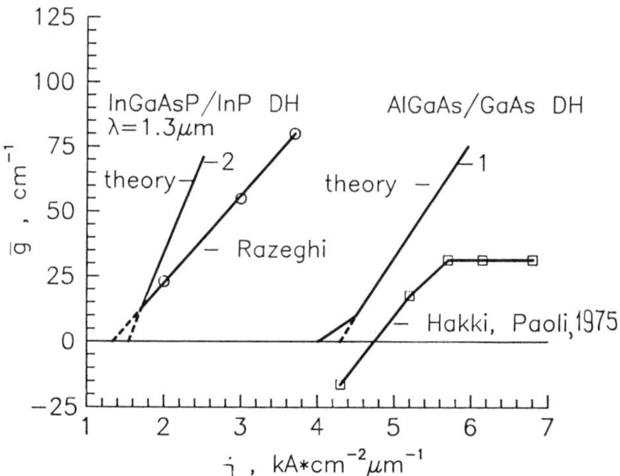


**Fig. 9.** The comparison of theoretical and experimental data on $\bar{g} = f(j)$ dependencies for InGaAsP/InP ($\lambda = 1.3$ $\mu$m) and AlGaAs/GaAs DH lasers with 3D active regions. Experimental data of Razeghi *et al.* [37] and Hakki and Paoli [38] are shown by circles and squares, respectively. The dashed lines drawn to the horizontal axis show the corresponding values of zero-gain current density.

from Fig. 9, in the case of InGaAsP/InP lasers the agreement between theory and experiment for $j_{tr}$ is likewise not bad ($j_{tr} = 1.3$–$1.8\ \mathrm{kA\ cm^{-2}\ \mu m^{-1}}$). The agreement between experimental and theoretical values for the slope of the $\bar{g}$ versus $j$ characteristics is somewhat worse for these lasers than for the AlGaAs/GaAs devices. An attempt at improving the agreement by introducing finite values of $\tau_{in}$ into the calculation has not met with success, just as in the case of lasers with a GaAs active region, since the slope of these characteristics does not depend strongly on $\tau_{in}$, while the value of $j_{tr}$ for $\tau_{in} = 10^{-13}$ s exceeds by more than a factor of two the experimental results for $j_{tr}$.

Considering on the whole the data of Fig. 9, it should be stressed that despite the higher losses in the active region and the effect of the nonradiative current component, both theory and experiment yield values of $j_{tr}$ for the InGaAsP/InP lasers at least two times lower than those for lasers with a GaAs active region.

Calculations suggest that the same relation between the threshold current densities should hold also for lasers with InGaAsP and GaAs quantum well active regions. For comparison, Fig. 10 shows calculated gains for the corresponding 50-Å quantum wells. As seen from Figs. 9 and 10, for InGaAsP/InP lasers, theory predicts not only lower zero-gain currents but also a steeper growth of gain with increasing pump current than is the case with the AlGaAs/GaAs diodes. We believe that this difference between the threshold characteristics, rather than being related to specific features of the active region materials in these two types of lasers, reflects a trend in the variation of these parameters with $E_g$ that is common for all $A^{III}B^{V}$ compounds. Two factors are of major importance here:

1. An increase of the energy density of the spontaneous radiation modes

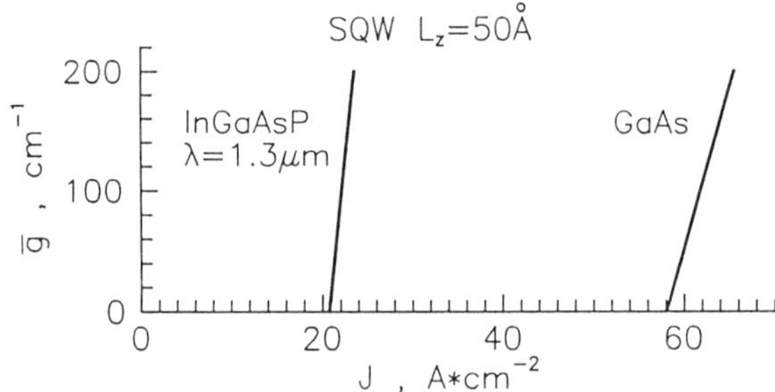


**Fig. 10.** Theoretical gain versus current dependencies for GaAs and InGaAsP 50- Å QW active regions in the low gain range.

with increasing $E_g$ of the active region ($\cong E_g^2$), which reduces the spontaneous radiative lifetime of nonequilibrium carriers.

2. The increase of the $m_c/m_0$ ratio with increasing $E_g$.

The first of these factors provides a dominant contribution to the decrease of the slope of the $\bar{g}$ versus $j$ characteristic in wider-gap materials and increases in them the zero-gain thresholds. The second factor dominates in the increase of the threshold current density in wider-bandgap materials.

The calculated values of the quantities $j_{tr}$ and $\beta$, which describe in the three dimensional case the initial linear regions in the $\bar{g}$ versus $j$ characteristics, are listed in the upper row of Table I.

A number of definitions should be specified prior to using similar quantities in describing gain versus current dependencies for quantum well lasers. Obviously, under strictly two dimensional conditions, the mode gain $G = \Gamma \times \bar{g}$, just as any other experimentally measured quantity, should not depend on active region thickness ($L_z$) for a fixed current density $J$; whence it follows that since $\Gamma \propto L_z$, $\bar{g}$ should be proportional to $1/L_z$. In other words, when extending our definition of $\bar{g}$ to the two dimensional case, the quantity $\bar{g}$ will turn out to depend in the 2D limit on quantum well thickness. With this in mind, and also taking into account that in the two dimensional limit the zero-gain threshold current density $J_{tr}$ becomes an $L_z$-independent quantity, one can approximate the initial part of the gain versus current dependencies by the relation

$$\bar{g} = \beta'[J - J_{tr}]/L_z. \tag{10}$$

The quantity $\beta'$ depends only on the parameters of the active region material and is in the 2D case a constant similar to the constant $\beta$ in the

**Table I.**

Parameters of the Linear Part of the Gain–Current Characteristics for GaAs and InGaAsP Active Layers in 3D and 2D Cases

| | GaAs | | $In_{0.7}Ga_{0.3}As_{0.65}P_{0.35}$ | |
|---|---|---|---|---|
| | kA cm$^{-2}$ $\mu$m$^{-1}$ | cm $\mu$m kA$^{-1}$ | kA cm$^{-2}$ $\mu$m$^{-1}$ | cm $\mu$m kA$^{-1}$ |
| 3D | $j_{tr}$ | $\beta$ | $j_{tr}$ | $\beta$ |
| | 4.3 | 50 | 1.5 | 70 |
| | A cm$^{-2}$ | cm $\mu$m kA$^{-1}$ | A cm$^{-2}$ | cm $\mu$m kA$^{-1}$ |
| 2D | $J_{tr}$ | $\beta'$ | $J_{tr}$ | $\beta'$ |
| | 60 | 180 | 20 | 350 |

three dimensional conditions. As seen from Table I, the calculated values of the $\beta'$ parameter for both the GaAs- and InGaAsP-active regions turn out to be larger by about a factor of four than the 3D values of the corresponding parameters. This result is a direct consequence of the ideal step-like distribution of the density of states assumed in the two dimensional version of the model under consideration.

As $L_z$ increases and one crosses over to quasi two dimensional conditions, the initial parts of the $\bar{g}$ versus $j$ dependencies should become smoother because of the effect of the higher lying subbands on the carrier distribution. Figure 11 shows the results of calculations of $\beta'$ carried out for a GaAs-active region structure (Fig. 2a) for several values of $L_z$. At $L_z = 200$ Å, the value of $\beta'$ differs already only by a factor of two from its three dimensional limit; while at $L_z = 500$ Å, the values of $\beta$ and $\beta'$ practically coincide.

### 3.4. Results of Calculations for the High Gain Region

The idea that the values of $\beta'$ in SCH SQW structures is larger than $\beta$ in conventional DH lasers has until recently been universally adopted and used frequently in interpretation of experimental data. Consider now their applicability to real operational conditions in such lasers.

Figure 12 presents the dependencies of the optical confinement factors ($\Gamma$) on active region thickness for the two types of laser structures discussed here. Also shown in the figure are the distributions of the values of the refractive

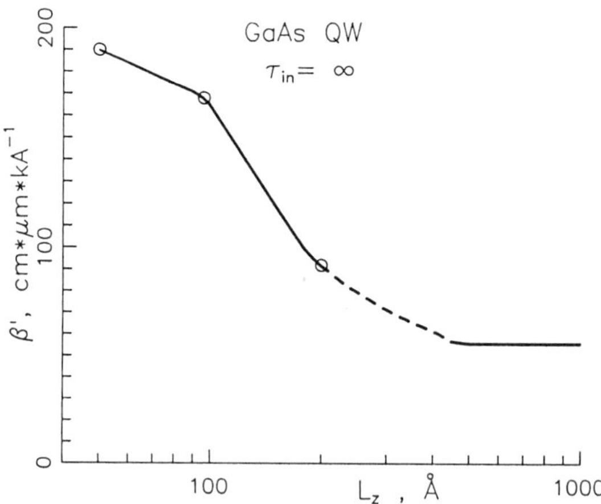


**Fig. 11.** Gain coefficient $\beta'$ as a function of GaAs active region thickness. The horizontal line represents 3D limit for $\beta'$.

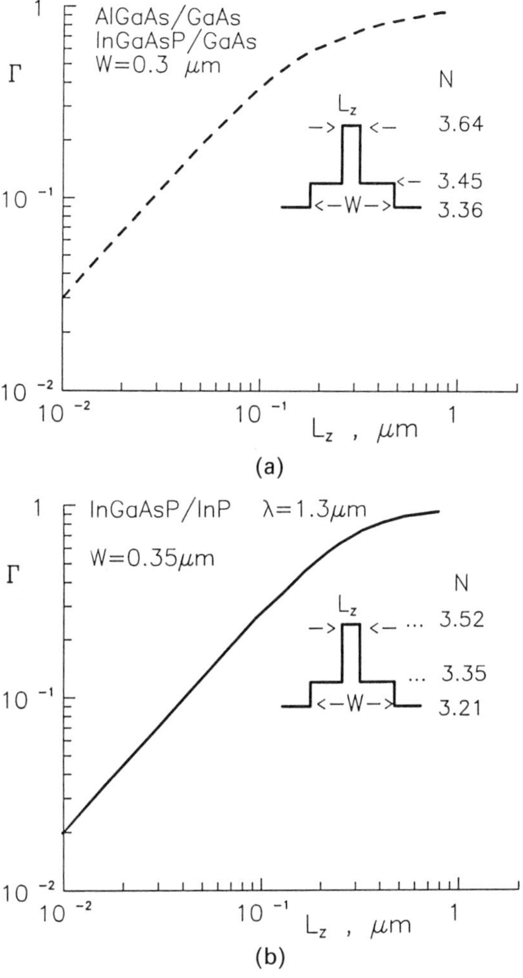


**Fig. 12.** The optical confinement factor dependencies on the active region thickness for InGaAsP/GaAs (a) and InGaAsP/InP (b) SCH SQW structures. The refractive index distributions are shown in the insets.

index adopted in the calculation of these dependencies. The values $\Gamma = 0.02$–$0.03$ for $L_z = 100$ Å derived from them are typical of the SCH SQW structures, and an increase of the refractive index differences to the maximum possible values for the AlGaAs/GaAs system can provide an increase of $\Gamma$ by not more than a factor of 1.5. Thus, the value of $\bar{g}$ corresponding to the onset of lasing, for example, in a diode with cavity length $L = 300$ $\mu$m and output losses $\alpha = 30$ cm$^{-1}$ should be not smaller than the product $\alpha\Gamma^{-1}$, i.e., should be near 1000 cm$^{-1}$, which exceeds by an order of magnitude the range of $\bar{g}$ considered in Fig. 9.

Figure 13 illustrates the calculation of the gain in narrow (50 Å) GaAs and InGaAsP quantum wells for $\bar{g}$ varying over a broad range of $10^2$–$10^4$ cm$^{-1}$. As seen from the calculated curves, for $\bar{g} > 10^3$ cm$^{-1}$, the rate of gain increase with increasing current density slows down rapidly, this trend manifesting itself nearly in the same way for both types of structures despite the difference in their energy diagrams and in active region material parameters. We believe that it is this phenomenon (to be called in what follows *the gain saturation effect*) that determines to a considerable extent the shape of the threshold characteristics of SCH SQW lasers. We turn now to a comprehensive discussion of this effect.

It can be shown that in an analysis of the gain saturation effect for both types of structures with $L_z = 50$ Å, the influence of the higher lying subbands may be neglected, the inclusion of transitions between the first electron and the first heavy hole subbands being sufficient for a qualitative consideration. As follows from calculations and straightforward estimates, for current densities close to $J_{tr}$, the electron Fermi level lies in the electron subband ($F_e \cong 2kT$), and the hole Fermi level above the top of the valence subband by the same amount ($F_p \cong 2kT$), in both the two and three dimensional cases (Fig. 14). For $J$ only slightly above $J_{tr}$, the fast growth of the peak spectral gain with increasing pumping is due to increasing occupancy by both holes

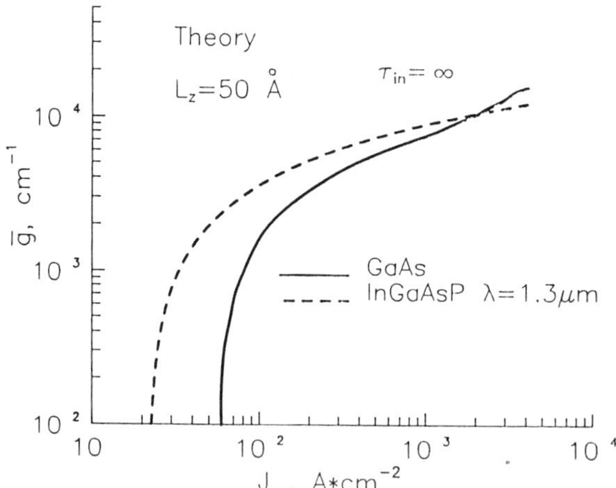


**Fig. 13.** The gain versus current density theoretical dependencies for SCH SQW lasers with GaAs and $In_{0.68}Ga_{0.32}As_{0.66}P_{0.34}$ 50-Å QW active regions. The calculation is stopped when the electron quasi-Fermi level comes to the top of the well.

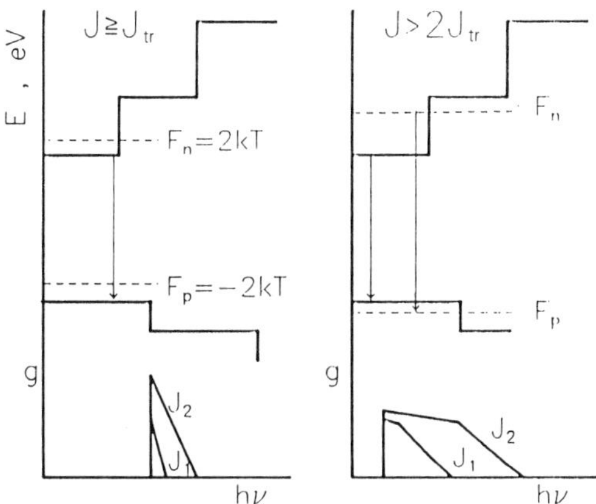


**Fig. 14.** Scheme of quasi-Fermi level positions and gain spectra for the near zero-gain current ($J \sim J_{tr}$) range and for the range of gain saturation ($J > 2J_{tr}$). ($J_2 > J_1$ in both cases.)

and electrons of states close to the subband bottom. However, starting with some values $J'$ ($J' \cong 2J_{tr}$) of the driving current, the states near the bottom of the electron subband become degenerated to such an extent that a further growth of gain for $h\nu_{max} = E_g + E_i + E_j$ with increasing driving current can occur only at the expense of a growth of the hole population. Obviously, at this pump level, one should expect $\bar{g} \propto J^{1/2}$. For still higher current densities, as hole degeneracy sets in, an increase of the pump level in the two dimensional limit should result only in a broadening of the gain spectrum, with no further increase of the maximum gain.

It is evident that such complete gain saturation cannot be reached in the 3D case, where, because of the density of states increasing as one moves into the band, the gain maximum under the conditions of complete electron and hole degeneracy should shift toward shorter wavelengths. At the same time, it is clear that under these conditions the maximum gain versus current dependence will be sublinear in the 3D case as well ($\bar{g} \propto j^{1/3}$ under total degeneracy of the electron–hole plasma), and, hence, the gain saturation effects are not a specific property of quantum well structures.

Figure 15 illustrates, for a 200-Å GaAs active region, the gain versus current dependencies calculated by means of Eq. (7), which takes into account quantum well effects, and of Eq. (6), which assumes the active region material to retain the three dimensional properties. As seen from Fig. 15, a slowed-down growth of gain with driving current manifests itself, on the average, in the same way in both dependencies and has the same cause in both cases,

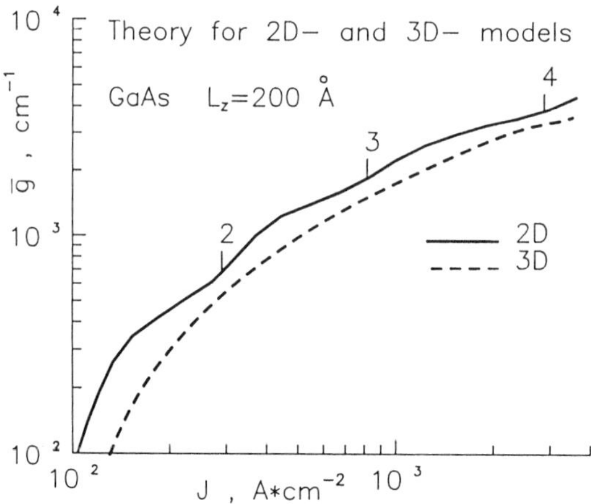


**Fig. 15.** The gain versus current density dependencies for SCH SQW with 200-Å GaAs active region calculated in 2D and 3D approximations.

namely, the degeneracy of the states responsible for the transitions that correspond to the maximum values of $\bar{g}$ in the spectrum. An analysis of the $\bar{g}$ versus $j$ dependence corresponding to the three dimensional conditions shows that here, just as in the 2D case, the population of the electron states should change dramatically in the current variation range of $j_{tr}$ to $2j_{tr}$. In the region of $j \geq 2j_{tr}$, where gain saturation sets in, and where the filling of the electron states no longer contributes significantly to the growth of $\bar{g}$, approximation in the three dimensional case yields a relation $\bar{g} \cong j^{2/3}$, which differs somewhat from the corresponding approximations in the 2D case. The previously mentioned conditions of total electron and hole degeneracy, under which in the three dimensional case $\bar{g} \cong j^{1/3}$, can occur only at volume current densities $\geq 3 \times 10^5$ A $cm^{-2}$ $\mu m^{-1}$, which are difficult to realize.

Summing up our discussion, one can say that the value of the parameter $\beta$ or $\beta'$ determining the threshold characteristics of lasers depends not only on the actual quantum well or bulk properties of the gain medium in their active region, but also on how much the driving current should exceed $j_{tr}$ for the threshold conditions set in real lasers.

For instance, if output loss compensation in an SCH SQW laser requires a current density of twice the value of $J_{tr}$, then the value of $\beta'$ in such a device may turn out to be smaller than that in a conventional DH laser with a 3D active region, if in this laser $j_{th} \sim j_{tr}$.

# 4. CHARACTERISTICS OF InGaAsP SCH SQW LASERS

## 4.1 InGaAsP/GaAs Laser (850 nm)

This subsection reports on studies of InGaAsP/GaAs SCH SQW lasers carried out in 1986 [22]. The studies were performed on broad-contact lasers made up of several InGaAsP/GaAs SCH SQW structures ($\lambda = 0.85\ \mu$m) with an active region about 100 Å thick. The cavity length in the laser diodes varied from 2 mm to 100 $\mu$m.

Figure 16 presents the current density dependencies of the external efficiency ($\eta_e$) of spontaneous radiation for three InGaAsP/GaAs diodes differing in cavity length. Calculations show that the values of $\eta_e \cong 1.8$–2% obtained for the shortest cavity diodes ($L = 100\ \mu$m) correspond to the quantum yield limit ($\eta_i = 100\%$) for spontaneous radiation propagating along the waveguide of the SCH SQW structure. An essential point is the nearly constant value of $\eta_e$ for the short-cavity diode throughout the current density region, where longer-cavity samples exhibit a transition to lasing (curves 2 and 3 in Fig. 16). In our opinion, this result permits one, in an analysis of the dependence of $J_{th}$ on output losses $\alpha$, not only to exclude from consideration the effect of the nonradiative recombination channels, as this could already be concluded from the photoluminescence experiments described in Section 2, but also to neglect the current leakage, which likewise could influence the dependencies discussed in what follows.

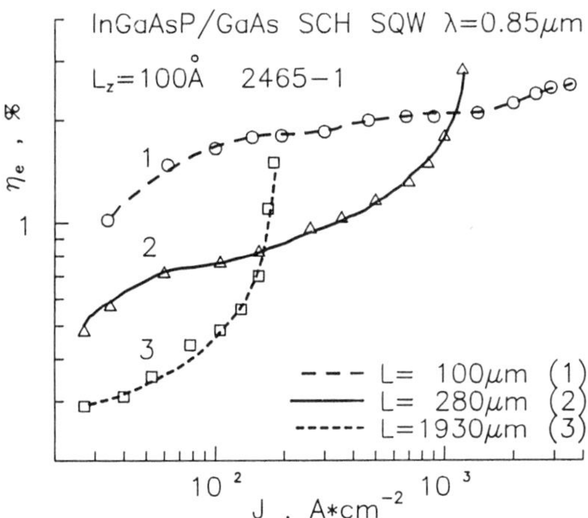


**Fig. 16.** The radiation efficiency as a function of driving current density for three broad area InGaAsP/GaAs diodes with different cavity lengths.

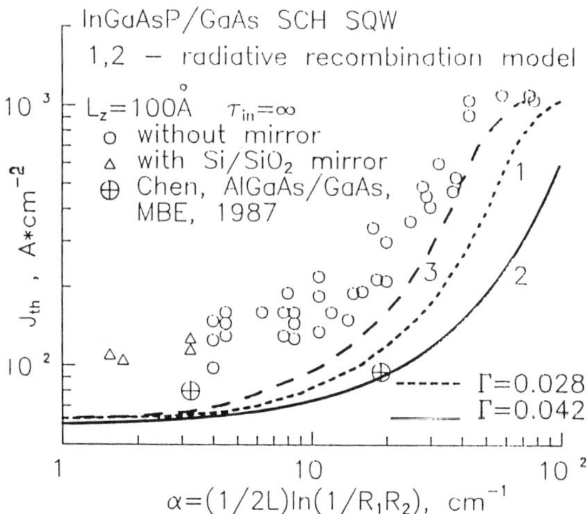


**Fig. 17.** Threshold current density versus output losses for InGaAsP/GaAs SCH SQW lasers ($\lambda = 0.85\ \mu$m). Circles and triangles represent our experimental results. Circles with crosses are the data of Chen *et al.* [40] for AlGaAs/GaAs diodes. Curves 1 and 2 demonstrate the results of the calculations for $L_z = 100$ Å ($\tau_{in} = \infty$) with $\Gamma = 0.028$ and 0.042, respectively. Curve 3 is obtained by a shift of the curve 1 along the $\alpha$ axis.

The $J_{th}$ versus $\alpha$ dependence for the diodes discussed here is shown in Fig. 17. The values of the threshold current density $\sim 100$ A/cm$^2$ obtained for InGaAsP/GaAs diodes with $L = 2$ mm were a record low for injection lasers at the time (1986). Utilizing strained quantum wells (see Chapters 1, 7, and 8 of this book), record $J_{th}$ values are now in the vicinity of 50 A/cm$^2$. To consider the general behavior of the $J_{th}$ versus $L$ dependence, we start with the results presented in Figs. 8 and 17.

As seen from Fig. 17, in the case of InGaAsP/GaAs SCH SQW lasers, fast superlinear increase of the threshold begins already at output losses in excess of 20 cm$^{-1}$, so that this dependence resembles, on the whole, those for AlGaAs/GaAs SCH SQW lasers shown in Fig. 8. However, the absolute values of $J_{th}$ presented in Fig. 8 exceed, on the average by a factor of 2.5 for the same output losses, those for InGaAsP/GaAs lasers. Therefore, the comparison with the ideal version of the theoretical model was carried out for InGaAsP/GaAs SCH SQW lasers.

The corresponding calculated current density dependence of the gain for lasers with a 100-Å active region obtained for $\tau_{in} = \infty$ is given in Fig. 18a. The second vertical axis in this figure corresponds to the mode gain

$\bar{G} = \Gamma \times \bar{g}$ in the laser diodes in question, for which the optical confinement factor was evaluated by us to be 0.03. The $J_{th}$ versus $\alpha$ dependence constructed with the data of Fig. 18a is shown in Fig. 17 with dashed curve 1. The agreement in the behavior of the experimental and calculated threshold current density dependencies on losses suggests that gain saturation is the

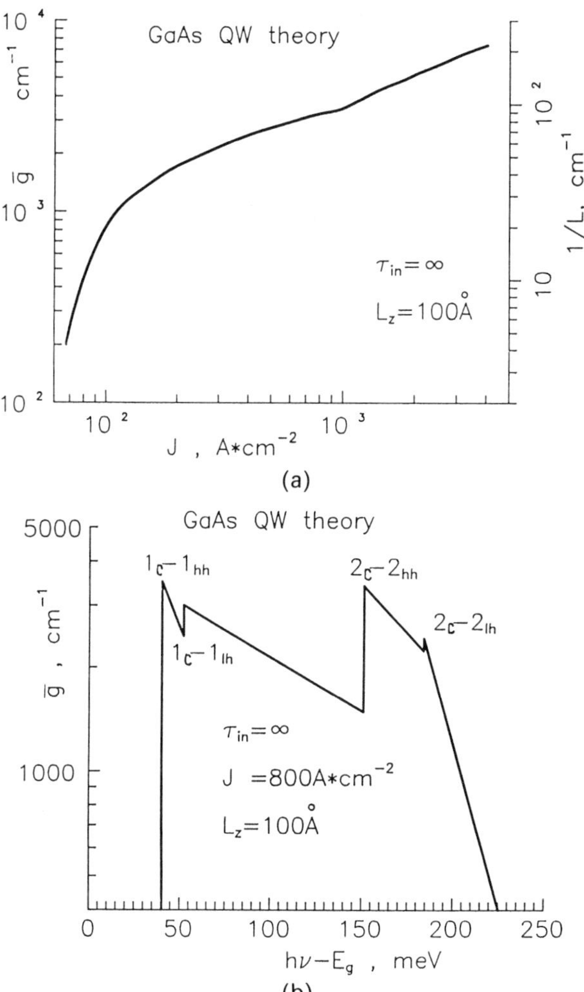


**Fig. 18.** Gain versus current density dependence calculated for a 100-Å GaAs QW. Singularity in the curve slope is due to the shift of transferring maximum gain position to the region of $2_c$–$2_{hh}$ transition energy range. (b) Gain spectrum for the same QW at current density in the transition region.

main cause of the superlinear behavior of the experimental $J_{th}$ versus $\alpha$ dependence.

An interesting calculation corresponding to $\Gamma = 0.042$ and obtained in the same way as curve 1 under the assumption $\tau_{in} = \infty$ is shown by the solid curve in Fig. 17. The close agreement of this calculated curve with the experimental data (crossed circles) of Chen *et al.* [40] is hardly accidental and may be considered as an indication that, at least for low losses where threshold current densities are not much different than $J_{tr}$ in the case of GaAs QW active regions, there are no fundamental reasons that could cause a substantial growth of $J_{th}$ compared with the ideal theoretical model. It should be stressed, however, that the effects resulting in a decrease of the gain at a given carrier density $N$ may become noticeable as one crosses over to the high gain region, where $J_{th} \gg J_{tr}$. One may expect that the broadening of the density of states distribution caused by fluctuations in the QW active region thickness and intraband relaxation should influence in exactly the same way the $g(J)$ dependence.

These effects account probably for the difference in the absolute experimental and calculated values of $J_{th}$ for InGaAsP/GaAs SCH SQW lasers with $\lambda = 0.85$ $\mu$m (Fig. 17). As seen from Fig. 17 (curve 3), the absolute values of the threshold current density for losses $\alpha > 0$ cm$^{-1}$ can be matched by shifting the calculated curve by an amount equivalent to a decrease of the gain by a factor of 1.5. The reasons for the disagreement between the values of $J_{th}$ in the region of $\alpha < 10$ cm$^{-1}$ remain unclear. One can note, however, that the excess of the experimental values of $J_{th}$ over the calculations cannot be attributed to nonradiative leakage (see Fig. 16) or internal losses that were not taken into account when presenting the experimental data. (As follows from the later discussion in Section 4.3, the magnitude of the current-independent losses in the InGaAsP/GaAs SCH SQW lasers does not exceed 1 cm$^{-1}$.)

We should mention here another phenomenon whose investigation was stimulated by calculations of the current dependence of the gain for the InGaAsP/GaAs SCH SQW lasers in question. The calculations accounting for the superlinear behavior of the $J_{th}$ versus $\alpha$ dependence revealed that, in such lasers with $L_z = 100$ Å, there should occur in the current density range studied a jump of the absolute maximum in the gain spectra from the position corresponding to the $1_c$–$1_{hh}$/$1_c$–$1_{lh}$ transitions to that for the $2_c$–$2_{hh}$. Figure 18b illustrates a calculated gain spectrum for the transitional current density. At the corresponding current densities, a step appears in the calculated $g$ versus $J$ plot (Fig. 18a).

In full agreement with the theoretical prediction, experimental studies of the lasing peak positions [22] have indeed revealed its jump toward shorter wavelengths with decreasing cavity length of the laser diodes in question (Fig. 19). The values of $1/L$ and $J_{th}$ at which the jump was observed to occur are lower than the calculated ones, a fact that perhaps can also be attributed to the previously mentioned processes, which result in a broadening of the density of states distribution in the active region of the laser diodes studied.

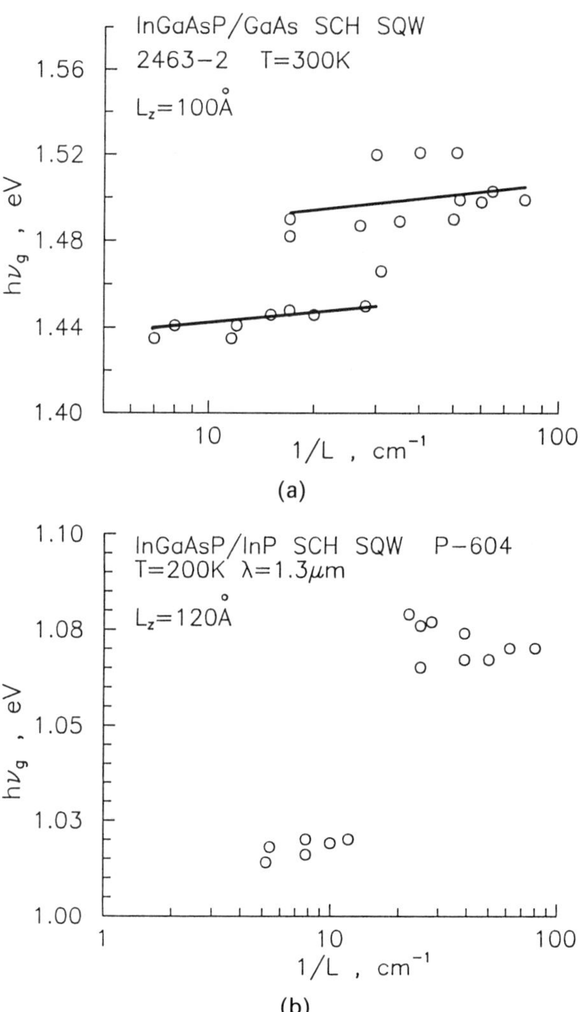


**Fig. 19.** Spectral position of laser peak as a function of reciprocal cavity length (a) for InGaAsP/GaAs and (b) for InGaAsP/InP laser diodes.

## 4.2 InGaAsP/InP Lasers (1300 nm)

Similar studies of the position of the lasing peak have been carried out also in 1986 for quantum well lasers in the InGaAsP/InP system. Figure 19b shows a plot of peak lasing energy $h\nu_g$ versus $1/L$ dependence for InGaAsP/InP SCH SQW diodes [41].

Somewhat earlier, similar results were obtained for SCH SQW lasers in the traditional AlGaAs/GaAs system, and in 1986, three research groups reported at about the same time on a jump of $h\nu_g$ in such lasers [1, 42, 43].

Postponing a detailed discussion of this subject to Section 5, we turn now to a consideration of the threshold characteristics of InGaAsP/InP-based laser diodes ($\lambda = 1.3$ $\mu$m). The results presented in what follows were obtained in an investigation of mesa stripe and buried mesa stripe diodes with a stripe width $W = 12$–$16$ $\mu$m. Several SCH SQW structures (Fig. 6b) with active region thickness varying from 150 to 250 Å were used to manufacture the lasers on which the data presented in Figs. 20 and 21 were obtained. Figure 20 shows the dependencies of the spontaneous radiation efficiency $\eta_e$ on current density for three diodes prepared from one of these structures and differing significantly in cavity length. In contrast to similar measurements made on InGaAsP/GaAs diodes (Fig. 16), the value of $\eta_i$ for the shortest cavity, nonlasing InGaAsP/InP laser diode, rather than remaining

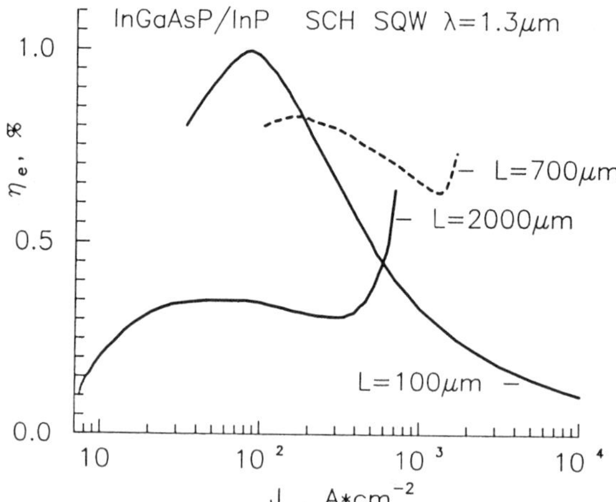


**Fig. 20.** The dependencies of radiation efficiency on current density for three mesa stripe InGaAsP/InP laser diodes ($L_z = 200$ Å) with various cavity lengths.

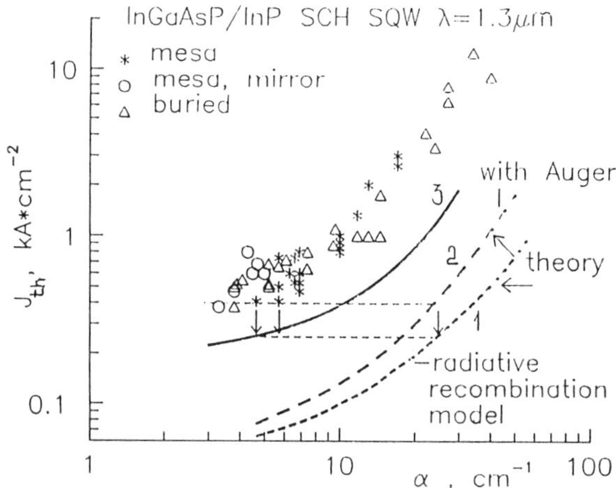


**Fig. 21.** Threshold current densities plotted against the value of output losses for mesa stripe and buried InGaAsP/InP laser diodes with $L_z \cong 200$ Å. Dashed curves are the results of calculation with and without Auger recombination. The solid curve is obtained by subtraction from experimental data of the Auger component of threshold current as it is shown in the figure.

constant, decreased with increasing current density in the range $10^2$–$10^3$ A/cm$^2$, where the diodes with greater $L$ exhibited a transition to lasing (Fig. 20).

Figure 21 presents the dependence of $J_{th}$ on output losses ($\alpha$) for the laser diodes in question. The corresponding studies show that for $W > 12$ $\mu$m the values of $J_{th}$ do not depend on the width of the stripe or on the method of its preparation. The value $J_{th} \cong 360$ A/cm$^2$ obtained for several diodes with the minimum output losses ($\alpha = 3$–$5$ cm$^{-1}$) seems to be the lowest published value for InGaAsP/InP lasers ($\lambda = 1.3$ $\mu$m). Nevertheless, they are substantially greater than the values of $J_{th}$ for lasers with a GaAs active region, or than those that could be expected based on the results of Razeghi *et al.* [37]. In the preceding work, the value $J_{th} = 430$ A/cm$^2$ was obtained for conventional DH InGaAsP/InP lasers ($\lambda = 1.3$ $\mu$m) with active region thickness an order of magnitude greater than that in the SCH SQW samples under study. Apart from this in the case of DH lasers, a linear, comparatively slow growth of $J_{th}$ occurred with $L$ decreasing from 1 down to 0.3 mm, while for SCH SQW lasers the threshold current density begins to increase in a superlinear way starting already from $L < 1$ mm (Fig. 21).

A comparison with the threshold calculations discussed in Section 3 and illustrated by curves 1, 2 in Fig. 21 shows that Auger recombination can only partially account for the higher values of $J_{th}$ of the additional current component that may be due to Auger recombination. A comparison of the obtained data suggests that other mechanisms resulting in an increase of $J_{th}$ must also be present. As will be shown in the next section, in the case of InGaAsP/InP SCH SQW lasers whose threshold current densities exceed by far the values of $J_{th}$ for InGaAsP/GaAs, SCH SQW lasers with $\lambda = 0.85\ \mu m$ (compare Figs. 21 and 17), these mechanisms in the region of $\alpha > 20\ cm^{-1}$ are the growth of losses associated with absorption involving the injected carriers, and electron leakage to the p-cladding layer. The effects resulting in a broadening of the density of states distribution that were discussed in the preceding section also undoubtedly contributed to the experimental values of $J_{th}$ being in excess of the predictions of the ideal model.

### 4.3 Differential Quantum Efficiency

In an analysis of the differential efficiency ($\eta_d$) of SCH SQW lasers, one can single out two features making them different from lasers based on conventional DH structures. The first feature, which has been known already for a long time, consists in the possibility of obtaining high values of $\eta_d$ in long cavity, low threshold SCH SQW laser diodes where the losses to absorption by free carriers can be substantially lower than those in conventional DH lasers and can be represented as $\alpha_{in}\Gamma$, where $\Gamma = 0.03$ is the optical confinement factor, and $\alpha_{in}$ are the internal losses in conventional DH lasers based on the corresponding heterostructures ($\alpha_{in} \sim 10\ cm^{-1}$ for lasers with a GaAs active region, and $\alpha_{in} \sim 20\ cm^{-1}$ for InGaAsP/InP lasers operating at $\lambda = 1.3\ \mu m$). The possibility of implementing this feature in the InGaAsP/GaAS and InGaAsP/InP, SCH SQW diodes studied by us is illustrated by Figs. 22 and 23. As seen from Fig. 22, the values of $\eta_d$ in excess of approximately 80% have been obtained for lasers with output losses $\sim 10\ cm^{-1}$, both for the InGaAsP/GaAs laser diodes with $\lambda = 0.85\ \mu m$, whose threshold characteristics were discussed in the preceding section, and for the InGaAsP/GaAs SCH SQW lasers with $\lambda = 0.81\ \mu m$ whose $\eta_d$ versus $\alpha$ dependence was investigated in more detail. Figure 23 shows that the values of $\eta_d \cong 70\%$ can be obtained for InGaAsP/InP SCH SQW laser diodes with output losses of 10–20 $cm^{-1}$.

The decrease of $\eta_d$ with increasing output losses seen from the data in Figs. 22 and 23 demonstrates another feature in the behavior of $\eta_d$ for the SCH SQW lasers mentioned in several recent publications [24, 44, 45] dealing with AlGaAs/GaAs SCH SQW lasers. They report on the observation of a decay of $\eta_d$ with decreasing $L$ for short cavity diodes, which exhibit a strong superlinear growth of threshold current densities.

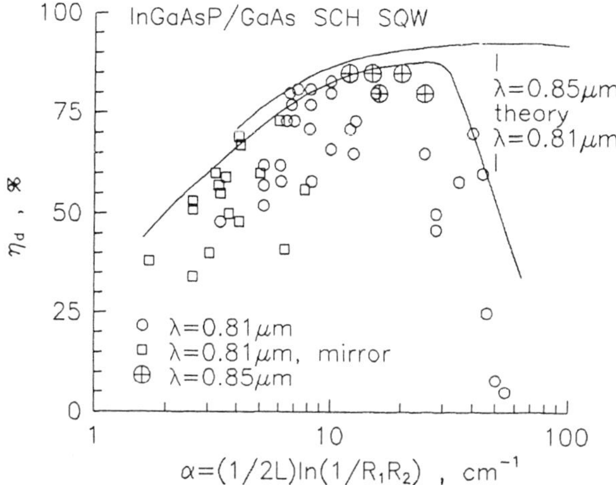


**Fig. 22.** The differential quantum efficiencies as a function of output losses for oxide stripe InGaAsP/GaAs laser diodes with $\lambda = 0.85\ \mu m$ ($L_z = 100$ Å) and $\lambda = 0.81\ \mu m$ ($L_z = 200$ Å). Solid curves represent the results of calculation based on $J_{th}$ versus $\alpha$ dependencies shown in Figs. 17 and 25.

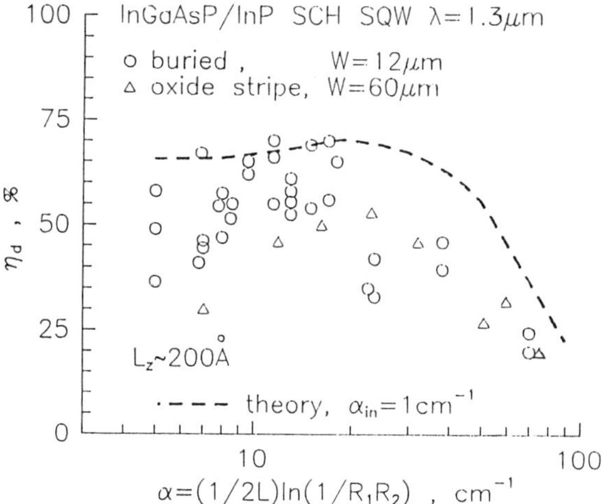


**Fig. 23.** The differential quantum efficiency versus values of output losses for buried and oxide stripe InGaAsP/InP laser diodes ($L_z \cong 200$ Å). The dashed curve is the result of calculation in which the $J_{th} = f(\alpha)$ dependence of Fig. 21 was used.

In the study especially dealing with this problem, Wilcox *et al.* [46] maintain that the anomalous behavior of $\eta_d$ in short cavity diodes can be accounted for by a growth of the threshold concentration of nonequilibrium carriers and the corresponding increase of losses associated with free carrier absorption. The main contribution to these losses is believed to come from nonequilibrium carriers in waveguide layers whose concentration grows dramatically with increasing driving current and filling of the quantum well. In our opinion, this interpretation raises a number of questions and objections, the most essential of them being the problem of lasing involving the transitions that correspond to recombination in the waveguide layers. As is well known from experiments on conventional DH laser diodes, such lasing should set in at the nonequilibrium carrier concentrations in the waveguide corresponding to losses of 10–20 $cm^{-1}$. Such losses cannot result in a strong decrease of $\eta_d$ in the diodes in question with cavity length of 100–200 $\mu$m. We believe that the calculations of Wilcox *et al.* [46] yield overevaluated figures for the nonequilibrium carrier concentrations in waveguides, which should be considered as an upper limit for the corresponding quantities. To interpret the $\eta_d$ versus $\alpha$ dependencies observed in real devices, one should take into account a number of other factors as well, which may be related with the driving current density not in such a direct way as the nonequilibrium carrier concentration is, and may depend on such parameters of the structure as, for instance, the doping level of the cladding layers.

Figure 24 presents a $\eta_d$ versus $1/L$ dependence for SCH SQW, AlGaAs/GaAs diodes that shows that a drop in $\eta_d$ observed as one crosses over to a very short–cavity diode may not occur at all. The corresponding structure was MOCVD-grown and consisted of a 200 Å thick GaAs active region, $Al_{0.2}Ga_{0.8}As$ waveguide layers of total thickness 0.3 $\mu$m, and $Al_{0.6}Ga_{0.4}As$ cladding layers. As the cavity length decreased from 1 mm to 100 $\mu$m, the threshold current densities for these diodes increased superlinearly from 0.45 to 2 $kA/cm^2$.

We turn back to the $\eta_d$ versus $\alpha$ dependencies for InGaAsP/GaAs and InGaAsP/InP, SCH SQW laser diodes shown in Figs. 22 and 23. No studies of the differential efficiency for short cavity diodes were performed in the case of InGaAsP/GaAs SCH SQW lasers with $\lambda = 0.85$ $\mu$m. Such studies were carried out on shorter wavelength ($\lambda = 0.81$ $\mu$m) InGaAsP/GaAs diodes designed for solid state lasers pumping. The corresponding structures differed from those shown in Fig. 6a in that the bandgaps of the solid solutions used as material for the active region and waveguide layers were larger by about 100 meV, 1.54, and 1.83 eV, respectively. The threshold current densities for wide stripe lasers were 200–300 $A/cm^2$ for a cavity length of $\sim$1 mm (Fig.

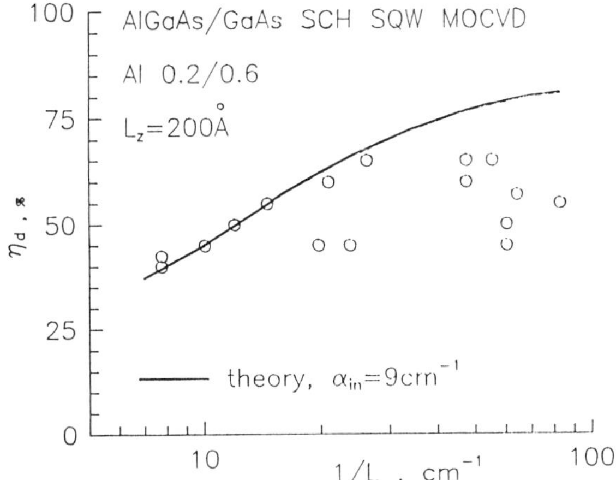


**Fig. 24.** The dependence of differential quantum efficiency on reciprocal cavity length for oxide stripe AlGaAs/GaAs MOCVD-grown laser diodes. The solid curve represents results of a calculation in which a value of 9 $cm^{-1}$ for internal losses was used.

25). As the cavity length was reduced, the values of $J_{th}$ grew rapidly, reaching to 10 kA/$cm^2$ for $L \sim 100$ $\mu$m. In contrast to the behavior of the spontaneous radiation efficiency for InGaAsP/GaAs diodes with $\lambda = 0.85$ $\mu$m presented in Fig. 16, the corresponding dependencies obtained for short cavity laser diodes with $\lambda = 0.81$ $\mu$m for current density $\geq 0.5$ kA/$cm^2$ revealed a

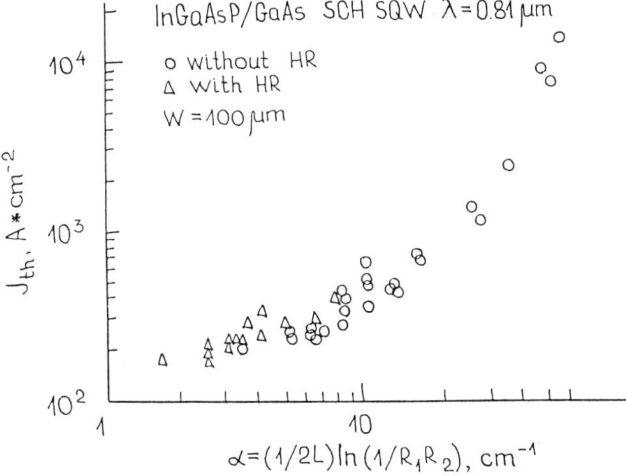


**Fig. 25.** Threshold current densities plotted against output losses for $\lambda = 0.81$ $\mu$m InGaAsP/GaAs oxide stripe laser diodes.

dropping region that was pronounced to a much lesser extent in photoluminescence studies of the corresponding structures.

This result has initiated our calculation of the $\eta_d$ versus $\alpha$ dependence, which included not only the absorption by free carriers but also the current leakage associated with electron injection from the active region into the p-cladding layer [39]. These calculations, to be briefly described in the following, were used also to explain the $\eta_d$ versus $\alpha$ dependencies for the InGaAsP/InP SCH SQW diodes (Fig. 23).

In contrast to Wilcox *et al.* [46], in our calculations we assumed the existence at the active region–waveguide and waveguide–cladding interfaces of space charge regions and of the corresponding additional potential barriers providing quasi-neutrality in the bulk of each layer. The conditions of quasi-neutrality and flatness of quasi-Fermi levels within the space charge regions were used to find the height of the barriers for each value of current. The concentration of nonequilibrium holes (and, respectively, of electrons) in the waveguide layer close to the cladding boundary was assumed to grow with increasing current density in such a way as to produce a hole component of the ambipolar diffusion current equal to the recombination current in the well ($J_R$).

For the current range $J > J_{th}$, the calculation of the total current increase ($\Delta J$), which corresponds to the recombination current growth by $\Delta J_R$, was carried out taking into account: (i) the increase of the hole concentration at the waveguide–cladding boundary (with the fixed value of the hole concentration near well boundary); (ii) the corresponding growth of the electron concentration at the cladding boundary; (iii) the increase of the recombination current in the waveguide $\Delta J_W$; (iv) the increase of the electron leakage current due to diffusion and drift in the p-cladding layer ($\Delta J_L$) as a result of both the just mentioned growth of electron concentration in the waveguide and of the increase of electric field in the cladding required for the growth of the hole current ($\Delta J_R + \Delta J_W$) to occur. The final calculation of the differential efficiency took into consideration: (i) the absorption by free carriers in the active region and the waveguides; (ii) the increase of the current leakage by $\Delta J_L + \Delta J_W$ accompanying the increase of the recombination current in the active region by $\Delta J_R$.

The experimental values of the threshold currents were used as a given parameter when calculating $\eta_d$ for SCH SQW lasers with different cavity lengths. The threshold concentration of nonequilibrium carriers in the active region was chosen such that the sum of the previously mentioned current components be equal to the experimental value of $J_{th}$. The current-independent background losses for both laser types were chosen equal to

1 $cm^{-1}$. The results of the calculations are presented in Figs. 22 and 23. The background losses and the losses associated with absorption by free carriers in InGaAsP/GaAs SCH SQW lasers represent the major reasons limiting the values of $\eta_d$ in diodes with external losses $< 30\ cm^{-1}$. As a result of the losses caused by additional absorption through transitions between the hole subbands [47], the maximum values of $\eta_d$ in InGaAsP/InP lasers are somewhat lower ($\sim 70\%$).

An analysis of these calculations shows that, within this model, the losses associated with absorption by free carriers in the active region and the waveguide layers limit the growth of $\eta_d$ with decreasing $L$ while not producing a dropping region for large $1/L$ in the $\eta_d$ versus $1/L$ dependencies. Both in InGaAsP/GaAs and InGaAsP/InP lasers the appearance of this anomalous region in the $\eta_d$ versus $\alpha$ plot originates from electron leakage to the p-cladding layer. The magnitude of this leakage in both laser types grows dramatically for $J > 2$–$3\ kA/cm^2$.

Besides the quantum well parameters, the value of the leakage current depends strongly on the doping level by acceptors of the cladding layer, particularly if the corresponding hole concentration is less than $10^{18}\ cm^{-3}$. In the calculation of the curves in Figs. 22 and 23, the hole concentration in the cladding layer was adopted to be $10^{18}\ cm^{-3}$. The similarity in the behavior of $\eta_d$ for large $\alpha$ between InGaAsP/GaAs and InGaAsP/InP diodes is determined by similarities in the $J_{th}$ versus $\alpha$ dependencies (see Figs. 21 and 25), as well as by the same value ($\sim 0.37$ eV) of the energy difference between the bandgap of the active region and of the cladding layer. This difference in our model, assuming the quasi-neutrality of each layer, determines primarily the distribution of the minority nonequilibrium carrier concentration between the layers making up the SCH SQW structure.

It should be stressed that the threshold concentrations of nonequilibrium carriers in the well derived from the calculations of this section, which are based on experimental values of $J_{th}$, exceed by far the threshold current densities corresponding to the theoretical model of gain considered in Section 2. If we had used for the calculation the theoretical values of the current density, then the threshold currents would be much lower, and no dropping regions would appear in the $\eta_d$ versus $\alpha$ dependencies up to $\alpha \sim 100\ cm^{-1}$. The values $J_{th} = 2\ kA/cm^2$ observed for InGaAsP/GaAs ($\lambda = 0.85\ \mu m$) (Fig. 22) lasers and AlGaAs/GaAs SCH SQW laser diodes (Fig. 24) with cavity length $\sim 100\ \mu m$ turn out also to be low enough that no anomalous regions appear in the calculated $\eta_d$ versus $\alpha$ relations for the corresponding structures (Figs. 22 and 24).

Summing up, we may conclude that the presence of anomalous regions in the $\eta_d$ versus $\alpha$ relations is not as fundamental a feature in the behavior of SCH SQW lasers as, for instance, the superlinear growth of threshold currents with increasing optical losses. Low threshold lasers with optimized parameters should not exhibit dropping regions in the $\eta_d$ versus $\alpha$ relations for the values of $\alpha$ of practical significance ($\alpha < 100\ \text{cm}^{-1}$).

### 4.4. High Power Characteristics

This subsection will discuss some considerations and results bearing on the practical application of SCH SQW lasers. It can apparently be considered as established that single quantum well structures are presently an optimal choice for the preparation of both lasers with super low (microampere scale) operating currents and super high power laser arrays radiating tens of watts. As follows from the preceding analysis, the major condition for implementing the potentialities inherent in SCH SQW lasers lies in maintaining their optical losses at a level of 5–20 $\text{cm}^{-1}$. In the case of low current microlasers (cf. [48] and Chapter 4, this book), this condition should be met if reflecting coatings are deposited on both mirrors of a very short–cavity laser, and for high power lasers, if long cavity diodes are used. Large cavity lengths are the necessary condition for obtaining high radiation powers, since by increasing $L$ one reduces the series and thermal resistance of devices and can obtain a high power conversion efficiency ($\eta$).

It is in the preparation of high power, long cavity devices that the most remarkable results in the investigation of InGaAsP SCH SQW lasers have been obtained. The light–current characteristics of three lasers with record high parameters are shown in Fig. 26. The rear mirrors of all three lasers were provided with high reflection coatings, while the front mirrors in these devices did not have additional coatings. Curve 1 in Fig. 26 relates to a buried InGaAsP/InP SCH SQW laser ($\lambda = 1.3\ \mu\text{m}$) with stripe width of 12 $\mu$m. The value of $\eta$ for this device was about 40% at a radiated power of over 350 mW, thus providing a record high radiation power at $\lambda = 1.3\ \mu\text{m}$ in a 50-$\mu$m multimode fiber (about 300 mW).

The other two characteristics presented in Fig. 26 relate to oxide stripe lasers with stripe widths of 200 and 100 $\mu$m. Curve 3 demonstrates a new result suggesting that, in the 1.3-$\mu$m spectral range, single-stripe injection lasers are capable of providing an output power $\sim 1$ W in cw mode. Curve 2 shows the best result obtained for an InGaAsP/GaAs SCH SQW laser with $\lambda = 0.85\ \mu\text{m}$ ($W = 200\ \mu\text{m}$). A very high value of power conversion efficiency ($\eta = 66\%$) was observed for this diode at a cw output power of about 1 W.

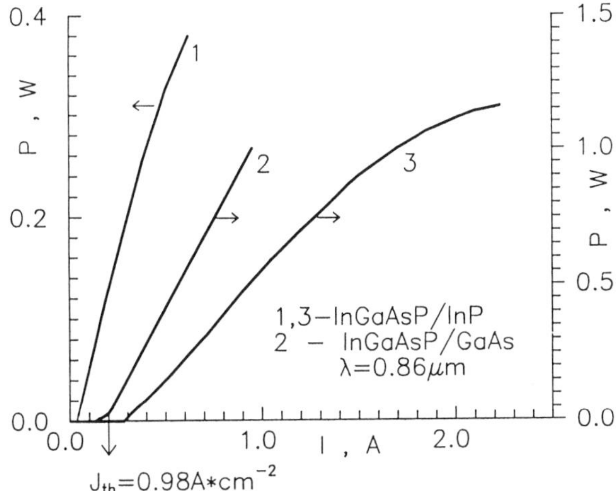


**Fig. 26.** Output power–current characteristics for three InGaAsP SCH SQW laser diodes: (1) 1.3 $\mu$m, buried laser with $W = 12$ $\mu$m; (2) 0.86 $\mu$m, oxide stripe laser diode with $W = 200$ $\mu$m; (3) 1.3 $\mu$m, oxide stripe laser with $W = 100$ $\mu$m.

Curve 2 in Fig. 27a relates to a high power InGaAsP/GaAs SCH SQW laser with $W = 100$ $\mu$m emitting in the region $\lambda = 0.81$ $\mu$m, which is of considerable interest in connection with the development of diode systems for driving YAG : $Nd^{3+}$ solid state lasers. The highest output power obtained with this laser (4.3 W) is only slightly inferior to the best figure reached in AlGaAs/GaAs SCH SQW lasers with an equivalent aperture [49]. The values of $\eta$ for this InGaAsP/GaAs laser are shown in Fig. 27a with a dashed line. In the power range 1–2 W, the obtained values of $\eta$ are higher than those for the AlGaAs/GaAs device discussed by Welsh *et al.* [49]. The decrease in $\eta$ down to 25% at higher output powers is to a considerable extent due to an overheating of the laser's active region, which causes saturation of the light–current relation (Fig. 27a).

The temperature of the active region in this laser was evaluated from the position of the radiation band maximum of the waveguide layer, which is clearly seen to exist (Fig. 27b) on the short-wavelength wing of the radiation band of the active region in both spontaneous and coherent operation. As seen from Fig. 27b, as one approaches the lasing threshold the overheating of the active region ($\Delta T$) reaches 8–10°C. After the threshold, the rate of increase of the active region temperature first slows down, and then begins to grow again. This behavior of the $\Delta T$ versus $J$ relation is in agreement with that expected for the bulk temperature of the active region, whose

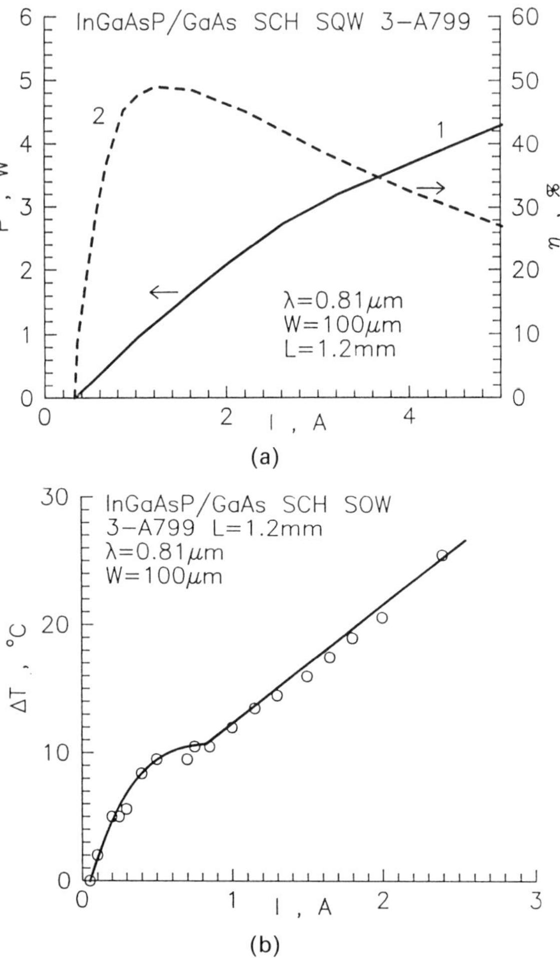


**Fig. 27.** (a) Output power (curve 1) and power conversion efficiency (curve 2) plotted against driving current for InGaAsP/GaAs laser diode. Average heatsink temperature was near 5°C. (b) The difference between laser active region and heatsink temperatures as a function of driving current.

overheating should be determined to a considerable extent by the conversion efficiency of electric power to light.

At the same time, it is obvious that the waveguide radiation is coupled out of crystal regions not more than a few microns in size that adjoin the mirror facet. Thus, the observed character of the $\Delta T$ versus $J$ relation permits a conclusion that the output mirror temperature in the laser in question does not differ much from the bulk temperature of the active region.

The question of the mirror temperature is very essential, since it is the catastrophic optical damage (COD) level of the mirrors that limits, as a rule, the ultimate power and lifetime of high power injection lasers [50]. Taking into account the absence of an antireflection coating on the output mirror facet of the InGaAsP/GaAs laser in question, one can maintain that the maximum optical power handled by the mirror facet of this laser was 1.5 times that of the best AlGaAs/GAs SCH SQW laser [49]. At the same time, there are data arguing for an extremely high COD level of mirror facets in InGaAsP/InP lasers [51]. It can thus be maintained that the high COD level is a common property of mirrors for lasers based on quaternary InGaAsP compounds. One of the reasons for the high COD level of mirror facets in such lasers may be the lower, compared with the Al–Ga–As system, rate of surface recombination for the InGaAsP compounds and, correspondingly, the lower overheating of the laser mirror facets.

## 5. TRANSITIONS FROM HIGHER LYING QUANTIZED STATES

This section discusses studies of the effect of optical losses ($\alpha$) on the position of the lasing peak in SCH SQW lasers. These studies represent a continuation of the investigation dealt with in Section 2 where a short jump of $h\nu_g$ with increasing $\alpha$ was observed to occur in InGaAsP/GaAs and InGaAsP/InP SCH SQW lasers.

Quantitative comparison of the corresponding experiments with calculations has become possible in studies of AlGaAs/GaAs SCH SQW lasers grown by MBE, a technique ensuring higher homogeneity and better control of the geometric parameters of the SQW structures compared with the liquid phase epitaxy method used to prepare InGaAsP-based lasers. As for the internal quantum efficiency of radiative recombination, however, the reproducibility of this parameter for the MBE-grown AlGaAs/GaAs structures turned out to be not as good as that typical of LPE-grown samples, as was pointed out in Section 2 dealing with the corresponding photoluminescence studies. It is probably because of this that in the studies of the external efficiency of short cavity length, MBE-grown SCH SQW diodes (Fig. 20) we have not succeeded in obtaining in the threshold current density region the values $\eta_e \sim 1.5\%$ typical of LPE-grown samples. In accordance with this, the threshold current densities for oxide stripe lasers with $W = 100$ $\mu$m and $L = 1$–$2$ mm made of MBE-grown AlGaAs/GaAs SCH SQW structures with $L_z = 100$–$200$ Å were higher than those for similar InGaAsP/GaAs lasers, the threshold for the MBE-grown lasers being 150–400 A/cm$^2$. We are not going to discuss here the $J_{th}$ versus $\alpha$ dependencies for MBE-grown AlGaAs/GaAs SCH SQW lasers, which are close to those shown in Fig. 8, since they are apparently affected substantially by the nonradiative current leakage, which can be removed by further improvements in the technology of preparation and optimization of the parameters of the structures. At the same time, this leakage should not affect the behavior of the $h\nu_g$ versus $\alpha$ relations discussed in the following.

As already mentioned (see Section 3.3), the gain spectrum of the active region broadens with increasing current density. If quasi two dimensional conditions take place, then additional maxima appear in the short wavelength part of the spectrum, which are associated with the contributions to the gain coming from transitions involving higher lying quantized subbands (Fig. 18b).

Denote by $\bar{g}_{ik}$ ($k = i + 1$) the transitional values of the gain at which the absolute maximum of the gain shifts with increasing driving current from the lower-energy band to the adjacent higher-energy one. The corresponding transitional values of output losses can be denoted by $\alpha_{ik}$ ($\alpha_{ik} \cong \Gamma\bar{g}_{ik}$). Consider now the effect of the quantum well thickness on the quantities $\bar{g}_{ik}$, $\alpha_{ik}$ and on the range of variation of $h\nu_g$ with varying output losses $\alpha$ ($\alpha \sim 1/L$).

In quasi two dimensional conditions, the gap between the adjacent electron subbands $\Delta E_{ik}$ is greater than $kT$, which makes it obvious that an increase of $L_z$ should result in a substantial increase in the nonequilibrium carrier concentration in the higher lying subbands and hence in a decrease of the driving current at which the gain maximum undergoes a jump and, accordingly, in a decrease of $\bar{g}_{ik}$. The decrease of $\bar{g}_{ik}$ with increasing $L_z$ plays a dominant role also in the behavior of the $\alpha_{ik}$ versus $L_z$ relation.

As for the range of the shift of the lasing peak ($\Delta h\nu_g$), the quantity $\Delta h\nu_g$ may be comparable with well depth for all $L_z$. However, in thicker wells containing a larger number of subbands, a lasing peak shift by $\Delta h\nu_g \sim \Delta E_c$ can occur at higher losses and current densities than in the case of wells with smaller $L_z$. For wells with $L_z > 300$ Å, where the three dimensional approach becomes valid ($\Delta E_{ik} \cong kT$), the weakening of the dependence of $h\nu_g$ on $\alpha$ with increasing $L_z$ may be interpreted as resulting from the increasing optical confinement factor.

We know of no publications before 1989 dealing with the effect of structure parameters on the $h\nu_g$ versus $\alpha$ dependence. The three pioneering studies of Zory *et al.* [2], Nagle *et al.* [42], and Mittelstein *et al.* [43], devoted to the lasing wavelength switching in AlGaAs/GaAs, SCH SQW lasers, were made on structures with relatively thin active layers. As a result, the value of $\alpha_{12}$

corresponding to losses $\sim 100\ \text{cm}^{-1}$ could be reached in such structures only in lasers with the shortest possible cavity length ($L \sim 100\ \mu\text{m}$), where the bottom of the second electron subband involved in lasing in short cavity lasers was already close to or in the waveguide conduction band. It thus follows that by increasing $L_z$ one could improve substantially the conditions for the corresponding experiments. [*Editor's note*: Temperature-induced wavelength switching studies in GaAs/AlGaAs QW lasers with thicker $L_z$ values have been reported by Zhu *et al.* [52]. The first report of temperature-induced wavelength switching appears to be that of Partin [53], who observed the phenomenon in lead-chalcogenide QW lasers.]

Figure 28b presents an $h\nu_g$ versus $\alpha$ dependence obtained in studies of lasers with different cavity lengths made of AlGaAs/GaAs SCH SQW structure with $L_z = 140$ Å, which is shown schematically in Fig. 28a [54]. As seen from the figure, the first jump in the position of the lasing peak occurs already at losses $\alpha_{12} = 40\ \text{cm}^{-1}$, corresponding to a cavity length of 250 $\mu$m, while at $\alpha_{23} = 100\ \text{cm}^{-1}$, a second jump appears, which is due to the gain peak shifting to an energy range that may correspond to the $3_c$–$3_{hh}$ transitions.

In the case of structure with a 200 Å thick active region (Fig. 29), lasing involving transitions between the $1_c$ and $1_{hh}$ subbands occurs only in lasers with the longest cavities, with $h\nu_g$ shifting already for $L \leq 1$ mm into the energy range corresponding to the $2_c$–$2_{hh}$ transitions. Two more jumps in the position of $h\nu_g$ are observed as the cavity length is decreased still further.

The solid lines in Figs. 28b and 29 present the $h\nu_g$ versus $\alpha$ relation calculated by means of Eqs. (2) and (7) in Section 3. The loss scales corresponding to the relations are given on top of the figures. As seen from the figures, good agreement between the experimental and calculated values of $\alpha_{ik}$ can be reached by shifting the corresponding log scales for $\alpha$, which is equivalent to an overevaluation of the theoretical values by 50–70% compared with the experimental data, proportionally for all $\alpha_{ik}$.

Note that such a disagreement between theory and experiment cannot be accounted for by internal losses, which are not included when plotting experimental data. It is most probably an indication that the maximum value of the gain that can be experimentally achieved in each band is by a factor of 1.5–2 less than theoretically predicted.

This disagreement is in no way unexpected, since there are a number of effects that are not included in the ideal theoretical model but should reduce the maximum value of the gain that can be reached for a given $h\nu$. Nearly the same overevaluation by a factor of 1.5–2 of the calculated values of $\bar{g}$ compared with experiment has been pointed out by us in the preceding

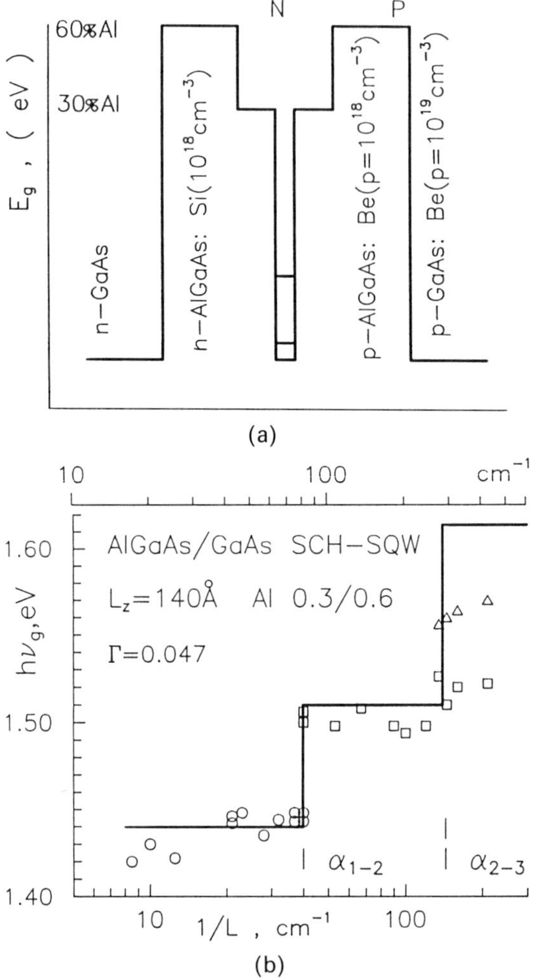


**Fig. 28.** (a) Band diagram of MBE-grown AlGaAs/GaAs SCH SQW structure with $L_z = 140$ Å. (b) Theoretical (solid curve) and experimental data on the spectral position of lasing peak versus the value of output losses for laser diodes fabricated from the structure of Fig. 28a. Loss axis for theoretical curve in the upper part of the Fig. 28b is shifted to fit experimental and theoretical data.

section when discussing the results relating to InGaAsP/GaAs SCH SQW lasers. Generally, such an effect could be caused by intraband relaxation, which reduces the gain particularly strongly in the spectral regions corresponding to the step in the density of states distribution (see Chapters 2 and 3, this book).

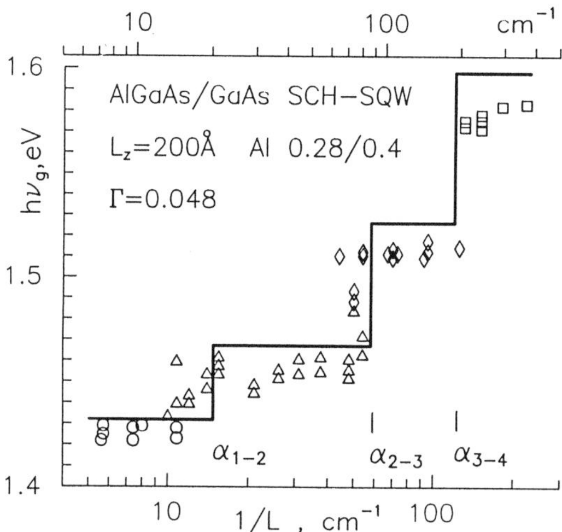


**Fig. 29.** The same as in Fig. 28b but for lasers fabricated from structures with $L_z = 200$ Å.

Figure 29 also exhibits an overestimation of the theoretical values of $h\nu_g$ compared with experimental data that increases with increasing transition number. This disagreement is not connected with any error in the calculations, since the calculated subband positions coincide with the well known experimental data for the corresponding AlGaAs/GaAs SQW structures. The reason for the disagreement lies in the renormalization of the bandgap energies associated with Coulomb interaction in a high density electron–hole plasma [24]. Despite the disagreements, both the experimental and theoretical data indicate that in SCH SQW lasers with $L_z \geq 140$ Å and through the range of cavity lengths of practical interest ($L = 1$–$0.2$ mm), the higher lying quantized subbands should participate in the lasing process, and that a change in output losses should result in a switching of the lasing wavelength.

## 6. CONCLUSIONS

1. Studies of thin active layer lasers based on InGaAsP/GaAs, InGaAsP/InP, and AlGaAs/GaAs systems show that, for losses in excess of 20–30 $cm^{-1}$, the threshold currents in such lasers start to grow superlinearly, the lasing peak shifts toward shorter wavelengths, and other laser parameters also undergo variation. These phenomena are connected with

the saturation of gain, and they originate from a transition to a degenerate distribution of first electrons and then also of holes in the active region of such lasers as the losses grow. These effects should be observed in lasers with active layer thickness scanning a fairly broad range (50–500 Å) irrespective of the conditions realized in their active regions: two dimensional, quasi two dimensional or three dimensional.

2. Calculations performed in terms of the simplest model, considering only radiative transitions with the rigorous selection rules, yield satisfactory quantitative agreement with the experimental values of threshold for lasers with three and two dimensional GaAs active layers in the low loss range, where $J_{th} \cong J_{tr}$, and offer a correct qualitative description of the behavior of the $J_{th}$ versus $\alpha$ and $h\nu_g$ versus $\alpha$ relations in the high loss region ($\alpha > 20\ \text{cm}^{-1}$). Quantitative description of experimental relations in the range of high losses requires inclusion into the calculations of corrections (for instance, for intraband scattering), reducing the gain at a fixed current density by about a factor of 1.5–2.

3. The preceding effects originating fundamentally from gain saturation may become strongly enhanced in lasers based on real structures because of the influence of additional factors, such as, for instance, carrier leakage from a finite depth quantum well and absorption by free carriers. The growing electron leakage to the cladding layer may be the reason for the anomalous decrease of differential efficiency, with increasing output losses observed to occur in some of the SCH SQW lasers studied. [*Editor's note*: Carrier transport effects of this type are also believed to be responsible for limiting the high speed capabilities of QW lasers—see Rideout *et al.* [55], Nagarajan *et al.* [56], and Chapter 5, this book.]

## ACKNOWLEDGMENT

The authors are grateful to academician Zh. I. Alferov for helpful discussions. We wish to acknowledge all of our colleagues from the Ioffe Physical-Technical Institute who fulfilled the work comprising the contents of this review.

## REFERENCES

1. Zory, P. S., Reisinger, A. R., Waters, R. G., Mawst, L. J., Zmudzinski, C. A., Emanuel, M. A., Givens, M. E., and Coleman, J. J. (1986). *Appl. Phys. Lett.* **49**, 16.

2. Zory, P. S., Reisinger, A. R., Mawst, L. J., Costrini, G., Zmudzinski, C. A., Emanuel, M. A., Givens, M. E., and Coleman, J. J. (1986). *Electron. Lett.* **22**, 475.
3. Tsang, W. T. (1982). *Appl. Phys. Lett.* **40**, 217.
4. Hersee, S., Baldy, M., Assenat, P., De Cremoux, B., and Duchemin, J. P. (1982). *Electron. Lett.* **18**, 618.
5. Alferov, Zh. I., and Garbuzov, D. Z. (1987). *18th international conference on the physics of semiconductors*, p. 136. Stockholm, Sweden.
6. Rezek, E. A., Holonyak, N., Jr., Vojak, B. A., Stillman, G. E., Rossi, J. A., Keune, D. L., and Fairing, J. D. (1977). *Appl. Phys. Lett.* **31**, 288.
7. Rezek, E. A., Vojak, B. A., Chin, R., Holonyak, N., Jr., and Samman, E. A. (1981). *J. Electron. Mater.* **10**, 255.
8. Alferov, Zh. I., Garbuzov, D. Z., Arsent'ev, I. N., Ber, B. I., Vavilova, L. S., Krasovskii, V. V., and Chudinov, A. V. (1985). *Sov. Phys. Semicond.* **19**, 679.
9. Alferov, Zh. I., Garbuzov, D. Z., Krasovskii, V. V., Nikishin, S. A., Sinyavskii, D. V., and Tikunov, A. V. (1985). *Sov. Tech. Phys. Lett.* **11**, 581.
10. Alferov, Zh. I., Garbuzov, D. Z., Denisov, A. G., Evtikhiev, V. P., Komissarov, A. B., Senichkin, A. P., Skorokhodov, V. N., and Tokranov, V. E. (1988). *Sov. Phys. Semicond.* **22**, 1331.
11. Alferov, Zh. I., Garbuzov, D. Z., Zhigulin, S. N., Kyz'min, I. A., Orlov, B. B., Sinitchin, M. A., Strugov, N. A., Tokranov, V. E., and Yavich, B. S. (1988). *Sov. Phys. Semicond.* **22**, 1334.
12. Alferov, Zh. I., Andreev, V. M., Aksenov, V. Yu., Larionov, V. R., Rumyantsev, V. D., and Khvostikov, V. P. (1988). *Sov. Phys. Semicond.* **22**, 1123.
13. Brunemeir, P. E., Hsieh, K. C., Deppe, D. G., Brown, J. M., and Holonyak, N., Jr. (1985). *J. Crystal Growth.* **71**, 705.
14. Bert, N. A., and Kosogov, A. O. (1990). *Proceedings XIIth international congress on electron microscopy*, Seattle, 1990 (in press).
15. Garbuzov, D. Z., Chalyi, V. P., Svelokuzov, A. E., Khalfin, V. B., and Ter-Martirosyan, A. L. (1988). *Sov. Phys. Semicond.* **22**, 410.
16. Alferov, Zh. I., Garbuzov, D. Z., and Arsent'ev, I. N. (1984). *Abstracts of the XVIth international conference on Solid State devices*, p. 48. Kobe, Japan.
17. Garbuzov, D. Z., Evtikhiev, V. P., Karpov, S. Yu., Sokolova, Z. N., and Khalfin, V. B. (1985). *Sov. Phys. Semicond.* **19**, 83.
18. Evtikhiev, V. P., Garbuzov, D. Z., Sokolova, Z. N., Tarasov, I. S., Khalfin, V. B., Chalyi, V. P., and Chydinov, A. V. (1985). *Sov. Phys. Semicond.* **19**, 8073.
19. Tarasov, I. S., Garbuzov, D. Z., Evtikhiev, V. P., Ovchinnikov, A. V., Sokolova, Z. N., and Chudinov, A. V. (1985). *Sov. Phys. Semicond.* **19**.
20. Garbuzov, D. Z. (1982). *J. Luminescence* **27**, 109.
21. Khalfin, V. B., Garbuzov, D. Z., and Krasovskii, V. V. (1986). *Sov. Phys. Semicond.* **20**, 1140.
22. Garbuzov, D. Z., and Khalfin, V. B. (1987). *Proceedings of the third binational U.S.A.–U.S.S.R. symposium on laser optics of condensed matter, Leningrad, 1987* (Birman, J. L., Cummins, H. Z., and Kaplyanskii, A. A., eds.), p. 103. Plenum Press, New York and London.

23. Garbuzov, D. Z., Tikunov, A. V., and Khalfin, V. B. (1987). *Sov. Phys. Semicond.* **27**, 662.
24. Chinn, S. R., Zory, P. S., and Reisinger, A. R. (1988). *IEEE J. Quantum Electron.* **QE-24**, 2191.
25. Chinn, S. R., Zory, P. S., and Reisinger, A. R. (1989). *SPIE* **143**, 157.
26. Garbuzov, D. Z., Arsent'ev, I. N., Chalyi, V. P., Chudinov, A. V., Evtikhiev, V. P., and Khalfin, V. B. (1984). *Sov. Phys. Semicond.* **18**, 1272.
27. Nelson, R. J., and Sobers, K. G. (1978). *J. Appl. Phys.* **49**, 6103.
28. Sermage, B., Eicher, H. J., Heribage, J. P., Nelson, R. J., and Dutta, M. N. (1983). *Appl. Phys. Lett.* **42**, 259.
29. Alferov, Zh. I., Garbuzov, D. Z., Evtikhiev, V. P., Komissarov, A. B., Sokolova, Z. N., Ter-Martirosjan, A. L., Tokranov, V. E., Chalyi, V. P., and Khalfin, V. B. (1988). *19th international conference on the physics of semiconductors, Warshawa* (Zawadzki, ed.), vol. 1, p. 271.
30. Matsusue, T., and Sakuki, H. (1987). *Appl. Phys. Lett.* **50**, 1929.
31. Yamanishi, M., and Suemune, I. (1984). *Japan. J. Appl. Phys.* **23**, L35.
32. Yamada, M., and Suematsu, Y. (1979). *IEEE J. Quantum Electron. QE-15*, 743.
33. Yamada, M., and Ishiguro, H. (1981). *Japan. J. Appl. Phys.* **20**, 1279.
34. Sugimura, A., Patzak, E., and Meissner, P. (1986). *J. Phys. D: Appl. Phys.* **19**, 7.
35. Asada, M. (1989). *IEEE J. Quantum Electron.* **QE-25**, 2019.
36. Ahn, D. (1989). *J. Appl. Phys.* **65**, 4517.
37. Razeghi, M., De Cremoux, B., and Duchemin, J. P. (1984). *J. Crystal Growth.* **68**, 389.
38. Hakki, B. W., and Paoli, T. L. (1975). *J. Appl. Phys.* **46**, 1299.
39. Casey, H. C., Jr., and Panish, M. B. (1978). Heterostructure lasers, Part B, Section 7. Academic Press, New York.
40. Chen, H. Z., Ghaffari, A., Morkoc, H., and Yariv, A. (1987). *Electron. Lett.* **23**, 1334.
41. Alferov, Zh. I., Garbuzov, D. Z., Zaitzev, S. V., Nivin, A. B., Ovchinnikov, A. V., and Tarasov, I. S. (1987). *Sov. Phys. Semicond.* **21**, 503.
42. Nagle, J., Hersee, S., Krakowski, M., Weil, T., and Weisbuch, C. (1986). *Appl. Phys. Lett.* **49**, 1325.
43. Mittelstein, M., Arakawa, Y., Larsson, A., and Yariv, A. (1986). *Appl. Phys. Lett.* **49**, 1689.
44. Reisinger, A. R., Zory, P. S., and Waters, R. G. (1987). *IEEE J. Quant. Electron.* **QE-23**, 993.
45. Wagner, D. K., Waters, R. G., Tihanyi, P. L., Hill, D. S., Rosa, A. J., Jr., Vollmer, H. J., and Leopold, M. M. (1988). *IEEE J. Quantum Electron.* **QE-24**, 1255.
46. Wilcox, J. Z., Ou, S., Yang, J. J., Jansen, M., and Peterson, G. L. (1989). *Appl. Phys. Lett.* **55**, 825.
47. Asada, M., Kameyama, A., and Suematsu, Y. (1984). *IEEE J. Quantum Electron.* **QE-20**, 745.
48. Lau, K. Y., Derry, P. L., and Yariv, A. (1988). *Appl. Phys. Lett.* **52**, 88.

49. Welsh, D. F., Chan, B., Streifer, W., and Scifres, D. R. (1988). *Electron. Lett.* **27**, 113.
50. Moser, A. (1991). *Appl. Phys. Lett.* **58**, 552.
51. Temkin, H., Mahajan, S., and Logan, R. A. (1983). *Inst. Phys. Conf. Ser.* **67** (Sect. 5), 279.
52. Zhu, L. D., Zheng, B. Z., and Feak, G. A. B. (1989). *IEEE J. Quantum Electron.* **QE-25,** 2007.
53. Partin, D. L. (1984). *Appl. Phys. Lett.* **45**, 487.
54. Alferov, Zh. I., Garbuzov, D. Z., Evtikhiev, V. P., Zhigulin, S. N., and Kudryashov, I. V. (1989). *Conference on lasers and electro-optics (CLEO-1989), Tech. Digest Series* **11**, p. 306.
55. Rideout, W., Sharfin, W. F., Koteles, E. S., Vassell, M. O., and Elman, B. (1991). *IEEE J. Photon. Lett.* **3**, 784.
56. Nagarajan, R., Fukushima, T., Ishikawa, M., Bowers, J. E., Geels, R. S., and Coldren, L. A. (1991). *Appl. Phys. Lett.* **59**, 1835.

# *Chapter 7*

# VALENCE BAND ENGINEERING IN QUANTUM WELL LASERS

**Eoin P. O'Reilly and Ali Ghiti**

*Department of Physics*
*University of Surrey*
*Guildford, United Kingdom*



## 1. INTRODUCTION

One of the major themes to emerge in the study of quantum well lasers is the key role of the valence band states in determining the lasing properties



ISBN 0-12-781890-1

of any quantum well structure. As an example, the internal gain mechanisms tend to favour transverse electric (TE) over transverse magnetic (TM) modes because of the modifications to the valence band structure upon formation of a quantum well. Also, the two major loss mechanisms in long wavelength (1.55 $\mu$m) lasers are both strongly influenced by the valence band structure, namely, Auger recombination and intervalence band absorption. An accurate description of the gain as a function of carrier or current density therefore requires a careful analysis of the valence band states in the quantum well structure, as discussed by Corzine *et al.* in the first chapter of this book [1]. Further, an understanding of the valence band structure allows the possibility of valence band engineering, whereby quantum well lasers can be grown with the valence states tailored to give particular beneficial properties, including reduced threshold current densities, improved temperature stability, and enhanced modulation response.

We focus in this chapter on the advantages that can be achieved by growing in particular strained layer, and also (111) quantum well lasers. It is now possible to grow high quality strained layer structures, in which the quantum well layer is composed of a semiconductor which would normally have a significantly different lattice constant than the barrier material. The lattice mismatch is accommodated by a tetragonal distortion of the quantum well layer, giving a built-in axial strain. This axial strain splits the degeneracy of the light- and heavy-hole zone centre states, accessing a wide range of subband structures, including the possibility of the highest valence subband being light-hole-like, of significant benefit for semiconductor lasers. The strain-induced band splittings and their consequences are of sufficient importance that a large part of this and four other chapters [1–4] are devoted to considering strained layer lasers.

Corzine *et al.* have discussed in Chapter 1 the role of the valence states in determining the gain in strained and unstrained quantum well structures. Much of the present chapter describes the benefits that should be achieved by growing long wavelength strained layer lasers, where the built-in strain may lead to the virtual elimination of the major loss mechanisms of Auger recombination and intervalence band absorption. In Chapter 8, Coleman gives a theoretical and experimental overview of GaAs-based strained layer lasers operating close to 1 $\mu$m [2], while Lau considers in Chapters 4 and 5 [3, 4] the low threshold and high speed dynamics of quantum well lasers and the speed enhancements that can be obtained from strained layer structures.

A second aspect of valence band engineering that we consider in this chapter is the potential benefits of growing unstrained lasers, using growth directions other than the conventional (001) direction. The highest valence band is highly anisotropic, even in unstrained bulk III-V semiconductors, where the heavy-hole mass is approximately twice as large along (111) as along (001) [5, 6]. This increases the number of heavy-hole confined states and the energy separation between the highest heavy-hole and light-hole states for a given well width. It has recently been shown that (111) GaAs–AlGaAs lasers can have a lower threshold current density than equivalent (001) lasers [7–9]. We describe laser gain calculations that show that this improvement is consistent with the different valence subband structures associated with the two growth directions.

We begin in the next section by reviewing the major loss mechanisms in long wavelength lasers and discussing how they might be eliminated by altering the valence band structure. We then give an overview in Section 3 of the theoretical description of the valence band structure in bulk unstrained and strained III-V compound semiconductors, and we describe the calculation of the valence subband structure in quantum wells as a function of strain and/or growth direction. This overlaps with some of the material in Chapter 1, where a more complete description of the band structure may be obtained. Section 4 considers the benefits of InP-based strained layer structures for long wavelength lasers. Finally, we consider in Section 5 the potential benefits of growing laser structures along the (111) direction, using the example of GaAs–AlGaAs laser structures. The examples in these last two sections emphasise the variety of means by which the valence subband structure can be modified in quantum well lasers and the advantages to be derived from such modifications. These advantages are further expounded elsewhere in this book [1–4].

## 2. LOSS MECHANISMS IN LONG WAVELENGTH LASERS

Much effort has been devoted to developing semiconductor lasers to operate at 1.3 and 1.55 $\mu$m, the wavelengths at which pulse dispersion and absorption losses are minimised in silica-based optical fibres. Despite these efforts, both bulk and unstrained quantum well lasers operating at these wavelengths suffer from significant intrinsic losses, and as a consequence both the threshold current and quantum efficiency are very sensitive to changes in temperature. The threshold current $I_{th}$ is often expressed in the form

$$I_{th}(T) = I_{th}(T') \exp\{(T - T')/T_0\}, \tag{1}$$

where $T_0$ is the characteristic temperature, which is used to express the temperature sensitivity of the threshold current, with values for long wavelength lasers typically in the range $T_0 = 55 \pm 20\ K$. The two most important mechanisms contributing to the intrinsic losses are Auger recombination [10] and intervalence band absorption [11]. The intervalence band absorption (IVBA) process is illustrated by the solid vertical arrows in Fig. 1, whereby an electron in the conduction band recombines with a hole near the valence band maximum, creating a photon of energy $h\nu$, close to the bandgap energy $E_g$. IVBA occurs when the photon is subsequently reabsorbed by lifting an electron from the spin–split-off band into an injected hole state in the heavy-hole band. This absorption loss contributes both to increasing the threshold current and to reducing the quantum efficiency above threshold. Because the bandgap $E_g$ is typically about $\frac{1}{2}$ eV greater than the spin–orbit splitting $\Delta$ in lasers used for optical communications, the reabsorption is expected well away from the Brillouin zone centre, at large wavevector **k**, to conserve energy and momentum (vertical arrow, Fig. 1a). The effect is therefore very temperature sensitive, because the number of

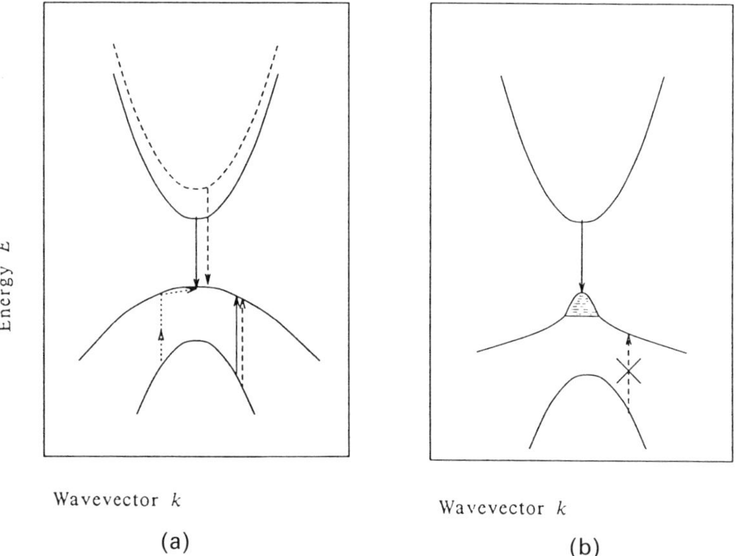


**Fig. 1.** Intervalence band absorption (IVBA) loss mechanisms in 1.55-$\mu$m lasers. (a) The solid vertical arrows indicate the direct IVBA process, while the dotted arrows indicate phonon-assisted IVBA. The dotted arrows indicate how IVBA moves to larger wavevector **k** and is thereby reduced with increasing bandgap. (b) The elimination of IVBA in strained structures, where holes are confined in a smaller region of **k** space.

holes available at such large wavevectors increases exponentially with temperature. Phonon-assisted IVBA is also possible (dotted line, Fig. 1a), and it has recently been suggested [12] that this process could be of comparable magnitude with the direct absorption process in long wavelength lasers.

In Auger recombination, an electron and hole again recombine across the bandgap but now, instead of emitting a photon, excite a third carrier to higher energy. The two most important band-to-band Auger processes are illustrated in Figs. 2a and b. In the first (CHCC) case, a Conduction electron and Heavy hole recombine, exciting a Conduction electron to a higher Conduction state. In the second (CHSH) case, the electron is excited from the Spin–split-off band into an empty Heavy-hole state. Because of the need to conserve energy and momentum, it is clear from Fig. 2b that, when $\Delta$ is significantly less than $E_g$, both these processes involve heavy holes at large wavevector **k**; so that the Auger rate also increases exponentially with temperature [13] with the CHCC rate, assuming parabolic bands and Boltzmann statistics, being given by

$$R = Cn^2pe^{-E_a/kT}, \tag{2}$$

with $E_a = m_c/(m_c + m_h)E_g$; while the CHSH rate varies as

$$R = Dnp^2e^{-E_a/kT}, \tag{3}$$

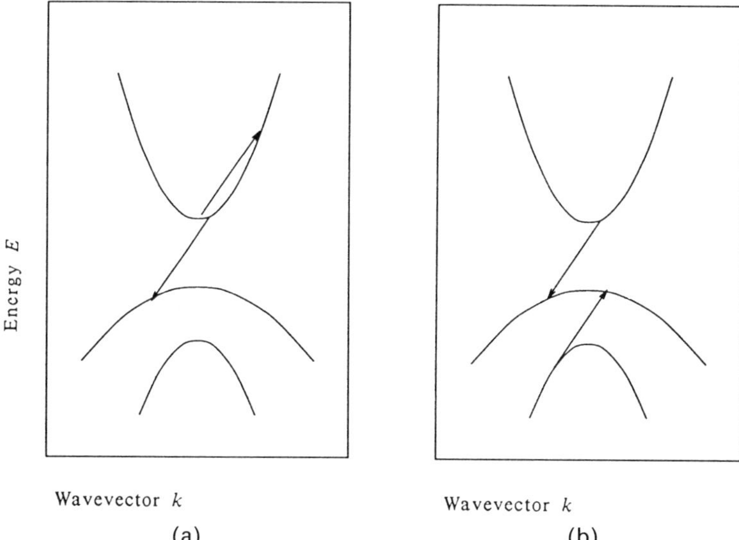


**Fig. 2.** CHCC (a) and CHSH (b) Auger processes.

with $E_a = m_s/(2m_h + m_c - m_s)(E_g - \Delta)$ for $E_g > \Delta$, as here, where $n$ and $p$ are the electron and hole densities, and $m_{c,h,s}$ the conduction band, heavy-hole, and split-off masses, respectively. Phonon-assisted and defect-assisted Auger recombination have a much weaker temperature dependence than the direct Auger processes [14] and so can be discounted as major contributors to the $T_0$ problem.

A range of studies have established the importance of Auger recombination [15–19] and of IVBA [11, 20–22] as intrinsic loss mechanisms in long wavelength lasers. As an example, the results of recent high pressure measurements of the threshold current of 1.55-$\mu$m quantum well lasers are shown in Fig. 3 [23]. As can be seen, $I_{th}$ decreases with increasing pressure. This is due to the fact that the direct bandgap, and hence the emission energy, increases with pressure at about 10 meV/kbar while leaving the valence band relatively unchanged [23]. Thus, at high pressure, IVBA occurs at larger **k** values, as shown by the dotted arrows in Fig. 1a. There are fewer holes here, and therefore the loss is decreased. Auger recombination processes also decrease with increasing pressure, because as the energy exchanged increases, so the momentum exchange must increase, and holes with a larger **k** are again required. There is a complicated interplay between IVBA and Auger

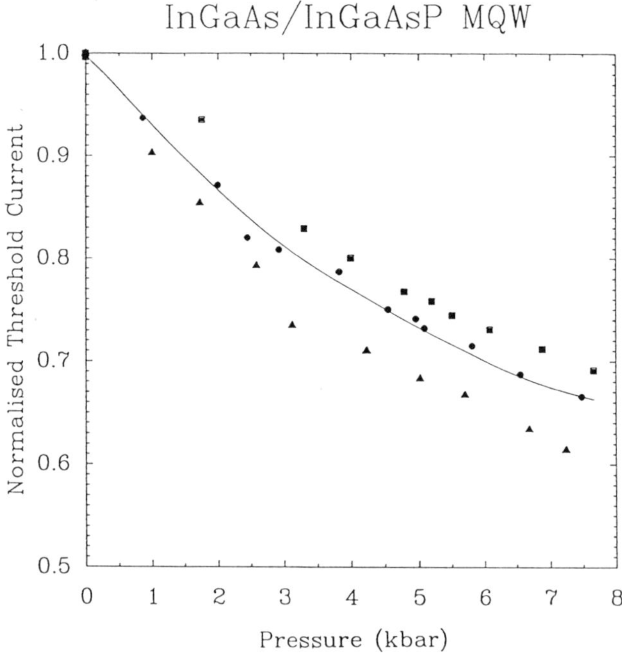


**Fig. 3.** Measured pressure dependence of the threshold current in 1.55-$\mu$m quantum well lasers together with theoretical fit assuming IVBA and Auger recombination.

recombination [20], since IVBA increases the threshold carrier density $n_{th}$ and hence also increases the Auger recombination due to its dependence on $n_{th}$. The theoretical curve through the points in Fig. 3 was obtained with an IVBA coefficient of 40 $cm^{-1}$ per $10^{18}$ $cm^{-3}$ of holes and an Auger coefficient of $1.4 \times 10^{-28}$ $cm^6/s$ at ambient pressure, values in general agreement with others in the literature [16].

Since the loss mechanisms decrease as the bandgap increases, neither IVBA nor Auger recombination present a problem in shorter wavelength GaAs lasers, where the bandgap $E_g$ is more than 1 eV greater than the spin–orbit splitting $\Delta$, and the Auger recombination coefficient is approximately $10^{-31}$ $cm^6/s$. We now discuss how efficient operation can be obtained if, instead of going to shorter wavelengths, we stay at 1.55 $\mu$m and modify the valence band structure to restrict the holes in **k** space.

We show in the next section how it is possible to engineer the valence band structure in strained quantum well lasers so that the in-plane effective mass $m_{\parallel}^*$ and the valence band density of states are significantly reduced. This occurs when the strained quantum well layer is under biaxial compression, as is the case for $In_xGa_{1-x}As$ on InP for $x > 0.53$, and on GaAs for all $x > 0$. The reduced density of valence states is predicted to bring significant benefits to lasers at all wavelengths [24, 25], reducing the carrier density $n_{th}$ and current density $J_{th}$ required to reach threshold and increasing the differential gain $dg/dn$ at threshold, with consequent improvements to the dynamic response and linewidth enhancement factor [1, 26–28].

The reduced density of valence states brings particular benefit to long wavelength lasers, as illustrated in Fig. 1b, where the hole distribution is being confined in a much smaller region of **k** space. This is predicted in the optimum case to virtually eliminate the direct IVBA and Auger processes [24]. Combining this with the reduced value of $n_{th}$, we see that valence band engineering in strained layer structures should lead to 1.3- and 1.5-$\mu$m lasers with reduced threshold current density, improved temperature stability, and enhanced differential quantum efficiency. We describe in more detail the dependence of the valence band structure on strain and on growth direction in the next section and return to consider the consequences for long wavelength lasers in the following section.

## 3. VALENCE BAND STRUCTURE

### 3.1. Bulk

The band structure of a bulk unstrained direct gap tetrahedral semiconductor is illustrated in Fig. 4a, while Figs. 4b and c show the band structure under

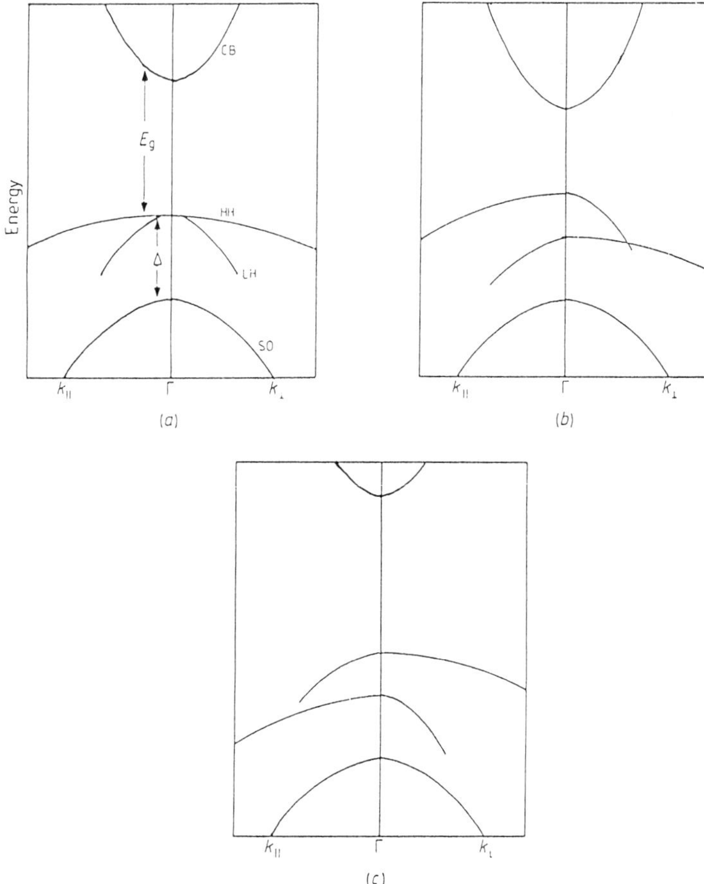


**Fig. 4.** (a) A schematic representation of the band structure of an unstrained direct gap tetrahedral semiconductor. The light-hole (LH) and heavy-hole (HH) bands are degenerate at the Brillouin zone centre Γ, and the spin–split-off (SO) band lies Δ lower in energy. The lowest conduction band (CB) is separated by the bandgap energy $E_g$ from the valence bands. (b) Under biaxial tension, the hydrostatic component of the tension reduces the mean bandgap, while the axial component splits the degeneracy of the valence band maximum and introduces an anisotropic valence band structure, with the highest band being light along the strain ($k_{\perp}$) axis and comparatively heavy perpendicular to that axis (along $k_{\parallel}$). Note that there is an ambiguity in the choice of notation, which leads us to the use here of $k_{\parallel}$ for the direction *perpendicular* to the strain axis. We define directions for SLS with respect to the growth plane: $k_{\parallel}$ lies in the growth plane and hence is perpendicular to the growth and strain direction. (c) Under biaxial compression the mean bandgap increases and the valence splitting is reversed, so the highest band is now heavy along the strain axis ($k_{\perp}$) and comparatively light along $k_{\parallel}$.

biaxial tension and compression, respectively. In each case, the description of the conduction band is relatively straightforward; as this band is approximately parabolic near the zone centre, and the electron dispersion at small **k** can be described by

$$E(k) = E_c + \frac{\hbar^2}{2}\left(\frac{k_{\|}^2}{m_{\|}^*} + \frac{k_{\perp}^2}{m_{\perp}^*}\right), \tag{4}$$

where $E_c$ is the conduction band edge energy, $m_{\|}^* = m_{\perp}^*$ in the unstrained band structure, and the variation $m_{\|}^*$ and $m_{\perp}^*$ with strain can be predicted using $\mathbf{k} \cdot \mathbf{p}$ theory [29].

By contrast, the description of the valence band is complicated even in the unstrained case (Fig. 4a), where the light- and heavy-hole bands are degenerate at the zone centre, of $\Gamma_8$ symmetry, and so mix with each other at small wavevectors **k**. The light-hole and split-off bands start to interact when the light-hole energy becomes comparable with the spin–orbit splitting $\Delta$, but this mixing can generally be ignored in the analysis of semiconductor lasers due to the low density of holes at such high energies in the light-hole band.

Axial strain breaks the cubic symmetry of the semiconductor and thus splits the degeneracy of the light- and heavy-hole states at $\Gamma$, typically by about 60–80 meV for 1% lattice mismatch. The resulting valence band structure is highly anisotropic (Figs. 4b and c), with the band that is heavy along the strain axis $k_{\perp}$ being comparatively light perpendicular to that direction, $k_{\|}$, and vice versa. The anisotropy is due directly to the splitting of the valence band maximum and can again be understood via $\mathbf{k} \cdot \mathbf{p}$ theory [30, 31].

A complete description of the unstrained bulk valence band dispersion over a wide energy range should include explicitly the interactions with the spin–split-off and the lowest conduction band [32]. For laser gain calculations, we are usually interested chiefly in the dispersion near the band maximum, which can be well described using the Luttinger–Kohn (LK) Hamiltonian [33], which is written in matrix notation as

$$H = -\tfrac{1}{2}\gamma_1 k^2 I + \gamma_2[k_x^2(J_x^2 - \tfrac{1}{3}J^2) + \text{cp}] + 2\gamma_3[k_x k_y\{J_x J_y\} + \text{cp}], \tag{5}$$

where $I$ is the $4 \times 4$ unit matrix, $J_x, J_y, J_z$ the angular momentum matrices for spin $\frac{3}{2}$, and $x, y$, and $z$ denote the crystallographic (001) directions; $\{J_x J_y\} = \frac{1}{2}(J_x J_y + J_y J_x)$, and cp stands for cyclic permutation of the preceding term. We have taken $\hbar = m = 1$. The LK Hamiltonian of Eq. (5) omits terms linear in **k**, which are present due to the lack of inversion symmetry in III-V semiconductors, as these terms are negligible [34, 35]. The valence band dispersion is determined by the Luttinger parameters $\gamma_1$, $\gamma_2$, and $\gamma_3$, which are related to inverse effective masses at the Brillouin zone centre as discussed in what follows. We take the axis of quantisation along the "3" direction, while the "1" and "2" directions are chosen such that "1," "2," and "3" form a cartesian coordinate set. The phase factors of the basis functions for $J = \frac{3}{2}$ are chosen such that the corresponding matrices are

$$
\begin{aligned}
J_1 &= \frac{1}{2}\begin{pmatrix} 0 & \sqrt{3} & 0 & 0 \\ \sqrt{3} & 0 & 2 & 0 \\ 0 & 2 & 0 & \sqrt{3} \\ 0 & 0 & \sqrt{3} & 0 \end{pmatrix}, \\
J &= \frac{1}{2}\begin{pmatrix} 0 & -i\sqrt{3} & 0 & 0 \\ i\sqrt{3} & 0 & -2i & 0 \\ 0 & 2i & 0 & -i\sqrt{3} \\ 0 & 0 & i\sqrt{3} & 0 \end{pmatrix}, \\
J_3 &= \frac{1}{2}\begin{pmatrix} 3 & 0 & 0 & 0 \\ 0 & 1 & 0 & 0 \\ 0 & 0 & -1 & 0 \\ 0 & 0 & 0 & -3 \end{pmatrix}.
\end{aligned}
\tag{6}
$$

The LK Hamiltonian of Eq. (5) is usually presented by choosing "1," "2," and "3" to coincide respectively with the $x$, $y$, and $z$ directions. We mainly follow this convention here, except when we consider (111) growth, where the "3" direction is along (111). In the former case (i.e., $1, 2, 3 = x, y, z$), combining Eqs. (5) and (6) leads to the bulk valence band Hamiltonian

$$
H = \begin{pmatrix} a_+ & b & c & 0 \\ b^* & a_- & 0 & c \\ c^* & 0 & a_- & -b \\ 0 & c^* & -b^* & a_+ \end{pmatrix}, \tag{7}
$$

where

$$
\begin{aligned}
a_\pm &= -\frac{1}{2}(\gamma_1 \mp 3\gamma_2)k_z^2 - \frac{1}{2}(\gamma_1 \pm \gamma_2)(k_x^2 + k_y^2), \\
b &= \sqrt{3}\gamma_3 k_z(k_x - ik_y), \\
c &= \frac{\sqrt{3}}{2}[\gamma_2(k_x^2 - k_y^2) - 2i\gamma_3 k_x k_y].
\end{aligned}
$$

Because we have neglected the linear **k** terms in Eq. (5), the valence band structure is doubly degenerate. The $4 \times 4$ Hamiltonian matrix of Eq. (7) can then be decoupled into two independent $2 \times 2$ matrices [36, 37]

$$H_{\pm} = \begin{pmatrix} a_{+} & |b| \pm i|c| \\ |b| \mp i|c| & a_{-} \end{pmatrix}, \tag{8}$$

where

$$|b| = \sqrt{3}\,\gamma_3 k_z k_t,$$
$$|c| = \frac{\sqrt{3}}{2} k_t^2 \sqrt{\gamma_2^2 + (\gamma_3^2 - \gamma_2^2)\sin^2 2\theta},$$

$k_t$ is the magnitude of the wavevector in the $x$–$y$ plane, and $\theta$ is the angle between the direction of $k_t$ and the [100] direction. The band dispersion $E(k)$ found by diagonalising Eq. (8) is

$$E(k) = -\frac{1}{2}\gamma_1 k^2 \pm \sqrt{\gamma_2^2 k^4 + 3(\gamma_3^2 - \gamma_2^2)(k_z^2 k_t^2 + (1/4)k_t^4 \sin^2(2\theta)}. \tag{9}$$

The LK Hamiltonian is often further simplified for specific applications, with the three most widely used modifications being the spherical approximation, the axial approximation, and the diagonal approximation. In the first (spherical) case, $\gamma_2$ and $\gamma_3$ are set equal to $\bar{\gamma} = (3\gamma_3 + 2\gamma_2)/5$ [38], so that the band dispersion in Eq. (9) simplifies to

$$E(k) = -\frac{1}{2}(\gamma_1 \pm 2\bar{\gamma})k^2. \tag{10}$$

The spherical approximation is particularly appealing in bulk lasers, as it provides analytic expressions for the density of states and the average transition matrix elements, which otherwise would have to be evaluated numerically. It is less successful when applied to quantum well structures, where it can give only a qualitative description of confinement and polarisation effects but does not accurately calculate confinement energies nor mixing effects. We note that the spherical approximation does not distinguish between (001) and (111) quantum well lasers, which we wish to discuss in more detail in Section 5.

The axial approximation describes the LK band structure along the "3" direction exactly, and it averages the band structure in the 1–2 plane, so that $a_{\pm}$ and $b$ remain unchanged in Eq. (8), while the $\theta$ dependence in $|c|$ is ignored by setting $\gamma_2 = \gamma_3 = \gamma'$ in this term only, so that $|c| = \sqrt{3}\gamma' k_t^2$, with $\gamma' = (\gamma_2 + \gamma_3)/2$ [39], or $\sqrt{\frac{1}{2}(\gamma_2^2 + \gamma_3^2)}$. The axial approximation is particularly suited to QW structures, as it gives the correct confinement energies and still allows a straightforward expression for the average matrix elements as a function of $k_t$. The gain curves agree closely with those calculated by retaining the angular dependence of $|c|$ [40], so that the axial approximation is now widely used to study QW lasers, as in this work. Finally, the diagonal approximation consists of ignoring the off-diagonal terms of $H$ and therefore mixing effects, so that the band dispersion is given by

$$E(k) = a_{\pm} = -\frac{1}{2}(\gamma_1 \mp 2\gamma_2)k_z^2 - \frac{1}{2}(\gamma_1 \pm \gamma_2)(k_x^2 + k_y^2). \tag{11}$$

The diagonal approximation gives a useful guide to the band dispersion in highly strained structures, where mixing effects are minimised, but is of less value in applications where band mixing is important, such as unstrained bulk and QW lasers.

In the presence of strain, the LK Hamiltonian of Eq. (5) must be modified to reproduce the band structure of Figs. 4b and c. This can be done by incorporating strain components in the Hamiltonian up to the desired order [29, 41]. When we consider growth of a high quality strained epilayer along the (001) direction, the epilayer is under a biaxial stress such that its in-plane lattice constant equals the substrate lattice constant $a_s$. The net strain in the layer plane, $\varepsilon_{\|}$, is given by

$$\varepsilon_{\|} = \varepsilon_{xx} = \varepsilon_{yy} = (a_s - a_0)/a_0, \tag{12}$$

where $a_0$ is the lattice constant of a free standing layer of the same composition. In response to the biaxial stress, the layer relaxes along the growth direction, the strain $\varepsilon_{\perp}$ $(=\varepsilon_{zz})$ being given by

$$\varepsilon_{zz} = \frac{-2\sigma}{1-\sigma}\varepsilon_{xx}, \tag{13}$$

where $\sigma$ is Poisson's ratio. All the off-diagonal elements of the strain tensor vanish for an (001) strain. The strain due to such a biaxial stress can be resolved into a hydrostatic and a purely axial component. The former is due to the fractional volume change $\Delta V/V = \mathrm{tr}(\varepsilon)$ $( = \varepsilon_{xx} + \varepsilon_{yy} + \varepsilon_{zz})$, and does not change the symmetry of the lattice. The latter changes the symmetry, introducing the valence band splittings in Figs. 4b and c. Both components introduce additional terms in the LK Hamiltonian, the most important of which are the terms that are linear in strain:

$$H_{\varepsilon} = a_v\, \mathrm{tr}(\varepsilon)I + \tfrac{2}{3}b[\varepsilon_{xx}(J_x^2 - \tfrac{1}{3}J^2) + \mathrm{cp}], \tag{14}$$

where $a_v$ and $b$ are the hydrostatic and uniaxial deformation potentials. The hydrostatic term (containing $\mathrm{tr}(\varepsilon)$), together with the conduction band strain

Hamiltonian $a_c$ tr($\varepsilon$), causes a change in mean bandgap, $\Delta E_g = (a_c - a_v)$ tr($\varepsilon$). The uniaxial component can be rewritten using the matrices of Eq. (6) as

$$H_\varepsilon = \begin{pmatrix} \delta\varepsilon_S & 0 & 0 & 0 \\ 0 & -\delta\varepsilon_S & 0 & 0 \\ 0 & 0 & -\delta\varepsilon_S & 0 \\ 0 & 0 & 0 & \delta\varepsilon_S \end{pmatrix}, \tag{15}$$

where $\delta\varepsilon_S = b(\varepsilon_{xx} - \varepsilon_{zz})$. The degeneracy of the heavy- ($m_J = \pm\frac{3}{2}$) and light- ($m_J = \pm\frac{1}{2}$) hole states is split, with the ordering of the splitting depending on the sign of $\delta\varepsilon_S$. Adding $H_\varepsilon$ to $H$ in Eq. (8) ($H_\varepsilon$ retains the same form after the Broido–Sham transformation), we see that in the diagonal approximation the masses become anisotropic with the introduction of axial strain:

$$E_\pm(k_z, k_t) = \pm\delta\varepsilon_s - \frac{1}{2}(\gamma_1 \mp 2\gamma_2)k_z^2 - \frac{1}{2}(\gamma_1 \pm \gamma_2)k_t^2. \tag{16}$$

The band that is heavy along the strain ($z$) direction ($m^* = (\gamma_1 - 2\gamma_2)^{-1}$) is comparatively light perpendicular to that direction ($m^* = (\gamma_1 + \gamma_2)^{-1}$), and vice versa. This anisotropy is the key to achieving light holes at the valence band maximum of strained layer structures.

### 3.2. 2D Structures

The description of the valence subband structure in quantum well systems is presented with excellent detail and clarity in Chapter 1 [1]; we therefore only summarise the most important features here.

For quantum wells grown along the (001) direction, the mean valence band energy and strain-induced splitting vary as $E_v(z)$ and $\delta\varepsilon_S(z)$ respectively, so that $k_z$ is no longer a good quantum number. We assume the structure remains periodic and that Bloch's theorem holds in the $x$–$y$ plane, so that we can use the axial approximation to describe the subband dispersion in the quantum well plane in terms of the in-plane wavevector $k_t$ of Eq. (8). The Hamiltonian of Eq. (8) is diagonalised at the Brillouin zone centre ($k_t = 0$), where we can calculate the zone centre heavy- and light-hole states by solving the independent Schroedinger equations

$$-\frac{1}{2}(\gamma_1 \mp 2\gamma_2)\frac{\partial^2 f_m^\pm(z)}{\partial z^2} + [E_v(z) \pm \delta\varepsilon_S(z)]f_m^\pm(z) = E_m^\pm f_m^\pm(z), \tag{17}$$

with appropriate boundary conditions [1]. $E_m^+$ ($E_m^-$) is the energy of the $m$th heavy-hole (light-hole) state and $f_m(z)$ is the $m$th zone centre envelope function. The resulting confinement energies are different for heavy- and light-hole states even in an unstrained quantum well ($\delta\varepsilon_S(z) = 0$); i.e., quantum size effects lift the degeneracy at the zone centre, with the confined state energies being determined by the effective masses $(\gamma_1 \pm 2\gamma_2)^{-1}$, the band offset, and the well width.

As opposed to the bulk case, it is the in-plane mass that determines the density of states in QW structures. In the diagonal approximation, this mass is light for "heavy-hole" states ($=(\gamma_1 + \gamma_2)^{-1}$), including in particular the highest subband, which plays an important role in lasers, as most of the injected holes populate this subband. The reduced in-plane mass, coupled with the more favourable shape of the DOS in 2D structures (step-like for parabolic bands) are the chief reasons why QW lasers have been predicted to have improved characteristics compared with their bulk counterparts. However, band-mixing effects reduce the energy over which the highest hole subband is light-hole-like, thus degrading the laser characteristics. Mixing effects depend both on the magnitude of the off-diagonal terms in Eq. (8) and on the energy separation between the heavy- and light-hole zone centre states. Strain offers an extra degrees of freedom for changing the zone centre energy separation. For a QW layer under biaxial compression (e.g., InGaAs on GaAs), the "heavy-hole" band shifts upwards in energy whereas the "light-hole" band shifts down ($\delta\varepsilon_S > 0$). This effect increases the zone centre energy separation between heavy- and light-hole subbands, and band mixing in the valence band become less important than in the unstrained case. The in-plane hole mass for the highest subband is then light-hole-like over a wider energy range than in the unstrained case, of potential benefit for semiconductor laser applications. In the case of biaxial tension in the $x$–$y$ plane, quantum size and strain effects compete against each other, and the average mass is generally heavy-hole-like or can even have an electron-like dispersion. The consequence of strain-induced changes in band structure form the main theme of Section 4.

### 3.3. (111) Valence Band Hamiltonian

In the case of (111) growth, we rotate the axes of quantisation to the (111) direction, with the "1" and "2" directions along $(11\bar{2})$ and $(\bar{1}10)$, respectively. In this way, $H$ in Eq. (5) can be written as a function of $k_1, k_2, k_3$ and $J_1, J_2, J_3$. The resulting Hamiltonian is

$$H = \begin{pmatrix} a_+ & b & c & 0 \\ b^* & a_- & 0 & c \\ c^* & 0 & a_- & -b \\ 0 & c^* & -b^* & a_+ \end{pmatrix}, \tag{18}$$

where

$$a_{\pm} = -\frac{1}{2}(\gamma_1 \mp 2\gamma_3)k_3^2 - \frac{1}{2}(\gamma_1 \pm \gamma_3)(k_1^2 + k_2^2),$$

$$b = -i\left[\frac{(2\gamma_2 + \gamma_3)}{\sqrt{3}}\right](k_1 - ik_2)k_3 + i\left[\frac{(\gamma_2 - \gamma_3)}{\sqrt{6}}\right](k_1 + ik_2)^2,$$

$$c = \frac{(\gamma_2 + 2\gamma_3)}{2\sqrt{3}}(k_1 - ik_2)^2 - \frac{2(\gamma_2 - \gamma_3)}{\sqrt{6}}(k_1 + ik_2)k_3.$$

In the axial approximation, we retain the exact form of the $a_{\pm}$ terms in Eq. (18) but omit the terms involving the anisotropy factor $\gamma_2 - \gamma_3$ in $b$ and $c$. The $4 \times 4$ Hamiltonian can be decoupled into two independent $2 \times 2$ matrices

$$H = \begin{pmatrix} a_+ & |b| \pm i|c| \\ |b| \mp i|c| & a_- \end{pmatrix}, \tag{19}$$

where

$$|b| = [(2\gamma_2 + \gamma_3)/\sqrt{3}]k_t k_3,$$

$$|c| = [(\gamma_2 + 2\gamma_3)/2\sqrt{3}]k_t^2,$$

and $k_t = \sqrt{k_1^2 + k_2^2}$. The bulk heavy- and light-hole masses along the (111) direction are given from the diagonal elements of Eq. (19) as $m_h = (\gamma_1 - 2\gamma_3)^{-1}$ and $m_l = (\gamma_1 + 2\gamma_3)^{-1}$, respectively. This compares with the values along the (001) direction of $m_{h,1} = (\gamma_1 \mp 2\gamma_2)^{-1}$. In most III-V semiconductors, $\gamma_3 > \gamma_2$ [42], reflecting the larger heavy-hole mass along (111) than along (001). This valence band anisotropy has important consequences for laser applications and can account for the measured differences in GaAs–AlGaAs laser characteristics. Detailed comparison of the two growth directions is considered in Section 5.

It is finally of interest to consider how the combination of (111) growth and uniaxial strain might affect laser applications. It was originally proposed theoretically [43, 44] and has since been confirmed experimentally [45] that (111) strained layer structures would have a built-in piezoelectric field due to the relative displacement of the cation and anion sublattices. The induced field is typically of order $1 \times 10^5$ V cm$^{-1}$ for a 1% lattice mismatch. Such large fields have a significant effect on the band edge energy and can lead to Stark shifts of order 50 meV [44] and to spatial separation of the electron and hole wavefunctions. Screening causes the built-in field to decrease with increasing carrier density, so that, above a critical density, the screened field is effectively zero. This carrier-dependent field effect has potential application for nonlinear devices [44], including systems that may incorporate lasers. The threshold carrier density in a (111) strained layer laser is expected to be sufficiently large to fully screen the built-in fields, so that the operation of an isolated broad area laser will not be influenced by the Stark-shifted band gap. If, however, the carrier density is reduced in part of the quantum well structure, either by confining the current laterally or else by only pumping a fraction of the total length, then the reduced bandgap in the region of low carrier density will lead to increased absorption losses [46]. While such effects are generally undesirable in conventional optical communication systems, they could be of benefit in such areas as optical switching applications.

## 4. LONG WAVELENGTH STRAINED LAYER LASERS

In this section, we consider the design of a strained layer quantum well laser to operate at 1.55 $\mu$m, the dominant wavelength in optical communications. In practice, such lasers can be formed using the same technology as conventional QW lasers, with the chief modifications being the replacement of the lattice-matched QW layers by QW layers under biaxial compression. The two most common III-V substrates are InP and GaAs. GaAs-based structures with strained InGaAs quantum wells have already demonstrated low threshold current densities and improved laser characteristics for short wavelength lasers operating near 1 $\mu$m (see Chapter 8 in this book [2]). However, such GaAs-based laser structures would require a substantial lattice mismatch between the wells and barrier to operate at 1.55 $\mu$m. We estimated a 1.55-$\mu$m laser would require $\sim$35 Å strained InAs wells between unstrained GaAs barriers [47]. Such a structure is likely to be beyond the limits of good quality growth. In the longer term, it may be possible to grow strained lasers off buffer layers grown on GaAs or any other substrate [48, 49], but at present, and for the foreseeable future, the most suitable 1.55-$\mu$m strained layer lasers are almost certainly those grown on InP substrates. A number of theoretical studies have considered this system [50–53], and several groups have recently reported promising experimental results on InP-based strained layer 1.55-$\mu$m lasers [54–60]. We first present a theoretical review of 1.55-$\mu$m strained lasers and then consider the experimental situation in Section 4.2.

### 4.1. Theoretical Analysis

We have considered a specific InP-based, strained layer, separate confinement heterostructure (SCH) laser operating at 1.55 $\mu$m, corresponding to a photon energy of 0.8 eV (Fig. 5). The structure in Fig. 5 is closely related to a conventional SCH laser with the well material replaced by strained layers. The wells consist of undoped $In_xGa_{1-x}As$, with the well width $L_z$ chosen for each value of $x$ to ensure an optical gap of 0.8 eV, so that $L_z$ decreases as $x$ (and $\varepsilon_{\|}$, the percentage lattice mismatch) increases. We have studied three different structures with well thicknesses of 20, 35, and 100 Å, respectively [50], while more recently Abram and coworkers [52, 53] have examined structures where $\varepsilon_{\|} = 0$, $-0.75$, and $-1.5\%$, corresponding to $x = 0.53, 0.63$, and 0.73, respectively. Different theoretical parameters were used in the two studies, so that direct comparison is difficult, but the general conclusions are similar. It was found in both cases for each laser structure that $\varepsilon_{\|} L_z < 100$ Å%. It is expected that the devices should then be stable, as Fritz *et al.* [61] showed good quality growth in the GaAs–InGaAs system for $\varepsilon_{\|} L_z$ up to at least 100 Å%, while Andersson *et al.* [62] achieved high quality growth up to of order 200 Å%. Quantum confinement is achieved by sandwiching the wells between quaternary $Ga_yIn_{1-y}As_zP_{1-z}$ barriers, lattice-matched to InP. Separate optical confinement is achieved with an InP cladding above and below the quaternary region. Two competing effects need to be considered in optimising the structure shown in Fig. 5. On the one hand, optical confinement is enhanced by increasing $y$ and $z$. On the other hand, this decreases the quantum confinement between the wells and quaternary barriers. We have not attempted to determine the

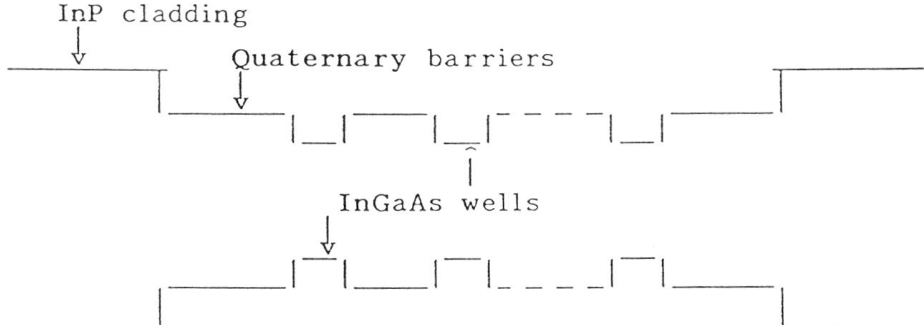


**Fig. 5.** InP-based strained layer MQW separate confinement heterostructure (SCH) laser. Carriers are confined in the strained $In_xGa_{1-x}As$ quantum wells, while optical confinement is due to the refractive index step between the quaternary barrier and InP cladding.

optimum quaternary composition for the barrier but find that, with $y = 0.14$ and $z = 0.3$, neither the optical nor the quantum confinement is significantly degraded. More recent experimental studies have employed graded index (GRIN) SCH structures [57, 59].

The quantum well valence subband structure was calculated using the LK $6 \times 6$ Hamiltonian [63] in the axial approximation for the three well widths we considered and is presented in Figs. 6a–c. The separation between the two highest valence band states at the zone centre decreases rapidly with increasing well width, falling from 164 meV in the 20-Å well to 123 and

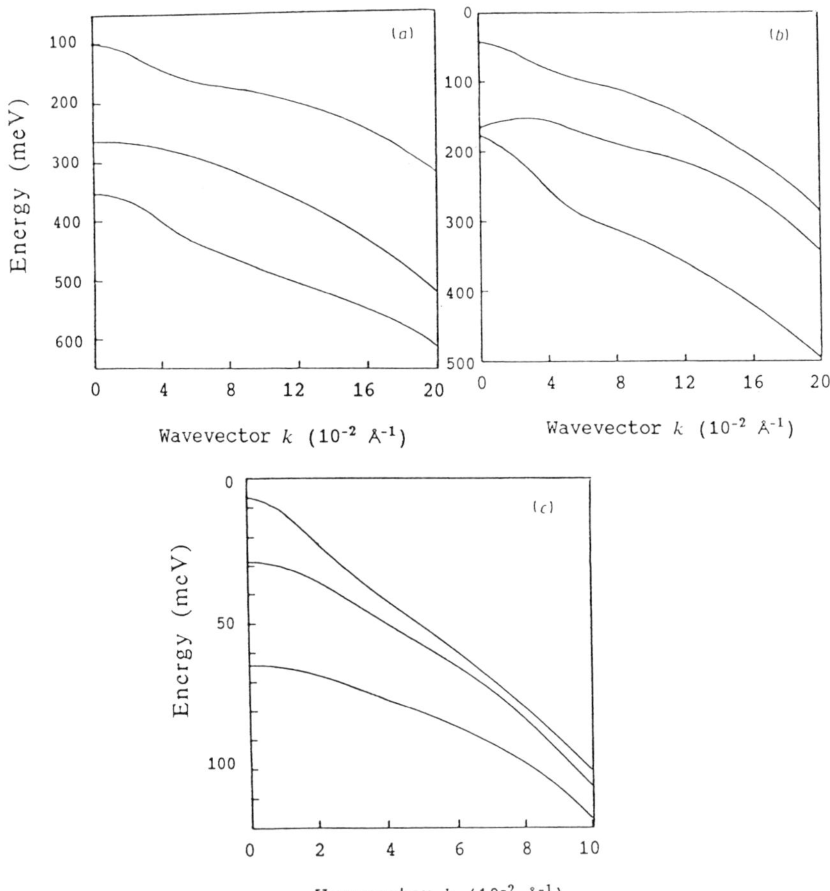


**Fig. 6.** Valence subband structure in the well plane for (a) 20-Å (b) 35-Å, and (c) 100-Å $Ga_xIn_{1-x}As$ quantum wells sandwiched between $Ga_{.14}In_{.86}As_{.3}P_{.7}$ barriers. In each case, $x$ is chosen to give an optical gap of 0.8 eV.

21 meV in the 35- and 100- Å wells, respectively. The decreasing separation is due to the reduction both in confinement energy and in net strain with increasing well width.

We calculated the gain spectra for the band structures of Fig. 6, following Asada [64]. We assumed an intraband relaxation time $\tau_{in} = 10^{-13}$ s, while the **k** dependence of the dipole matrix elements were taken directly from the band structure calculations, as described by Corzine *et al.* in Chapter 1 [1]. The most promising results were obtained for the case of 35-Å wells, which we now summarise.

Transparency is achieved at a carrier concentration $n_0$ of less than $2 \times 10^{18}$ cm$^{-3}$. The differential gain above transparency, $\beta = dg/dn$, is large, implying an improved dynamic response and enhanced modulation bandwidth [26–28], and the gain exceeds 1000 cm$^{-1}$ by a carrier density of $3 \times 10^{18}$ cm$^{-3}$. Lasing occurs when the maximum gain equals the threshold gain $g_{th}$, given by

$$g_{th} = \alpha_{ac} + \Gamma^{-1}(1 - \Gamma)\alpha_{ex} + (\Gamma L)^{-1} \ln(1/R). \tag{20}$$

We assume typical values for the cavity length $L = 300\ \mu$m, reflectivity $R = 0.31$, and external losses $\alpha_{ex} = 10$ cm$^{-1}$. The internal losses are expected to be considerably reduced in comparison with bulk lasers, due to the virtual elimination of IVBA, so we take $\alpha_{ac} = 10$ cm$^{-1}$. The optical confinement factor for a SCH QW laser is approximately proportional to the product of the well thickness times the number of wells $N_w$, i.e., $\Gamma = \gamma N_w L_z$ [65]. We choose $\gamma = 2.09 \times 10^{-4}$ Å$^{-1}$.

The threshold current density $J_{th}$, assuming only radiative recombination, can be derived from the spontaneous emission rate [1], and, setting the carrier density to its threshold value, $J_{th}$ can be written as

$$J_{th} = eBN_w L_z n_{th}^2, \tag{21}$$

where $B$ is the radiative recombination coefficient, which is band structure and carrier density dependent. Here, $B$ is set equal to $10^{-10}$ cm$^3$ s$^{-1}$. For a small number of wells $N_w$, the low value of $\Gamma$ implies a high value of the threshold carrier density $n_{th}$, and gain saturation effects can become important. For larger $N_w$ values, $n_{th}$ approaches $n_0$, the inversion carrier density, and, from Eq. (21), the threshold current density then increases linearly with $N_w$.

The lower curve in Fig. 7 shows the calculated current density $J_{th}^{rad}$ at 300 K as a function of the number of wells. Assuming negligible Auger recombination, the threshold current density is down by nearly an order of magnitude over conventional long wavelength lasers and is comparable with the best

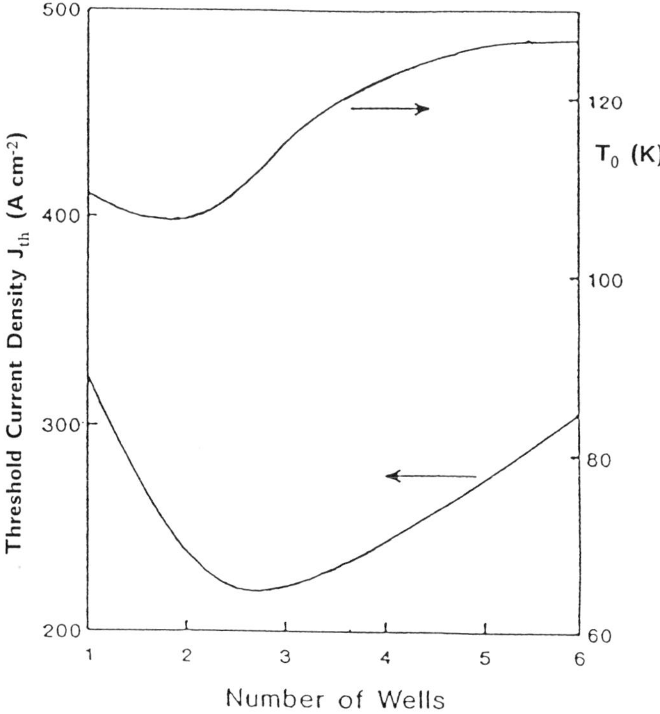


**Fig. 7.** Calculated threshold current density $J_{th}$, in A/cm$^2$, and $T_0$ value in K, using the band structure of Fig. 6b.

short wavelength lasers. In addition, the calculated $T_0$ values, shown in the upper curve of Fig. 7, are over 100 K, implying much improved temperature stability over conventional lasers, where $T_0 \sim 60$ K.

Our results in Fig. 7, where Auger recombination is set to zero, present the most optimistic view of the predicted improvement in threshold current density in strained layer lasers. The more recent work of Robbins and coworkers [52, 53] takes a more pessimistic view and still predicts a significant reduction in threshold current density. From Eqs. (2) and (3), the Auger rate $R$ varies approximately as

$$R \sim Cn_{\text{th}}^3, \tag{22}$$

where $C$ depends on details of the band structure and is predicted to be reduced in a strained layer laser [24]. Robbins *et al.* [53] assume $C$ remains unchanged between the strained and unstrained structures that they consider and find a minimum total threshold current density of $\sim 1000$ A/cm$^2$, which is still a considerable reduction over their unstrained case (Fig. 8). The cubic dependence of the Auger current on carrier density favours a higher number of wells than in our calculations, enabling the threshold current density $n_{th}$ to approach closer to the transparency density $n_0$. Interestingly, they find that little or no advantage is obtained by increasing the magnitude of the lattice mismatch $\varepsilon_{\|}$ from 0.75% ($x = 0.63$, $L_z = 53$ Å) to 1.5% ($x = 0.73$, $L_z = 43$ Å) and so predict that useful improvements can be obtained in long wavelength (1.55 $\mu$m) quantum well lasers by the incorporation of a relatively modest amount of strain in the quantum well layers.

## 4.2. Experimental Results

The first experimental work on 1.55-$\mu$m strained layer lasers demonstrated that such structures could be grown but showed little or no improvement over comparable lattice-matched MQW lasers [66]. Since then, there has been a dramatic increase in interest in strained lasers, with promising results being reported by groups from many laboratories. Thijs and coworkers [54, 55] have used low pressure organometallic vapour phase epitaxy to grow high quality strained layer lasers, similar to the structure in Fig. 5, discussed

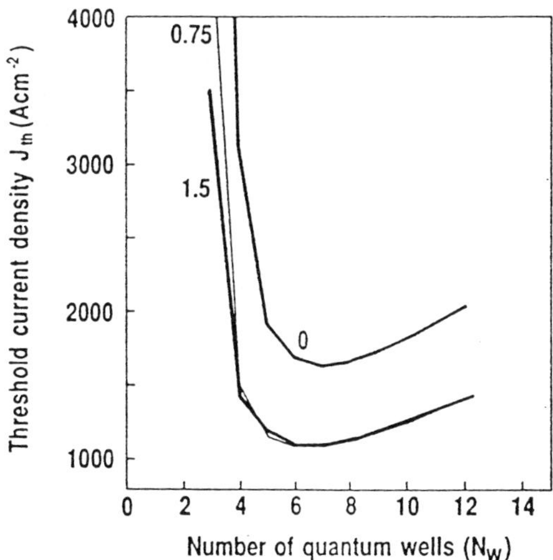


**Fig. 8.** Calculated threshold total current density as a function of the number of $In_xGa_{1-x}As$ quantum wells for 1.55-$\mu$m lasers in the three cases of lattice match ($\varepsilon_{\|} = 0$, $x = 0.53$); a mismatch of magnitude 0.75% ($x = 0.63$), and a mismatch of magnitude 1.5% ($x = 0.73$).

previously. Excellent properties were demonstrated for structures containing four quantum wells. All showed very high differential external efficiency $\eta$, with a value as high as 82% being demonstrated for a 256 $\mu$m long device with as-cleaved mirrors. The internal efficiency $\eta_i$ was deduced to be 100%, considerably improved over the typical lattice-matched MQW value of 60–80% [67–70] and indicative of a significant reduction in nonradiative recombination processes such as IVBA and Auger recombination. A $T_0$ value as high as 97 K between 20–50°C was reported for 1000 $\mu$m long strained devices, which compares favourably with the theoretical predictions and with room temperature $T_0$ values of 50–80 K for 1000 $\mu$m long lattice-matched 1.55-$\mu$m wavelength MQW lasers [71, 72]. Such characteristics are excellent prerequisites for high power operation, and a maximum CW output power as high as 200 mW was obtained from 1500 $\mu$m long DCPBH devices that had antireflection and high reflection mirrors on opposite facets ($R_f = 4\%$, $R_r = 98\%$). What is more, lifetime tests performed at a heatsink temperature of 60°C and a CW output power of 5 mW showed almost no degradation after 4,000 hours, which, taken with the lifetime data on 1-$\mu$m wavelength strained layer lasers in Chapter 8 [2], is most encouraging for future device applications.

A number of groups have succeeded in demonstrating reduced threshold current densities in strained layer 1.55-$\mu$m lasers. Koren *et al.* [60] have reported a threshold current density as low as 440 A/cm$^2$ in a structure with four 25 Å wide $In_{.77}Ga_{.23}As$ wells and a high reflectivity mirror ($R_r = 97\%$) on one facet. Both Yamada *et al.* [58] and Tsang *et al.* [56] have made direct comparison of lattice-matched and strained MQW laser structures, with similar results. Yamada *et al.* considered lasers with four $In_xGa_{1-x}As$ quantum wells, with $x = 0.53, 0.62$, and 0.71 and well widths chosen to achieve lasing at about 1.55 $\mu$m. The strained MQW lasers, with $x = 0.62$, showed the minimum threshold current densities in the measured cavity length range, with the lowest $J_{th}$ value (398 A/cm$^2$) being obtained in a 1850 $\mu$m long device. The extrapolated $J_{th}$ values for infinite cavity lengths $L$ (equivalent to zero mirror loss in Eq. (20)) were 540, 340, and 660 A/cm$^2$ for $x = 0.53, 0.62$, and 0.71, respectively. Tsang *et al.* studied a wider range of well compositions $x$, and they again found the minimum threshold current densities for In composition $x$ between 0.6 and 0.65, with the lowest value of $J_{th} = 370$ A/cm$^2$ being reported for $x = 0.65$ and cavity lengths $\geq 1500$ $\mu$m. Tsang *et al.* found the $T_0$ value to be improved, but not dramatically, in the strained lasers studied, rising from 54 K in the lattice-matched structures ($x = 0.53$) to 71 K for $x = 0.65$.

The lowest threshold current densities should be achieved in structures with few quantum wells, where both the cavity losses and the mirror losses of Eq. (20) are minimized. The threshold carrier density $n_{th}$ should then approach the transparency density $n_0$, and for a single quantum well ($N_w = 1$), the radiative current density of Eq. (21) can be approximated by

$$J_{th}^{rad} = eBL_z n_0^2, \tag{23}$$

with the minimum value of $J_{th}^{rad}$ predicted to be of order 10 A/cm$^2$ [25] for strained layer 1.55-$\mu$m lasers. Zah *et al.* [59] have sought such a structure by growing a single quantum well graded index SCH laser with high reflection coatings to minimise the mirror losses. The laser was optically pumped, and an upper limit of 92 A/cm$^2$ was estimated for $J_{th}$, based on 100% collection efficiency. Allowing for carrier losses, the actual threshold current density could be as low as 70 A/cm$^2$. In practice, the output power for such a laser is very low ($\sim$1 mW) because of the high reflectivity mirrors.

In summary, significant effort has recently been devoted to developing 1.55-$\mu$m strained layer MQW lasers with improved device characteristics associated with the modified valence band structure of the strained quantum wells. The results to date are very promising, and while no single laser has yet demonstrated all of the predicted advantages of strained growth, it has nevertheless been possible to show that strained layer structures can lead to reduced threshold current densities [56–58], improved temperature stability [55], enhanced internal efficiency [55], and higher output powers [55]. The devices have also been shown to be stable [55].

We finally mention in this section the increasing interest in QW strained layer lasers and devices where parts of the structure are deliberately under biaxial *tension* [73, 82, 83]. Such a tensile stress shifts the light-hole band edge upwards in energy ($\delta\varepsilon_S < 0$ in Eq. (15)), so that the highest heavy-hole and light-hole states can become degenerate with each other, or the light-hole state can even be shifted above the heavy-hole state. As the heavy-hole states tend to favour TE gain, while the light-holes favour TM gain, tensile stress can be used to tailor the ratio of the peak TE to TM gain, including the possibility of identical peak gains for moderate tensile stress or of the TM gain dominating the TE gain for larger tensile stress and/or wider wells. The case of equal TE and TM gain is of potential benefit for optical amplifier applications, where the intensity of the output beam can then be independent of the polarisation of the incident beam [73]. For tensile-strained lasers, it has recently been shown experimentally that the threshold current density decreases with increasing tensile strain. A threshold current density as low as 197 A cm$^{-2}$ was obtained with a single 125 Å wide $In_{0.3}Ga_{0.7}As$ quantum well GRINSCH laser grown on an InP substrate [83]. This reduction in threshold current density is due to a largely different mechanism than in the case of compressive-strained lasers, namely, the suppression here of the spontaneous emission associated with light polarised in the plane of the quantum well structure [84]. This suppression of unwanted emission is more efficient in tensile- than compressive-strained lasers, as biaxial tension removes the $\pm\frac{3}{2}$ state from the valence band edge, which make no contribution to the TM gain, while under biaxial compression the $\pm\frac{1}{2}$ states removed from the band edge had contributed both to TE and TM gain [1]. This is discussed in more detail elsewhere [84].

With the range of band structures and effects that can be accessed in strained layer structures under biaxial compression and tension, it is clear that the growth of strained layer lasers and devices should remain an exciting and fruitful research area in the coming years.

## 5. COMPARISON OF (001) AND (111) QUANTUM WELL LASERS

### 5.1. Introduction

Because the lasers that were considered in the previous section were in general designed to emit light at 1.55 $\mu$m, it is difficult to separate quantitatively the influence of the different factors such as well width, built-in strain, and alloy composition on the overall emission efficiency of the laser. In this section, we try to emphasise the overall importance of valence subband structure on quantum well laser operation by comparing lasers that are grown along different crystallographic directions but are in every other respect equivalent. High quality semiconductor epistructures have been grown almost exclusively along the (001) crystallographic direction. There has occasionally been interest in growth along other directions, in particular along (111). The best GaAs–AlGaAs (111) layers are now of device quality, comparable with conventional (001) structures [5–9, 74–76]. Hayakawa and coworkers have been particularly successful in fabricating (111) quantum well lasers [7–9], which for well widths $L_w < 100$ Å have consistently lower threshold current densities $J_{th}$ than the equivalent (001) lasers. They also measured the polarisation dependence of gain–current characteristics [9] and found marked differences between the (111) and (001) results, including strong gain saturation of the $n = 1$ transition for a (111) laser with $L_w =$ 100 Å, and more pronounced transverse electric (TE) mode discrimination than in the equivalent (001) case.

In this section, we are mainly concerned with the effects of valence band anisotropy on the subband structure of (111) quantum wells compared with (001) wells, and on its consequences for laser applications. The heavy- and light-hole confinement energies have been shown experimentally to be very different in (001) and (111) GaAs–AlGaAs quantum wells [5, 6], reflecting the anisotropy of the bulk valence bands, where the heavy-hole mass, for example, is approximately twice as large along (111) as along (001). We present envelope function calculations of the valence subband structure of (111) and (001) wells in Section 5.2, and of the optical matrix elements associated with the two growth directions in Section 5.3, including band-mixing effects. The calculated matrix elements are then used as input to laser gain calculations in Section 5.4. We demonstrate that the highest valence subband has a low in-plane effective mass (light-hole cap) in both (001) and (111) structures, and the light-hole cap extends over a greater energy range in (111) wells for $L_w < 100$ Å. This reduces the carrier density required for population inversion in thin (111) quantum wells. The changes in valence subband structure due to band anisotropy can fully account for the measured differences in $J_{th}$ [8] between (001) and (111) growth. We also show how the strong gain saturation of the $n = 1$ transition in (111) lasers with $L_w = 100$ Å and the gain switching to the $n = 2$ transition at high injected carrier density follow from the valence subband structure. Finally, we find that the TE to TM mode discrimination is more pronounced for thin (111) wells than (001) wells as a result of the different subband ordering.

## 5.2. Electronic Structure

The valence band structure is again described using the Luttinger–Kohn (LK) Hamiltonian, as explained in Section 3. The valence band dispersion is then determined by the Luttinger parameters $\gamma_1$, $\gamma_2$, and $\gamma_3$, with the bulk heavy- and light-hole masses along the (111) direction being given by $m_h^* = (\gamma_1 - 2\gamma_3)^{-1}$ and $m_l^* = (\gamma_1 + 2\gamma_3)^{-1}$, respectively, compared with the values along the (001) direction $m_{l,h}^* = (\gamma_1 \pm 2\gamma_2)^{-1}$. The $\gamma$ values used are shown in Table I, along with the corresponding heavy- and light-hole masses along (001) and (111). The GaAs $\gamma$ values were taken from [6], while those for GaAlAs were calculated by linear interpolation of effective masses between GaAs and AlAs [77]. We assume the conventional 65:35 band offset ratio and the bandgap of $Al_xGa_{1-x}As$ to increase as $E_g(x) - E_g(\text{GaAs}) = 1.34x$ (eV) [78]. Details of the calculation method used to determine the subband structure can be found in [79] and [80]. We first consider how the zone centre confinement energies vary with well width for (001) and (111) growth. These are plotted in Figs. 9a and b, respectively, where we take as

**Table I.**

Luttinger Parameters and Valence Band Effective Masses Used

| | | | | (001) | | (111) | |
|---|---|---|---|---|---|---|---|
| | $\gamma_1$ | $\gamma_2$ | $\gamma_3$ | $m_{hh}$ | $m_{lh}$ | $m_{hh}$ | $m_{lh}$ |
| GaAs | 6.78 | 1.92 | 2.70 | 0.340 | 0.094 | 0.725 | 0.082 |
| $Al_{.2}Ga_{.2}As$ | 5.64 | 1.46 | 2.20 | 0.368 | 0.110 | 0.810 | 0.100 |
| $Al_{.3}Ga_{.7}As$ | 5.21 | 1.29 | 2.02 | 0.382 | 0.128 | 0.852 | 0.108 |

our cxample GaAs wells between $Ga_{.7}Al_{.3}As$ barriers, using the effective masses in Table I. The light-hole confinement energies are very similar for the two growth directions. By contrast, the heavy-hole confinement energies are much reduced for (111) growth, reflecting the twofold increase in effective mass. This has three important consequences, all of which influence subband dispersion and laser applications.

(i) The splitting between the highest heavy- and light-hole states, E(H1) − E(L1), is always enhanced in (111) growth. This enhancement increases with decreasing well width, with the splitting increasing from 30 to 43 meV in a 50-Å well.

(ii) There are significantly more heavy-hole confined states in a (111) well of given width than in the equivalent (001) well; and the splitting between the two highest heavy-hole states, E(H1) − E(H2), decreases much more rapidly with increasing well width in the (111) case.

(iii) For sufficiently wide wells, the L1 state must always lie below the H2 state for (111) growth, as $m_h^* > 4m_l^*$ for the (111) direction. In thin wells, only one light and one heavy bound state are expected, so L1 must rise above H2 at some intermediate well width $L_w$. We calculate the crossover to be about 50 Å for (111) growth between $Ga_{.7}Al_{.3}As$ barriers. As $m_h^* < 4m_l^*$ in (001) growth, the L1 state will always be above the H2 state in this case. For most well widths of interest, the second confined value state is therefore H2 in (111) growth, while it is L1 in the (001) case.

The calculated valence subband structures for 50-, 100-, and 150-Å wells between $Ga_{.7}Al_{.3}As$ barriers are shown in Fig. 10. There are marked differences between the (001) (a–c) and (111) (d–f) results. In each case, the highest valence subband, H1, has a low in-plane effective mass over a small energy range. In thin wells, the light-hole behaviour persists over a greater energy range in the (111) than in the (001) case. This light-hole-like behaviour can be understood by treating the LK Hamiltonian of Eq. (19) in the diagonal

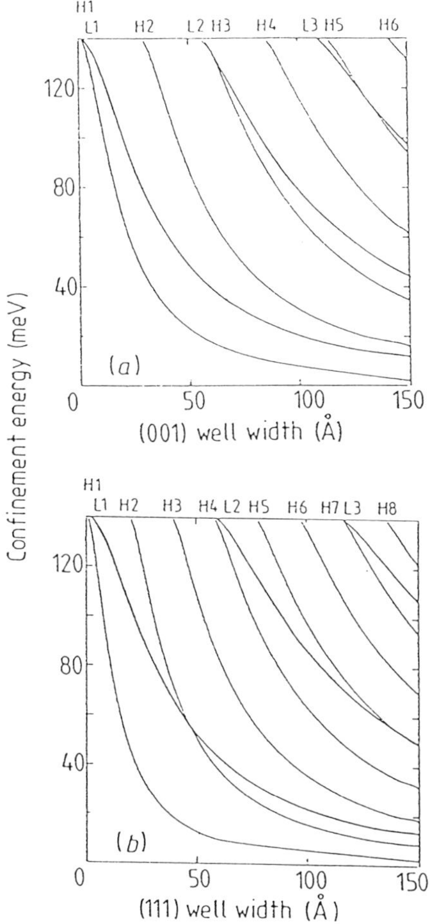


**Fig. 9.** Calculated valence confined state energies as a function of well width for (001) (a) and (111) (b) growth GaAs wells between $Ga_{.7}Al_{.3}As$ barriers.

approximation, considering only the $a_{\pm}$ terms and ignoring band-mixing effects. From these diagonal elements, the states that are heavy along the growth direction ($m^* = (\gamma_1 - 2\gamma_3)^{-1}$) are comparatively light in the well plane ($m^* = (\gamma_1 + \gamma_3)^{-1}$). This light-hole-like behaviour persists in the well plane until mixing with lower lying bands becomes important.

Turning to the 50-Å well, for instance, we estimate that the (111) valence band density of states is approximately halved over the first 25 meV ($kT$ at room temperature) compared with the (001) case. This (111) light-hole cap covers a smaller energy range than in the optimum strained layer cases

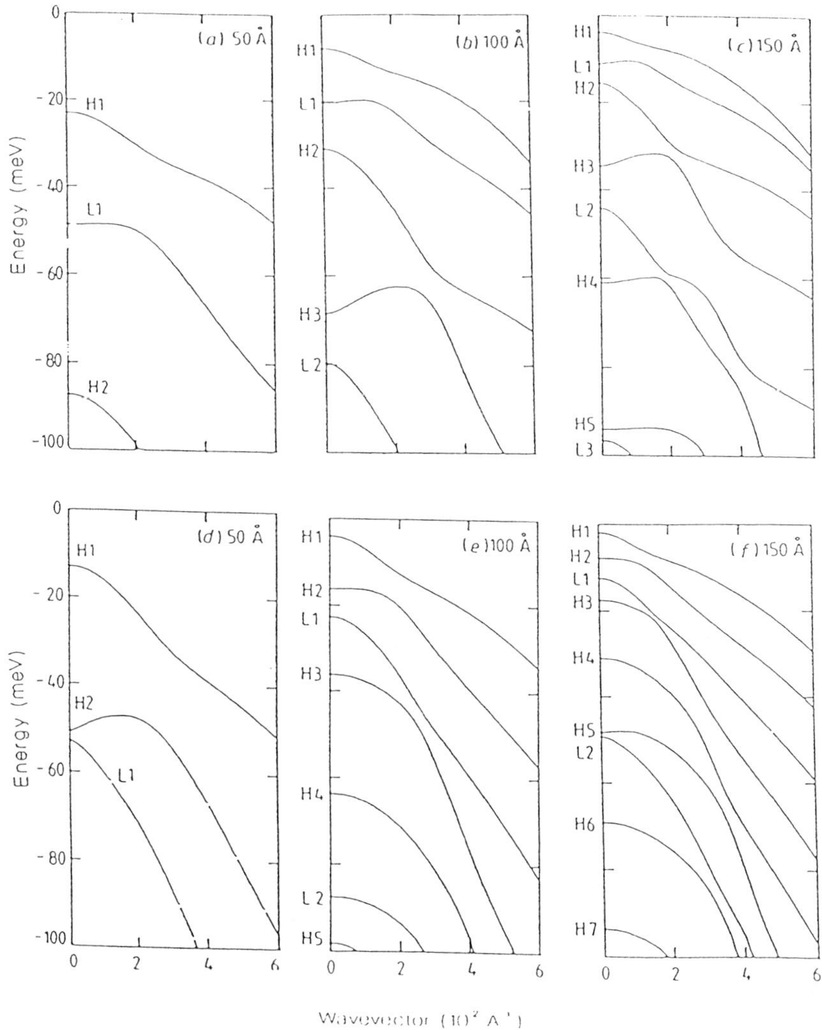


**Fig. 10.** Valence subband structure of (001) (a–c) and (111) (d–f) GaAs wells between $Ga_{.7}Al_{.3}As$ barriers. The wells are 50, 100, and 150 Å wide.

discussed in Section 4; nevertheless, (111) growth of thin quantum wells could be of benefit for laser applications [24], providing an alternative means of achieving a valence light-hole cap.

There are significant band-mixing effects in the (001) subbands, but these are generally absent from the (111) subbands. In particular, many of the lower subbands have an electron-like effective mass at the zone centre for (001) growth (e.g., L1, H3, and H4 in Figs. 1(a–c), whereas less upward dispersion is found in the (111) structures considered (Figs. 10d–f). The sharp reduction in heavy-hole confinement energies in (111) growth leads to an increased energy splitting between the relevant heavy and light zone centre states and consequent weakening of mixing effects [79, 81].

The (111) quantum well laser of [7] consisted of a 70-Å GaAs well between parabolically graded barriers, the barrier composition varying from $Ga_{.8}Al_{.2}As$ to $Ga_{.3}Al_{.7}As$ over 1500 Å. We modeled this laser structure by taking a barrier of constant composition, $Ga_{.8}Al_{.2}As$ [79]. The calculated band structures for 70-Å (001) and (111) quantum wells showed the same features as in Fig. 10: an enhanced light-hole cap at the valence band maximum in the (111) case, and the second subband being HH2 for (111) growth and LH1 for (001) growth.

### 5.3. Optical Matrix Elements

The valence subband structures presented in Fig. 10 suggest that the threshold current density may be reduced in thin (111) quantum well lasers compared with equivalent (001) lasers. A detailed comparison of the two types of structures, however, requires consideration also of the optical matrix elements between confined valence and conduction states, as discussed in Chapter 1 [1]. The optical matrix element describing a vertical transition between a conduction state and a valence state each with wavevector $k_t$ depends on the field direction (TE or TM modes), the subbands being considered, and on $k_t$. Transitions at the zone centre ($k_t = 0$) from the $n$th electron band E$n$ to the $n$th heavy-hole band H$n$ favour TE modes, while transitions to the $n$th light-hole band L$n$ favour TM modes. Zone centre transitions are forbidden between states of opposite parity (e.g. E1–H2). The situation becomes more complicated due to the mixing of different zone centre wavefunctions at finite $k_t$ [1].

Figure 11 shows the calculated average optical matrix elements as a function of $k_t$, normalised to the zone centre E1–H1 transition, for transitions between the first electron subband (E1) and the first and second hole subbands, for $L_w = 100$ Å and for both directions of light polarisation (TE and TM). The highest hole subband in both cases is H1, for which the optical matrix element is large for TE modes (Fig. 11a), and vanishing for TM modes (Fig. 11b) near the zone centre. At low carrier densities, H1 is the only populated subband, so that TE modes are favoured with respect to TM. The second hole subband is H2 in (111) growth. From parity considerations, this contributes little to transitions associated with the lowest conduction band E1 (Figs. 11a,b) but contributes most strongly to TE transitions associated

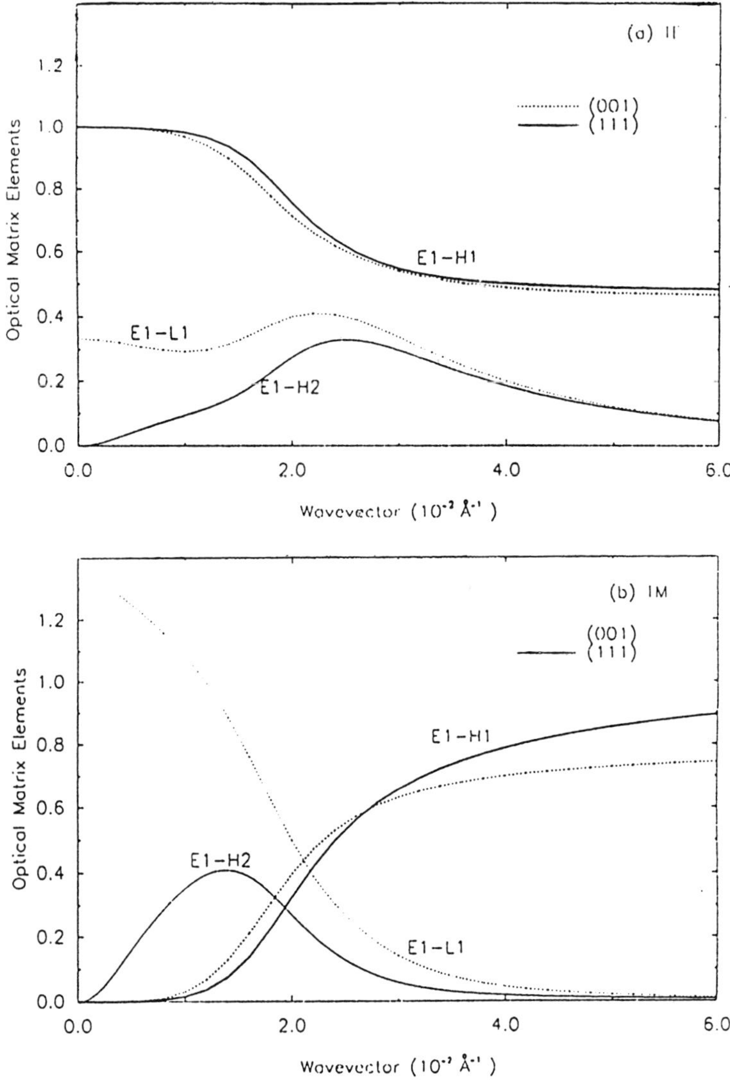


**Fig. 11.** Optical matrix elements for 100-Å GaAs wells between $G_{.8}Al_{.2}As$ barriers.

with the second conduction subband E2. By contrast, the second hole subband is L1 in (001) wells, which contributes more strongly to E1-related TM transitions (Fig. 11b). We describe later how the changed subband ordering can lead to different mode-hopping behaviour in (111) compared with (001) quantum well structures. Finally, with increasing well width, the third and lower hole subbands play an important role due to the reduced subband splitting and to strong mixing effects, and we find that by $L_w =$ 150 Å the TE and TM modes have similar gains as for bulk lasers, and the two growth directions are equivalent [79].

### 5.4. Laser Gain

Figure 12 shows the peak TE gain $g_m$ calculated as a function of carrier density at 300 K for the (111) quantum wells (solid lines) and the (001) quantum wells (dashed lines) [79]. The influence of the valence subband structure is clearly evident in the gain curves. For the 50-, 70- and 100-Å wells, transparency is achieved at a lower carrier density for the (111) well, because of the reduced density of states at the valence band maximum: fewer carriers are required for population inversion. At higher carrier densities, a crossover occurs in the 70- and 100-Å wells, whereby greater gain is achieved in the (001) well than the (111) well at a given carrier density $n$. This crossover occurs when the occupation of the second and further subbands becomes important, and it is due to the different subband ordering in the two cases: carriers in the second subband in (001) semiconductors (L1) contribute more strongly to the gain than do carriers in the second (111) subband (H2). The crossover does not occur for the carrier density range considered in the 50-Å well for two reasons: Firstly, because L1 and H2 are now approximately degenerate in the (111) case; and secondly, as the splitting is larger than in the other structures considered, the lower subbands do not play such an important role. Little difference is found between the maximum gain spectra of the two 150-Å wells, reflecting the fact that the subband splittings are now comparable with the assumed level broadening of 6.7 meV.

The experimentally determined differences in threshold current density between (001) and (111) growth are all consistent with the calculated peak gain curves of Fig. 12. Lasing occurs when the peak gain $g_m$ equals the total losses in the laser. The threshold current density $J_{th}$, assuming radiative recombination only, varies with the threshold carrier density $n_{th}$ approximately as $J_{th} \sim n_{th}^2$. From Fig. 12, a low loss (111) quantum well laser with thin wells should have a lower threshold carrier and therefore lower current density than an equivalent (001) quantum well laser. We calculate, assuming typical loss values, that $J_{th}$ would be reduced by about 22% in a (111) structure with 50-Å wells compared with a (001) structure with equal low losses. A 12% improvement should be found in 70-Å wells, while little difference is expected between similar structures for $L_w > 100$ Å. These reductions and variations compare very favourably with those found experimentally in [8] (see Fig. 13).

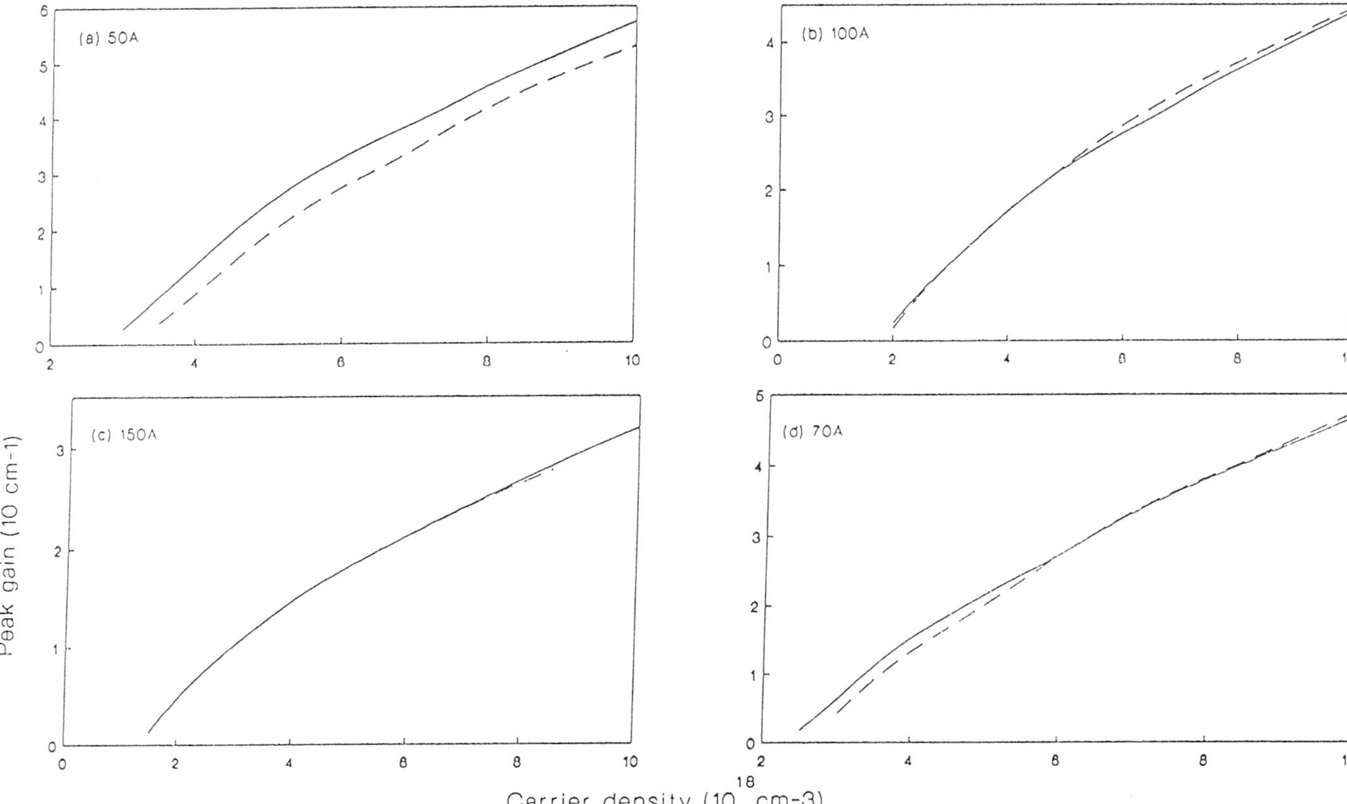


**Fig. 12.** The variation of peak gain with carrier density for the band structures calculated in Section 5. Full curves: (111) quantum wells; broken curves: (001) quantum wells.

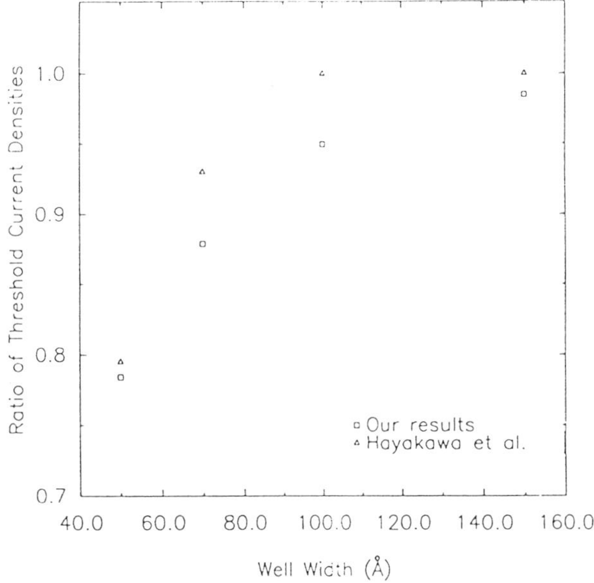


**Fig. 13.** Ratio of (111) to (001) threshold current densities as a function of well width.

Hayakawa *et al.* [9] also measured the gain spectrum as a function of current and found strong gain saturation of the $n = 1$ transition for a (111) laser with 100-Å wells. They demonstrated the peak gain switching to the $n = 2$ transition at higher current densities (see Fig. 14a). The calculated gain spectra in Fig. 14b confirm that this switching is favoured in such a (111) laser. The $n = 1$ transition (at about 0.85 $\mu$m) saturates with increasing carrier density, and the peak gain switches to the E2–H2 transition. In the (001) case, the H1–H2 energy separation is larger and the $n = 1$ TM transition (E1–L1) then lases before the $n = 2$ TE transition (E2–H2).

We thus conclude that the valence subband structure plays a key role in determining both the gain mechanisms and the threshold current density in high quality, low loss (001) and (111) GaAs–AlGaAs quantum well lasers and that the experimentally determined difference between (001) and (111) gain structures can be fully accounted for by the differences in the valence subband structure associated with the two growth directions.

## 6. CONCLUSIONS

We have emphasised in this chapter how the characteristics of quantum well lasers can be improved by engineering the valence band structure, either by

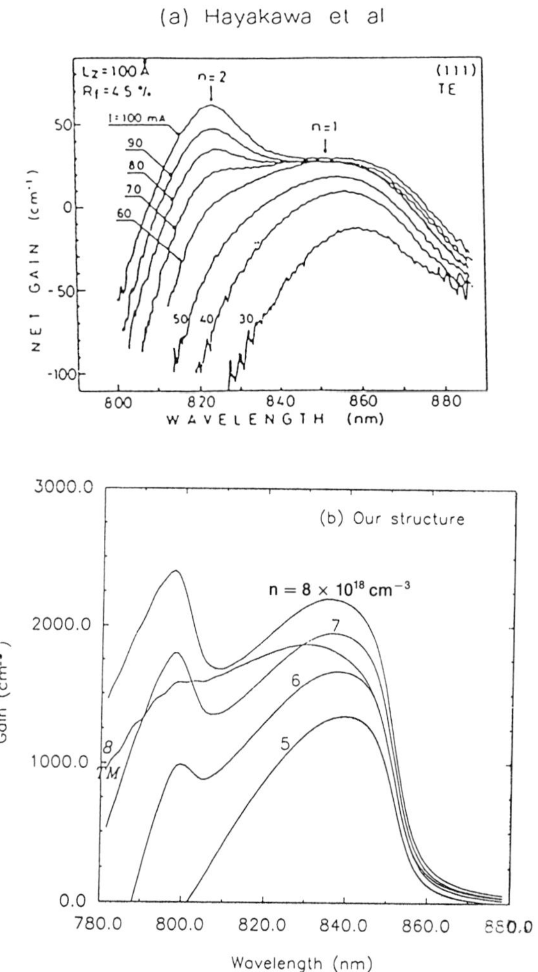


**Fig. 14.** The variation of gain with wavelength for 100-Å GaAs (111) wells (a) as measured for Hayakawa's structure and (b) as calculated for our structure with $Ga_{.8}Al_{.2}As$ barriers. All curves are for TE modes except one for TM modes.

the growth of strained layer structures or by unstrained growth of thin wells along the (111) direction. Both methods have been predicted to give a reduced density of states near the valence band maximum, leading to lower threshold current density and improved gain characteristics at all wavelengths, with the further benefit for long wavelength (1.55 $\mu$m) lasers of reduced intrinsic losses, implying an improved temperature stability and improved differential quantum efficiency. Experimental evidence becoming available not only supports the theoretical proposals but also indicates that such laser structures are stable even under quite extreme operating conditions. It is clear that valence band engineering, in particular via strained layer growth, will remain an exciting and fruitful area of research over the coming years.

## ACKNOWLEDGMENTS

We are particularly grateful to A. R. Adams for many valuable discussions on strained layer lasers; to W. Batty, U. Ekenberg, K. C. Heasman, and G. P. Witchlow for their contributions to the theoretical work and to our colleagues in the Strained Layer Structures Research group at Surrey and RSRE for valuable interactions. We are grateful to the Science and Engineering Research Council (EOR, AG) and to the Algerian Ministry of Higher Education (AG) for financial support.

## REFERENCES

1. Corzine, S. W., Yan, R.-H., and Coldren, L. A. Chapter 1 of this book.
2. Coleman, J. J. Chapter 8 of this book.
3. Lau, K. Y. Chapter 4 of this book.
4. Lau, K. Y. Chapter 5 of this book.
5. Molenkamp, L. W., Eppenga, R., 'T Hooft, G. W., Dawson, P., Foxon, C. T., and Moore, K. (1988). *Phys. Rev. B* **38**, 4314.
6. Shanabrook, B. V., Glembocki, O. J., Broido, D. A., and Wang, W. I. (1989). *Superlatt. Microstruct.* **5**, 503.
7. Hayakawa, T., Kondo, M., Suyama, T., Takahashi, K., Yamamoto, S., and Hijikata, T. (1987). *Japan. J. Appl. Phys.* **26**, L302.
8. Hayakawa, T., Suyama, T., Takahashi, K., Kondo, M., Yamamoto, S., and Hijikata, T. (1988). *Appl. Phys. Lett.* **52**, 339.
9. Hayakawa, T., Suyama, T., Tahakashi, K., Kondo, M., Yamamoto, S., and Hijikata, T. (1988). *J. Appl. Phys.* **64**, 297.
10. Beattie, A. R., and Landsberg, P. T. (1958). *Proc. R. Soc. A* **249**, 16.
11. Asada, M., Adams, A. R., Stubkjaer, K. E., Suematsu, Y., Itaha, Y., and Arai, S. (1981). *IEEE J. Quantum Electron.* **QE-17**, 611.
12. Haug, A. (1990). *Semicond. Sci. Technol.* **5**, 557.
13. Taylor, R. I., Abram, R. A., Burt, M. G., and Smith, C. (1985). *IEE Proc. Part J, Optoelectron.* **132**, 364.

14. HAUG, A. (1983). *Appl. Phys. Lett.* **42**, 512.
15. TAYLOR, R. I., ABRAM, R. A., BURT, M. G., and SMITH, C. (1990). *Semicond Sci. Technol.* **5**, 90.
16. HAUSSER, S., FUCHS, G., HANGLEITER, A., STREUBEL, K., and TSANG, W. T. (1990). *Appl. Phys. Lett.* **56**, 913.
17. CHIN, L. C., and YARIV, A. (1982). *IEEE J. Quantum Electron.* **QE-18**, 1406.
18. SUGIMURA, A. (1981). *IEEE J. Quantum Electron.* **QE-17**, 627.
19. DUTTA, N. K. (1983). *J. Appl. Phys.* **54**, 1236.
20. ADAMS, A. R., HEASMAN, K. C., and HILTON, J. (1987). *Semicond. Sci. Technol.* **2**, 761.
21. MOZER, A., ROMANELI, K. M., SCHMID, W., and PILKUHN, M. J. (1982). *Appl. Phys. Lett.* **41**, 964.
22. HENRY, C. H., LOGAN, A. R., MERRIT, F. R., and LUARGO, J. P. (1983). *IEEE J. Quantum Electron.* **QE-19**, 947.
23. ADAMS, A. R., ALLAM, J., CZAJKOWSKI, I. K., GHITI, A., O'REILLY, E. P., and RING, W. S. (1991). *Condensed Systems of Low Dimensionality*, ed. J. L. BEEBY, Plenum Press, New York, p. 623.
24. ADAMS, A. R. (1986). *Electron. Lett.* **22**, 249.
25. YABLONOVITCH, E., and KANE, E. O. (1986). *J. Lightwave Technol.* **LT-4**, 504.
26. SUEMUNE, I., COLDREN, L. A., YAMANISHI, M., and KAU, Y. (1988). *Appl. Phys. Lett.* **53**, 1378.
27. GHITI, A., O'REILLY, E. P., and ADAMS, A. R. (1989). *Electron Lett.* **25**, 821.
28. LAU, K. Y., XIN, S., WANG, W. I., BAR-CHAIM, N., and MITTELSTEIN, M. (1989). *Appl. Phys. Lett.* **55**, 1173.
29. PIKUS, G. E., and BIR, G. L. (1974). *Symmetry and strain induced effects in semiconductors.* Wiley.
30. PIKUS, G. E., and BIR, G. L. (1959). *Sov. Phys. Solid State* **1**, 1502.
31. O'REILLY, E. P. (1986). *Semicond. Sci. Technol.* **1**, 128.
32. EPPENGA, R., SCHUURMANS, M. F. H., and COLAK, S. (1987). *Phys. Rev. B* **36**, 1554.
33. LUTTINGER, J. M., and KOHN, W. (1955). *Phys. Rev.* **97**, 869.
34. CARDONA, M., CHRISTENSEN, N. E., and FASOL, G. (1986). *Phys. Rev. Lett.* **56**, 2831.
35. KANE, E. O. (1957). *J. Phys. Chem. Solids* **1**, 249.
36. BROIDO, D. A., and SHAM, L. J. (1985). *Phys. Rev. B* **31**, 888.
37. AHN, D., and CHUANG, S. (1988). *IEEE J. Quantum Electron.* **QE-24**, 2400.
38. BALDERESCHI, A., and LIPARI, N. O. (1973). *Phys. Rev. B* **8**, 2697.
39. ALTARELLI, M., EKENBERG, U., and FASALINO, A. (1985). *Phys. Rev. B* **32**, 5138.
40. COLAK, S., EPPENGA, R., and SCHUURMANS, M. F. H. (1987). *IEEE J. Quantum Electron.* **QE-23**, 960.
41. TREBIN, H. R., ROSSLER, U., and RANVAUD, R. (1979). *Phys. Rev. B* **20**, 686.
42. LAEWETZ, P. (1971). *Phys. Rev. B* **4**, 3460.
43. SMITH, D. L. (1986). *Solid State Comm.* **57**, 919.
44. MAILHIOT, C., and SMITH, D. L. (1987). *Phys. Rev. B* **35**, 1242.
45. LAURICH, B. K., ELCESS, K., FONSTAD, C. G., BEERY, J. G., MAILHIOT, C., and SMITH, D. L. (1989). *Phys. Rev. Lett.* **62**, 649.

46. ADAMS, A. R. (1990). Private communication.
47. HEASMAN, K. C., O'REILLY, E. P., WITCHLOW, G. P., BATTY, W., and ADAMS, A. R. (1987). *Proc. SPIE* **800**, 50.
48. O'REILLY, E. P., HEASMAN, K. C., ADAMS, A. R., and WITCHLOW, G. P. (1987). *Superlatt. Microstruct.* **3**, 99.
49. DUNSTAN, D. J., and ADAMS, A. R. (1990). *Semicond. Sci. Technol.* **5**, 1202.
50. GHITI, A., BATTY, W. B., EKENBERG, U., and O'REILLY, E. P. (1987). *Proc. SPIE* **861**, 96.
51. YABLONOVITCH, E., and KANE, E. O. (1988). *IEEE J. Lightwave Technol.* **LT-6**, 1292.
52. ROBBINS, D. J., WOOD, A. C. G., and ABRAM, R. A. (1990). *Proceedings 17th international symposium on GaAs and related compounds, IoP Conference Series* **112**, 533.
53. ABRAM, R. A., WOOD, A. C. G., and ROBBINS, D. J. (1991). *Proc. SPIE* **1361**, 69 Bellingham WA.
54. THIJS, P. J. A., and VAN DONGEN, T. (1989). *Electron. Lett.* **25**, 1735.
55. THIJS, P. J. A., MONTIE, E. A., VAN DONGEN, T., and BULLE-LIEUWMA, C. W. T. (1990). *J. Crystal Growth* **105**, 339.
56. TSANG, W. T., WU, M. C., YANG, L., CHEN, Y. K., and SERGENT, A. M. (1990). *Electron. Lett.* **26**, 2053.
57. TEMKIN, H., TANBUN-EK, T., and LOGAN, R. A. (1990). *Appl. Phys. Lett.* **56**, 1210.
58. YAMADA, H., KITAMURA, M., and MITO, I. (1990). *Device research conference, Santa Barbara.*
59. ZAH, C. E., BHAT, R., CHEUNG, K. W., ANDREADAKIS, N. C., FAVIRE, F. J., MENOCAL, S. G., YABLONOVITCH, E., HWANG, D. M., KOZA, M., GWITTER, T. J., and LEE, T. P. (1990). *Appl. Phys. Lett.* **57**, 1608.
60. KOREN, U., ORON, M., YOUNG, M. G., MILLER, B. I., DEMIGUEL, J. L., RAYBON, G., and CHIEN, M. (1990). *Electron. Lett.* **26**, 466.
61. FRITZ, I. J., PICRAUX, S. T., DAWSON, L. R., DRUMMOND, T. J., LAIDIG, W. D., and ANDERSON, N. G. (1985). *Appl. Phys. Lett.* **46**, 967.
62. ANDERSSON, T. J., CHEN, Z. G., KULAKOVSKII, V. D., UDDIN, A., and VALLIN, J. T. (1987). *Appl. Phys. Lett.* **51**, 752.
63. EKENBERG, U., BATTY, W., and O'REILLY, E. P. (1987). In *Proceedings of the 3rd international conference on modulated semiconductor structures* (RAYMOND, A., and VOISON, P., eds.), *J. Physique* **48** C5, 553.
64. ASADA, M., KAMEYAMA, A., and SUEMATSU, Y. (1984). *IEEE J. Quantum Electron.* **QE-20**, 745.
65. ELISEEV, P. E., and DRAKIN, A. E. (1984). *Sov. J. Quantum Electron* **14**, 119.
66. MONSERRAT, K. J., and TOTHILL, J. N. (1989). *J. Electron. Mater.* **18**, 475.
67. THIJS, P. J. A., TIEMEYER, L. F., VAN DONGEN, T., KUINDERSMA, P. I., BINSMA, J. J. M., VAN DE HOFSTAD, G. L., and NIJMAN, W. (1988). *Proceedings of the European conference on optical communication, Brighton, U.K.*, vol. **1**, 372.
68. KOREN, U., MILLER, B. I., SU, Y. K., KOCH, T. L., and BOWERS, J. E. (1987). *Appl. Phys. Lett.* **51**, 1744.

69. KITAMURA, M., TAKANO, S., SASAKI, T., YAMADA, H., and MITO, I. (1988). *Appl. Phys. Lett.* **53**, 1.
70. COOPER, D. M., SELTZER, C. P., AYLETT, M., ELTON, D. J., HARLOW, M., WICKES, H., and SPILLET, R. E. (1989). *Proceedings of the European conference on optical communication, Gotenburg, Sweden*, vol. **3**, 82.
71. MITO, I., YAMAZAKI, H., YAMADA, H., SASAKI, T., TAKANO, S., AOKI, Y., and KITAMURA, M. (1989). *Proceedings of the conference on integrated optics and optical fiber communication*, Kobe, Japan, paper **20PDB-12**.
72. AGRAWAL, G. P., and DUTTA, N. K. (1986). *Long wavelength semiconductor lasers*. Van Nostrand Reinhold Company, Inc.
73. MAGARI, K., OKAMOTO, M., YASAKA, H., SATO, K., NOGUCHI, Y., and MIKAMI, O. (1990). *Photon. Technol. Lett.* **2**, 556.
74. VINA, L. I., and WANG, W. I. (1986). *Appl. Phys. Lett.* **48**, 36.
75. FUKUNAGA, T., and NAKASHIMA, H. (1986). *Japan. J. Appl. Phys.* **25**, L856.
76. HAYAKAWA, T., TAKAHASHI, K., KONDO, M., SUYAMA, T., YAMAMOTO, S. F., and HIJIKATA, T. (1988). *Phys. Rev. Lett.* **60**, 349.
77. HESS, E., TOPOL, I., SCHULZE, K.-R., NEUMANN, H., and UNGER, K. (1973). *Phys. Status. Solidi b* **55**, 187.
78. LAMBERT, B., CAULET, J., REGRENY, A., BAUDET, M., DEVEAUD, B., and CHOMETTE, A. (1987). *Semicond. Sci. Technol.* **2**, 491.
79. BATTY, W., EKENBERG, U., GHITI, A., and O'REILLY, E. P. (1989). *Semicond. Sci. Technol.* **4**, 904.
80. BATTY, W. (1991). Ph.D. Thesis, University of Surrey.
81. GHITI, A., BATTY, W. and O'REILLY, E. P. (1990). *Superlatt. Microstruct.* **7**, 353.
82. THIJS, P. J. A. (1991). In *Technical digest, OFC91*, paper **WB2**, p. 67.
83. ZAH, C. E., BHAT, R., PATHAK, B., CANEAU, C., FAVIRE, F. J., ANDREAKIS, N. C., HWANG, D. M., KOZA, M. A., CHEN, C. Y. and LEE, T. P. (1991). *Electron. Lett.* **27**, 1414.
84. O'REILLY, E. P., JONES, G., GHITI, A., and ADAMS, A. R. (1991). *Electron. Lett.* **27**, 1417.

# Chapter 8

# STRAINED LAYER QUANTUM WELL HETEROSTRUCTURE LASERS

**James J. Coleman**

*Microelectronics Laboratory and Materials Research Laboratory*
*University of Illinois*
*Urbana, Illinois*



## 1. INTRODUCTION

The development of a semiconductor heterostructure laser materials system is driven by one of two expedients. For example, the oldest, best developed, and still most sophisticated heterostructure laser system is the AlGaAs–GaAs system. The expedient for development of this system is a unique combination of practical physical, electrical, optical, and chemical features that has





ISBN 0-12-781890-1

withstood the test of time. These features include a tractable chemical system that lends itself equally well to basic epitaxial growth methods, such as liquid phase epitaxy (LPE), and to more sophisticated thin layer epitaxial growth methods, such as metalorganic chemical vapor deposition (MOCVD) and molecular beam epitaxy (MBE). Controlled thin layer epitaxy has, in turn, allowed the development of AlGaAs–GaAs quantum well heterostructure lasers. Reasonably high quality GaAs substrates are available. Looking toward the future, sophisticated heterostructure electronic devices and complex MESFET integrated circuitry on GaAs substrates that presently exist can form, with heterostructure lasers, the basis for optoelectronic integrated circuitry. This is enhanced by the suitability of silicon detectors for the wavelength range of lasers made from AlGaAs–GaAs. Probably the single most important attractive feature of the AlGaAs–GaAs heterostructure system is the fortunate coincidence that AlAs and GaAs have, for all practical purposes, the same lattice constant ($\Delta a/a < 0.12\%$). This allows design of any structure, with any combination of layer compositions, without regard to lattice mismatch or the associated dislocation formation.

After AlGaAs–GaAs, the best studied semiconductor heterostructure laser materials system is InGaAsP–InP. And in many ways this is a much more limited system from a practical point of view. It presents a much less tractable chemical system that, thus far, lends itself well only to LPE growth. The more sophisticated growth methods of MOCVD and MBE do not easily provide for handling of difficult elements like phosphorus. Correspondingly, InGaAsP–InP quantum well heterostructure lasers have been reported but are far less common. High quality InP substrates are less readily available and more expensive. Heterostructure electronic devices and integrated circuitry on InP substrates are far less advanced. The wavelength range of lasers made from InGaAsP–InP is beyond the long wavelength cutoff of silicon detectors, necessitating the use of less well developed photodetectors. Obtaining a lattice match between InGaAsP and InP is straightforward in principle, but requires more stringent control over the individual constituents of the alloy during growth. A very strong expedient for developing this system, however, arises from its application as a source in the 1.3–1.5 $\mu$m wavelength range for low loss, minimum dispersion optical fibers.

The range of wavelengths available from lattice-matched AlGaAs–GaAs conventional double heterostructure lasers and quantum well heterostructure lasers is from $\lambda \approx 0.88$–$0.65\ \mu$m. The long wavelength limit is defined by the band edge for GaAs; and AlGaAs active layers or the quantum size effect, or both, are used to shift the emission to shorter wavelengths. The short wavelength limit is not well defined, but arises from the intrusion of the large mass indirect conduction band minima at Γ and $L$ on direct recombination processes. The range of wavelengths available from lattice-matched InGaAsP–InP conventional double heterostructure lasers and quantum well heterostructure lasers is from $\lambda \approx 1.1$–$1.6\ \mu$m. The long wavelength limit is defined by the band edge for the ternary endpoint InGaAs. The approximate short wavelength limit results from an insufficient heterostructure conduction band discontinuity between InGaAsP and InP, which results in poor carrier collection and higher laser threshold current densities for wavelengths shorter than 1.1 $\mu$m.

These wavelength ranges are sufficient to cover many important applications such as the use of modulated InGaAsP–InP lasers at $\lambda \approx 1.55\ \mu$m as sources for low loss, minimum dispersion optical fiber communications links, and the use of high power AlGaAs–GaAs laser arrays at $\lambda \approx 0.82\ \mu$m for diode-pumped Nd:YAG solid state lasers. This is shown schematically in Fig 1, which is simply the conversion of wavelength versus energy given by

$$\lambda = \frac{hc}{E}, \tag{1}$$

where $h$ is Planck's constant, and $c$ is the speed of light. In Fig. 1, the wavelength ranges supported by the lattice-matched heterostructure laser systems just described are shown as solid lines, with an obvious gap in the

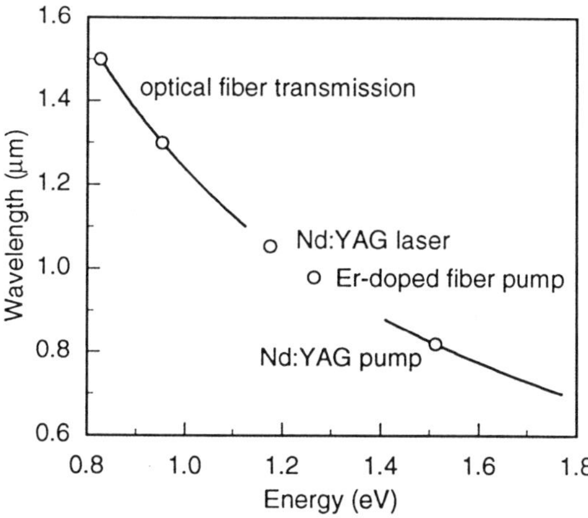


**Fig. 1.** The wavelength ranges supported by lattice-matched heterostructure laser systems (solid lines), with an obvious gap in the wavelength range of $\lambda \approx 0.88$–$1.1\ \mu$m. The data points correspond to the wavelengths of various important applications of heterostructure lasers.

wavelength range of $\lambda \approx 0.88$–$1.1\ \mu$m. This range is unavailable in any III-V lattice-matched heterostructure laser materials system. There are, however, a number of important applications that require laser emission in this range, including frequency doubling of wavelengths near the 1.06-$\mu$m emission wavelength of Nd:YAG solid state lasers and pumping the upper states of rare earth–doped silica fiber amplifiers.

Shown schematically in Fig. 2 are the $^4I_{n/2}$ energy levels ($n = 11, 13, 15$) of the rare earth erbium. Light at a pump wavelength of $\lambda \sim 0.98\ \mu$m provides excitation from the ground state to the $^4I_{11/2}$ states followed by rapid internal relaxation to the relatively long-lived $^4I_{13/2}$ states. Incoming $\lambda \sim 1.55\ \mu$m laser light can be amplified by stimulation of the transition to empty states in the upper $^4I_{15/2}$ level. This process can in principle be employed in a simple system containing a laser diode pump at $\lambda \sim 0.98\ \mu$m, an optical coupler, and a length of $Er^{3+}$-doped optical fiber to form a traveling wave

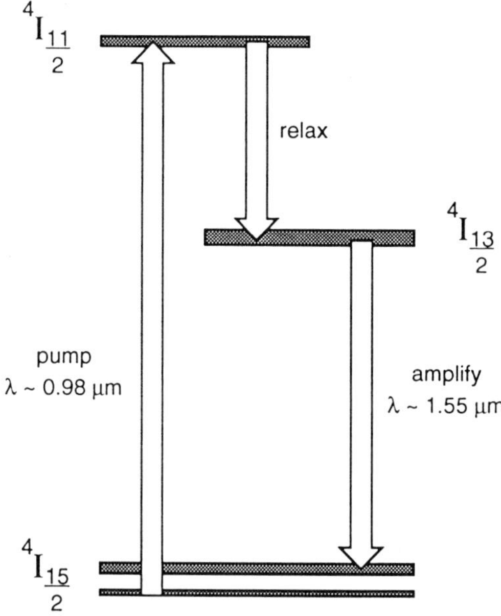


**Fig. 2.** The $^4I_{n/2}$ energy levels ($n = 11, 13, 15$) of the rare earth erbium. Absorption at $\lambda \sim 0.98\ \mu$m provides excitation from the ground state to the $^4I_{11/2}$ states followed by rapid internal relaxation to the relatively long-lived $^4I_{13/2}$ states. 1.55-$\mu$m laser light can be amplified by stimulation of the transition to empty states in the upper $^4I_{15/2}$ level.

optical amplifier ([1–3]; see also the reviews by Urquhart [4] and Shimada [5]).

A simple examination of the available direct gap III-V compound semiconductor alloys (see [6]) suggests that perhaps the best material system for obtaining emission wavelengths in the range $\lambda \approx 0.88–1.1\ \mu m$ is $In_xGa_{1-x}As$. There is, however, no suitable binary substrate material that allows lattice-matched compositions of $In_xGa_{1-x}As$ in the wavelength range of interest here. Thus, only a heterostructure materials system in which a very large mismatch must be accommodated ($\Delta a/a$ as great as 3%) may be considered. One's intuition would perhaps argue, however, that strained layer materials must be unsuitable, at least in terms of device reliability.

The metallurgical implications of accommodating strain elastically with layers thinner than some misfit-dependent critical thickness were described in the seminal work of Matthews and Blakeslee [7]. In the early 1980s, workers at the Sandia National Laboratories (see the review by Osbourn *et al.* [8]) first considered some of the effects of biaxial strain on the energy band structure, optical properties, and electrical transport in these structures. Studies of photopumped InGaAs–GaAs strained layer superlattices [9, 10] demonstrated cw 300 K laser operation, but indicated that the superlattices were unstable under high excitation, with failure occurring in under an hour. An important key in the continued development of strained layer InGaAs–GaAs heterostructure lasers was the work by Laidig and coworkers on injection laser diodes [11, 12]. This work showed the suitability of InGaAs–GaAs strained layer heterostructures for diode lasers at $\lambda \sim 1\ \mu m$ and were the first reports of reliable laser operation from strained layer lasers.

Since 1986, the number of research laboratories studying various attributes of strained layer InGaAs quantum well heterostructure lasers has risen dramatically. In this review, we will consider the metallurgical aspects of critical thickness in strained layer heterostructure systems, the effects of strain and quantum size effect on emission wavelength, and briefly outline considerations of optical gain and threshold current density in strained layer lasers. We will briefly describe the MOCVD growth of InGaAs–GaAs strained layer heterostructure materials and consider the characteristics of broad area lasers under conditions of varying quantum well composition, thickness with respect to the critical thickness, growth conditions, stripe width, and other structural parameters. We will describe more complicated InGaAs–GaAs strained layer quantum well heterostructure lasers such as index guided laser structures and laser arrays. Finally, we will discuss the reliability of these laser structures.

## 2. CRITICAL THICKNESS IN STRAINED LAYER LASERS

In lattice-matched heterostructure systems, the number and thickness of quantum wells in the structure is not a design constraint. In a strained layer, lattice-mismatched system, however, the elastic accommodation of the strain energy associated with the mismatch, without the formation of misfit dislocations, must be considered [7, 13]. Shown in Fig. 3a is a representation, at an exaggerated scale, of the unit cells of $In_xGa_{1-x}As$ and GaAs. The unit cell of $In_xGa_{1-x}As$ can be as much as 3.6% larger ($x = 0.50$) than GaAs, in contrast to $Al_xGa_{1-x}As$, which has a unit cell that is never more than about 0.13% larger than the GaAs unit cell. In the case of a single layer, as shown schematically assuming a relatively thick GaAs host in Fig. 3b, the $In_xGa_{1-x}As$ cell is shortened in both directions parallel to the interface (biaxial compression) and elongated in the direction normal to the interface (uniaxial tension). The strain energy that results is approximately equal to the misfit and produces a force $F_\varepsilon$ at the interface. If this force exceeds the tension $F_1$ in a dislocation line, migration of a threading dislocation results in formation of a single misfit dislocation [7].

If the $In_xGa_{1-x}As$ cell is inserted between layers of the host as in a quantum well, assuming that the host is relatively thick on both top and bottom, both interfaces are under biaxial compression, as shown in Fig. 3c. The strain energy that results is again approximately equal to the misfit and, if the force $F_\varepsilon$ exceeds twice the tension $F_1$ in a dislocation line, migration of a threading dislocation results in formation of two misfit dislocations [7]. In the case of a superlattice, where the concept of a host does not apply and

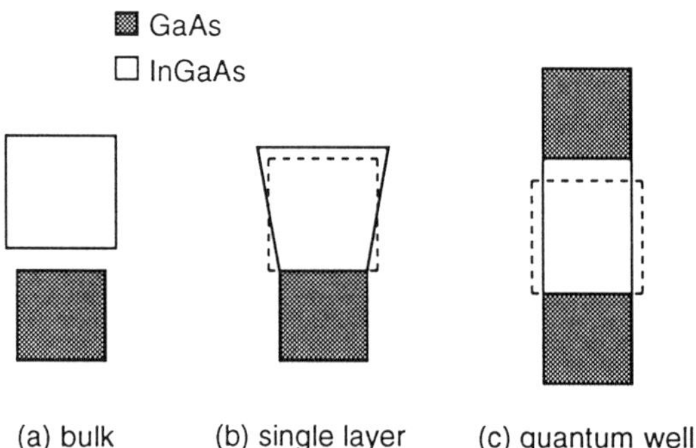


**Fig. 3.** Schematic diagram (a) of separate $In_xGa_{1-x}As$ and GaAs unit cells, (b) a single layer, strained layer $In_xGa_{1-x}As$–GaAs structure, and (c) an inserted $In_xGa_{1-x}As$–GaAs strained layer structure. Shown for reference as a dashed line is the original $In_xGa_{1-x}As$ cell shape.

alternating layers are under either biaxial compression or biaxial tension, the misfit strain is distributed among all of the layers, and the strain energy is equal to approximately half of the misfit.

For each of these cases, a critical layer thickness $h_c$, below which the misfit strain is accommodated without formation of misfit dislocations, can be defined in terms of the elastic constants of the materials. Matthews and Blakeslee [7] calculate the critical thickness for layers in a superlattice having as equal elastic constants

$$h_c = \frac{a}{\kappa\sqrt{2}\pi f}\frac{1-0.25v}{1+v}\left(\ln\frac{h_c\sqrt{2}}{a}+1\right), \tag{2}$$

where $h_c$ is the critical thickness, and $a$ is the lattice constant of the strained layer. The misfit $f$ is defined simply as

$$f = \frac{\Delta a}{a}, \tag{3}$$

and $v$ is Poisson's ratio, which is defined as

$$v = \frac{C_{12}}{C_{11}+C_{12}}. \tag{4}$$

The coefficient $\kappa$ has a value of 1 for a strained layer superlattice, 2 for a single quantum well, and 4 for a single strained layer. The critical thickness of Eq. (2) for $In_xGa_{1-x}As$–GaAs as a function of indium composition in the range $x < 0.50$ can be solved simply by numerical methods and is shown in Fig. 4 for (a) superlattice, (b) quantum well, and (c) single strained layer structures.

The definition of a critical thickness, particularly in terms of its effect on the characteristics of strained layer quantum well heterostructure lasers, is somewhat less clear than indicated by the description thus far. For example, People and Bean [14] have proposed an alternative mathematical description for the critical thickness of a single layer by defining a balance between the strain energy in a layer and the energy required for formation of an isolated dislocation. They assume that there are no threading dislocations present in the structure and argue that screw dislocations have the lowest energy density and thus will form first. The expression thus obtained for the critical thickness for a single strained layer is given by

$$h_c = \frac{a}{32\sqrt{2}\pi f^2}\frac{1-v}{1+v}\left(\ln\frac{h_c\sqrt{2}}{a}\right), \tag{5}$$

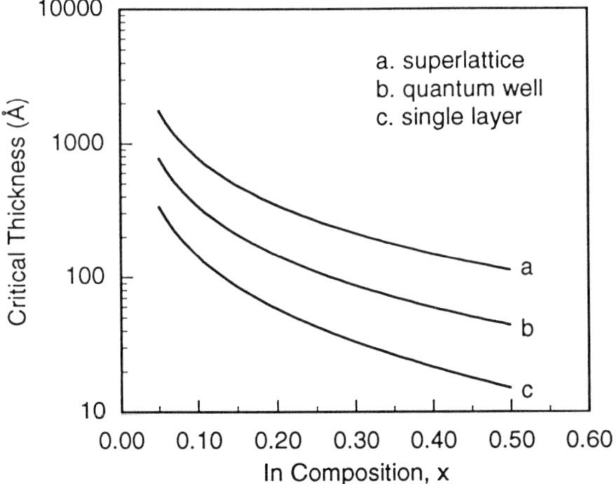


**Fig. 4.** The critical thickness according to the model of Matthews and Blakeslee [7] as a function of indium composition in the range $x < 0.50$ for (a) superlattice, (b) quantum well, and (c) single strained layer structures.

where the terms are defined as in Eq. (2). Figure 5 is a comparison of the computed critical thicknesses of Eqs. (2) and (5) for $In_xGa_{1-x}As$–GaAs single layers as a function of indium composition in the range $x < 0.50$. The models markedly differ, especially at lower indium compositions.

Others have reported variations of these models for critical thickness.

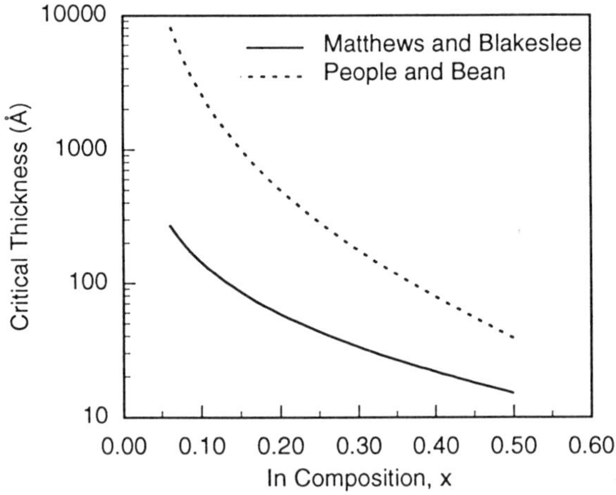


**Fig. 5.** Comparison of the computed critical thicknesses for $In_xGa_{1-x}As$–GaAs single layers as a function of indium composition in the range $x < 0.50$ according to the models of Matthews and Blakeslee [7] and People and Bean [14].

Dodson and Tsao [15] applied a model for dislocation dynamics and plastic flow to account for the experimentally observed behavior in SiGe–Si strained layers. Vawter and Myers [16] have described an equivalent strained layer model for design purposes, which utilizes a reduced effective strain and a reduced effective thickness, and applied this model to single and multiple quantum wells and strained layer superlattices.

Most workers report reasonable agreement between experimentally determined critical thickness data for $In_xGa_{1-x}As$–GaAs quantum well structures with the Matthews and Blakeslee model of Eq. (2) using $\kappa = 2$ [17–21], although departures, sometimes large, have been reported [22, 23]. A number of explanations for smaller deviations from the Matthews and Blakeslee model of Eq. (2) have been suggested, including residual strain resulting from incomplete relaxation [24] and differing mechanisms for strain relaxation for different substrate orientations [25].

An interesting dilemma that arises as a consequence of Eq. (2) is that, since the critical thickness for a single strained layer is half that of a quantum well of the same composition, the limiting thickness for both structures should be that of the single layer. This, of course, must be the case, because conventional epitaxial growth requires that the strained layer be grown prior to the growth of an overlayer. Price [26] has suggested that the reported agreement of quantum well critical thickness with the Matthews and

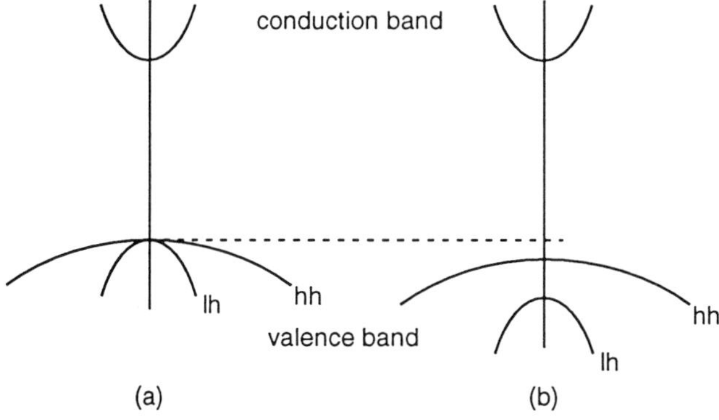


**Fig. 6.** Schematic *E–k* diagram for (a) an unstrained direct semiconductor such as GaAs, showing the approximately parabolic conduction band and degenerate light-hole and heavy-hole valence bands. The same semiconductor under biaxial compression (b) shows an increase in the energies of both light- and heavy-hole valence bands, relative to the conduction band edge. There is also a net separation of the bands, removing the degeneracy, because of the different effective masses in the valence bands.

Blakeslee model of Eq. (2) for $\kappa = 2$ implies the existence of a mechanism that allows re-establishment of misfit by the growth of an overlayer. They subsequently presented experimental evidence [27] that, for $In_xGa_{1-x}As$–GaAs quantum wells thicker than the single layer critical thickness ($\kappa = 4$) but thinner than the quantum well critical thickness ($\kappa = 2$), dislocations can indeed move in and out of the structure reversibly during growth.

## 3. STRAIN AND QUANTUM SIZE EFFECTS ON EMISSION WAVELENGTH

The conduction and valence energy bands of strained layer lattice-mismatched systems are altered by the presence of biaxial strain. This includes changes in the effective masses of the various energy bands and shifts in energy of the band edges relative to each other. In an unstrained compound semiconductor quantum well heterostructure, the band edges of interest are the conduction band edge and the degenerate light-hole (spin $\pm\frac{1}{2}$) and heavy-hole (spin $\pm\frac{3}{2}$) valence band edges. Although there is also a split-off band in these materials resulting from spin–orbit splitting, the energy difference between this band edge and the energy of the light- and heavy-hole band edge is sufficiently large that it doesn't contribute significantly to recombination in quantum well heterostructures. For the purposes of this discussion, the conduction and valence bands can be treated as parabolic near zone center, as shown in Fig. 6a, with the dispersion relation

$$E(k) = \frac{\hbar^2 k^2}{2m^*}, \tag{6}$$

where $m^*$ is the effective mass.

In the strained InGaAs–GaAs heterostructure system, biaxial compression defeats the cubic symmetry of the semiconductor [28]. This results in removal of the degeneracy in the valence band edge because, although both heavy-hole and light-hole band edge energies increase in energy with respect to the conduction band edge with increasing strain, the increase in the light-hole band edge energy is greater, as shown schematically in Fig. 6b [29–33]. An orbital strain Hamiltonian for a given band at $k = 0$ can be written [34, 35] in terms of the components of the strain tensor $\varepsilon_{ij}$, the angular momentum operator **L**, the hydrostatic deformation potential $\Xi$, and the tetragonal and rhombohedral shear deformation potentials ($b$ and $d$, respectively). Assuming biaxial stress in the growth plane, the Hamiltonian can be simplified and eigenvalues calculated [36]. This results in energy differences between the conduction band and valence bands at $k = 0$, given to first order by

$$\Delta E_{\text{hh}} = -2\Xi\varepsilon\left(\frac{C_{11} - C_{12}}{C_{11}}\right) + b\varepsilon\left(\frac{C_{11} + C_{12}}{C_{11}}\right), \tag{7}$$

$$\Delta E_{\text{lh}} = -2\Xi\varepsilon\left(\frac{C_{11} - C_{12}}{C_{11}}\right) - b\varepsilon\left(\frac{C_{11} + 2C_{12}}{C_{11}}\right), \tag{8}$$

where $\Delta E_{\text{hh}}$ is the shift in the heavy-hole valence band edge with respect to the conduction band edge, $\Delta E_{\text{lh}}$ is the shift in the light-hole valence band edge with respect to the conduction band edge, $C_{ij}$ are elastic stiffness coefficients, and the strain $\varepsilon$ is given by

$$\varepsilon = \frac{\Delta a}{a_0}. \tag{9}$$

The dependence of strain (9) on indium composition in the range $0 < x < 0.50$ for strained $In_xGa_{1-x}As$–GaAs is shown in Fig. 7.

The parameters necessary to calculate Eqs. (7) and (8) for $In_xGa_{1-x}As$ strained layers on GaAs can be interpolated from values [37–42] for the endpoint binary semiconductors, InAs and GaAs. The values used here are given in Table I. The results of calculating the increase in energy for the light-hole and heavy-hole valence band edges, Eqs. (7) and (8), are shown in

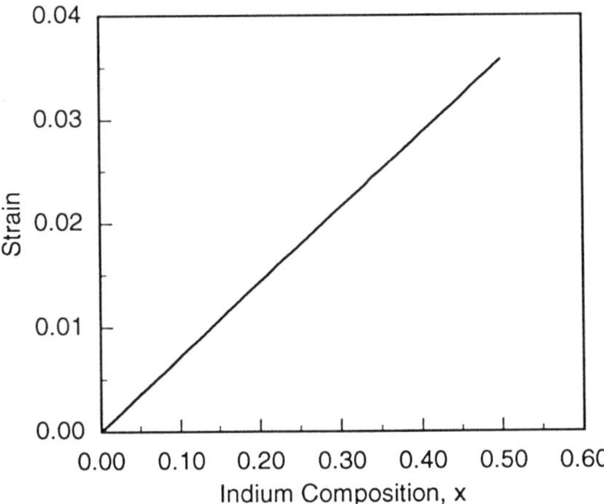


**Fig. 7.** The dependence of strain on In composition in the range $0 < x < 0.50$ for the $In_xGa_{1-x}As$–GaAs strained layer system. For reference, the strain present in the $Al_xGa_{1-x}As$–GaAs heterostructure system, considered to be lattice-matched for all compositions, is less than $1.5 \times 10^{-3}$.

**Table I.**

List of Material parameters for InAs, GaAs, and $In_xGa_{1-x}As$

| | | InAs | GaAs |
|---|---|---|---|
| Lattice constant (Å) | $a_0$ | 6.0585 | 5.6535 |
| Elastic coefficient ($10^{12}$ dyne/cm$^2$ | $C_{11}$ | 0.865 | 0.538 |
| | $C_{12}$ | 0.485 | 1.188 |
| Hydrostatic deformation potential (eV) | $\Xi$ | −6.0 | −8.2 |
| Shear deformation potential (eV) | $b$ | −1.8 | −2.0 |
| Electron effective mass | $m_e$ | 0.023 | 0.069 |
| Heavy-hole effective mass | $m_h$ | 0.41 | 0.47 |
| Light-hole effective mass | $m_l$ | 0.027 | 0.074 |
| | | $In_xGa_{1-x}As$ | |
| Energy gap (eV) | $E_g$ | $1.424 - 1.62x - 0.56x^2$ | |
| Heterostructure discontinuity | $\Delta E_c/\Delta E_g$ | 0.65 | |

[References: [37–42].

Fig. 8. Since the shift in the light-hole valence band edge with composition is much greater than the shift in the heavy-hole valence band edge, recombination in bulk strained layer $In_xGa_{1-x}As$ is dominated by transitions from the conduction band to the heavy-hole valence band. Thus, there is a strain-induced increase in the bulk bandgap energy with indium composition, given by $\Delta E_{hh}$ in Eq. (7) and Fig. 8, that partly offsets the decrease in bandgap energy with composition in unstrained $In_xGa_{1-x}As$. This net shift is indicated in Fig. 9, which shows the unstrained bulk band edge for $In_xGa_{1-x}As$ as a function of composition from Table I and the strained corrected bulk band edge given by

$$E_g(In_xGa_{1-x}As) + \Delta E_{hh}(x). \tag{10}$$

The inclusion of strain effects results in a decrease in the range of band edge energies available from $In_xGa_{1-x}As$ when grown as a strained layer on a GaAs host, as shown in Fig. 9. Because of the limits on strained layer thickness imposed by the critical thickness, and because of the advantages associated with the use of quantum wells in semiconductor lasers in general, an additional correction to Eq. (10) is necessary to account for energy shifts in thin layers resulting from the quantum size effect. The shifts in energy associated with the quantum size effect can be determined from solutions to the finite square potential well problem of modern physics. Solutions to this problem for both confined electrons in the conduction band and confined

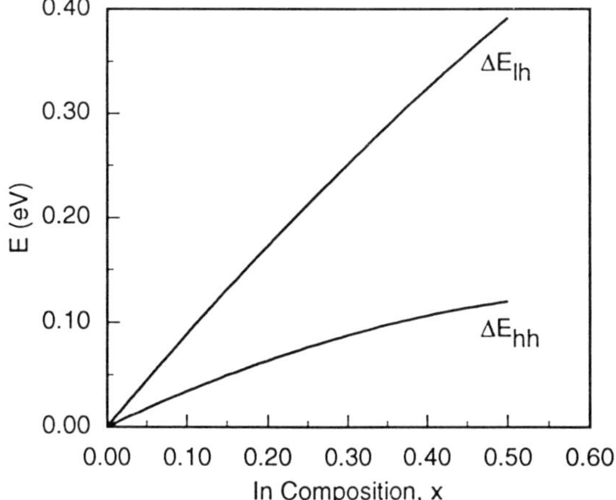


**Fig. 8.** The shift in the heavy-hole valence band edge ($\Delta E_{\mathrm{hh}}$) and light-hole valence band edge ($\Delta E_{\mathrm{lh}}$) with respect to the conduction band edge calculated from Eqs. (7) and (8) as a function of indium composition in the range $0 < x < 0.50$ for $\mathrm{In}_x\mathrm{Ga}_{1-x}\mathrm{As}$–GaAs strained layers.

holes in the heavy-hole valence band can be found given the heights of the potential barriers and the effective masses of the particles. Here, we take the effective masses for electrons and heavy holes for $\mathrm{In}_x\mathrm{Ga}_{1-x}\mathrm{As}$ to be interpolated from the bulk values for the endpoint binary semiconductors, InAs and GaAs, given in Table I.

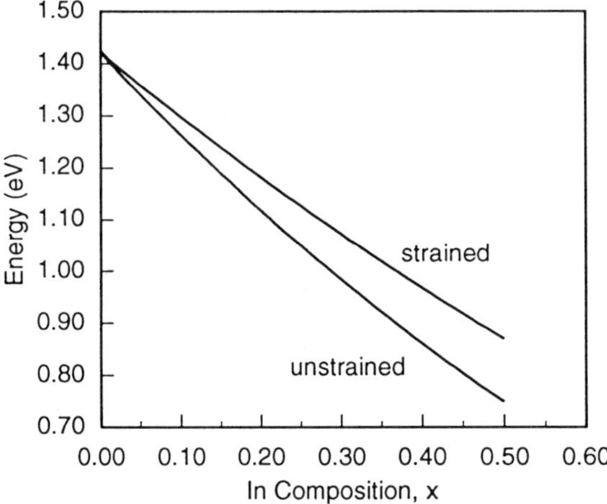


**Fig. 9.** Bulk bandgap energy for strained and unstrained $\mathrm{In}_x\mathrm{Ga}_{1-x}\mathrm{As}$ on GaAs as a function of indium composition in the range $0 < x < 0.50$.

The energy of the electron potential barrier is determined from the difference $\Delta E_g$ between the bulk bandgap energy of the barrier layers and the strained bandgap energy for $In_xGa_{1-x}As$, as given by Eq. (10) and shown in Fig. 9, multiplied by a dimensionless conduction band heterostructure discontinuity fraction of $\Delta E_c/\Delta E_g$. The barrier layers here are taken to be GaAs in order to be consistent with laser data presented later. The energy of the heavy-hole potential barrier is determined from the same bulk energy difference multiplied by $1 - \Delta E_c/\Delta E_g$. Thus,

$$\Delta E_c = \frac{\Delta E_c}{\Delta E_g}\,[E_g(\mathrm{GaAs}) - E_g(\mathrm{In}_x\mathrm{Ga}_{1-x}\mathrm{As}) - \Delta E_{hh}(x)], \tag{11}$$

$$\Delta E_v = \left[1 - \frac{\Delta E_c}{\Delta E_g}\right][E_g(\mathrm{GaAs}) - E_g(\mathrm{In}_x\mathrm{Ga}_{1-x}\mathrm{As}) - \Delta E_{hh}(x)]. \tag{12}$$

There is some controversy over what is an appropriate value for the conduction band heterostructure discontinuity fraction $\Delta E_c/\Delta E_g$. A summary of reported values is given in Table II. We have chosen to use a value $\Delta E_c/\Delta E_g = 0.65$ for purposes of demonstration. In any case, the overall transition energy from a bound state in the conduction band to a bound heavy-hole valence band state for a quantum well heterostructure laser is not particularly sensitive to the choice of $\Delta E_c/\Delta E_g$. The choice of $\Delta E_c/\Delta E_g$ becomes more important in consideration of bound and unbound light-hole valence band states [46, 47, 49, 53] or confinement in heterostructure systems under biaxial tension [46, 47].

**Table II.**

Reported Values of $\Delta E_c/\Delta E_g$ for $In_xGa_{1-x}As$–GaAs Strained layers

| Reference | $\Delta E_c/\Delta E_g$ |
|---|---|
| Ramberg *et al.* [43] | 0.35 |
| Menéndez *et al.* [44] | 0.40 |
| Priester *et al.* [45] | 0.41 |
| Koteles *et al.* [46. 47] | 0.55 |
| Niki *et al.* [41] | 0.65 |
| Zou *et al.* [48] | 0.65 |
| Marzin *et al.* [49] | 0.68 |
| Ji *et al* [50] | 0.70 |
| Andersson *et al.* [51] | 0.83 |
| Xinghau and Laiho, [52] | 0.83 |

The range of sizes for a quantum well in an $In_xGa_{1-x}As$ laser structure is bounded on the lower end by the practicality of growing ultrathin layers and the requirement for at least one bound state in both conduction and heavy-hole valence bands. The range is limited on the upper end by the critical thickness for elastic accommodation of strain as described previously. In order to illustrate how the quantum size effect impacts the transition energy for a laser diode, we have chosen a value for the well thickness equal to the Matthews and Blakeslee critical thickness at the composition of interest. These values have appropriate bound states for compositions in the range of interest, are sufficiently thin that the quantum size effect is significant, and are comfortably in the range of practical thicknesses for modern epitaxial methods. The total transition energy for recombination from an $n = 1$ electron state to an $n = 1$ heavy-hole state in a strained layer quantum well heterostructure then consists of the bulk in strained energy gap for the well composition of interest, a correction for the shift in the heavy-hole valence band edge with strain, and corrections for shifts in both the conduction band and heavy-hole valence band associated with the quantum size effect. Thus, for $In_xGa_{1-x}As$,

$$E(x, L_z) = E_g(x) + \Delta E_{hh}(x)|_{strain} + \Delta E_c(x)|_{L_z} + \Delta E_{hh}(x)|_{L_z}; \tag{13}$$

and the wavelength for this transition energy is given, of course, by

$$\lambda(x, L_z) = \frac{hc}{E(x, L_z)}, \tag{14}$$

where $h$ is Planck's constant and $c$ is the speed of light. Shown in Fig. 10 is the transition wavelength as a function of composition for a strained layer $In_xGa_{1-x}As$–GaAs quantum well heterostructure laser having a well thickness equal to the Matthews and Blakeslee critical thickness, and for a structure in which quantum size effects can be neglected (bulk). The correction for strain is included in both curves. From Fig. 10, it is clear that, even with corrections for strain and quantum size effect, the emission wavelength range of $\lambda = 0.9$–$1.1\ \mu m$ is adequately covered by $In_xGa_{1-x}As$–GaAs quantum well heterostructure lasers in the composition range of $x = 0.10$–$0.40$.

## 4. OPTICAL GAIN AND THRESHOLD CURRENT DENSITY IN STRAINED LAYER LASERS

In a strained layer lattice-mismatched system, biaxial strain can be expected to strongly affect more than just the positions of the valence band edges at

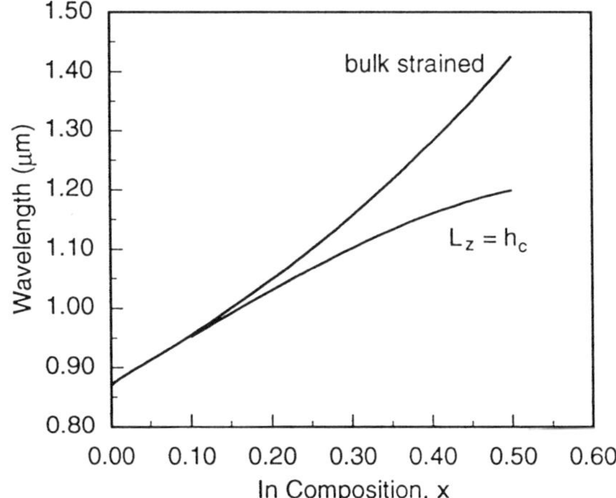


**Fig. 10.** Total transition wavelength for recombination from an $n = 1$ electron state to an $n = 1$ heavy-hole state as a function of composition for a strained layer $In_xGa_{1-x}As$–GaAs quantum well heterostructure laser having a well thickness equal to the Matthews and Blakeslee critical thickness, and for a structure in which quantum size effects can be neglected (bulk). A correction for strain is included in both curves.

$k = 0$. In fact, the form of the valence bands shown in Fig. 6 is overly simplified and is only correct in the direction perpendicular to the plane of the heterostructure interface. A more accurate representation of the changes in the valence band structure resulting from biaxial compression is shown in Fig. 11. This figure shows the heavy-hole and light-hole valence bands as a function of $k$ vector, normalized to $2\pi/a_0$, where $a_0$ is the GaAs lattice constant. This figure is calculated for $In_xGa_{1-x}As$ at a composition of $x = 0.30$, but the light-hole band edge has been shifted with respect to the heavy-hole band edge for illustration. The salient feature of Fig. 11 is that the upper valence band (hh) has the heavy-hole effective mass in the direction perpendicular ($\perp$) to the plane of the interfaces, but it is no longer symmetry and has a much lighter effective mass in the direction parallel ($\|$) to the growth interfaces [34, 35, 54–56]. It is, of course, the effective masses in the direction parallel ($\|$) to the growth interfaces that affect the density of states function and, hence, optical gain and threshold current density in strained layer quantum well heterostructure lasers.

Calculations of the gain in general for strained layer semiconductor lasers using the density matrix formalism have been reported [55, 57]. Shown in Fig. 12 [55] are transverse electric (TE) and transverse magnetic (TM)

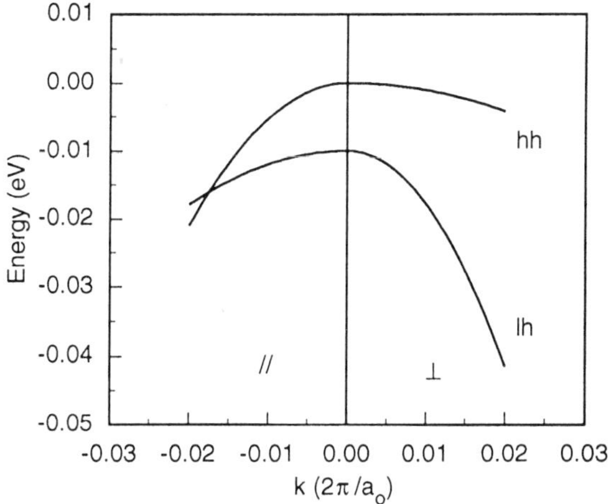


**Fig. 11.** The heavy-hole and light-hole valence band structure resulting from biaxial compression as a function of $k$ vector, normalized to $2\pi/a_0$, where $a_0$ is the GaAs lattice constant, in the directions perpendicular (⊥) and parallel (∥) to the growth interfaces, calculated for $In_xGa_{1-x}As$ at a composition of $x = 0.30$. The light-hole band edge has been shifted with respect to the heavy-hole band edge for illustration.

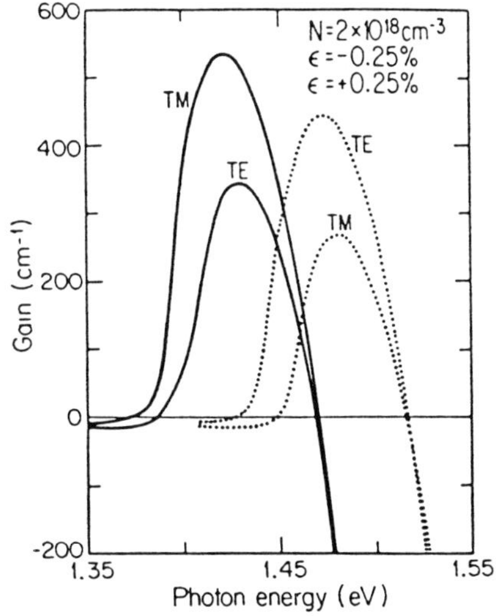


**Fig. 12.** Polarization-resolved gain spectra for strained layer lasers under both biaxial compression (dotted) and biaxial tension (solid). The magnitude of the stress is 0.25% and the injected carrier density is $2 \times 10^{18}$ cm$^{-3}$ (after [55]; © 1989 *IEEE*).

polarization-resolved linear gain spectra for strained layer lasers under both biaxial compression (dotted lines) and biaxial tension (solid lines). The magnitude of the stress is 0.25%, the injected carrier density is $2 \times 10^{18}\ \mathrm{cm}^{-3}$, and an intraband relaxation time of 0.2 ps is assumed. For strained layer lasers under biaxial compression, as is the case for $In_xGa_{1-x}As$–GaAs strained layer lasers, the peak emission wavelength is shifted to higher energies because of strain, as described in the preceding, and the TE mode is favored. Measured optical gain spectra [58] for In-rich $In_xGa_{1-x}As$ multiple quantum well heterostructure lasers (under biaxial compression) show relatively strong suppression of the TM mode.

There are significant performance parameters that are expected in strained layer lasers including modulation bandwidth in excess of 90 GHz, enhanced differential gain $\delta G/\delta N$, increased carrier dependence of the real part of the index of refraction $\delta n/\delta N$, and a reduced linewidth enhancement factor $\alpha$ [56, 59–61]. These enhanced parameters offer the possibility for reduced chirp, narrower linewidth, and extended single longitudinal mode tuning range. Rideout *et al.* [62] have reported the experimental measurement of $\delta G/\delta N$, $\delta n/\delta N$, and $\alpha$ for strained layer InGaAs–GaAs quantum well heterostructure lasers. They report a strong dependence of these parameters on carrier density with significantly enhanced values at certain carrier densities, compared with unstrained bulk or quantum well heterostructure lasers.

Perhaps more interesting is that the changes in the valence band structure of strained layer $In_xGa_{1-x}As$–GaAs quantum well heterostructure lasers result in the expectation of reduced laser threshold current density. Yablonovitch and Kane [42, 63] have argued that the large asymmetry between the light conduction band effective mass and the much heavier valence band effective mass is a fundamental limitation to low threshold laser performance in conventional double heterostructure or quantum well heterostructure lasers. At threshold, the quasi-Fermi levels for electrons and holes must satisfy the condition [64] that

$$E_{F_n} - E_{F_p} > \hbar\omega, \tag{15}$$

where $\hbar\omega$ is greater than the bandgap energy or, in the case of a quantum well heterostructure laser, where $\hbar\omega$ is greater than the transition energy. Assuming lightly doped active regions, detailed balance results in the requirement that the injected electron density $\delta n$ and the injected hole density $\delta p$ are equal. Satisfying simultaneously the equation

$$\delta n = \delta p \tag{16}$$

and Eq. (15) defines the values of $E_{F_n}$ and $E_{F_p}$. In unstrained conventional lasers, the hole effective mass is so much greater than the electron effective mass that the electron quasi-Fermi level is degenerate, while the hole quasi-Fermi level is not. The large differences in occupation probability result in a greater injection level at threshold than would be required if the conduction band and valence band densities of states were the same.

The reduction in hole effective mass in the direction parallel to the growth interfaces associated with strain results in a corresponding reduction in the injected carrier density required to reach threshold. The decrease in injected carrier density with decreasing hole effective mass for 50-Å quantum wells having an electron effective mass corresponding to either a GaAs or $In_xGa_{1-x}As$ ($x = 0.25$) quantum well is shown in Fig. 13. A reduction in the hole effective mass to equal the electron effective mass results in a reduction of the injected carrier density by more than a factor of two, with a further reduction resulting from the reduced InGaAs electron effective mass. Any reduction in the injected carrier densities manifests itself: (1) linearly in reduced nonradiative surface recombination losses; and (2) to the third power in reduced losses from Auger recombination. As shown in Fig. 11, the hh valence band in strained layer lasers has a much smaller effective mass in the direction parallel to the growth plane, perhaps as small as 1.5 $m_l$ [54].

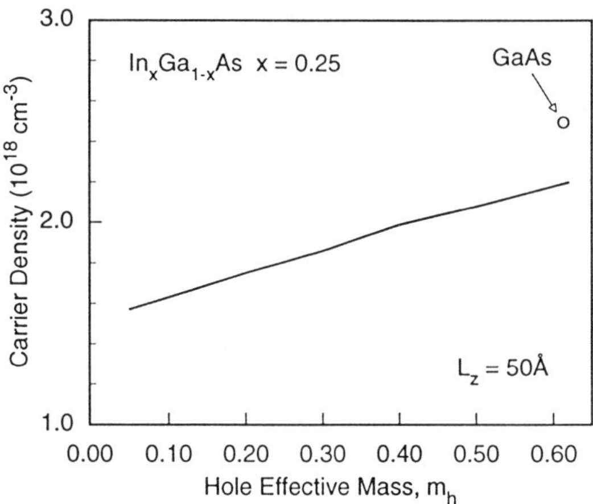


**Fig. 13.** The decrease in injected carrier density with decreasing hole effective mass for 50-Å quantum wells having electron effective mass corresponding to GaAs and $In_xGa_{1-x}As$ ($x = 0.25$). A reduction in the hole effective mass to values approaching the electron effective mass results in a reduction of the injected carrier density by more than a factor of two.

This results in significant shift of the gain in a strained layer laser to lower injected carrier densities [55].

Adams [54] has argued that direct intervalence band absorption (IVBA) would also be eliminated by the strain-induced removal of the valence band degeneracy and that the activation energies for Auger processes would be much larger in the presence of strain, resulting in an additional reduction of Auger recombination rates by many orders of magnitude. A numerical estimate of the extent of these combined effects has been made by O'Reilly *et al.* [65]. Chong and Fonstad [55] have also pointed out that an additional reduction in laser threshold can be expected from strained layer quantum well heterostructure lasers under biaxial compression, since TE polarization is dominant. This results from reduced end losses associated with greater facet reflectivity for TE modes than for TM modes [66].

## 5. MOCVD GROWTH OF STRAINED LAYER LASER STRUCTURES

Nearly all of the InGaAs–GaAs strained layer laser structures reported have been grown by metalorganic chemical vapor deposition (MOCVD). (For a detailed description of the MOCVD process see Coleman and Dapkus [67], Miller and Coleman [68], Dapkus and Coleman [69], and references cited.) The only process modification necessary is the addition of one of several relatively low volatility In-precursors to an otherwise well established conventional AlGaAs–GaAs MOCVD growth technology. The consistency with which these lasers can be grown by MOCVD is limited only by the stability of the vapor pressure of the indium precursor. In much early work on In-compound semiconductors by MOCVD, the liquid alkyl triethylindium (TEIn) was utilized [70, 71] with some problems associated with prereaction products. Many groups [72–75] have been successful in the use of trimethylindium (TMIn), which is a solid, at both atmospheric and low pressure. However, TMIn converts from a powder to very large crystals over the period of a few months, thereby continually reducing the surface area available for sublimation [76]. This reduction in surface area results in an ever lower effective vapor pressure, leading to a reduction in growth efficiency as well as poor composition control and irreproducibility. Ethyldimethylindium (EDMIn) is a relatively new organometallic source that has been reported [77–79] as a potential replacement for TMIn. The compound is comprised of an indium atom bonded to two methyl radicals ($CH_3$) and a single ethyl radical ($C_2H_5$), and it has a vapor pressure similar to TMIn. In contrast to TMIn, EDMIn is a liquid at room temperature, presumably free of the problems associated with solid sources.

Later, we describe data from a series of InGaAs–GaAs strained layer laser structures grown by MOCVD. The epitaxial growth of these structures, typical of the structures grown in many laboratories, was carried out in a vertical reaction chamber at atmospheric pressure using trimethylgallium (TMGa), trimethylaluminum (TMAl), and EDMIn. Diethylzinc and silane were used for p- and n-type dopants, respectively. As an example, the $In_{0.25}Ga_{0.75}As$ layers are grown with TMGa and EDMIn mass flow rates of 2 and 44.2 sccm, respectively, and a V/III ratio of 147. Typically, buffer and confining layers are grown at 720°C, while the temperature is lowered during the growth of the GaAs separate confinement region to 625°C for the InGaAs quantum well, as discussed in what follows.

Bulk crystal measurements of composition and growth rate for strained InGaAs layers are not possible because of the limitation imposed by the critical thickness. Transmission electron microscopy (TEM) can be used to accurately determine the quantum well thicknesses for even single well quantum well heterostructure laser structures. Then accurate knowledge of the InGaAs growth rate $g$, the GaAs growth rate $g_{\mathrm{GaAs}}$ and the TMGa vapor pressure $p_{\mathrm{G}}$ allows calculation [80] of the vapor pressure of EDMIn, $p_{\mathrm{I}}$, from

$$g = g_{\mathrm{GaAs}}(1 + p_{\mathrm{I}}F_{\mathrm{I}}/p_{\mathrm{G}}F_{\mathrm{G}}), \tag{17}$$

where $F_{\mathrm{G}}$ and $F_{\mathrm{I}}$ are the flow rates of TMGa and EDMIn, respectively. The vapor pressure for EDMIn has been reported to be 0.56 Torr at 11°C [79] and 0.85 Torr at 17°C [77]. The composition for $In_xGa_{1-x}As$ strained layers can then be extracted from the molar ratio of the column III constituents:

$$x = p_{\mathrm{I}}F_{\mathrm{I}}/(p_{\mathrm{I}}F_{\mathrm{I}} + p_{\mathrm{G}}F_{\mathrm{G}}). \tag{18}$$

There has been some speculation that EDMIn may rapidly decompose into its constituents within the bubbler over time. Fry *et al.* [77] observed with mass spectrometry, however, that EDMIn consists of only a single compound, with no trace of TMIn or TEIn, and York *et al.* [78, 79] reported stable InGaAs growth rates over a six month period, indicating that EDMIn probably has a stability similar to that observed for TMGa and TMAl. In general, the growth rate of $In_xGa_{1-x}As$ is linear with the column III precursor mass flow rate [80] in the diffusion-limited growth regime ($T > 575$°C). Beernink *et al.* [82] have relied on this to form strained layer InGaAs–GaAs quantum well heterostructure lasers in which the quantum well thickness was kept constant, while the composition was varied from $In_{0.07}Ga_{0.93}As$ to $In_{0.42}Ga_{0.58}As$ by changing the EDMIn mass flow rate.

The emission wavelengths for those lasers matched values predicted from composition and thickness data using the preceding EDMIn vapor pressure data.

The microscopic quality and atomic abruptness of interfaces [83–86] in III-V heterostructures are important issues in optical and electronic devices where the interface plays a critical role in device performance. For InGaAs strained layer lasers, this manifests itself in a somewhat narrower and lower range of acceptable growth temperatures than for AlGaAs–GaAs heterostructures. The heats of formation [87] for GaAs, InAs, and AlAs are

| | $\Delta f$ (kcal/mole) |
|---|---|
| InAs | 114 |
| GaAs | 128 |
| AlAs | 150 |

The heat of formation is an indication of the column III to arsenic bond strength, with the In–As bond being the weakest. A weaker bond suggests increased surface mobility during growth. York *et al.* [88] reported TEM studies of InAs–GaAs multilayer structures grown at 625 and 800°C. Symmetric abrupt InAs–GaAs interfaces were observed at lower growth temperatures, and an asymmetry was observed at higher growth temperatures. Previous studies of AlAs–GaAs superlattices grown at 750°C indicate the presence of a similar asymmetry in the interfaces [86] and is thought to be caused by surface roughness associated with the lower surface mobility and faster reaction kinetics of aluminum atoms as compared with gallium atoms [89]. Indium atoms are more mobile than gallium atoms [89], however, so it is likely that the cause of the intermixing of the GaAs on InAs interface at higher temperatures is the result of increased surface mobility of the indium atom relative to the surface mobility at 625°C coupled with slower reaction kinetics. At higher temperatures, incoming gallium atoms have the opportunity to mix with the mobile indium atoms and, as the GaAs lattice can be considered the host for these structures, mixing to form a less mismatched alloy layer is energetically favorable.

The combining of indium and aluminum alloys in structures of InGaAs–GaAs–AlGaAs requires temperature variation during the growth, since AlGaAs is generally optimized at much higher temperatures than InGaAs [91]. The effects of temperature excursions and long pauses on either side of the InGaAs quantum well that would be required to lower the temperature can be minimized by the insertion of the InGaAs quantum well into a separate confinement laser structure with only low composition (20%)

aluminum confining layers and GaAs barrier layers, where much lower growth temperatures would be feasible while still maintaining the optical quality of the material. A schematic drawing of the laser structure and growth temperature cycle is shown in Fig. 14. Many of the lasers described in the following sections are based on this design and, despite the unconventionally low cladding layer aluminum composition, threshold current densities of 140–150 A/cm$^2$ have been observed in structures with several different types of active regions. Subsequently, otherwise identical lasers, excepting aluminum confining layer compositions as high as 85%, have also been grown [92] with a 800–625°C temperature cycle. These lasers have time zero characteristics that are better than lasers with $Al_{0.2}Ga_{0.8}As$ cladding layers. Increased confining layer aluminum compositions generally necessitate higher growth temperatures, although the quantum well can still be grown at a lower temperature.

## 6. BROAD AREA AND OXIDE DEFINED STRIPE LASERS

The first reports of injection laser diodes from strained layer InGaAs–GaAs quantum well heterostructures [11] were for MBE-grown triple quantum

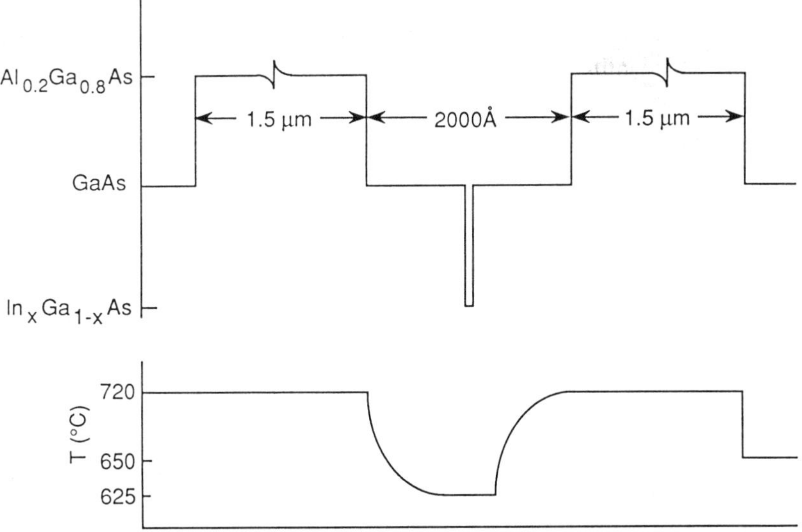


**Fig. 14.** Schematic diagram of a separate confinement strained layer InGaAs–GaAs–AlGaAs heterostructure laser structure and optimized growth temperature cycle.

well devices emitting at $\sim 1.0\ \mu m$. These early lasers clearly demonstrated the capability for injection lasers at long wavelengths but were characterized by somewhat high threshold current densities ($\sim 1.2\ kA/cm^2$) and low output powers ($\sim 5$ mW/facet). Perhaps somewhat of concern, especially coupled with reports of rapid degradation in photopumped samples [9, 10], was their preliminary report of a 10% increase in threshold current in 25 hours of operation. This report was quickly followed by reports of lower threshold current densities (465 $A/cm^2$) and somewhat better reliability [12]. The issue of reliability will be discussed in detail later. After 1986, research activity on this materials system for broad area lasers increased rapidly, with other groups reporting low broad area threshold current densities at $\lambda \sim 1\ \mu m$ [93, 94], high power cw operation at $\lambda \sim 0.93\ \mu m$ [95], high power conversion efficiency at $\lambda \sim 0.93\ \mu m$ [96–98], and details of the operating characteristics, such as threshold current density $J_{th}$, differential quantum efficiency of $\eta$, and characteristic temperature of threshold current $T_0$ in the range $\lambda \sim 0.93$–$1.0\ \mu m$ [96–98]. In this section, we describe details of several studies [81, 82, 90–92] in which various structural parameters of strained layer $In_xGa_{1-x}As$–GaAs–$Al_yGa_{1-y}As$ quantum well heterostructure lasers have been varied over a wide range and compared with resulting laser characteristics.

In comparison with the design of conventional $Al_xGa_{1-x}As$–GaAs quantum well heterostructure lasers, the design of strained layer $In_xGa_{1-x}As$–GaAs–$Al_yGa_{1-y}As$ has a larger parameter matrix to consider. In addition to the quantum size effect associated with well thickness in strained layer lasers, consideration must be given to that thickness with respect to the critical thickness. Composition of the well must be considered in terms of bulk emission wavelength, the shift in energy associated with strain, and the compositional dependence of critical thickness. Also, extension to longer wavelengths increases the useable range of aluminum composition in both barrier layers and outer confining layers. We describe here a series of strained layer laser structures with design parameters adjusted to traverse portions of this parameter matrix. The laser structures, shown schematically in Fig. 15, consist of a 0.25 $\mu m$ thick GaAs:$n^+$ buffer layer grown on a (100) GaAs:$n^+$ substrate, n- and p-type 1.5 $\mu m$ thick $Al_yGa_{1-y}As$ outer confining layers surrounding an active region consisting of a single $In_xGa_{1-x}As$ undoped strained layer quantum well of thickness $L_z$ centered in a 0.2 $\mu m$ thick GaAs barrier, or carrier collection layer. Each structure has a 0.2-$\mu m$ GaAs : $P^+$ contact layer. The combinations of quantum well thickness, quantum well indium mole fraction, and outer confining layer aluminum composition considered here are shown in Table III.

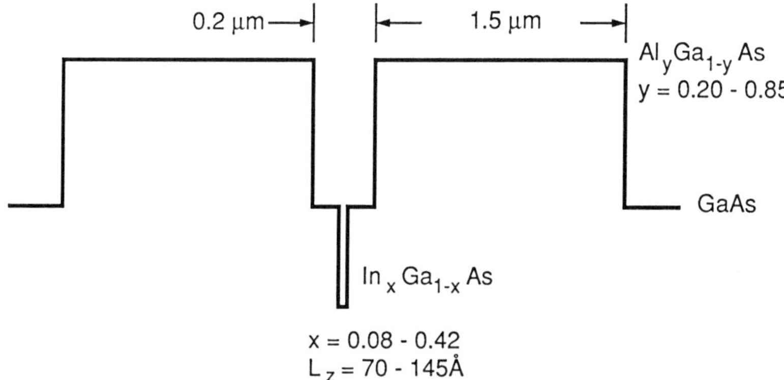


**Fig. 15.** Schematic diagram for a series of strained layer laser structures with design parameters adjusted to traverse portions of the quantum well thickness, quantum well composition, and confining layer composition parameter matrix for the combinations indicated in Table III.

Standard photolithographic processing was used to fabricate 150 $\mu$m wide oxide defined stripes on each sample. The substrates were lapped and polished, n-type contacts were formed by alloyed Ge/Au, and nonalloyed Cr/Au contacts were evaporated on the p-side. Various cavity lengths were formed by cleaving, and the resulting bars were cleaved into individual diodes for testing under pulsed conditions (1.5 $\mu$s pulse width, 2 kHz repetition rate) at room temperature.

For the first series of strained layer lasers, the quantum well thickness was held at 70 Å, the outer confining layer $Al_yGa_{1-y}As$ composition was held at $y = 0.20$, and the $In_xGa_{1-x}As$ well composition was varied from $x =$ 0.08–0.42. Shown in Fig. 16 for reference is the critical thickness as a function of indium composition for the models of Matthews and Blakeslee and People

**Table III.**

Parameter marix $Al_yGa_{1-y}As$–GaAs–$In_xGa_{1-x}As$ Strained Layer, Broad Area Laser Study

| $L_z$ (Å) | $x$ | $y$ |
|---|---|---|
| 70 | 0.08–0.42 | 0.20 |
| 100, 125, 143 | 0.25 | 0.20 |
| 70, 100 | 0.25 | 0.20–0.85 |

[References: [81, 82, 90–92].

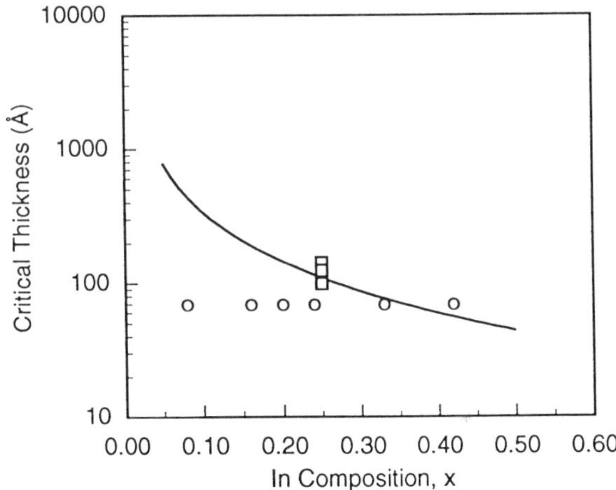


**Fig. 16.** The critical thickness as a function of indium composition for the models of Matthews and Blakeslee and of People and Bean. Shown as open squares are a series of strained layer lasers in which the quantum well thickness was held at 70 Å, the outer confining layer $Al_yGa_{1-y}As$ composition was held at $y = 0.20$, and the $In_xGa_{1-x}As$ well composition was varied from $x = 0.08$ to 0.42. Shown as open circles are a series of lasers having a fixed $In_xGa_{1-x}As$ composition of $x \sim 0.25$ and varying well thicknesses approaching the critical thickness as outlined in Table III.

and Bean. The series of laser structures considered here are shown as open squares. The measured threshold current density $J_{th}$ as a function of well composition for 815 $\mu$m long devices is shown in Fig. 17. All of the devices with $In_{0.42}Ga_{0.58}As$ wells failed to reach threshold below 1.6 kA/cm$^2$. The trend in this figure is the same for all cavity lengths tested.

The minimum in $J_{th}$ versus composition of less 150 A/cm$^2$ at an indium fraction in the range $0.20 \leq x \leq 0.30$ in Fig. 17 is explained by the consideration of two effects. For low indium fraction in the wells, the confining energy $\Delta E_{conf}$ between the bound state of electrons in the conduction band and the band edge of the GaAs barrier layers is small. A confining energy of at least a few $kT$ ($\sim$75 meV) is necessary to have good carrier confinement in the well. Smaller confining energies result in a significant portion of the electron population in the active region having energies greater than the barrier height and, thus, unconfined in terms of the quantum well. Similar considerations apply to the confinement of holes in the valence band, but the higher density of states reduces the effect. Since the barriers have a much larger volume than the well, a significantly larger current density would then be required

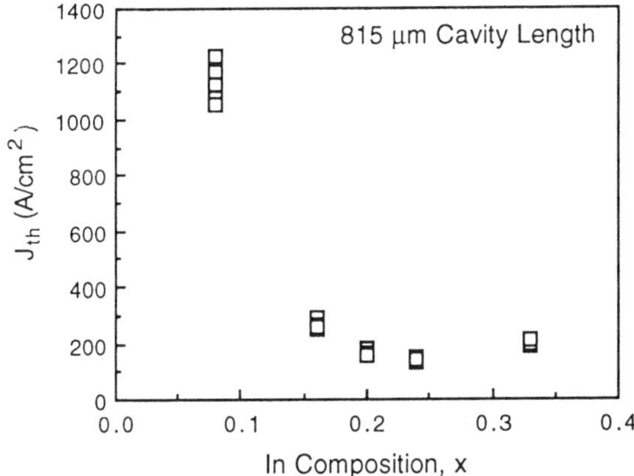


**Fig. 17.** The measured threshold current density $J_{th}$ as a function of well composition for 815 $\mu$m long devices from a series of strained layer lasers in which the quantum well thickness was held at 70 Å, the outer confining layer $Al_yGa_{1-y}As$ composition was held at $y = 0.20$, and the $In_xGa_{1-x}As$ well composition was varied from $x = 0.08$ to 0.42.

to supply the necessary carriers to the barriers to reach a threshold carrier density in the well. For devices with $In_xGa_{1-x}As$ wells with $x < 0.15$, the electrons are poorly confined. As the indium fraction in the wells is increased and the well is made deeper, the state is better confined, resulting in decreased $J_{th}$. A similar decrease in $J_{th}$ might be expected if the barrier layers were instead $Al_yGa_{1-y}As$, and the aluminum composition $y$ was increased. The results for low indium fraction are similar to those observed for AlGaAs–GaAs quantum well heterostructure lasers with low aluminum composition barrier layers [99].

As the indium fraction of the wells increases beyond 0.30, the effects of strain begin to adversely affect the performance of the lasers of this study. For 70-Å wells with indium fraction below $x \sim 0.30$, the strained active region thickness is well below the critical thickness as calculated from either model. For higher indium mole fractions in the well, the thickness of the active region approaches the Matthews and Blakeslee calculated critical thickness, and the deleterious effects of the strain become evident as $J_{th}$ increases. For all devices with well compositions in the range $0.16 \leq x \leq 0.33$, $\Delta E_{conf}$ is greater than 75 meV, and the well thickness is below the calculated critical thickness; the performance of devices in this range does not vary dramatically.

A second series of lasers, having a fixed $In_xGa_{1-x}As$ composition of $x \sim 0.25$, varying well thicknesses as outlined in Table III, and approaching the critical thickness, shown as open circles in Fig. 16, were studied. The laser characteristics temperature dependence of threshold ($T_0$), temperature dependence of wavelength ($\partial\lambda/\partial T$), and total differential quantum efficiency ($\eta$) are found to have no observable dependence on well size for the range of thicknesses considered. $T_0$ was measured to be between 49 and 62 K. The shift in emission wavelength, $\partial\lambda/\partial T$, ranged from 4.0–5.1 Å/°C for temperatures above 300 K, and the total eternal differential efficiencies were approximately 40%. Figure 18 shows the average threshold current density for four cavity lengths of each well thickness. As observed at higher compositions in Fig. 17, there is a modest increase in the laser threshold current density as the quantum well thickness approaches the Matthews and Blakeslee critical thickness. A much stronger dependence of laser reliability on quantum well thickness is observed for these samples and is described in what follows.

Figure 19 shows the variation in laser threshold current density with cavity length for $In_xGa_{1-x}As$–GaAs quantum well heterostructure lasers having a well thickness of 100 Å, composition $x = 0.25$, and confining layer composition of $y = 0.20$. These very low threshold lasers show the expected short cavity effect for thin single quantum well structures [100] at cavity lengths below 500 $\mu$m and approach 100 A/cm$^2$ in the long cavity limit.

For the laser structures we have described, the index of refraction differences in the structures are not large and, as a consequence, optical confine-

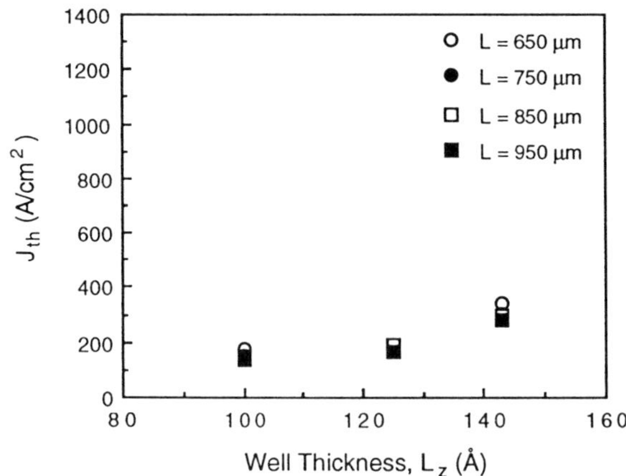


**Fig. 18.** The average threshold current density for four cavity lengths of each well thickness for a series of lasers having a fixed $In_xGa_{1-x}As$ composition of $x \sim 0.25$ and an outer confining layer $Al_yGa_{1-y}As$ composition of $y = 0.20$.

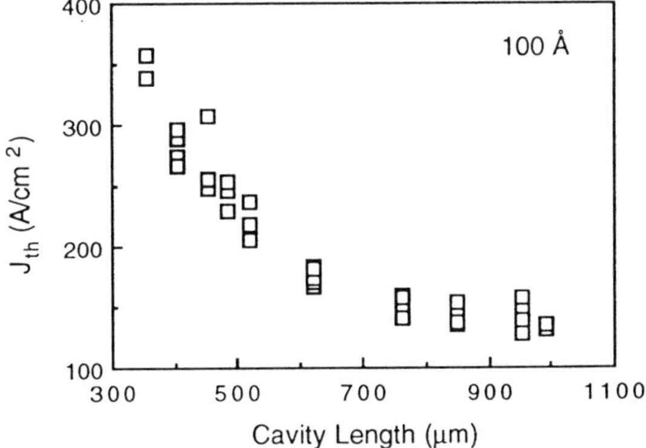


**Fig. 19.** The variation in laser threshold current density with cavity length for $In_xGa_{1-x}As$ quantum well heterostructure lasers having a well thickness of 100 Å, composition $x = 0.25$, and confining layer composition of $y = 0.2$, showing the expected short cavity effect for thin single quantum well structures and approaching 100 A/cm$^2$ in the long cavity limit.

ment factor Γ is also small, typically less than 0.02. In a third series of lasers, two sets of $In_{0.25}Ga_{0.75}As$ single quantum well laser structures, with 70- and 100-Å wells, were studied [92], each set with confining layer aluminum composition varying from $y = 0.20$ to 0.85. The threshold current densities $J_{th}$ for both the 70-Å (circles) and the 100-Å (squares) $In_{0.25}Ga_{0.75}As$ single quantum well lasers are plotted as a function of the aluminum composition

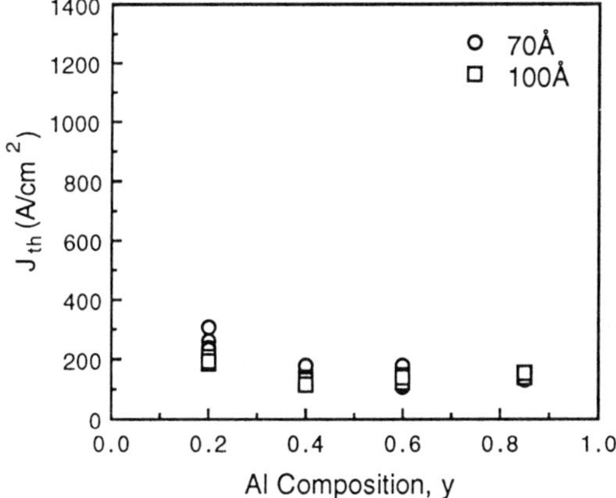


**Fig. 20.** The threshold current density $J_{th}$ for both the 70-Å (circles) and the 100-Å (squares) $In_{0.25}Ga_{0.75}As$ single quantum well lasers as a function of the aluminum composition in the confining layers.

in the confining layers in Fig. 20. The threshold current density $J_{th}$ decreases as the composition is increased from 0.20 to 0.85 in the confining layers, corresponding to the increase in the optical confinement factor $\Gamma$ due to the lower refractive index in layers with higher aluminum composition. This is evidenced by the usual expression for the threshold current density in the low current range, where the gain is a logarithmic function of current density:

$$J_{th} = \frac{J_{tr}}{\eta} \exp\left[\frac{\frac{1}{L} \ln \frac{1}{R}}{\Gamma \beta J_{tr}}\right]. \tag{19}$$

Here, $J_{tr}$ is the transparency current (A/cm$^2$), $\eta_i$ is the internal quantum efficiency, $\beta$ is the gain coefficient (cm/A), $L$ is the laser diode cavity length, $R$ is the fact reflectivity, and $\alpha_i$ represents the internal optical loss (cm$^{-1}$). The optical confinement factors were calculated for each of the eight structures using a five-slab waveguide model [101], taking into account the index of refraction of each of the confining and barrier layers at the appropriate emission wavelength. The decrease in the threshold current density clearly follows the decrease in $\Gamma^{-1}$. The change in threshold current density associated with changes in confinement factor results in a corresponding well defined change in emission wavelength [92]. The electron effective mass $m_e^*(x)$ in $In_xGa_{1-x}As$ decreases for increasing indium content, which brings about a reduced conduction band density of states as compared with GaAs. Consequently, band filling becomes pronounced [102] for increasing carrier injection levels, shifting the laser transition energy up from the quantum well bound state energy. A decrease in threshold current density is thus accompanied in these structures by an increase in the laser emission wavelength.

The behavior of oxide defined stripe geometry InGaAs–GaAs strained layer lasers with narrower stripe widths ($< 50$ $\mu$m) is markedly poorer than for wide stripe broad area devices. Kolbas *et al.* [103, 104] reported on 6 $\mu$m wide oxide stripe lasers with an emission wavelength of $\sim$920 nm. These lasers had a threshold current density of $\sim$4600 A/cm$^2$, calculated from the actual oxide stripe width, threshold current, and cavity length. Vawter *et al.* [105] reported a threshold current density for proton isolated 12 $\mu$m wide stripe lasers of 1333 A/cm$^2$. Shieh *et al.* [106, 107] reported good laser threshold current densities for single- and multiple-well strained layer quantum well heterostructures with 50-$\mu$m oxide defined stripe widths. For single well structures, as the stripe width was reduced, however, they observed [106, 107] an increase in threshold current, a large shift in wavelength to a higher energy quantum transition, and evidence of index antiguiding in the

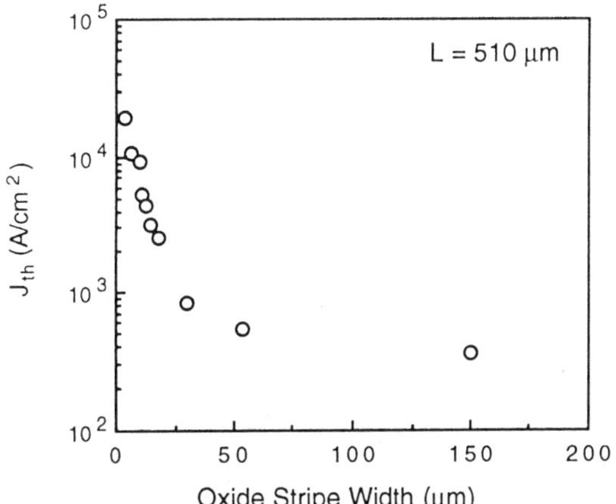


**Fig. 21.** Calculated threshold current density versus stripe width for oxide defined stripe $In_{0.175}Ga_{0.825}As$ single quantum well strained layer lasers ($L_z$ = 150 Å, cavity length 510 μm).

form of prominent side lobes in the narrow stripe laser far field emission patterns.

Shown in Fig. 21 are data similar to those reported by Shieh *et al.* [106, 107] on the threshold current density versus stripe width calculated from the actual oxide stripe width for a series of $Al_{0.20}Ga_{0.80}As$–GaAs–$In_{0.175}Ga_{0.825}As$ strained layer quantum well heterostructure lasers ($L_z$ = 150 Å, cavity length 510 μm) grown in our laboratory. These lasers are characterized by a gradual increase in laser threshold current density as the stripe width is decreased from 150 to 30 μm, followed by a much larger increase as the stripe width is made smaller than 30 μm. In the intermediate stripe width range, current spreading becomes more important as the stripe width is reduced, similar to the behavior of oxide defined stripe AlGaAs–GaAs lasers. As the losses become larger in narrower stripe quantum well heterostructure lasers, however, the gain peak is shifted first to higher energy quantum transitions [100] and finally to the GaAs barrier layers [108]. Figure 22 is the emission energy as a function of stripe width for the same series of lasers, and it clearly shows the jump to higher energy transitions for stripe widths below 30 μm. Thus, an additional increase in the threshold current density arises from the larger transparency current required in order for higher order transitions to reach threshold.

The basis for the unusually large losses apparent in narrow width oxide defined stripe lasers is, as suggested by Shieh *et al.* [106, 107], index

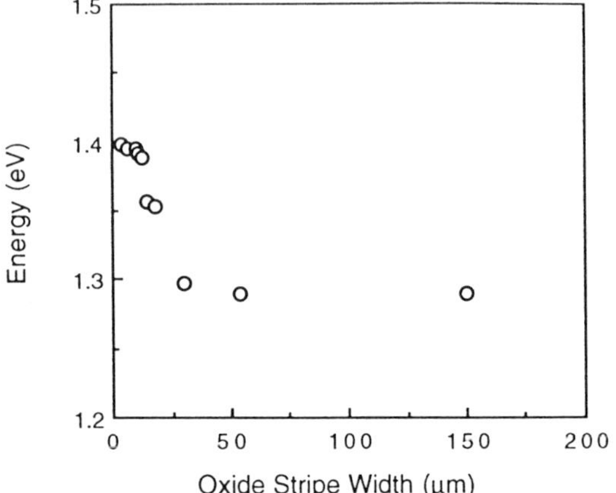


**Fig. 22.** Emission energy versus stripe width for oxide defined stripe $In_{0.175}Ga_{0.825}As$ single quantum well strained layer lasers ($L_z = 150$ Å, cavity length 510 $\mu$m).

antiguiding in these structures. Figure 23 shows the far field emission patterns for an 8-$\mu$m oxide stripe $Al_{0.20}Ga_{0.80}As$–GaAs–$In_{0.175}Ga_{0.825}As$ strained layer quantum well heterostructure laser ($L_z = 70$ Å) in the drive current range of 1.1–4 times laser threshold. As the current increases, the angular beam divergence also increases, indicating a net increasing defocusing as a result of a decreasing real index in the stripe region.

A number of workers have considered gain and thermal guiding effects in stripe geometry lasers [109–111]. An antiguiding factor $b$ can be defined [111] as the ratio of the decrease in the real refractive index $\Delta n_r$ to the increase in the imaginary refractive index $\Delta \bar{n}_i$ by

$$b = \frac{-\Delta \bar{n}_r}{\Delta \bar{n}_i}, \tag{20}$$

where the imaginary refractive index is related to the gain $g$ by

$$\Delta \bar{n}_i = \frac{g}{2k_0}, \tag{21}$$

and where $k_0 = 2\pi/\lambda_0$, with $\lambda_0$ the free space wavelength. The decrease in the real refractive index is a result of injected carriers under the stripe, offset somewhat by an increase from the thermal contribution. Thus, $b$ is a measure of the strength of the carrier-induced index depression relative to the gain that results from the same carrier density with positive values resulting in

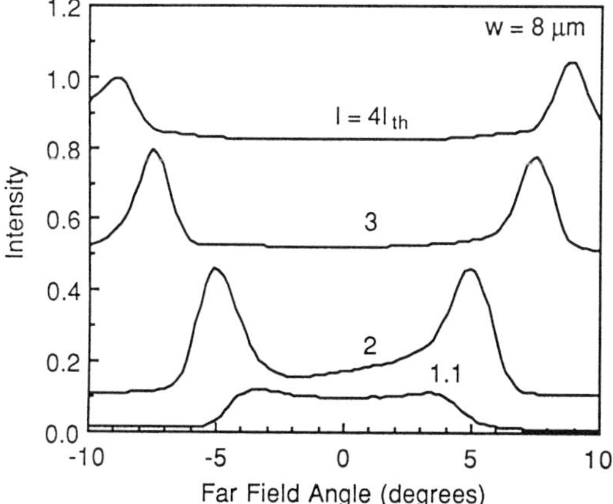


**Fig. 23.** Double lobed far field emission patterns for an 8 $\mu$m wide oxide defined stripe $In_{0.175}Ga_{0.825}As$ single quantum well strained layer laser ($L_z = 70$ Å) for drive currents in the range of 1.1–4 times threshold.

antiguiding, with increased curvature of the wavefronts outside the stripe. For narrow stripes, the antiguiding results [111] in a double-lobed far field pattern, as observed for the narrow stripe lasers of Fig. 23. Antiguiding also results in a large diffraction loss due to light propagating out from under the stripe. The very high threshold current densities of the narrow stripe devices are a result of this high diffraction loss [106, 107].

The Petermann $K$ factor [112] is a measure of the astigmatism in the transverse mode, and is related to the antiguiding factor $b$. Petermann $K$ factors of 12.9 [108] and 20 have been reported [106, 107] for similar InGaAs strained layer lasers. These values are much larger than the values of 3–5 reported [113] for similar GaAs–AlGaAs lasers. A Petermann $K$ factor of 12.9 gives a value of $b = 6.4$, which is considerably larger than the values 0.5–3 reported [114] for GaAs–AlGaAs lasers, indicating the strong antiguiding present in these InGaAs strained layer lasers.

An alternate way to view the origin of the twin lobes in the far field of Fig. 23 is to neglect the gain guiding and thermal effect and model the waveguiding in the transverse direction, considering only the real index depression caused by the injected carriers under the stripe [108]. Assuming a step index profile in the transverse direction, leaky waves within the device produce a lobe in the far field at an angle on either side of the facet normal given by [149, 150].

$$\sin\theta \sim (2|\Delta\bar{n}|\bar{n}_{\text{eff}})^{1/2} \tag{22}$$

where $\theta$ is the far field angle, $\Delta n$ is the magnitude of the index depression, and $n_{\text{eff}}$ is the effective index of the region outside the stripe. From Eq. (22) and the optical confinement factor $\Gamma$, the approximate carrier-induced index depression in the well can be calculated as a function of drive current. Large values, ranging from 0.032 to 0.41 [108] and as high as 0.57 [106, 107] have been reported.

## 7. INDEX GUIDED LASERS AND LASER ARRAYS

Low threshold current or controlled optical mode applications require a modal stability that cannot be obtained from either gain guided or antiguided devices. For most of these applications, structural changes forcing a large current-independent lateral effective index variation are necessary. A number of workers have reported index guided lasers from InGaAs–GaAs strained layer materials. For example, low threshold transverse junction stripe (TJS) lasers formed by zinc diffusion have been reported [103, 104, 115]. Following the early work by Fischer *et al.* [116] on ridge waveguide single mode InGaAs–GaAs strained layer lasers, there have been a number of reports including growth by MBE [117–119] and high power, high efficiency, high reliability structures by MOCVD [120–124]. Silicon–oxygen impurity induced disordering has been used to form buried heterostructure strained layer lasers [125, 126]. Recently, presumably for low threshold current and reliability as much as for the 0.9–1.1-$\mu$m wavelength range, there is interest in using strained layer InGaAs–GaAs heterostructures for surface emitting laser diodes [38, 39, 127]. Various forms of buried heterostructure InGaAs–GaAs strained layer lasers formed by liquid phase epitaxy (LPE) regrowth [128], by ion implantation and rapid thermal anneal [129], and by growth on nonplanar substrates [130] have been reported. All of these index guided lasers have in common that the structures were well developed earlier for use with AlGaAs–GaAs quantum well heterostructure lasers and that the inclusion of strained layers of InGaAs is straightforward and incidental.

One structure that benefits, albeit indirectly, from the use of long wavelength strained InGaAs layers for the active region of the laser is the buried heterostructure laser reported by York *et al.* [78, 90]. Although, the buried heterostructure (BH) laser is one of the most attractive index guided stripe geometry semiconductor lasers because of the combination of strong lateral index guiding and absolute current confinement provided by a heterostructure discontinuity in the lateral direction, the structure is difficult to fabricate because of the need for processing high quality narrow stripe etched mesas having an etched depth of typically greater than 1 $\mu$m and the requirement for a high quality regrowth interface at the edges of the active region. The regrowth becomes especially difficult for AlGaAs–GaAs BH lasers having higher aluminum composition confining layers because of the relative stability of the native oxide that forms at the interface after etching. Long wavelength ($\lambda > 1$ $\mu$m) strained layer InGaAs–GaAs–AlGaAs quantum well buried heterostructure lasers, such as the structure shown in Fig. 24, can be more easily formed by wet chemical etching and a two step MOCVD growth process, because the relatively low aluminum composition of the confining layers ($x = 0.20$) allows for high quality regrowth interfaces and effective use of a silicon dioxide mask for selective epitaxy limited to the etched regions. The longer emission wavelength and, hence, lower energy of the InGaAs strained layer active region allows the use of lower Al-composition confining layers than for AlGaAs–GaAs quantum well heterostructure lasers, with a modest penalty in threshold current density as described previously. The structures reported by York *et al.* [78] have active region stripe widths of ~3.5 $\mu$m and an emission wavelength of $\lambda \sim 1.074$ $\mu$m. Threshold currents of less than 7 mA have been observed for devices having a cavity length of 405 $\mu$m. The threshold current density for these devices is comparable with oxide defined stripe broad area lasers made from the same material, indicating that current spreading is negligible. Output powers in excess of 130 mW per uncoated facet have been observed with total differential quantum efficiencies of greater than 60%. Near field patterns indicate

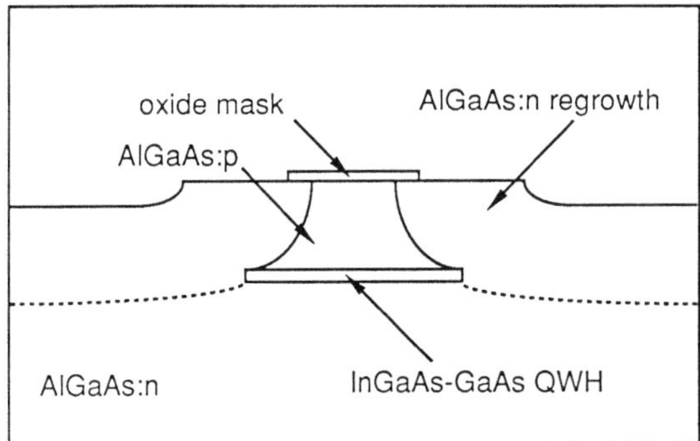


**Fig. 24.** Schematic diagram of the cross section of an index guided $Al_{0.20}Ga_{0.80}As$–GaAs–$In_xGa_{1-x}As$ buried heterostructure strained layer quantum well laser formed by wet chemical etching and a two step MOCVD growth process. Minimal polycrystalline deposition of the $y = 0.20$ $Al_yGa_{1-y}As$ takes place on the oxide mask, and the regrowth interface is of high quality.

that the lasers are operating on a fundamental lateral mode and are stable to more than 30 times laser threshold.

A number of multiple stripe laser array structures utilizing InGaAs strained layer active regions have been reported. Baillargeon *et al.* [131] reported a high power, phase-locked nonplanar corrugated substrate array with an InGaAs strained layer active region. Ten stripe arrays formed by proton isolation [120, 132] or hydrogenation [126] have been reported, yielding up to 3 W of continuous power [120]. Two dimensional InGaAs strained layer arrays operating at very high power levels have been reported using both the stacked array approach [133] and the grating surface emitter approach [98]. Similar to the index guided lasers described in the preceding, most of these array structures were well developed earlier for use with AlGaAs–GaAs quantum well heterostructure lasers.

Beernink *et al.* [134] have made use of the antiguiding behavior associated with InGaAs strained layer lasers (Fig. 23) to form multiple element, oxide defined stripe phase-locked diode arrays operating in the in-phase fundamental array mode. Five element oxide stripe arrays of 8 $\mu$m wide stripes with center to center spacing between 14 and 30 $\mu$m were fabricated from previously characterized $In_xGa_{1-x}As$ ($x \sim 0.25$ and 0.33) strained layer laser material. In contrast to broad area lasers with stripe width comparable with the total width of the pumped region in the array 64 $\mu$m), the far field of the array, shown in Fig. 25, is much narrower (2–3°) and is stable to greater

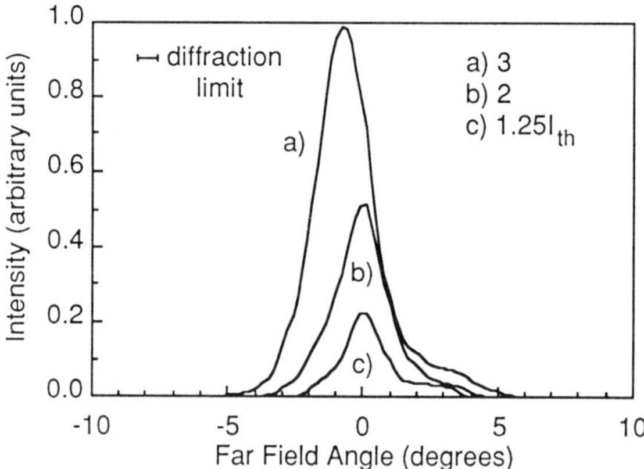


**Fig. 25.** Far field emission patterns for a five element oxide defined stripe InGaAs–GaAs strained layer laser array. The far field patterns are stable to greater than $3 \times I_{th}$ and much narrower (2–3°) than similar broad area lasers. The single central lobe in the far field is suggestive of in-phase fundamental array mode operation.

than $3 \times I_{th}$. The single central lobe in the far field is suggestive of in-phase fundamental array mode operation and counter to what is observed for similar arrays made from AlGaAs–GaAs quantum well heterostructures.

## 8. RELIABILITY

It is of little practical consequence if a semiconductor diode laser system has the low threshold current, high differential gain, and attractive spectral properties described in the preceding for strained layer InGaAs–GaAs quantum well heterostructure lasers if the devices are unreliable. Perhaps one's intuition would support the early studies of photopumped InGaAs–GaAs strained layer superlattices [9, 10] which demonstrated cw 300 K laser operation but suggested that these strained materials are inherently unreliable. The first limited reports of reliable laser operation from strained layer InGaAs–GaAs heterostructure lasers appeared in 1985 [12], followed by reports of cw operation for a few hundred hours in 1988 [132]. Fischer *et al.* [135] reported 5000 hour cw operation of strained layer InGaAs–GaAs lasers with 1.8% per kh degradation rates and 100% survival. Bour *et al.* [121, 122] recently reported 10,000 hour operation with a degradation rate under 1% per kh. Both of these degradation rates are significantly better than for comparable AlGaAs–GaAs quantum well heterostructure laser diodes [136, 137].

The thickness of the InGaAs quantum well with respect to the critical thickness plays an important role [81]. The results of reliability testing for the InGaAs–GaAs strained layer lasers of Fig. 18 are shown in Fig. 26. Standard processing was used to define 60 $\mu$m wide oxide defined stripes with facets formed by cleaving, and 600 $\mu$m long uncoated devices were mounted junction up on a copper heat sink. Lifetesting was performed at constant power on devices operating cw at 60 mW per uncoated facet. All of the devices fabricated from material with the thinnest well, 100 Å, were still operating after 2800 hours with only a gradual increase in current. In stark contrast, the performance of the 143 Å thick quantum well devices degraded rapidly, and the current doubling limit was reached for all devices in under 100 hours. The performance of devices with 125-Å wells was intermediate between these two extremes. The 100-Å quantum well lasers have reliabilities comparable with other InGaAs strained layer lasers [121, 122, 132, 135, 138, 139]. There is a lack of sudden failure in these devices, which can be attributed [135, 138, 139] to resistance of the InGaAs strained layer lasers to fabrication-induced damage. The devices showed gradual degradation, with the 143-Å quantum well devices degrading at a much faster

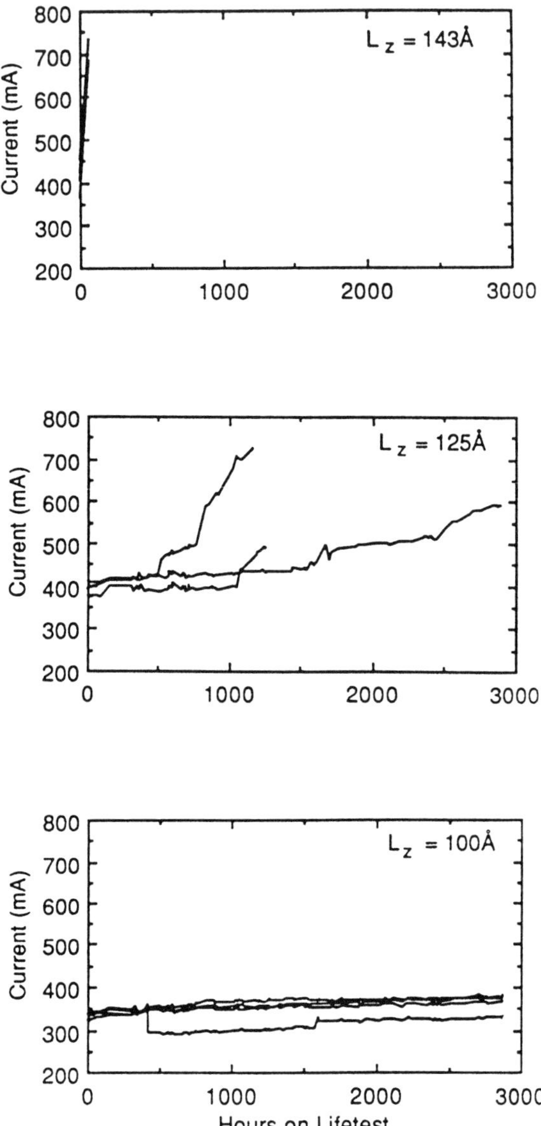


**Fig. 26.** Results of laser reliability testing for a series of lasers having a fixed $In_xGa_{1-x}As$ composition of $x \sim 0.25$, three quantum well thicknesses, and an outer confining layer $Al_yGa_{1-y}As$ composition of $y = 0.20$. All of the devices fabricated from material with the thinnest well, 100 Å, were still operating after 2800 hours, while the performance of the 143 Å thick quantum well devices degraded rapidly. The performance of devices with 125-Å wells was intermediate between these two extremes.

rate than those with thinner wells. These data suggest that the poorer reliability of the 143 Å thick quantum well devices is directly related to the thickness of the material itself. These results also indicate that the maximum *practical* thickness for $In_{0.25}Ga_{0.75}As$ quantum well active regions is less than 120 Å, which lies slightly above the critical thickness of ~105 Å predicted by the mechanical equilibrium model of Matthews and Blakeslee, and well below the value of ~265 Å given by the energy balance model of People and Bean for this composition.

Provided the quantum well thickness of strained layer InGaAs–GaAs lasers is kept below some practical thickness, the lasers show reliability characteristics that are more like those observed for 1.3- and 1.5-μm InGaAsP–InP lasers [140]. These characteristics include very low gradual degradation rates and the absence of sudden failures, even for devices that have no facet coatings and were not screened or burned-in [135, 138, 139]. Waters *et al.* [138, 139] have reported on the reliability of InGaAs–GaAs strained layer quantum well heterostructure lasers having a step graded strained layer active region and emitting at 1.1 μm. In addition to the characteristics just described, these lasers are also similar to InGaAsP–InP lasers in that they have a lower temperature dependence of laser threshold ($T_0 \sim 90$ K), and the cw output power for these devices is thermally limited rather than limited by catastrophic optical degradation (COD). Shown in Fig. 27 are the light–current characteristics for three strained layer

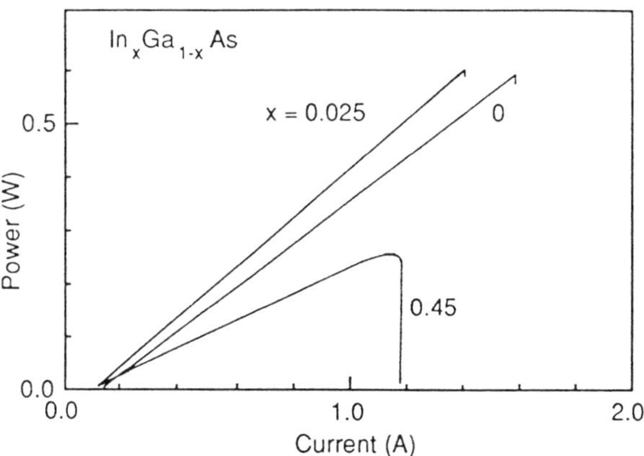


**Fig. 27.** Cw light–current characteristic for a step graded $x = 0.45$ $In_xGa_{1-x}As$–GaAs strained layer quantum well heterostructure laser. The output power of these devices is thermally limited. Shown for comparison are similar characteristics for single well graded index $In_xGa_{1-x}As$ quantum well heterostructure lasers ($x = 0, 0.025$), which are limited by catastrophical optical degradation.

$In_xGa_{1-x}As$–GaAs quantum well heterostructure lasers structures having $x = 0, 0.025$, and 0.45 [138, 139]. The lasers with little or no indium in the quantum well fail catastrophically, while the laser with $x = 0.45$ is thermally limited. Nevertheless, these lasers have operated for more than 10,000 hours with very low gradual degradation.

## 9. OTHER STRAINED LAYER MATERIAL SYSTEMS

As many of the attractive features of strained layer quantum well heterostructure lasers have developed, and the concerns about reliability have proven unfounded, it is natural to expect that strained layers will bc incorporated into otherwise lattice-matched systems, such as the InGaAsP–InP heterostructure system. In addition to lower threshold currents and higher differential gain, the likelihood that Auger recombination and intervalence band absorption can be greatly diminished in strained layer systems provides strong incentive to apply strain to InGaAsP–InP quantum well heterostructure lasers, which have thus far been disappointingly slow in developing. Of course, the applications requiring 1.3–1.5 $\mu$m laser emission are an important factor. Melman *et al.* [141] reported photoluminescence emission to 1.3 $\mu$m from strained layer InGaAs–GaAs heterostructures. More recently, high efficiency, high power, and low threshold currents have been obtained for $In_xGa_{1-x}As$–InP with $x$ adjusted for biaxial compressive strain. Although Tothill *et al.* [142] saw no improvement in the threshold current density or $T_0$ of strained layer InGaAs–InP lasers emitting at 1.5 $\mu$m, others have reported substantial improvements in the laser characteristics of these devices, including lower threshold, higher efficiency, higher output power, higher $T_0$, and other interesting optical properties [143–147]. It is fair to say that these excellent results on such an important laser system presage a dramatic increase in effort toward the continued development of strained layer 1.5-$\mu$m lasers. The use of strained layers in separate confinement AlGaInP–GaInP quantum well heterostructure lasers has resulted [148] in low threshold currents at visible wavelengths ($\lambda \sim 691$ nm).

## ACKNOWLEDGMENTS

It is a distinct pleasure to acknowledge the many hours of hard work, the long enjoyable discussions, and the creative imagination of many bright students at the University of Illinois, without whom much of this work would not have been possible, including J. J. Alwan, R. P. Bryan, T. M. Cockerill, M. E. Favaro, J. Kim, L. M. Miller, and especially K. J. Beernink and P. K. York. I greatly appreciate the expertise and effort contributed by R. C. Waters and his coworkers at the McDonnell Douglas Electronics Systems Company in determining the reliability of these structures. I would also like to thank Prof. P. D. Dapkus at the University of Southern California for occasional glances into his lab notebook, which is characterized by both enormous breadth and depth. This work has been supported by the National Science Foundation, the Charles Stark Draper Laboratory, the McDonnell Douglas Electronics Systems Company, the Naval Research Laboratory, the Joint Service Electronics Program, and GTE.

## REFERENCES

1. Whitley, T. J., Creamer, M. J., Steele, R. C., Brain, M. C., and Millar, C. A. (1989). *IEEE Photon. Tech. Lett.* **1**, 425.
2. Vodhanel, R. S., Laming, R. I., Shah, V., Curtis, L., Bour, D. P., Barnes, W. L., Minelly, J. D., Tarbox, E. J., and Favire, F. J. (1989). *Electron. Lett.* **25**, 1386.
3. Yamada, M., Shimizu, M., Takeshita, T., Okayasu, M., Horiguchi, M., Uehara, S., and Sugita, E. (1989). *IEEE Photon. Tech. Lett.* **1**, 422.
4. Urquhart, P. (1988). *IEE Proc.* **135**, pt. J, no. 6.
5. Shimada, S. (1990). *Optics Photon. News* **1**, 6.
6. Casey, H. C., Jr. and Panish, M. B. (1978). *Heterostructure Lasers, Part B: Materials and Operating Characteristics*, p. 3. Academic Press, New York.
7. Matthews, J. W., and Blakeslee, A. E. (1974). *J. Crystal Growth* **27**, 118.
8. Osbourn, G. C., Gourley, P. L., Fritz, I. J., Biefeld, R. M., Dawson, L. R., and Zipperian, T. E. (1987). In *Semiconductor and semimetals* (Willardson, R. K., and Beer, A. C., eds.), vol. 24, p. 459. Academic Press, New York.
9. Camras, M. D., Brown, J. M., Holonyak, N., Jr., Nixon, M. A., Kaliski, R. W., Ludowise, M. J., Dietze, W. T., and Lewis, C. R. (1983). *J. Appl. Phys.* 6183.
10. Ludowise, M. J., Dietze, W. T., Lewis, C. R., Camras, M. D., Holonyak, N., Jr., Fuller, B. K., and Nixon, M. A. (1983). *Appl. Phys. Lett.* **42**, 487.
11. Laidig, W. D., Caldwell, P. J., Lin, Y. F., and Peng, C. K. (1984). *Appl. Phys. Lett.* **44**, 653.
12. Laidig, W. D., Lin, Y. F., and Caldwell, P. J. (1985). *J. Appl. Phys.* **57**, 33.
13. Frank, F. C., and van der Merwe, J. H. (1949). *Proc. Roy. Soc. (London) A* **198**, 216.
14. People, R., and Bean, J. C. (1985). *Appl. Phys. Lett.* **47**, 322.
15. Dodson, B. W., and Tsao, J. Y. (1987). *Appl. Phys. Lett.* **51**, 1325.
16. Vawter, G. A., and Myers, D. R. (1989). *J. Appl. Phys.* **65**, 4769.
17. Fritz, I. J., Picraux, S. T., Dawson, L. R., Drummond, T. J., Laidig, W. D., and Anderson, N. G. (1985). *Appl. Phys. Lett.* **46**, 967.

18. Andersson, T. G., Chen, Z. G., Kulakovskii, V. D., Uddin, A., Vallin, J. T. (1987). *Appl. Phys. Lett.* **51**, 752.
19. Yao, J. Y., Andersson, T. G., and Dunlop, G. L. (1988). *Appl. Phys. Lett.* **53**, 1420.
20. Weng, S. L. (1989). *J. Appl. Phys.* **66**, 2217.
21. Bertolet, D. C., Hsu, J. K., Agahi, F., and Lau, K. M. (1990). *J. Electron. Mater.* **19**, 967.
22. Elman, B., Koteles, E. S., Melman, P., Jagannath, C., Lee, J., and Dugger, D. (1989). *Appl. Phys. Lett.* **55**, 1659.
23. Reithmaier, J.-P., Cerva, H., Lösch, R. (1989). *Appl. Phys. Lett.* **54**, 48.
24. Drigo, A. V., Aydinli, A., Carnera, A., Genova, F., Rigo, C., Ferrari, C., Franzosi, P., and Salviati, G. (1989). *J. Appl. Phys. Lett.* **66**, 1975.
25. Morris, D., Roth, A. P., Masut, R. A., Lacelle, C., and Brebner, J. L. (1988). *J. Appl. Phys.* **64**, 4135.
26. Price, G. L. (1988). *Appl. Phys. Lett.* **53**, 1288.
27. Price, G. L., and Usher, B. F. (1989). *Appl. Phys. Lett.* **55**, 1984.
28. Olsen, G. H., Nuese, C. J., and Smith, R. T. (1978). *J. Appl. Phys.* **49**, 5523.
29. Asai, H., and Oe, K. (1983). *J. Appl. Phys.* **54**, 2052.
30. Schirber, J. E., Fritz, I. J., and Dawson, L. R. (1985). *Appl. Phys. Lett.* **46**, 187.
31. Jones, E. D., Ackermann, H., Schirber, J. E., and Drummond, T. J. (1985). *Solid State Comm.* **55**, 525.
32. Anderson, N. G., Laidig, W. D., Kolbas, R. M., and Lo, Y. C. (1986). *J. Appl. Phys.* **60**, 2361.
33. O'Reilly, E. P. (1989). *Semicond. Sci. Technol.* **4**, 121.
34. Pikus, G. E., and Bir, G. L. (1959). *Sov. Phys. Solid State* **1**, 136.
35. Pikus, G. E., and Bir, G. L. (1960). *Sov. Phys. Solid State* **1**, 1502.
36. Gavini, A., and Cardona, M., (1970). *Phys. Rev. B* **1**, 672.
37. Adachi, S. (1982). *J. Appl. Phys.* **53**, 8775.
38. Huang, K. F., Tai, K., Chu, S. N. G., and Cho, A. Y. (1989). *Appl. Phys. Lett.* **54**, 2026.
39. Huang, K. F., Tai, K., Jewell, J. L., Fischer, R. J., McCall, S. L., and Cho, A. Y. (1989). *Appl. Phys. Lett.* **54**, 2192.
40. Madelung, O., ed. (1982). *Numerical data and functional relationships in science and technology*, Group III, **17**. Springer-Verlag, Berlin.
41. Niki, S., Lin, C. L., Chang, W. S. C., and Wieder, H. H. (1989). *Appl. Phys. Lett.* **55**, 1339.
42. Yablonovitch, E., and Kane, E. O. (1988). *J. Lightwave Tech.* **LT-6**, 1292.
43. Ramberg, L. P., Enquist, P. M., Chen, Y.-K., Najjar, F. E., Eastman, L. F., Fitzgerald, A., and Kavanagh, K. L. (1987). *J. Appl. Phys.* **61**, 1234.
44. Menéndez, J., Pinczuk, A., Werder, D. J., Sputz, S. K., Miller, R. C., Sivco, D. L., and Cho, A. Y. (1987). *Phys. Rev. B* **36**, 8165.
45. Priester, C., Allan, G., and Lanoo, M. (1988). *Phys. Rev. B* **38**, 9870.
46. Koteles, E. S., Owens, D. A., Bertolet, D. C., Hsu, J. K., and Lau, K. M. (1990). *Surface Sci.* **228**, 314.

47. Koteles, E. S., Owens, D., Elman, B., Melman, P., Bertolet, D. C., and Lau, K. M. (1990). *Mater. Res. Soc. Symp. Proc.* **160**, 649.
48. Zou, Y., Grodzinski, P., Menu, E. P., Jeong, W. G., Dapkus, P. D., Alwan, J. J., and Coleman, J. J. (1991). *Appl. Phys. Lett.* **58**, 601.
49. Marzin, J.-Y., Charasse, M. N., and Sermage, B. (1985). *Phys. Rev. B* **31**, 8298.
50. Ji, G., Huang, D., Reddy, U. K., Henderson, T. S., Houdre, R., and Morkoç, H. (1987). *J. Appl. Phys.* **62**, 3366.
51. Andersson, T. G., Chen, Z. G., Kulakovskii, V. D., Uddin, A., and Vallin, J. T. (1988). *Phys. Rev. B* **37**, 4032.
52. Xinghua, W., and Laiho, R. (1989). *Superlatt. Microstruct.* **5**, 79.
53. Reithmaier, J.-P., Höger, R., Riechert, H., Hiergeist, P., and Abstreiter, G. (1990). *Appl. Phys. Lett.* **57**, 957.
54. Adams, A. R. (1986). *Electron. Lett.* **22**, 249.
55. Chong, T. C., and Fonstad, C. G. (1989). *IEEE J. Quantum Electron.* **QE-25**, 171.
56. Ohtoshi, T., and Chinone, N. (1989). *IEEE Photon. Tech. Lett.* **1**, 117.
57. Ahn, D., and Chuang, S. L. (1988). *IEEE J. Quantum Electron.* **QE-24**, 2400.
58. Zielinski, E., Keppler, F., Streubel, Scholz, F., Sauer, R., and Tsang, W. T. (1989). *Superlatt. Microstruct.* **5**, 555.
59. Suemune, I., Coldren, L. A., Yamanishi, M., and Kan, Y. (1988). *Appl. Phys. Lett.* **53**, 1378.
60. Ghiti, A., O'Reilly, E. P., and Adams, A. R. (1989). *Electron. Lett.* **25**, 821.
61. Dutta, N. K., Wynn, J., Sivco, D. L., and Cho, A. Y. (1990). *Appl. Phys. Lett.* **56**, 2293.
62. Rideout, B., Yu, B., LaCourse, J., York, P. K., Beernink, K. J., and Coleman, J. J. (1990). *Appl. Phys. Lett.* **56**, 706.
63. Yablonovitch, E., and Kane, E. O. (1986). *J. Lightwave Tech.* **LT-4**, 504.
64. Bernard, M. G. A., and Duraffourg, G. (1961). *Phys. Status. Solidi.* **1**, 699.
65. O'Reilly, E. P., Heasman, K. C., Adams, A. R., and Witchlow, G. P. (1987). *Superlatt. Microstruct.* **3**, 99.
66. Ikegami, T. (1972). *IEEE J. Quantum Electron.* **QE-8**, 470.
67. Coleman, J. J., and Dapkus, P. D. (1985). In *Gallium arsenide technology* (Ferry, D. K., ed.), p. 79. Howard W. Sams & Co., Inc., Indianapolis.
68. Miller, L. M., and Coleman, J. J. (1988). *CRC critical reviews in solid state and materials sciences* **15**, 1.
69. Dapkus, P. D., and Coleman, J. J. (1989). In *III-V semiconductor materials and devices* (Malik, R. J., ed.), p. 147. North-Holland Co., Amsterdam.
70. Manasevit, H. M., and Simpson, W. I. (1973). *J. Electrochem. Soc.* **120**, 135.
71. Duchemin, J. P., Bonnet, M., Beuchet, G., and Koelsch, F. (1979). *Inst. Phys. Conf. Ser.* **45**, Chapter 1, 10.
72. Mircea, A., Azoulay, R., Dugrand, L., Mellet, R., Rao, K., and Sacilotti, M. (1984). *J. Electron. Mater.* **13**, 603.
73. Kuo, C. P., Jan, J. S., Cohen, R. M., Dunn, J., and Stringfellow, G. B. (1984). *Appl. Phys. Lett.* **44**, 550.

74. Carey, K. W. (1985). *Appl. Phys. Lett.* **46**, 89.
75. Meyer, R., Grützmacher, D., Jürgensen, H., and Balk, P. (1988). *J. Crystal Growth* **93**, 285.
76. Knauf, J., Schmitz, D., Strauch, G., Jürgensen, H., and Heyen, M. (1988). *J. Crystal Growth* **93**, 34.
77. Fry, K. L., Kuo, C. P., Larsen, C. A., Cohen, R. M., and Stringfellow, G. B. (1986). *J. Electron. Mater.* **15**, 91.
78. York P. K., Beernink, K. J., Fernandez, G. E., and Coleman, J. J. (1989). *Appl. Phys. Lett.* **54**, 499.
79. York, P. K., Beernink, K. J., Kim, J., Coleman, J. J., Fernandez, G. E., and Wayman, C. M. (1989). *Appl. Phys. Lett.* **55**, 2476.
80. Reep, D. H., and Ghandi, S. K. (1983). *J. Electrochem. Soc.* **130**, 675.
81. Beernink, K. J., York, P. K., Coleman, J. J., Waters, R. G., Kim, J., and Wayman, C. M. (1989). *Appl. Phys. Lett.* **55**, 2167.
82. Beernink, K. J., York, P. K., and Coleman, J. J. (1989). *Appl. Phys. Lett.* **55**, 2585.
83. Jeng, S. J., Wayman, C. M., Costrini, G., and Coleman, J. J. (1984). *Mater. Lett.* **2**, 359.
84. Jeng, S. J., Wayman, C. M., Costrini, G., and Coleman, J. J. (1985). *Mater. Lett.* **3**, 89.
85. Jeng, S. J., Wayman, C. M., Costrini, G., and Coleman, J. J. (1985). *Mater. Lett.* **3**, 331.
86. Coleman, J. J., Costrini, G., Jeng, S. J., and Wayman, C. M. (1986). *J. Appl. Phys.* **59**, 428.
87. Weast, R. C., ed. (1984). *Handbook of chemistry and physics*, E-89. CRC Press, Boca Raton, Florida.
88. York, P. K., Kiely, C. J., Fernandez, G. E., Baillargeon, J. N., and Coleman, J. J. (1988). *J. Crystal Growth* **93**, 512.
89. Newman, P. G., Cho, N. M., Kim, D. J., Madhukar, A., Smith, D. D., Aucoin, T. R., and Iafrate, G. J. (1988). *J. Vac. Sci. Technol.* **B6**, 1483.
90. York, P. K., Beernink, K. J., Fernandez, G. E., and Coleman, J. J. (1990). *Semicond. Sci. Tech.* **5**, 508.
91. York, P. K., Beernink, K. J., Kim, J., Alwan, J. J., Coleman, J. J., and Wayman, C. M. (1990). *J. Crystal Growth* **107**, 741.
92. York, P. K., Langsjoen, S. M., Miller, L. M., Beernink, K. J., Alwan, J. J., and Coleman, J. J. (1990). *Appl. Phys. Lett.* **57**, 843.
93. Feketa, D., Chan, K. T., Ballantyane, J. M., and Eastman, L. F. (1986). *Appl. Phys. Lett.* **49**, 1659.
94. Choi, H. K., and Wang, C. A. (1990). *Appl. Phys. Lett.* **57**, 321.
95. Bour, D. P., Gilbert, D. B., Elbaum, L., and Harvey, M. G. (1988). *Appl. Phys. Lett.* **53**, 2371.
96. Bour, D. P., Ramon, U., Martinelli, D. B., Elbaum, Gilbert, L., and Harvey, M. G. (1989). *Appl. Phys. Lett.* **55**, 1501.

97. Bour, D. P., Evans, G. A., and Gilbert, D. B. (1989). *J. Appl. Phys.* **65**, 3340.
98. Bour, D. P., Sstabile, P., Rosen, A., Janton, W., Elbaum, L., and Holmes, D. J. (1989). *Appl. Phys. Lett.* **54**, 2637.
99. Hersee, S. D., de Cremoux, B., and Duchemin, J. P. (1984). *Appl. Phys. Lett.* **44**, 476.
100. Zory, P. S., Reisinger, A. R., Mawst, L. J., Costrini, G., Zmudzinski, C. A., Emanuel, M. A., Givens, M. E., and Coleman, J. J. (1986). *Electron. Lett.* **22**, 475.
101. Casey, H. C., Jr., Panish, M. B., Schlosser, W. O., and Paoli, T. L. (1974). *J. Appl. Phys.* **45**, 322.
102. Hall, D., Major, J. S., Jr., Holonyak, N., Jr., Gavrilovic, P., Meehan, K., Stutius, W., and Williams, J. E. (1989). *Appl. Phys. Lett.* **55**, 752.
103. Kolbas, R. M., Anderson, N. G., Laidig, W. D., Sin, Y., Lo, Y. C., Hsieh, K. Y., and Yang, Y. J. (1988). *IEEE J. Quantum Electron.* **QE-24**, 1605.
104. Kolbas, R. M., Yang, Y. J., and Hsieh, K. Y. (1988). *Superlatt. Microstruct.* **4**, 603.
105. Vawter, G. A., Myers, D. R., Brennan, T. M., Hammons, B. E., and Hohimer, J. P. (1989). *Electron. Lett.* **25**, 243.
106. Shieh, C., Mantz, J., Lee, H., Ackley, D., and Engelmann, R. (1989). *Appl. Phys. Lett.* **54**, 2521.
107. Shieh, C., Lee, H., Mantz, J., Ackley, D., and Engelmann, R. (1989). *Electron. Lett.* **25**, 1226.
108. Beernink, K. J., Alwan, J. J., and Coleman, J. J. (1991). *J. Appl. Phys.* **69**, 56.
109. Paoli, T. L. (1977). *IEEE J. Quantum Electron.* **QE-13**, 662.
110. Nash, F. R. (1973). *J. Appl. Phys.* **44**, 4696.
111. Streifer, W., Burnham, R. D., and Scifres, D. R. (1982). *IEEE J. Quantum Electron.* **QE-18**, 856.
112. Petermann, K. (1979). *IEEE J. Quantum Electron.* **QE-15**, 566.
113. Idler, W., Hausser, S., Schweizer, H., Weimann, G., and Schlapp, W. (1988). *Electron. Lett.* **24**, 787.
114. Kirkby, P. A., Goodwin, A. R., Thompson, G. H. B., and Selway, P. R. (1977). *IEEE J. Quantum Electron.* **QE-13**, 705.
115. Yang, Y. J., Hsieh, K. Y., and Kolbas, R. M. (1987). *Appl. Phys. Lett.* **51**, 215.
116. Fischer, S. E., Fekete, D., Feak, G. B., and Ballantyne, J. M. (1987). *Appl. Phys. Lett.* **50**, 714.
117. Offsey, S. D., Schaff, W. J., Tasker, P. J., Ennen, H., and Eastman, L. F. (1989). *Appl. Phys. Lett.* **54**, 2527.
118. Offsey, S. D., Schaff, W. J., Tasker, P. J., and Eastman, L. F. (1990). *IEEE Photon. Technol. Lett.* **2**, 9.
119. Wu, M. C., Olsson, N. A., Sivco, D., and Cho, A. Y. (1990). *Appl. Phys. Lett.* **56**, 221.
120. Welch, D. F., Steifer, W., Schaus, C. F., Sun, S., and Gourley, P. L. (1990). *Appl. Phys. Lett.* **56**, 10.

121. Bour, D. P., Gilbert, D. B., Fabian, K. B., Bednarz, J. P., and Ettenberg, M. (1990). *IEEE Photon. Tech. Lett.* **2**, 173.
122. Bour, D. P., Dinkel, N. A., Gilbert, D. B., Fabian, K. B., and Harvey, M. G. (1990). *IEEE Photon. Tech. Lett.* **2**, 153.
123. Larsson, A., Cody, J., Forouhar, S., and Lang, R. J. (1990). *Appl. Phys. Lett.* **56**, 1731.
124. Larsson, A., Forouhar, S., Cody, J., and Lang, R. J. (1990). *IEEE Photon. Tech. Lett.* **2**, 307.
125. Major, J. S., Jr., Guido, L. J., Hsieh, K. C., and Holonyak, N., Jr. (1989). *Appl. Phys. Lett.* **54**, 913.
126. Major, J. S., Jr., Guido, L. J., Holonyak, N., Jr., Hsieh, K. C., Vesely, E. J., Nam, D. W., Hall, D. C., and Baker, J. E. (1990). *J. Electron. Mater.* **19**, 59.
127. Gourley, P. L., Lyo, S. K., and Dawson, L. R. (1989). *Appl. Phys. Lett.* **54**, 1397.
128. Eng, L. E., Chen, T. R., Sanders, S., Zhuang, Y. H., Zhao, B., and Yariv, A. (1989). *Appl. Phys. Lett.* **55**, 1378.
129. Vawter, G. A., Myers, D. R., Brennan, T. M., and Hammons, B. E. (1990). *Appl. Phys. Lett.* **56**, 1945.
130. Arent, D. J., Brovelli, L., Jackel, H., Marclay, E., and Meier, H. P. (1990). *Appl. Phys. Lett.* **56**, 1939.
131. Baillargeon, J. N., York, P. K., Zmudzinski, C. A., Fernandez, G. E., Beernink, K. J., and Coleman, J. J. (1988). *Appl. Phys. Lett.* **53**, 457.
132. Stutius, W., Gavrilovic, P., Williams, J. E., Meehan, K., and Zarrabi, J. H. (1988). *Electron. Lett.* **24**, 1494.
133. Evans, G. A., Bour, D. P., Carlson, N. W., Hammer, J. M., Lurie, M., Butler, J. K., Palfrey, S. L., Amantea, R., Carr, L. A., Hawrylo, F. Z., James, E. A., Kirk, J. B., Liew, S. K., and Reichert, W. F. (1989). *Appl. Phys. Lett.* **55**, 2721.
134. Beernink, K. J., Alwan, J. J., and Coleman, J. J. (1990). *Appl. Phys. Lett.* **58**, 2076.
135. Fischer, S. E., Waters, R. G., Fekete, D., and Ballantyne, J. M. (1989). *Appl. Phys. Lett.* **54**, 1861.
136. Harnagel, G. L., Paoli, T. L., Thornton, R. L., Burnham, R. D., and Smith, D. L. (1985). *Appl. Phys. Lett.* **46**, 118.
137. Waters, R. G., and Bertaska, R. K. (1988). *Appl. Phys. Lett.* **52**, 179.
138. Waters, R. G., York, P. K., Beernink, K. J., and Coleman, J. J. (1990). *J. Appl. Phys.* **67**, 1132.
139. Waters, R. G., Bour, D. P., Yellen, S. L., and Ruggieri, N. F. (1990). *IEEE Photon. Technol. Lett.* **2**, 531.
140. Fukuda, M. (1988). *IEEE J. Lightwave Technol.* **LT-6**, 1488.
141. Melman, P., Elman, B., Jagannath, C., Koteles, E. S., Silletti, A., and Dugger, D. (1989). *Appl. Phys. Lett.* **55**, 1436.
142. Tothill, J. N., Westbrook, L., Hatch, C. B., and Wilkie, J. H. (1989). *Electron. Lett.* **25**, 580.
143. Thijs, P. J. A., and VanDongen, T. (1989). *Electron. Lett.* **25**, 1737.

144. TEMKIN, H., TANBUN-EK, T., and LOGAN, R. A. (1990). *Appl. Phys. Lett.* **56**, 1210.
145. KOREN, U., ORON, M., YOUNG, M. G., MILLER, B. I., DEMIGUEL, J. L., RAYBON, G., and CHIEN, M. (1990). *Electron. Lett.* **26**, 465.
146. GERSHONI, D., and TEMKIN, H. (1989). *J. Luminescence* **44**, 381.
147. TANBUN-EK, T., LOGAN, R. A., OLSSON, N. A., TEMKIN, H., SERGENT, A. M., and WECHT, K. W. (1990). *Appl. Phys. Lett.* **57**, 224.
148. KATSUYAMA, T., YOSHIDA, I., SHINKAI, J., HASHIMOTO, J., and HAYASHI, H. (1990). *Electron. Lett.* **26**, 1375.
149. ACKLEY, D. E., and ENGELMANN, R. W. H. (1980). *Appl. Phys. Lett.* **37**, 866.
150. ACKLEY, D. E. (1982). *IEEE J. Quantum Electron.* **QE-18**, 1910.

# Chapter 9

# AlGaInP QUANTUM WELL LASERS

**David P. Bour**

*Xerox PARC*
*Palo Alto, California*






ISBN 0-12-781890-1

# 1. INTRODUCTION

In order to directly realize visible semiconductor lasers, several approaches are under investigation, including the high bandgap III-V and II-VI compounds. Among these, the greatest success has been achieved with the lattice-matched (to GaAs) AlGaInP materials system, with wavelength in the 600-nm band. Room temperature, cw operation was first accomplished in 1985. Presently, low power AlGaInP lasers are available commercially, and AlGaInP quantum well structures are being investigated.

In order to understand the operation of AlGaInP lasers, familiarity with the alloy's properties is essential. Therefore, we first describe the lattice, optical, electrical, and epitaxial growth characteristics of AlGaInP, highlighting comparisons with AlGaAs. Historically, the incorporation of this alloy into double heterojunction (DH) lasers preceded its use in quantum well lasers by about three years. A description of DH laser design and characteristics introduces some of the important issues that are also relevant to the more complex quantum-confined laser structures, including carrier and optical confinement and sensitivity to operating temperature.

Performance improvements, particularly lower threshold currents and shorter wavelengths, have been obtained with quantum well active regions. Furthermore, the characteristic temperature, maximum operating temperature, and power conversion efficiency are also increased in quantum-confined laser structures. These improvements are similar to those first demonstrated in the AlGaAs material system and can be explained well using the theories developed for AlGaAs quantum well lasers. Similarly, the more recent introduction of strained quantum wells has enhanced the performance even further. Compared with AlGaAs quantum well lasers, however, several of the alloy properties limit the performance of AlGaInP quantum well lasers. For instance, the conduction band offsets are relatively small, and it is difficult to P-dope alloys with high AlP-content. Thus, there are greater tradeoffs in designing structures that simultaneously achieve both strong electron and optical confinement. Likewise, the carrier effective masses are greater than in AlGaAs, increasing the transparency carrier density.

# 2. MATERIAL CHARACTERISTICS

## 2.1. Lattice Properties

Semiconductor heterojunction lasers are most often grown lattice-matched to binary compound substrates, in order to reduce the density of mismatch-related crystal defects. In the case of AlGaInP lasers, the composition is adjusted to lattice-match GaAs. The room temperature lattice parameter ($a_{300\,K}$) and thermal expansion coefficient ($\alpha$) of the binary phosphide compounds and GaAs are shown in Table I. Using Vegard's law, the composition of the ternary alloys GaInP and AlInP can be adjusted to lattice-match GaAs. The composition ($y$) dependence of the $Ga_yIn_{1-y}P$ lattice parameter is shown in Fig. 1, at both room temperature and at a typical growth temperature of 700°C [1]. The GaAs substrate's lattice parameter is also shown; it undergoes a greater thermal expansion than the GaInP alloy. Therefore, the lattice-matching condition is slightly different at room temperature ($y \sim 0.52$) than at the growth temperature ($y \sim 0.50$).

A GaInP layer that is lattice-matched at the growth temperature is under slight biaxial compression ($\Delta a/a \sim 0.1\%$) when cooled to room temperature; likewise, a layer that is lattice-matched at room temperature is under tension during growth. This tension, although slight, is sufficient to cause cracking (striations along $\langle 0\bar{1}1 \rangle$) of thick ($> 1\ \mu m$) layers, such as are required in laser structures [1, 2]. In contrast, such thick layers are capable of supporting compressive strains of similar magnitude. Therefore, the lattice-matching condition is usually specified at the growth temperature.

An equivalent situation exists for the $Al_yIn_{1-y}P$ ternary alloy. In fact, since the AlP and GaP lattice parameters are nearly equal, the lattice-matching compositions are essentially the same (AlP alloy content $y = 0.52$ at room

**Table I.**

Room Temperature Lattice Parameter ($a_{300K}$) and Thermal Expansion Coefficient ($\alpha$) of Relevant Binary Compounds

| Material | $a_{300K}$ (Å) | $\alpha$ ($10^{-6}$/K) |
|---|---|---|
| AlP | 5.4511[a] | 4.50[b] |
| GaP | 5.4512[a] | 5.91[c] |
| InP | 5.8686[a] | 4.75[c] |
| GaAs | 5.6533[a] | 6.86[d] |

[a] S. M. Sze, *Physics of semiconductor devices*, 2nd ed., John Wiley, New York, 1981.

[b] M. G. Grimmeis and B. Monemar, *Phys. Stat. Solidi* (*a*) **5**, 109 (1971).

[c] I. Kudman and R. J. Paff, *J. Appl. Phys.* **43**, 3760 (1972).

[d] E. D. Pierron, D. L. Parker, and J. B. McNeely, *Acta Crystallographica* **21**, 290 (1966).

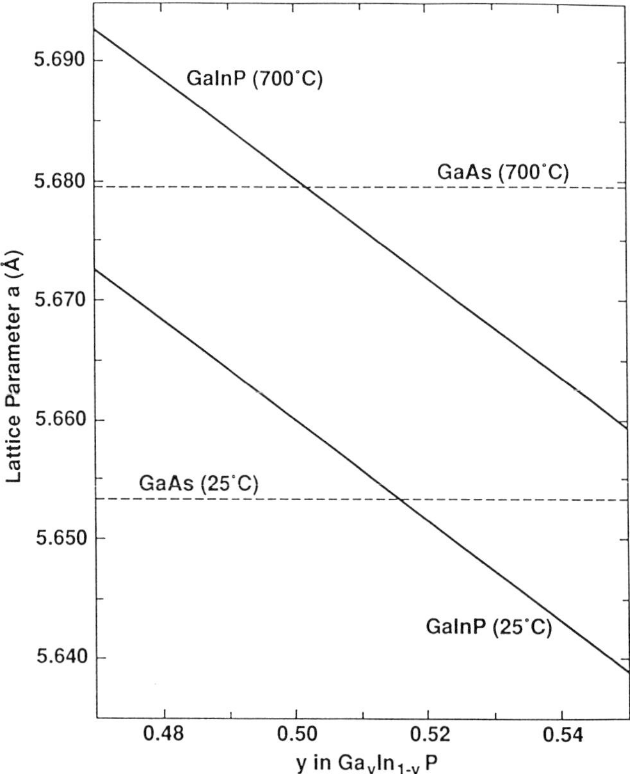


**Fig. 1.** Composition dependence of $Ga_yIn_{1-y}P$ lattice parameter at room temperature and at a typical growth temperature of 700°C. For lattice matching, the GaAs substrate's lattice parameter is also shown (dashed lines). From Ref. [1], with permission.

temperature and $y = 0.50$ at growth temperature). It is fortunate that these two lattice-matched ternaries each contain the same InP mole fraction of 0.50. Consequently, by starting with GaInP, the lattice-matched quaternary alloy $(Al_xGa_{1-x})_{0.5}In_{0.5}P$ is simply formed by exchanging Ga for Al in equal amounts. This procedure is fairly straightforward in gas-source crystal growth, where accurate gas flow control is accomplished, and graded composition profiles can be grown.

The lattice thermal resistivity ($W$) of AlGaInP is quite high. The large mass difference between Ga and In atoms creates a lattice anharmonicity, which increases the phonon scattering rate and thereby increases thermal resistivity [3]. For $Ga_{0.5}In_{0.5}P$, a value $W \sim 19$ cm °C/W is estimated, compared with $W < 10$ cm K/W for any AlGaAs composition [4]. Since the lattice anharmonicity is even greater in $Al_{0.5}In_{0.5}P$, the thermal resistivity of AlGaInP is also necessarily increased.

## 2.2. Band Structure

The bandgap energies of $(Al_xGa_{1-x})_{0.5}In_{0.5}P$ increase roughly linearly with composition $x$, as shown in Fig. 2 [5–7]. Anomalous behavior is observed in the direct (Γ) bandgap of $Ga_{0.5}In_{0.5}P$, due to atomic ordering of Ga and In on the group III sublattice. Extensive investigations of the ordered alloy have revealed that it consists of a monolayer GaP–InP superlattice on a $\{111\}$ plane, occurring naturally under certain growth conditions. Such an ordered alloy has a bandgap energy $E_\Gamma \sim 1.84$ eV, while the bandgap of the random alloy is $E_\Gamma - 1.91$ eV [8–11]. Intermediate energy values have

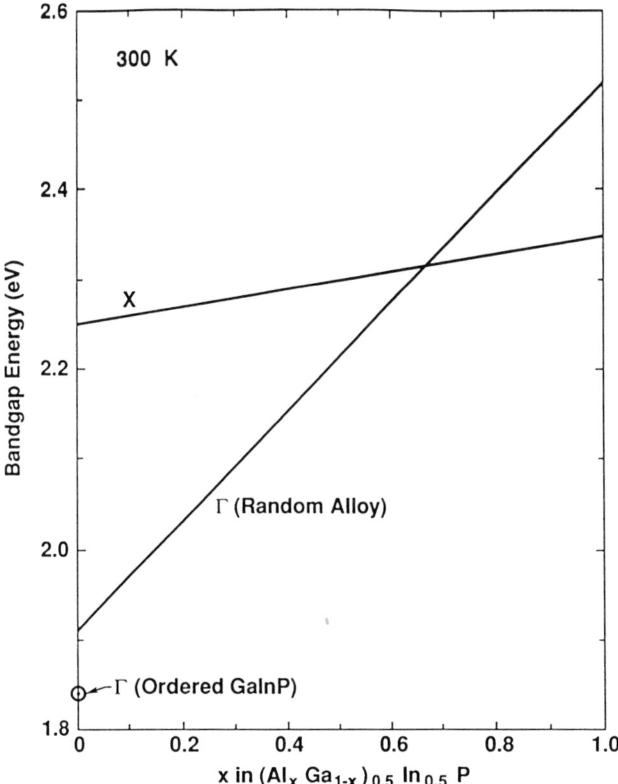


**Fig. 2.** Composition dependence of bandgap energies for $(Al_xGa_{1-x})_{0.5}In_{0.5}P$, lattice-matched to GaAs.

also been observed, depending upon the degree of ordering. Although ordering also occurs in AlInP, it does not influence its bandgap energy [5]. The composition dependence of the various bandgap energies in $(Al_xGa_{1-x})_{0.5}In_{0.5}P$ are summarized in Table II.

The indirect ($X$) bandgap energy increases slowly with composition, such that the direct–indirect crossover occurs near $x \sim 0.7$, for either ordered or random alloys. The split-off valence band energy ($\Delta_0$) is estimated to be

**Table II.**

Important Bandgap Energies in $(Al_xGa_{1-x})_{0.5}In_{0.5}P$

| Energy | GaInP | AlInP | $(Al_xGa_{1-x})_{0.5}In_{0.5}P$ |
|---|---|---|---|
| $E_\Gamma$ (eV) | 1.91[a–c] | 2.52[a–d] | $1.91 + 0.61x$[a] (random alloy) |
| | 1.84[a–c] | 2.52[a] | (ordered alloy) |
| $E_X$ (eV) | 2.25[e] | 2.35[f] | $2.25 + 0.10x$ |
| $\Delta_0$ (eV) | 0.1[g,h] | 0.1[h] | 0.1 |

[a] C. Nozaki, Y. Ohba, H. Sugawara, S. Yasuami, and T. Nakanisi, *J. Crystal Growth* **93**, 406 (1988).
[b] M. Kondow and S. Minagawa, *J. Appl. Phys.* **64**, 793 (1988).
[c] T. Suzuki, A. Gomyo, S. Iijima, K. Kobayashi, S. Kawata, I. Hino, and T. Yuasa, *Japan. J. Appl. Phys.* **27**, 2098 (1988).
[d] T. Hayakawa, K. Takahashi, K. Sasaki, M. Hosada, S. Yamamoto, and T. Hijikata, *Japan. J. Appl. Phys.* **27**, L968 (1988).
[e] A. Onton, M. R. Lorentz, and W. Reuter, *J. Appl. Phys.* **43**, 3420 (1971).
[f] A. Onton and R. J. Chicotka, *J. Appl. Phys.* **41**, 4205 (1970).
[g] C. Alibert, G. Bordure, A. Laugier, and J. Chevallier, *Phys. Rev. B* **6**, 1301 (1972).
[h] P. Lawaetz, *Phys. Rev. B* **4**, 3460 (1971).

**Table III.**

Carrier Effective Masses in $(Al_xGa_{1-x})_{0.5}In_{0.5}P$

| Effective mass | GaInP | AlInP | $(Al_xGa_{1-x})_{0.5}In_{0.5}P$ |
|---|---|---|---|
| $m_e^*/m_0$ | 0.11[a] | 0.35[b] | 0.11 ($x < 0.7$) |
| | | | 0.35 ($x > 0.7$) |
| $m_{hh}^*/m_0$ | 0.62[c] | 0.67[c] | $0.62 + 0.05x$[c] |
| $m_{lh}^*/m_0$ | 0.11[d] | 0.14[d] | $0.11 + 0.03x$[d] |

[a] C. Alibert, G. Bordure, A. Laugier, and J. Chevallier, *Phys. Rev. B* **6**, 1301 (1972).
[b] M. O. Watanabe, and Y. Ohba, *Appl. Phys. Lett.* **50**, 906 (1987).
[c] M. Honda, M. Ikeda, Y. Mori, K. Kaneko, and N. Watanabe, *Japan. J. Appl. Phys.* **34** , L187 (1985).
[d] Linear interpolation from binary properties in P. Laewaetz, *Phys. Rev. B* **4**, 3460 (1971).

approximately 0.1 eV for all $x$ in $(Al_xGa_{1-x})_{0.5}In_{0.5}P$ [12]. For these materials, with $E_\Gamma \gg \Delta_0$, both intervalence band absorption and Auger recombination [13] are considered relatively unimportant.

The carrier effective masses are fundamentally important in semiconductor lasers, as they determine the density of states; in a quantum well structure, the density of states in any subband is linearly proportional to the effective mass. The carrier effective masses are greater in AlGaInP than in AlGaAs. Generally, greater effective masses lead to higher lasing threshold currents, since the larger density of carrier states necessarily requires higher injected carrier densities in order to invert the population. Table III lists the electron, heavy hole, and light hole effective masses in AlGaInP.

## 2.3. Heterojunction Band Offsets

Photoluminescence studies of GaInP/AlGaInP quantum well heterostructures have shown that the direct bandgap energy difference ($\Delta E_g$) is split between the Γ conduction band offset $\Delta E_\Gamma = 0.43\ \Delta E_g$ and the valence band offset $\Delta E_v = 0.57\ \Delta E_g$ [14–16]. Moreover, the 43/57 band offset ratio is also consistent with other experiments, such as capacitance–voltage profiling across heterojunctions [17], that have measured the band alignments with

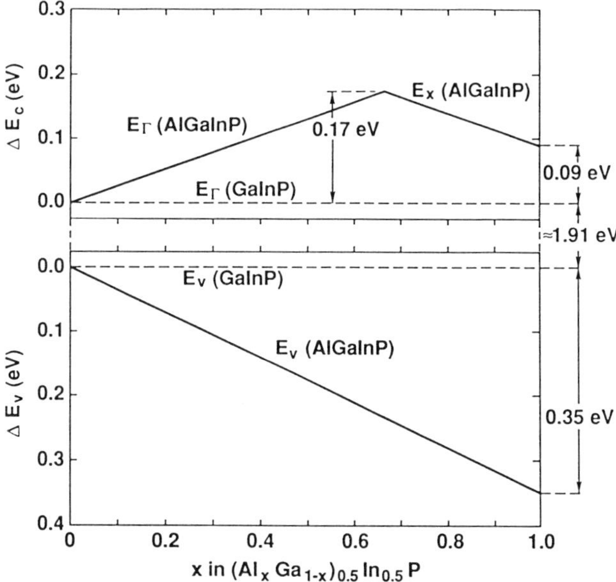


**Fig. 3.** Composition dependence of band offsets in $Ga_{0.5}In_{0.5}P$/$(Al_xGa_{1-x})_{0.5}In_{0.5}P$ heterojunctions.

respect to the GaAs band edges. These investigations of GaAs/AlGaInP interfaces are directed toward understanding selectively doped heterojunction transistor structures.

For a $Ga_{0.5}In_{0.5}P$ quantum well (random alloy; $E_{\Gamma} = 1.91$ eV) with $(Al_xGa_{1-x})_{0.5}In_{0.5}P$ confining layers, the conduction band offset ($\Delta E_c$) and valence band offset ($\Delta E_v$) are shown as a function of the confinement barrier composition ($x$) in Fig. 3. The maximum valence band confining potential is 0.35 eV, provided by $Al_{0.5}In_{0.5}P$. The greatest achievable electron confining potential is only $\Delta E_c \sim 0.17$ eV, however, when the barriers have alloy composition $x \sim 0.7$, for which the $\Gamma$ and $X$ bands are crossing over. The maximum electron confinement available in the GaInP/AlGaInP material system is therefore approximately half that available in the GaAs/AlGaAs system.

## 2.4. Refractive Indices

The refractive indices of $(Al_xGa_{1-x})_{0.5}In_{0.5}P$ have been described by Tanaka

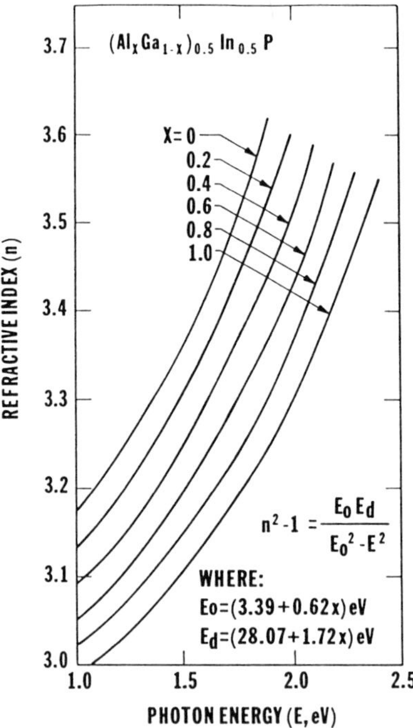


**Fig. 4.** Refractive indices of various $(Al_xGa_{1-x})_{0.5}In_{0.5}P$ alloys, as a function of photon energy. From Ref. [18], with permission.

*et al.* [18], and follow the relation

$$n^2 - 1 = \frac{E_0 E_d}{E_0^2 - E^2}, \tag{1}$$

where $n$ = refractive index, $E$ = photon energy (eV), $E_0 = 3.39 + 0.62x$ (eV), and $E_d = 28.07 + 1.72x$ (eV). The energy-dependent refractive index is shown for various alloy compositions in Fig. 4.

### 2.5. Electrical Properties

In order to realize $(Al_xGa_{1-x})_{0.5}In_{0.5}P$ diode lasers, n- and p-type doping of the high-bandgap cladding layers is required. For all $x$ values, controllable electron concentrations in the $10^{17}$–$10^{18}$ cm$^{-3}$ range have been accomplished using Se, Te, or Si donors. The Se-donor binding energy, shown as a function of alloy composition in Fig. 5 (solid curve) [19], displays behavior similar to n-doped AlGaAs. At $x \sim 0.3$, it begins to become a deep donor, likely due to the donor state's association with a higher lying conduction band minimum [20]. For $x \sim 0.4$, the donor binding energy reaches a maximum value of $\sim$95 meV, then drops slowly to 72 meV for $x = 1.0$. Donor-related deep levels, similar to the DX center in n-type AlGaAs, have also been examined for selenium-doped AlGaInP [21].

Most p-type doping has been accomplished using zinc, magnesium, or beryllium acceptors. High-aluminum alloys ($x \geq 0.8$) have presented serious difficulties, however, because the hole conductivity is often unacceptably low. This is due to the increasingly great acceptor binding energy [19], shown for zinc in Fig. 5 (dashed curve), along with the very low hole mobility in AlGaInP [22–24]. Together, the poor activation and solid-solubility limit of zinc [23, 25–28] seriously limits the maximum hole concentration that can be achieved in AlInP. Therefore, high conductivity AlInP has been very difficult to achieve, and many lasers contain $(Al_xGa_{1-x})_{0.5}In_{0.5}P$ cladding layers with $0.6 \leq x \leq 0.8$. By growing at lower temperatures to improve the zinc incorporation efficiency, adequate zinc-doping levels have been accomplished [29, 30]. Alternatively, beryllium [31–34] and magnesium [24, 35–38] acceptors have been used successfully to attain hole concentrations of $p > 4 \times 10^{17}$ cm$^{-3}$ in AlInP. In some cases, however, magnesium is not so well behaved, and the Mg-doping mechanism is complicated [23, 39, 40].

The electrical transport properties of AlGaInP alloys with various doping and composition are summarized in Table IV.

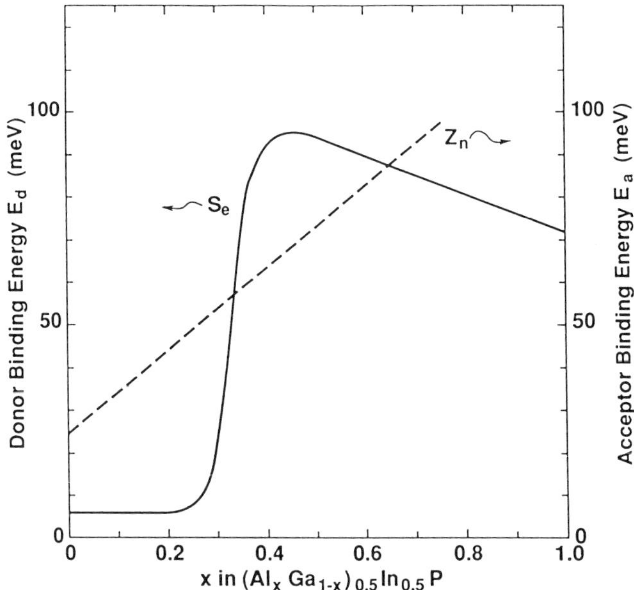


**Fig. 5.** Composition dependence of selenium donor binding energy (solid curve) and zinc acceptor binding energy (dashed curve) in $(Al_xGa_{1-x})_{0.5}In_{0.5}P$. From Ref. [19], with permission.

**Table IV.**

Summary of Transport Characteristics of $(Al_xGa_{1-x})_{0.5}In_{0.5}P$

| $x$ | Dopant | 300 K carrier concentration ($cm^{-3}$) | 300 K mobility ($cm^2$/V-sec) |
|---|---|---|---|
| 0 | undoped | $n \sim 2 \times 10^{15}$ | 3000[a] |
| 0 | Se | $n \sim 1 \times 10^{18}$ | 700[a] |
| 0 | Zn | $p \sim 1 \times 10^{18}$ | 40[a] |
| 0.5 | Se | $n \sim 1 \times 10^{18}$ | 600[b] |
| 0.5 | Zn | $p \sim 1 \times 10^{18}$ | 20[b] |
| 1 | Se | $n \sim 1 \times 10^{17}$ | 400[a] |
| 1 | Mg | $p \sim 1 \times 10^{18}$ | 7–9[c] |

[a] Y. Ohba, M. Ishikawa, H. Sugawara, M. Yamamoto, and T. Nakanisi, *J. Crystal Growth* **77**, 374 (1986).

[b] M. Ikeda and K. Kaneko, *J. Appl. Phys.* **66**, 5285 (1989).

[c] Y. Ohba, Y. Nishikawa, C. Nozaki, H. Sugawara, and T. Nakanisi, *J. Crystal Growth* **93**, 613 (1988).

## 2.6. Epitaxial Growth

The growth of device quality $(Al_xGa_{1-x})_{0.5}In_{0.5}P$ is a relatively recent development, due to difficulties with liquid phase epitaxial growth (LPE) and hydride vapor phase epitaxy (VPE). In the case of LPE, high quality GaInP films have been achieved [41], but aluminum segregation in the melt has prevented growth of AlGaInP [42]. Similarly, VPE growth has demonstrated high purity GaInP [43], yet the formation of highly corrosive aluminum chlorides is a practical limitation that discourages VPE growth of any aluminum-containing alloys. To date, all AlGaInP lasers have been prepared by organometallic vapor phase epitaxy (OMVPE) or molecular beam epitaxial (MBE) growth. Both these techniques are ideally suited for producing quantum well laser structures.

There are several significant difficulties associated with the growth of AlGaInP lasers. Foremost among these is lattice matching. Since typical laser structures contain approximately 3 $\mu$m of epitaxial material, even slight lattice mismatch can cause cracking or misfit dislocations. Thus, composition control is extremely critical. The films are especially sensitive to tensile strain (when composition is indium-deficient), which causes cracks (striations) along $\langle 0\bar{1}1 \rangle$ [1, 2]. In contrast, thick films can more easily support slight compressive strains ($\Delta a/a \leq 0.2\%$) without generating serious defects.

Like AlGaAs, AlGaInP quality is seriously degraded by even very low levels of oxygen-containing species ($O_2$ or $H_2O$) in the growth ambient. Thus, for any growth technique, the source purity and the system's leak integrity are critical. For example, OMVPE growth requires high phosphine ($PH_3$) flows, so $PH_3$ purity is a significant factor [39]. Likewise, increased luminescence efficiency of MBE-grown AlGaInP has been observed when the growth is performed in a sparse $H_2$ environment [16]. Photoluminescence (PL) spectroscopy is commonly used to assess the quality of epitaxial AlGaInP; the narrowest room temperature PL linewidths are 35–37 meV for GaInP [15, 16, 33, 41, 44–46].

Finally, in any III-V alloy containing both aluminum and indium, growth temperature is a critical parameter [47]. Due to the disparity in bonding strength between AlP (strong) and InP (weak), only a narrow temperature window (usually around 700°C for OMVPE, 500°C for MBE) is available for growth of AlGaInP [2, 22, 29, 48]. Although high temperatures are preferred for growth of AlP alloys, indium is thermally desorbed from the surface at elevated temperatures [45], so that composition control is difficult. Alternatively, low temperature growth is preferred for InP alloys, but the resultant short diffusion length of aluminum leads to three dimensional (hillocked) growth [22, 29]. The GaInP ternary is much more forgiving in this respect, with a wide temperature range available for growth of high quality material.

## 2.7. Ordering

Normally, in a ternary alloy such as GaInP, the group III atoms are expected to occupy their sublattice sites in a random arrangement. An ordered phase of $Ga_{0.5}In_{0.5}P$ occurs naturally under certain crystal growth conditions, however, and has been studied extensively [8–10, 49–52]. The ordered alloy consists of alternating monolayers of GaP and InP, stacked on the $\{111\}$ planes, as shown in Fig. 6. The ordering occurs along two of the four $\{111\}$ variants, $(\bar{1}11)$ and $(1\bar{1}1)$. This long range ordering has a periodicity twice that of the normal lattice parameter, and so it causes superstructure peaks in the electron diffraction pattern, halfway between the normal $\{111\}$ Bragg

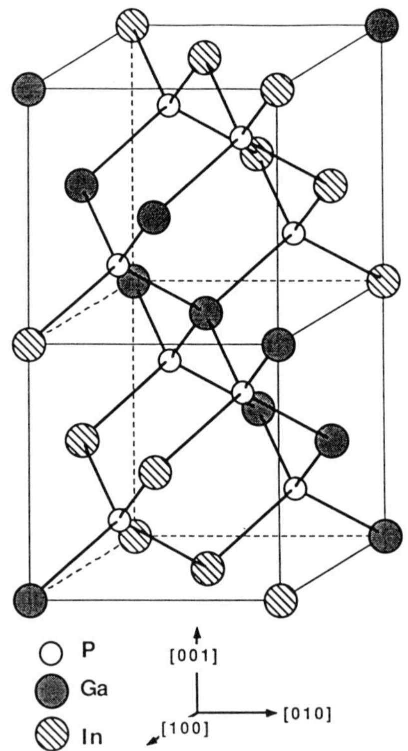


**Fig. 6.** Unit cell of ordered $Ga_{0.5}In_{0.5}P$ alloy.

reflections. In addition, the effects of ordering have also been observed in the Raman scattering [53, 54] and electroreflectance [51, 55] spectra of GaInP. Nearly all studies of ordered material have been performed on OMVPE-grown films. In this case, the growth temperature has a strong influence on ordering, while the group V overpressure has a weaker effect [8, 50, 54]. Atomic ordering on the group III sublattice also reduces the bandgap energy (see Fig. 2). A random $Ga_{0.5}In_{0.5}P$ alloy, such as is obtained in LPE growth, has a bandgap 1.91 eV, while an ordered alloy has a reduced bandgap of 1.84 eV [8, 50, 54].

The same ordered structure also appears in OMVPE-grown $Al_{0.5}In_{0.5}P$ [56], as well as in the $(Al_xGa_{1-x})_{0.5}In_{0.5}P$ quaternary [5, 11, 53, 57]. For these aluminum-containing alloys, ordering has a similar effect upon the bandgap [53]. For low AlP alloy content ($x \leq 0.5$), the random alloy still has a greater bandgap than the ordered alloy. Despite this, the two phases of AlInP ($x = 1$) have the same Γ bandgap [5].

Substrate orientation also influences ordering in AlGaInP. Those materials grown on (001) GaAs surfaces demonstrate strong ordering. Inclining the substrate surface toward (111) [58] or (110) [59] reduces the tendency to order. Indeed, the ordered structure is not observed for OMVPE growth on either (311)B, (111)B, or (110) substrates [60–62].

The ordered structure of AlGaInP has been intentionally disordered by several techniques. For example, heavy p-doping during growth [2, 8, 29, 63, 64], impurity diffusion following growth [65, 66], and (group III) vacancy-enhanced intermixing [67] have been used to randomize the atomic arrangements on the group III sublattice. Selective disordering is also useful for making nonabsorbing laser facet windows, to increase the COD maximum power limit [68, 69].

## 3. DOUBLE HETEROJUNCTION LASERS

### 3.1. Design Considerations

A typical AlGaInP double heterojunction (DH) laser is shown schematically in Fig. 7a. Similar layer structures are also used in high efficiency red, orange, and yellow LEDs [70–72]. Room temperature, pulsed operation of AlGaInP double heterostructure (DH) lasers was first achieved in 1983 [31, 73, 74]. Following this, continuous wave (cw) operation at 77 K was demonstrated in 1984 [37, 75]. Continuous operation at room temperature was first announced in 1985, at the Japanese research laboratories of NEC Corp.

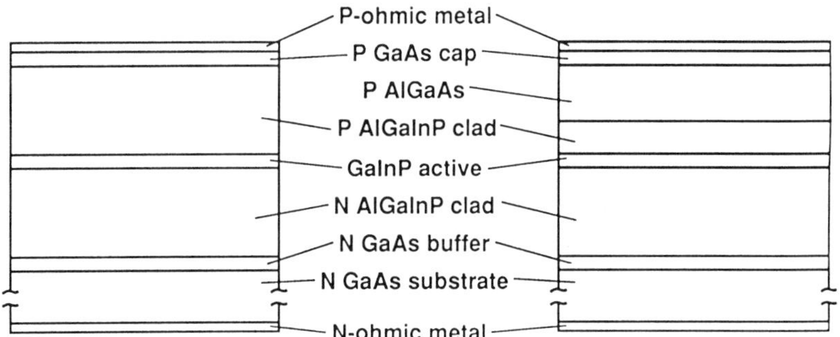


**Fig. 7.** $(Al_xGa_{1-x})_{0.5}In_{0.5}P$ double heterojunction (DH) laser structures: (a) conventional layer structure; and (b) with a dually stacked AlGaInP–AlGaAs p-cladding layer for greater thermal conductivity.

[76, 77], Sony Corp. [78, 79], and Toshiba Corp. [80]. Presently, cw operation at temperatures as high as 90–130°C are possible for AlGaInP DH lasers [39, 81–83].

AlGaInP DH lasers have higher threshold currents than AlGaAs DH lasers. Despite this, their differential quantum efficiencies are roughly comparable with those in AlGaAs lasers; external quantum efficiencies exceeding 50% have been demonstrated [2, 35, 84–86]. The higher thresholds arise from both the larger carrier effective masses in AlGaInP, since a greater density of states increases transparency current; and to the comparatively weak carrier confinement available in the AlGaInP materials system, which can lead to a greater rate of carrier leakage over the heterobarrier.

For DH lasers with AlGaInP cladding layers, reported good threshold current densities are 2–4 $kA/cm^2$ [2, 29, 32, 33, 38, 48, 76, 85, 87]. Using AlInP cladding layers, threshold current densities as low as 1.7 $kA/cm^2$ have been achieved in DH lasers [24, 35]. The record low $J_{th}$ for AlGaInP DH lasers is 860 $A/cm^2$, for a long cavity device [88]. It is not entirely clear whether $(Al_{0.7}Ga_{0.3})_{0.5}In_{0.5}P$ or AlInP is the best cladding layer material. While the $x = 0.7$ alloy provides the greatest electron confinement, AlInP is effective in increasing both the optical confinement factor and the hole confinement barrier. Consequently, AlInP cladding layers are expected to increase the electron leakage current, since the electron confining potential is reduced compared with $(Al_xGa_{1-x})_{0.5}In_{0.5}P$ $(0.5 \leq x)$ barriers (see Fig. 3), and their poor electrical conductivity contributes to a greater drift component of the electron leakage current [13, 89]. Nevertheless, reasonably low thresholds have been obtained with AlInP cladding layers.

The p-doping difficulties associated with AlInP (grown by OMVPE) have been overcome by either using Mg instead of Zn acceptors [24, 35, 37, 38] or growing the p-cladding layer at lower temperature in order to incorporate very high concentrations of Zn [29, 30]. For MBE growth, Be-doping yields AlInP of acceptably low resistivity for laser cladding layers [31–33]. Performance characteristics are improved with high p-cladding layer doping densities, due to a reduction in the electron leakage current, and lower series resistance [81, 82, 88, 90]. In the case of zinc-doped diodes, however, acceptor diffusion into the active region imposes an upper limit on the hole density in the p-cladding layer, resulting in an optimum concentration of p = 3–4 × $10^{17}$ $cm^{-3}$ when $x = 0.6$ [87].

The high threshold current densities along with the high thermal resistivity of AlGaInP make cw operation a challenge. In order to increase the maximum operating temperature, structures with thinner AlGaInP p-cladding layers have been investigated. For example, the structure shown in Fig. 7b contains a "dually stacked" AlGaInP/$Al_yGa_{1-y}As$ ($y \geq 0.7$) cladding layer. The lower thermal resistance of AlGaAs reduces the diode's overall thermal resistance by a factor of two, thereby raising the maximum operating temperature [78, 79]. Similarly, improvements have been noted in simpler structures like that in Fig. 7a, with thin ($\sim 0.5$ $\mu$m) p-cladding layers [81, 82].

Strong optical confinement can be obtained in AlGaInP lasers, due to the reasonably large refractive index differences between typical alloy compositions used to form the waveguide's cladding and optical cavity region (see Fig. 4), along with the short wavelength. At low drive current densities ($J$), the modal gain ($g$) versus current relationship for DH lasers is

$$g = \left(\frac{\Gamma}{h}\right) A(J - J_0), \tag{2}$$

where $\Gamma$ is the optical confinement factor, $h$ is the active layer thickness, $A$ is a gain coefficient, and $J_0/h$ is the normalized transparency current density [88, 91]. Thus, the optimum active layer thickness in GaInP/$(Al_xGa_{1-x})_{0.5}In_{0.5}P$ DH lasers is that for which the modal gain is greatest, and occurs when the quantity $\Gamma/h$ is maximized. The results of calculations for DH structure optimization are shown in Fig. 8 for two cladding layer compositions ($x = 0.6$ and 1), and $\lambda = 670$ nm. Optimum thickness with $(Al_{0.6}Ga_{0.4})_{0.5}In_{0.5}P$ cladding layers is $\sim 1200$ Å, whereas with $Al_{0.5}In_{0.5}P$ cladding layers it is 1000 Å. Experimental studies of GaInP/AlInP DH lasers with various active layer thicknesses follow these predictions well [35].

Since the carrier confinement available in the AlGaInP material system is somewhat less than in AlGaAs, achieving adequate carrier confinement is crucial to optimizing the performance of AlGaInP DH lasers. Figure 9 shows

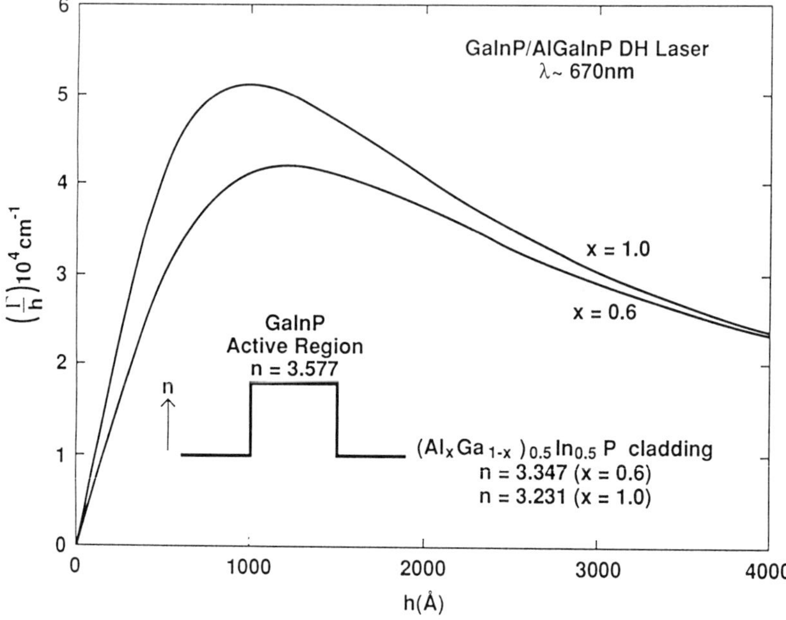


**Fig. 8.** Optical cavity thickness ($h$) dependence of $\Gamma/h$ for GaInP/$(Al_xGa_{1-x})_{0.5}In_{0.5}P$ DH lasers, for two cladding layer compositions; $x = 0.6$, 1.0.

the cladding layer composition ($x$) dependence of threshold current densities and characteristic temperatures ($T_0$) for GaInP/$(Al_xGa_{1-x})_{0.5}In_{0.5}P$ DH lasers [82]. Both parameters are optimized for $x \sim 0.7$, near the point where the $\Gamma$ and $X$ bandgaps cross over, and where carrier confinement is maximized (see Fig. 3).

A high characteristic temperature suggests strong carrier confinement, where thermal energy is insufficient to excite a large fraction of carriers over the confinement barrier. Moreover, a high $T_0$ also implies lack of nonradiative recombination, which is often a very temperature sensitive carrier loss [13, 92–94]. Values of $T_0$ for AlGaInP DH lasers usually fall in the range $T_0 = 80$–140 K, but values as high as 222 K have been reported in structures optimized for maximum carrier confinement, and with separate optical confinement layers [29]. Usually, however, the threshold currents of AlGaInP lasers do not follow a simple exponential increase with temperature. In many cases, the $T_0$ value is reduced at higher junction temperatures, due to the heterojunction leakage current, which starts to become significant near room temperature [88, 95]. Another mechanism, recombination in the

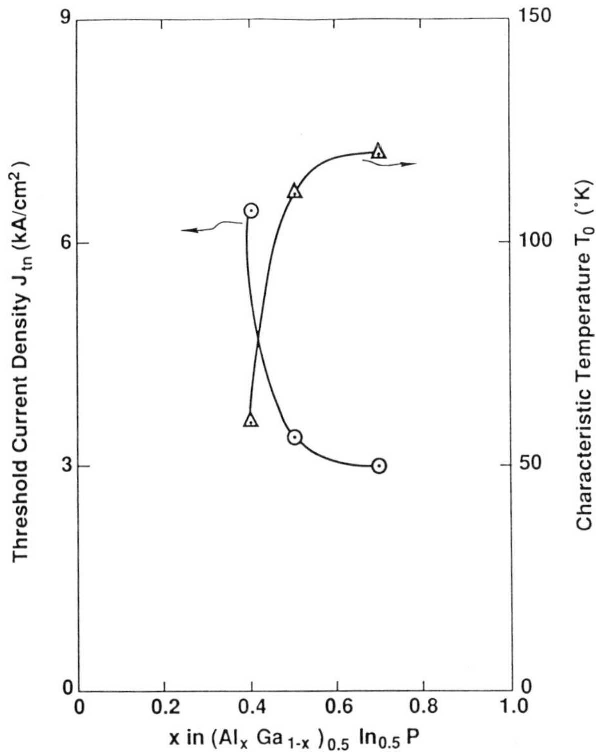


**Fig. 9.** Characteristic temperature ($T_0$, triangles) and threshold current density ($J_{th}$, circles) as a function of cladding layer composition ($x$) in GaInP/$(Al_xGa_{1-x})_{0.5}In_{0.5}P$ DH lasers (from Ref. [82], © 1989 IEEE).

depletion layers around the active region, may also contribute to greater temperature sensitivity [161].

Several experiments have also shown that two characteristic temperatures are required to describe AlGaInP lasers, much like InGaAsP lasers [13]. Generally, in the threshold versus temperature curve, a high $T_0$ describes the low temperature operation, while a lower $T_0$ describes the operation at higher temperatures; the break point temperature at which $T_0$ changes is typically $20 \leq T_b \leq 80°C$ [29, 78, 81, 82, 84, 88, 95]. This behavior can be attributed to the onset of electron heterobarrier leakage current, which increases rapidly with temperature, thus reducing $T_0$. Similarly, values of $T_0$ for AlGaInP lasers are often lower for cw than for pulsed operation. This occurs because the threshold current is high and the thermal conductivity is low, leading to a relatively large junction temperature rise during cw operation.

Heterojunction leakage current also reduces the internal quantum efficiency $\eta_i$. In cavity length studies performed at various temperatures, the internal quantum efficiency is independent of temperature up until some point, where it begins to fall roughly linearly with temperature [88]. This fall off begins near room temperature, and coincides with the break point temperature for the characteristic temperature $T_0$ and the onset of heterojunction electron leakage.

A distributed electron reflector has been proposed as a means of effectively increasing the electron barrier height. Such a "multiquantum barrier" (MQB) consists of a short-period AlGaInP superlattice with the layer thicknesses and compositions designed to reflect the electron wave [96, 97]. The MQB is situated at the active/cladding layer interface, thereby improving carrier confinement. A low threshold current density of 840 A/cm$^2$ and a high $T_0$ of 167 K have been demonstated for a 660-nm laser with MQB superlattice confining layers along with short-period superlattice confining layers [98]. Short-period superlattice confining layers are beneficial for increasing the carrier lifetime and lowering threshold [99, 100].

### 3.2. Short Wavelength Operation

Lasing wavelength of most OMVPE-grown DH lasers is 670–680 nm rather than 650 nm, due to the natural ordering of the GaInP active region, which lowers the bandgap. Nonetheless, devices with a random alloy active region, oscillating at ~650–660 nm, have been demonstrated. The random alloy can be selected by growing under the proper conditions [8]; alternatively, the ordered atomic arrangement can be randomized through the out-diffusion of the zinc p-dopant from the cladding layer during growth [2, 8, 29]. Growth on off-axis substrates (tilted at least 5° from (001), toward [110]) is a convenient and simple way to suppress ordering. Lasers grown on off-axis [101, 102], (311)B [62], and (111)B [103] substrates operate near 650 nm, since the ordering does not occur for growth on such surfaces.

The shortest wavelengths achieved with AlGaInP DH lasers are ~580 nm (yellow), for 77 K operation of diodes containing quaternary $(Al_{0.3}Ga_{0.7})_{0.5}In_{0.5}P$ active regions [37, 104]. Room temperature operation of $(Al_xGa_{1-x})_{0.5}In_{0.5}P$ injection lasers with quaternary active regions is difficult because of the limits imposed by poor carrier confinement, but, under pulsed conditions, this has been achieved for $x = 0.17$, with $\lambda = 626$ nm [105]. For cw operation near room temperature, lasing wavelengths in the 630-nm band have been established, and in the 620-nm band at slightly lower temperatures [30, 38, 90, 196–110].

Strained, lattice-mismatched active regions are also possible for DH lasers, as long as the active region is thin enough that misfit dislocations are not generated at the interfaces. Lasers with compressively strained ($Ga_yIn_{1-y}P$ with $y < 0.5$) active regions have improved carrier confinement for high temperature operation [111]. Conversely, those with tensile-strained ($Ga_yIn_{1-y}P$ with $y > 0.5$) active regions have shorter wavelength. Since the strain distorts the band structure, it also influences the gain spectra and the polarization of the emission [112].

### 3.3. Reliability

Reliability is a critical determinant of a diode laser's utility; and some of the lifetime issues in AlGaInP lasers are similar to those in AlGaAs lasers. For example, facet oxidation is expected to be a serious problem, due to the highly reactive nature of AlP-containing materials in air. Indeed, it has been shown that oxidation of the light-emitting region of the facets is a significant mode of gradual degradation in AlGaInP lasers, with activation energy 0.65 eV [113]. Facet coatings are effective in suppressing such degradation [82, 114]. Other important considerations are the higher drive current requirements and poor thermal conductivity of AlGaInP lasers, relative to AlGaAs lasers. This leads to excessive heating, which accelerates degradation of AlGaInP DH lasers [113].

At low power levels (several mW, for diodes aimed at optical recording applications), long lifetime has been established [49, 82, 113, 114]. The results of a contant-power lifetest of AlGaInP inner stripe lasers operating at 40°C are shown in Fig. 10 [82], a record of the current required to maintain a constant output of 3 mW. In addition to the facet corrosion, another slow degradation mechanism has been identified in these devices, one that is very sensitive to drive current. Lifetesting performed at elevated temperatures reveals an activation energy greater than 1.75 eV for this degradation mode [113]. In these experiments, the mean time to failure exceeds $10^4$ hours at 50°C; this extrapolates to a lifetime greater than $10^5$ hours at room temperature [109, 113]. Most lifetesting experiments have been performed on gain guided inner stripe lasers, since they were the first devices to become available

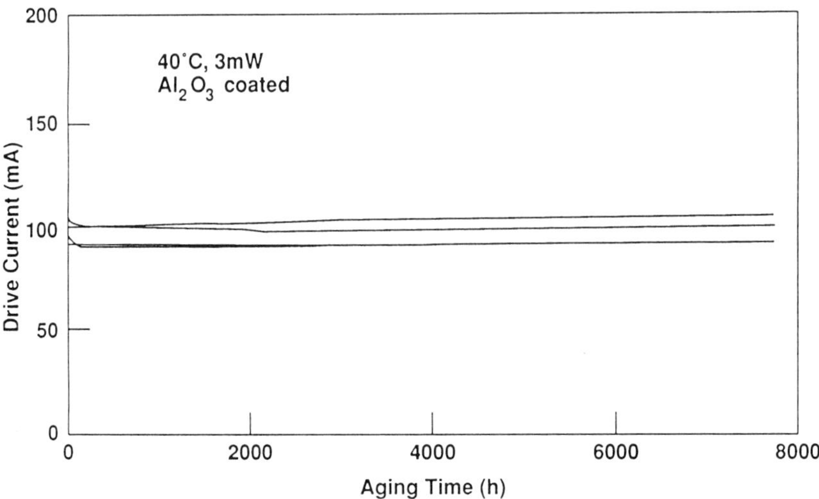


**Fig. 10.** Pictured is the result of a 40°C, 3-mW, constant-power lifetest characteristic for facet-coated, GaInP/$(Al_xGa_{1-x})_{0.5}In_{0.5}P$ DH, inner stripe laser (from Ref. [82], © 1989 IEEE).

commercially, and because they are simpler to fabricate than index guided structures.

The (uncoated) facet power level at which catastrophic optical damage (COD) occurs in AlGaInP DH lasers is $\sim 5$ MW/cm$^2$ [35, 109]. This limit is comparable with that of AlGaAs lasers [89]. An increase in the COD limit has been achieved with sulfur treatment of the facets, by immersion in $(NH_4)_2S_x$ [115]; and by selectively disordering the ordered-alloy GaInP active region in the vicinity of the facets, to increase the bandgap and thereby reduce the facet absorption [68, 69].

## 4. QUANTUM WELL LASERS

### 4.1. AlGaInP Quantum Wells

In the AlGaAs materials system, the development of quantum well (QW) laser structures has led to shorter wavelength emission and drastically reduced thresholds. Likewise, compared with DH lasers, lower thresholds are expected for AlGaInP QW lasers, because quantum confinement reduces the density of states in the active region and thereby lowers the number of injected carriers required to invert the population.

The energy band diagram for a $Ga_{0.5}In_{0.5}P$ (random alloy) single quantum well of thickness $L_z = 100$ Å, and with $(Al_{0.6}Ga_{0.4})_{0.5}In_{0.5}P$ barriers, is shown in Fig. 11. The conduction band offset is ~0.16 eV, while the valence band offset is ~0.21 eV. The energies ($E_n$) of the confined electron and hole states are calculated in the standard manner, using a simple envelope wavefunction approximation that incorporates the effective masses and heterojunction band offsets described in Table III and Fig. 3, respectively. Within each subband, the areal density of carrier states per unit energy is $\rho_n = m^*/\pi\hbar^2$.

Experimental studies of GaInP/AlGaInP quantum wells and superlattices reveal quantum confinement effects in good agreement with those predicted by the square well potential model [14, 116–118]. The measured quantum shifts of $Ga_{0.5}In_{0.5}P/(Al_{0.6}Ga_{0.4})_{0.5}In_{0.5}P$ single quantum wells are shown in Fig. 12a; they agree quite well with the predicted values, which are shown by the curve [14]. Likewise, room temperature absorption spectra of a (MBE-grown) 35-period $Ga_{0.5}In_{0.5}P$ (70 Å)/$(Al_{0.6}Ga_{0.4})_{0.5}In_{0.5}P$ (140 Å) superlattice (Fig. 12b) have shown the $n = 1$ and $n = 2$ excitonic resonances.

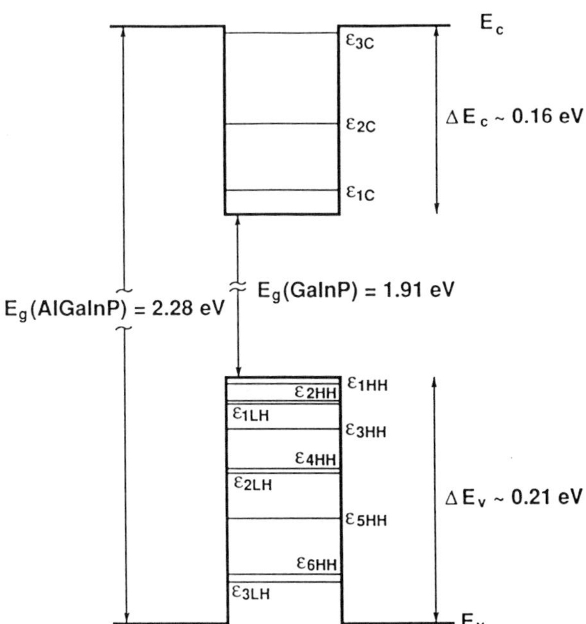


**Fig. 11.** Energy band diagram and quantum shifts of 100-Å GaInP/$(Al_{0.6}Ga_{0.4})_{0.5}In_{0.5}P$ quantum well.

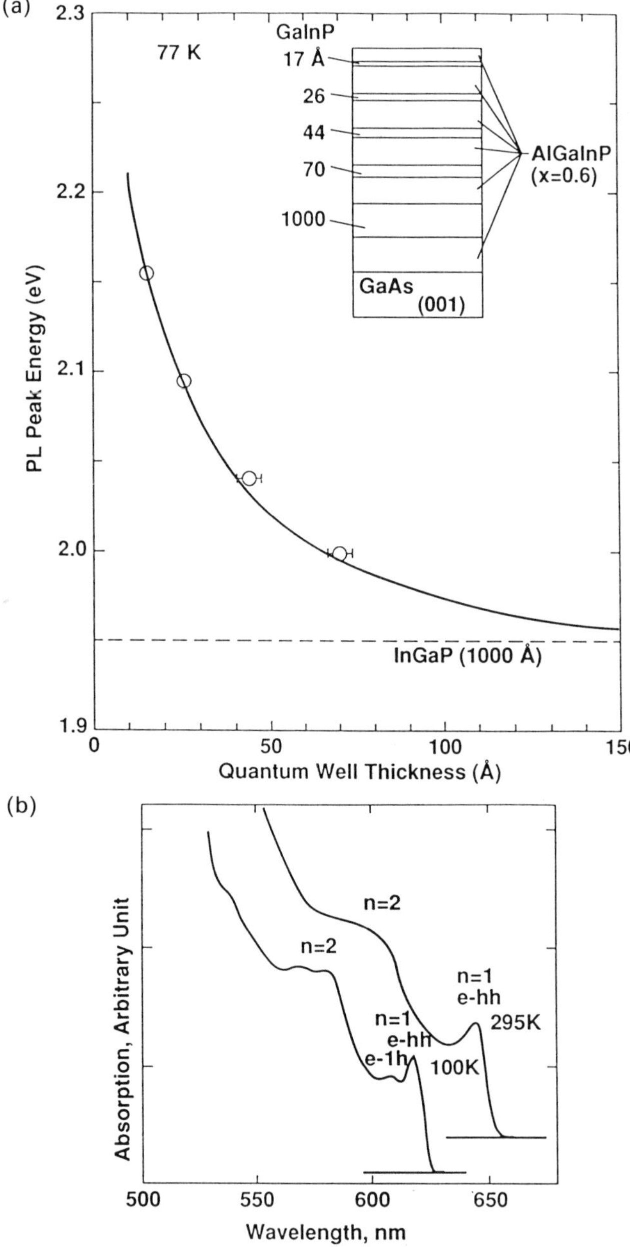


**Fig. 12.** (a) Experimental (open circles) and predicted (curve) quantum shifts for $GaInP/(Al_{0.6}Ga_{0.4})_{0.5}In_{0.5}P$ single quantum well, from Ref. [14]; and (b) room temperature and 100 K absorption spectra of GaInP (70 Å)/$(Al_{0.6}Ga_{0.4})_{0.5}P$ (140 Å) superlattice, showing excitonic resonances, from Ref. [15].

At lower temperature (100 K), the $n = 1$ electron to heavy hole and electron to light hole transitions were clearly distinguished at the band edge [15]. These transitions have also been observed in room temperature photoluminescence of strained GaInP/AlGaInP single QWs [119].

## 4.2. AlGaInP Quantum Well Lasers

Composition profiles of several representative $(Al_xGa_{1-x})_{0.5}In_{0.5}P$ QW laser structures are illustrated schematically in Fig. 13. Both single quantum well (SQW, Figs. 13a,b) and multiple quantum well (MQW, Fig. 13c) laser structures have been demonstrated successfully, grown by either MBE or OMVPE. Operation of an AlGaInP QW laser was first reported in 1986 [14]. Like in AlGaAs QW lasers, optical confinement is provided by a step index separate confinement heterostructure (SCH, Figs. 13a,c) or a graded index (GRIN) SCH (Fig. 13b).

For a given composition difference between the cladding layer and waveguide layer, the optimum SCH thickness is that which maximizes the quantum well optical confinement factor (Γ, defined as the spatial overlap between the QW gain and the optical intensity profiles) [120]. For example, the confinement factor's dependence upon the waveguide's half-thickness ($d/2$) is shown in Fig. 14, for both step index and GRIN waveguides. In each case, the cladding layers are $Al_{0.5}In_{0.5}P$ ($x = 1$), for maximum optical confinement; and the QW barrier is $(Al_{0.6}Ga_{0.4})_{0.5}In_{0.5}P$, for strong carrier confinement to the 100-Å $Ga_{0.5}In_{0.5}P$ SQW. Refractive indices are from Fig. 4, while the lasing wavelength is assumed to be 650 nm. An analogous

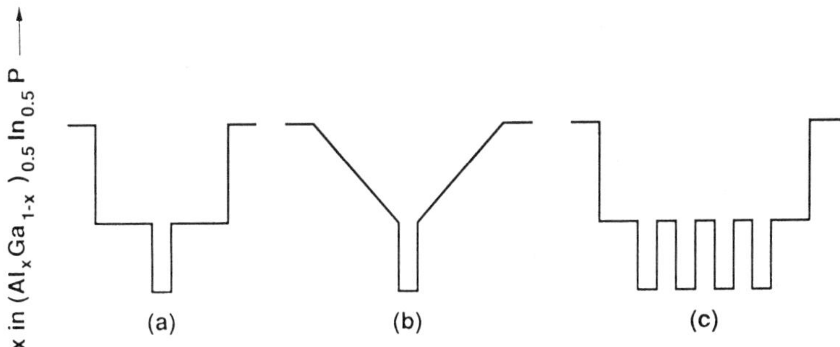


**Fig. 13.** Composition profiles of $(Al_xGa_{1-x})_{0.5}In_{0.5}P$ quantum well (QW) laser structures: (a) single QW with step index separate confinement heterostructure (SCH); (b) single QW with graded index (GRIN) SCH: and (c) multiple QW with step index SCH.

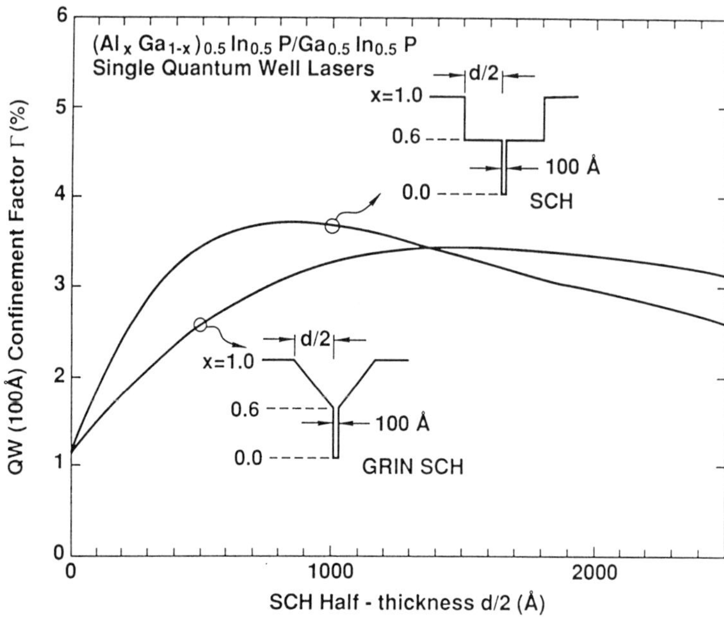


**Fig. 14.** Quantum well confinement factor (Γ) as a function of SCH half-thickness ($d/2$), comparing a graded index SCH with a step index SCH, each with a 100-Å $GaInP/(Al_{0.6}Ga_{0.4})_{0.5}In_{0.5}P$ QW ($\lambda = 650$ nm); the $(Al_xGa_{1-x})_{0.5}In_{0.5}P$ SCH is formed with $0.6 \leq x \leq 1.0$.

DH laser, only with a thicker, bulk-like active region and a step index SCH, resulted in the highest $T_0$ ($= 222$ K) for AlGaInP lasers [29]. Therefore, these QW structures are also expected to yield good performance.

The optimum SCH half-thickness is ~850 Å for the step index SCH and ~1500 Å for the GRIN SCH. These are the thicknesses that should be chosen to produce the greatest modal gain for any injection current, and therefore the lowest threshold current. The maximum Γ values that can be attained in AlGaInP QW lasers are roughly comparable with those in AlGaAs QW lasers. Although the magnitude of the refractive index difference is somewhat lower, it is offset by the shorter wavelength, which tends to produce a more tightly confined mode. Also, despite the slightly higher Γ that can be achieved with a step index SCH, the GRIN SCH contains a reduced density of states in the confining layers. This advantage can be important, especially in structures with weak QW carrier confinement [121].

Lower threshold currents are a major goal in QW lasers, and, as in the case of AlGaAs QW lasers, they have been achieved in the AlGaInP material

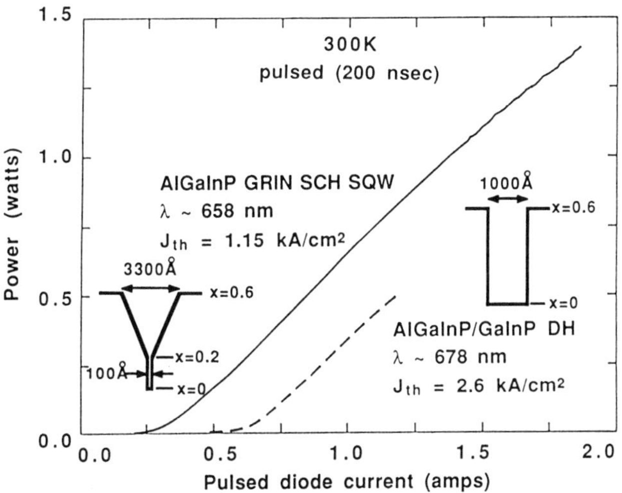


**Fig. 15.** Pulsed power versus current characteristics of GaInP/AlGaInP lasers, comparing DH and single QW, GRIN SCH devices, (from Ref. [122], © 1988 IEEE).

system by using a single quantum well active region, as shown in Fig. 15 [2, 85]. Here, the pulsed (200 ns pulse, 1 kHz rep. rate) power versus current characteristics of a GaInP/$(Al_xGa_{1-x})_{0.5}In_{0.5}P$ ($0.2 \leq x \leq 0.6$) GRIN SCH SQW laser are compared with a GaInP/$(Al_{0.6}Ga_{0.4})_{0.5}In_{0.5}P$ DH laser. The threshold current density of the SQW laser ($J_{th} = 1.15$ kA/cm$^2$) is less than half that of the DH laser ($J_{th} = 2.6$ kA/cm$^2$), thus demonstrating the primary benefit of a quantum-confined active region. Moreover, from the diode's current–voltage characteristic, the power conversion efficiency of the SQW laser exceeds 30% [122]. Thresholds as low as 250 A/cm$^2$ have been obtained for $Ga_{0.5}In_{0.5}P$/AlGaInP QW lasers, and even lower for strained QW devices [123]. Still, the reduction in threshold is not quite as dramatic as in AlGaAs SQW lasers, where threshold current densities around 0.2 kA/cm$^2$ are common, and have even approached 0.1 kA/cm$^2$ [124].

This $0.2 \leq x \leq 0.6$ GRIN SCH laser also serves to illustrate the difficulty of designing a structure that simultaneously provides both strong carrier and optical confinement. For example, although optimized structures like those in Fig. 13 can perform both these functions well, they are difficult to realize. Often the $(Al_xGa_{1-x})_{0.5}In_{0.5}P$ cladding layer composition is limited to $x \leq 0.8$, due to p-doping problems associated with alloys of high AlInP content. In many cases, quaternary $(Al_xGa_{1-x})_{0.5}In_{0.5}P$ cladding layers with

$0.6 \leq x \leq 0.8$ are used. Consequently, if the structure maintains maximum carrier confinement (barrier $x \sim 0.6$–0.7), the optical confinement becomes weak. Alternatively, strong optical confinement can be obtained with lower barrier composition $x$ at the expense of reduced carrier confinement.

This latter case describes the GRIN SCH laser in Fig. 15. The $(Al_xGa_{1-x})_{0.5}In_{0.5}P$ GRIN waveguide has $0.2 \leq x \leq 0.6$, providing strong optical confinement. Yet carrier confinement is weak with barriers of $x = 0.2$. In spite of its low threshold, this structure's poor carrier confinement is manifest in a low characteristic temperature [125] of $T_0 = 70$ K, compared with more typical values of $80 \leq T_0 \leq 140$ K for DH and MQW lasers with better electron confinement. In this structure, then, the reduced density of states in the GRIN SCH confining layers might provide an advantage over a step index SCH [121].

Although lattice-matched AlGaInP SQW lasers having thresholds approaching 0.2 kA/cm$^2$ have been demonstrated, MQW structures have not (yet) resulted in such an improvement. This is not surprising, since for a given QW thickness the transparency current density is roughly proportional to the number of QWs [126], and structures with 4–8 QWs are typical. The best threshold current densities of unstrained AlGaInP MQW lasers are ~3–5 kA/cm$^2$, roughly comparable with DH lasers [14, 16, 84, 127–130]. In a comparison of MQW and DH lasers with similar cladding layers, however, an increase in the characteristic temperature has been noted for MQW lasers [84]. MQW active regions have been highly effective in achieving shorter wavelengths [62, 99, 129, 130]. The incorporation of strained QWs has led to even lower thresholds for short wavelength lasers [131, 132].

Generally, the characteristic temperature of AlGaInP SQW and MQW lasers follows the same trend as do DH lasers. For lattice-matched QWs, Fig. 16 shows the characteristic temperature as a function of the confining barrier's composition ($x$ in $(Al_xGa_{1-x})_{0.5}In_{0.5}P$). There is a clear increase in $T_0$ as the barrier composition increases from $x = 0.2$ to 0.6; this reflects the increase in carrier confining potential (as shown in Fig. 3). Based on electron confinement considerations, the best values of $T_0$ are expected for barriers with $x = 0.6$–0.7. Indeed, the highest $T_0$ (222 K) has been obtained for a DH laser structure with $x = 0.6$ barriers and a separate optical confinement heterostructure [29]. For structures with higher barrier composition, there is only limited data, but these suggest that values of $T_0$ begin to decrease slightly for $x > 0.7$. Such behavior is expected on the basis of weaker electron confinement. Table V contains a summary of unstrained AlGaInP laser results.

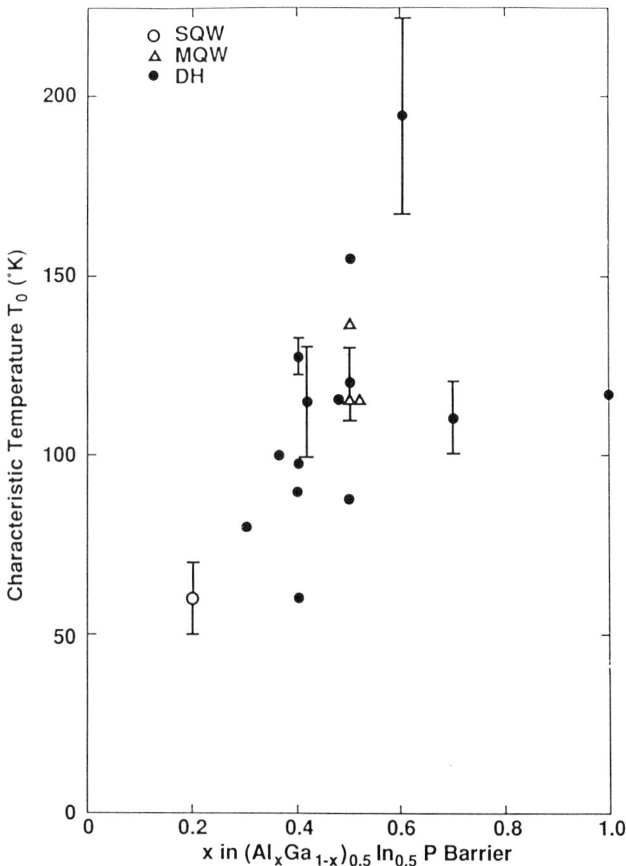


**Fig. 16.** Characteristic temperature ($T_0$) of GaInP/AlGaInP lasers, as a function of $(Al_xGa_{1-x})_{0.5}In_{0.5}P$ barrier composition ($x$), for single QW, multiple QW, and DH lasers.

## 4.3. Comparison with AlGaAs Quantum Well Lasers

In comparing the performance of AlGaInP and AlGaAs lasers, the most important feature is the higher threshold currents of AlGaInP lasers. These are caused by a number of factors; among the most important are the larger carrier effective masses in the higher bandgap alloy. Therefore, the density of states is greater, as must be the density of injected carriers required to reach transparency. Moreover, since the quantum energy shifts are inversely proportional to the carrier effective masses, the quantum shifts in AlGaInP QWs are reduced. Consequently, there is less separation (energy difference)

**Table V.**

Summary of Room Temperature, Unstrained $(Al_xGa_{1-x})_{0.5}In_{.5}P$ Quantum Well Lasers

| Reference | Structure | $\lambda$ | $J_{th}$ | $T_0$ |
|---|---|---|---|---|
| Tanaka *et al.* [14, 16] | 7QW, 100 Å | 658 nm | 7.6 kA/cm$^2$ | 115 K |
| Ikeda *et al.* [84] | 4QW, 100 Å. $0.5 \leq x \leq 0.7$ SCH | 668 | 3.5 kA/cm$^2$ | 138 K |
| Bout *et al.* [2, 48, 85, 125] | 1QW, 100 Å, $0.2 \leq x \leq 0.6$ GRIN SCH | 658 | 1.1 kA/cm$^2$ | 70 K |
| Kawata *et al.* [127] | 8QW, 35 Å | 646 | 4.4 kA/cm$^2$ | |
| Kuo *et al.* [133] | 1QW, 200 Å $x_{QW} = 0.22$ | 625 | photopumped | |
| | $x_{QW} = 0.38$ | 593 | photopumped | |
| Nam *et al.* [134] | 1QW, 200 Å, $x_{QW} = 0.20$ | 625 | | |
| Dallassasse *et al.* [128] | 4QW, 200 Å, $x_{QW} = 0.20$ | 640 | 3.7 kA/cm$^2$ | |
| Dallassasse *et al.* [135] | 3QW, 200 Å, $x_{QW} = 0.20$ | 625 | 5.2 kA/cm$^2$ | |
| Valster *et al.* [129, 130] | 8QW, 30 Å, $0.5 \leq x \leq 0.7$ SCH | 626 | 7 kA/cm$^2$ | 40 K |
| Kaneko *et al.* [99] | 20QW, 30 Å, all ternary | 601 | | |
| Valster *et al.* [62] | 8QW, 50 Å, $0.5 \leq x \leq 0.7$ SCH | 633 | 4.7 kA/cm$^2$ | 50 K |

between the individual subbands. Therefore, injected carriers are not as well confined (energetically) to the $n = 1$ subband, and a greater fraction of injected carriers are likely to occupy the higher lying subbands through thermal excitation. In this way, the larger carrier effective masses contribute to a further increase in the transparency carrier density. Significant thermal occupation of the higher subbands can also increase the temperature sensitivity of the lasing threshold current [94, 136].

In order to estimate the influence of AlGaInP's larger effective masses on threshold, Fig. 17 shows calculated peak modal gain ($g$) versus injected carrier sheet density ($\sigma$) for both AlGaAs and AlGaInP lasers. To provide the best comparison, each structure contains a 100-Å QW active region and a step index SCH whose thickness is optimized for maximum optical confinement factor. For the $Al_yGa_{1-y}As$ laser, the cladding and SCH layers have composition $y = 0.7$ and 0.3, respectively; whereas the corresponding compositions for the $(Al_xGa_{1-x})_{0.5}In_{0.5}P$ laser are $x = 1.0$ and 0.6. These calculations follow those prescribed by Chinn *et al.* [94]; for simplicity, the $L$ and $X$ band occupations are ignored (a good approximation for AlGaAs at low gain [137], but not necessarily for AlGaInP). The gain spectra are

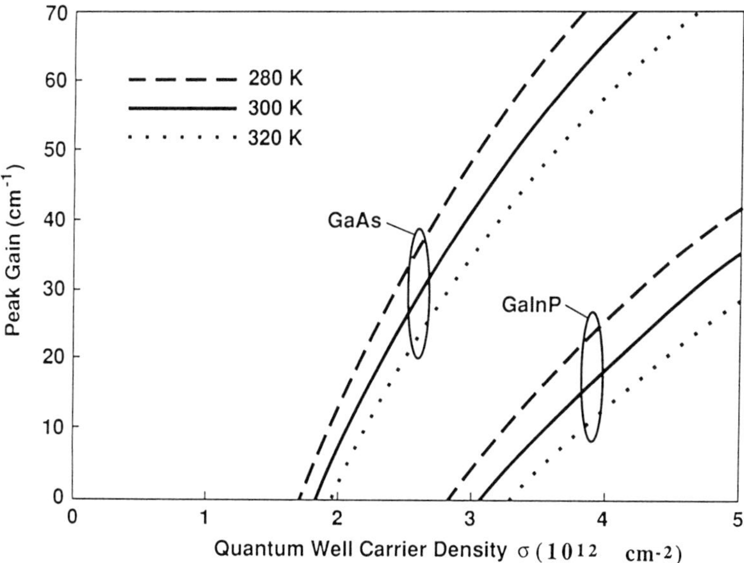


**Fig. 17.** Calculated gain ($g$) versus injected carrier sheet density ($\sigma$) for GaAs/AlGaAs and GaInP/AlGaInP single QW (100-Å) lasers, at temperatures of 280, 300, and 320 K.

broadened by convolution with a simple Lorentzian lineshape, characteristic of an intraband scattering time of $\tau = 0.1$ ps. Shown in Fig. 18 are the calculated 300 K gain spectra for the AlGaInP QW laser, at sheet carrier densities of 3, 4, and $5 \times 10^{12}$ cm$^{-2}$. Unbroadened gain spectra, with sharp features at the $n = 1$ e–hh and $n = 1$ e–lh band edges, are shown by the solid curves, while the broadened spectra are shown by the dashed curves. Both the relaxation broadening and band filling cause the gain peak to shift to higher energies at increasing injection levels. But these effects are offset by the bandgap renormalization, arising from the electron–hole plasma's coulombic interaction, which effectively decreases the active region's bandgap.

For both lasers, Fig. 17 shows that the gain increases sublinearly with carrier density. Such a gain characteristic is understood well for AlGaAs lasers, having been predicted and observed experimentally [94, 126, 137, 138]; it is roughly described by $g = g_0 \ln(\sigma/\sigma_0)$, where $g_0$ is a gain constant and $\sigma_0$ is the transparency density. At 300 K, and for typical threshold gains of 30 cm$^{-1}$, the threshold carrier density for an AlGaInP device is approximately 1.7–1.8 times that for an AlGaAs laser.

A more complete gain computation, accounting for leakage currents as well as for occupation of the indirect band minima and states associated

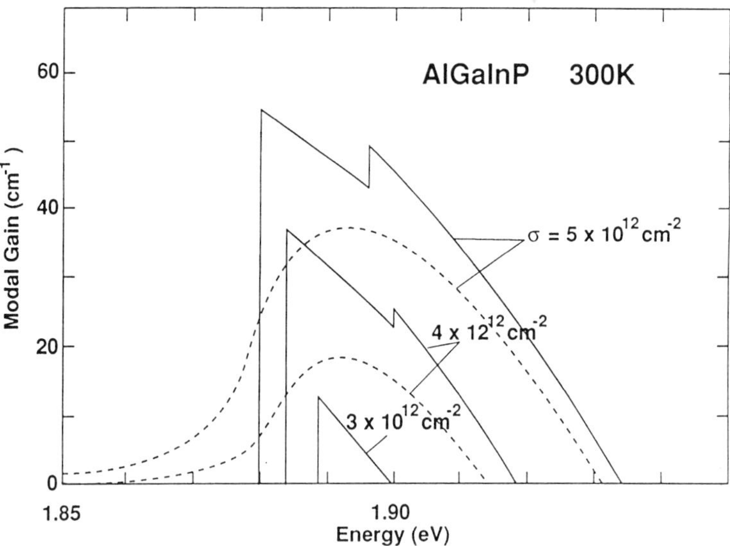


**Fig. 18.** Calculated 300 K gain spectra of 100-Å $Ga_{0.5}In_{0.5}P/(Al_xGa_{1-x})_{0.5}In_{0.5}P$ QW laser with step index SCH ($0.6 \leq x \leq 1.0$; 1700 Å thickness) at various injected carrier concentrations. Solid curves are unbroadened spectra, while dashed curves include relaxation broadening.

with the SCH, would likely result in an even greater difference in threshold carrier density and threshold current. With the strong carrier confinement afforded by the AlGaAs structure, these effects can be ignored except in the case of high threshold gain or high temperatures [94, 137]. On the other hand, the importance of these mechanisms is increased in AlGaInP, since the indirect band edges in GaInP do not lie so far above the Γ band edge, and also it is generally more difficult to achieve a large conduction band offset in GaInP/AlGaInP QW heterostructures.

Most important, a limited electron-confining potential increases the heterojunction electron leakage current in AlGaInP lasers, compared with AlGaAs lasers. Not only is the diffusive leakage current important, but the drift component of the electron leakage current can also be large; this because the poor electrical conductivity of the p-cladding layer implies a high voltage drop across the layer, when injection current is great. The associated electric field sweeps any electrons that leak over the confining heterobarrier toward the p-contact, contributing to the leakage current [13, 89]. In strongly confined AlGaAs QW lasers, electron leakage currents are important only under high drive current or temperature conditions [94, 137], while in AlGaInP lasers they can be a significant component of the diode current [29] even under less extreme conditions. In this case, the leakage current leads to lower $T_0$, reduced internal quantum efficiency, and higher threshold.

Carrier confinement effects dominate the temperature sensitivity of AlGaInP lasers, as shown in Fig. 16. This is an indication that Auger recombination is not significant in AlGaInP. In cases where Auger processes are important, they typically limit $T_0$ to lower values, because the Auger rate increases so rapidly with temperature [13, 94]. In fact, the dominant CHSH Auger recombination coefficient is expected to be negligibly small in AlGaInP, mainly because the fundamental bandgap is much greater than the spin–orbit splitting of the valence bands [139]. For GaAs QW lasers, Auger recombination becomes important at high operating temperatures, or in the short cavity regime, where threshold carrier densities are large. This leads to a stronger cavity length dependence of $T_0$ in GaAs QW lasers [92, 93, 137]. In contrast, no such Auger-related characteristics have been observed in AlGaInP SQW lasers [125].

### 4.4. 633-nm AlGaInP Quantum Well Lasers

In the wavelength range $600 \leq \lambda \leq 700$ nm, the eye's spectral sensitivity increases sharply with decreasing wavelength. There is, therefore, a strong incentive to fabricate AlGaInP lasers with wavelengths shorter than

650–680 nm, which are normally achieved. Among their numerous applications, 633-nm diode lasers are especially desirable, as a substitute for the He–Ne lasers that are already used in bar code scanners. Shorter wavelength lasers would further increase the eye safety of such systems, since lower power devices could be used while still maintaining visibility. In a manner that is analogous to that used in the AlGaAs material system, AlGaInP quantum well structures have also been used to attain shorter lasing wavelengths through either very narrow GaInP wells or higher bandgap, quaternary AlGaInP QWs. In addition, growth on misoriented substrates, in order to overcome the $\{111\}$ ordering, is also a technique for reaching shorter wavelengths.

From structures incorporating a quaternary quantum well active region, green emission has been obtained. Pulsed photopumping at 77 K of a structure with a 400-Å $(Al_{0.56}Ga_{0.44})_{0.5}In_{0.5}P$ active region, resulted in oscillation at wavelengths as short as 543 nm. Likewise, continuous 77 K photopumped operation at 553 nm was achieved in a similar structure, containing a 250-Å $(Al_{0.4}Ga_{0.6})_{0.5}In_{0.5}P$ active region [133]. Carrier confinement, thermal population of the indirect valleys, and leakage currents all impose a severe limitation on higher temperature operation of such structures. At room temperature, for example, the shortest wavelengths for photopumped operation of a structure containing an $(Al_xGa_{1-x})_{0.5}In_{0.5}P$ QW (with $x \sim 0.2$) are currently 593 nm (pulsed) and 625 nm (cw) [133, 134]. For p–n diode injection, 300 K lasing wavelengths as short as 625 nm (pulsed) [134] and 638 nm (cw) [128] have been demonstrated with similar structures. Impurity-induced intermixing of AlGaInP QW laser structures has also been shown effective in shifting emission to higher energies [140].

Short wavelength operation has also been demonstrated using structures with very narrow GaInP QWs. For practical, low threshold devices, this approach might be preferable over AlGaInP QWs, because the luminescence efficiency of alloys containing no AlP is greatest. Also, for lowest threshold, these structures usually contain a MQW active region, since the modal gain saturates at a low value for a narrow single well. For best performance in the 630-nm band, strained QW lasers are most promising.

At low temperature (109 K), 576-nm emission has been produced by an all-ternary laser structure containing twenty 30-Å GaInP QWs, separated by 20-Å AlInP barriers, grown by gas source MBE [99]. By modifying the structure to include short-period superlattice (7 Å/7 Å) GaInP/AlInP (instead of alloy) confining layers, the threshold was dramatically lowered, by a factor of $\sim 5$ at 125 K. This improvement led to near-room temperature, pulsed operation at $\lambda = 601$ nm. Likewise, similar improvements have been noted in AlGaAs QWs with short-period superlattice confining layers, due to enhanced carrier lifetimes [141].

The first cw, room temperature 633-nm diode laser was demonstrated in 1990, at Philips Laboratories (Netherlands) [129, 130]. An even shorter wavelength, 626 nm, was achieved for cw operation slightly below room temperature, 5°C, still within the range of thermoelectric cooling. As shown in Fig. 19, the structure consists of eight 30-Å GaInP QWs with 40-Å $(Al_{0.5}Ga_{0.5})_{0.5}In_{0.5}P$ barriers and step index SCH layers, and $(Al_{0.7}Ga_{0.3})_{0.5}In_{0.5}P$ cladding layers. A pulsed lasing spectrum at 20°C is shown inset, with $\lambda = 626$ nm. By increasing the QW thickness to ~50 Å, room temperature, cw operation at 633 nm was obtained. This achievement represents a significant advance in practical, short wavelength diode lasers.

A more recent advance is the use of either compressively strained ($y \leq 0.5$) and tensile-strained ($y \geq 0.5$) $Ga_yIn_{1-y}P$ QWs to achieve 630-nm band lasers

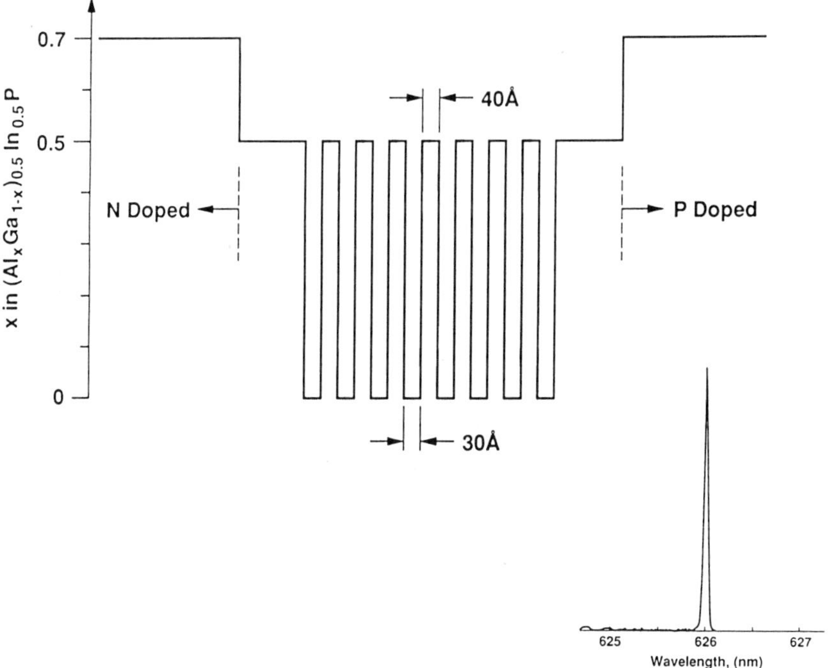


**Fig. 19.** Composition profile of short wavelength GaInP/AlGaInP MQW laser; 20°C pulsed spectrum is shown inset (from Ref. [129], © 1990 OSA). A slight increase in the QW thickness resulted in the first 633-nm cw, room temperature injection laser.

with low threshold. For devices with four 25-Å $Ga_{0.38}In_{0.62}P$ QWs, a wavelength of 634 nm and threshold current density of 1.7 kA/cm$^2$ have been demonstrated [131]. The lowest threshold for 630-nm band injection lasers is 1.2 kA/cm$^2$ for a structure with three 70-Å $Ga_{0.6}In_{0.4}P$ QWs (under tension), for $\lambda = 636$ nm. Overall, QW heterostructures, and especially strained QWs, have improved the short wavelength limits of the AlGaInP material system, bringing diode laser wavelengths further down into the visible range and also into the realm of He–Ne lasers.

### 4.5. Strained Quantum Well Lasers

For applications requiring high power, cw lasers, such as optical pumping of chromium-doped solid state laser materials [142], lower threshold current is a tremendous advantage. One possibility for reducing the threshold current of AlGaInP QW lasers is to use compressively strained $Ga_yIn_{1-y}P$ ($y < 0.5$) QWs. Although this alloy is no longer lattice-matched, if the QW thickness is less than some critical value, the biaxial compression will not cause interfacial misfit dislocations; but rather an elastic, tetragonal deformation of the GaInP unit cell. In the plane of the QW, the lattice parameter is compressed to match the lattice of GaAs and $(Al_xGa_{1-x})_{0.5}In_{0.5}P$, while in the quantization direction the unit cell stretches through Poisson's ratio. Overall, the unit cell volume is reduced, and the deformation has a pronounced effect upon the band structure. In addition to increasing the bandgap, the valence band structure is changed, so that the heavy- and light-hole bands are split at $k = 0$, and, most importantly, the (in-plane) heavy-hole effective mass is lower. The resultant decrease in the valence band density of states reduces the transparency density and increases the differential gain ($dg/dN$) [143–145; Chapters 1, 7, and 8, this book]. Both these benefits have been observed in strained InGaAs/AlGaAs lasers [146], and they are also applicable to GaInP/AlGaInP lasers. A simpler, but more radical approach could be to grow a strained GaInP/AlGaP QW structure on a (transparent) GaP substrate, in a phosphide analog of the high performance InGaAs/AlGaAs structures [48].

An In-rich alloy has a lower bandgap than the lattice-matched alloy. Although the compressive strain increases the bandgap somewhat, the net effect is still a lower bandgap QW and therefore a longer wavelength laser. For lasers with a single 100-Å $Ga_{0.43}In_{0.57}P$ QW, lasing wavelength is 690–705 nm, with threshold current density as low as 191 A/cm$^2$ [123, 147]. This represents as much as a two- to threefold decrease with respect to unstrained QW lasers [148]. With thinner wells (70 Å instead of 100 Å), the wavelength is reduced to 665 nm, while the threshold current remains very low [149, 150].

Obtaining short wavelengths from compressively strained devices requires very thin wells. For instance, with four 25-Å $G_{0.38}In_{0.62}P$ QWs, grown on a 6° off-axis substrate, 610 mW pulsed power at 634 nm was achieved, with a threshold current density of 1.7 kA/cm$^2$ [131]. An alternative means of achieving 630-nm band lasers is with tensile-strained QWs. In this case, a Ga-rich alloy QW has higher bandgap energy, so that such thin wells are not needed. For example, a device with three 70-Å $Ga_{0.6}In_{0.4}P$ QWs had a wavelength $\lambda = 636$ nm and threshold current density $J_{th} = 1.2$ kA/cm$^2$. To date, this is the lowest threshold for 630-nm band injection lasers [132].

In a QW under biaxial tensile strain, the light hole is involved in the fundamental transition, so that the polarization is transverse magnetic (TM). Unlike compressive strain, however, tensile strain does not lead to lower threshold [123]. This could be due to the strain influencing the band structure, the reduced carrier confinement resulting from a higher bandgap active region, or the reduced splitting between the $\Gamma$ and indirect valley in $Ga_yIn_{1-y}P$ with $y > 0.5$. In a comparison of strained and unstrained QW lasers, compressively strained QW lasers had the lowest threshold (191 A/cm$^2$) and highest $T_0$ (110 K). Conversely, the lasers with QWs under tensile strain had high $J_{th}$ (430 A/cm$^2$) and low $T_0$ (50 K) characteristics, that were attributed to their reduced barrier height [123]. Table VI summarizes strained $Ga_yIn_{1-y}P/(Al_xGa_{1-x})_{0.5}In_{0.5}P$ QW laser results.

## 5. STRUCTURES

### 5.1. Gain and Index Guides

The two structures that are most often used for AlGaInP lasers are illustrated in Fig. 20. They are the selectively buried ridge (SBR) waveguide laser (Fig. 20a, also called a transverse-mode stabilized laser) [24, 30, 35, 127, 151], and the inner stripe (IS) laser (Fig. 20b) [29, 76, 77, 80, 82, 95, 105]. Other promising techniques for fabricating index guided AlGaInP lasers are impurity-induced disordering [135], impurity-free vacancy diffusion [67], ion implantation [115, 152], and ridge waveguides [153, 154]. Both the SBR and IS structures have been used to produce diodes with good beam quality at low power, for incorporation in near-diffraction-limited optical systems such as optical data reading heads. Unfortunately, although growth on patterned substrates is a highly effective technique for fabricating index guided AlGaAs

**Table VI.**

Summary of Room Temperature, $Ga_yIn_{1-y}P/(Al_xGa_{1-x})_{0.5}In_{0.5}P$ Strained QW Lasers

| Reference | Structure | $y_{QW}$ | $\lambda$ | $J_{th}$ | $T_0$ | Other |
|---|---|---|---|---|---|---|
| Katsuyama *et al.* [147] | 1QW, 100 Å, $0.2 \le x \le 0.7$ SCH | 0.43 | 691 nm | 0.9 $kA/cm^2$ | 76 K | |
| Change-Hasnain *et al.* [131] | 4QW, 25 Å, $0.5 \le x \le 0.7$ SCH | 0.38 | 634 nm | 1.7 $kA/cm^2$ | | 610 mW |
| Hashimoto *et al.* [148] | 1QW, 100 Å, $0.2 \le x \le 0.7$ SCH | 0.51 | 674 nm | 0.55 $kA/cm^2$ | | unstrained |
| | | 0.43 | 705 nm | 0.22 $kA/cm^2$ | | |
| Katsuyama *et al.* [123] | 1QW, 100 Å, $0.4 \le x \le 0.7$ SCH | 0.61 | 652 nm | 0.43 $kA/cm^2$ | 50 | tension, TM |
| | | 0.51 | 673 nm | 0.25 $kA/cm^2$ | | unstrained |
| | | 0.43 | 697 nm | 0.19 $kA/cm^2$ | 110 | |
| Serreze *et al.* [149] | 1QW, 70 Å, GRIN SCH | 0.43 | 665 nm | 0.43 $kA/cm^2$ | 78 | 680 mW cw |
| Serreze *et al.* [150] | 1QW, 70 Å, GRIN SCH | 0.43 | 665 nm | 0.38 $kA/cm^2$ | 95 | 47 mW cw |
| Welch *et al.* [132] | 3QW, 70 Å, $0.3 \le x \le 0.6$ SCH | 0.60 | 636 nm | 1.2 $kA/cm^2$ | | tension, TM |

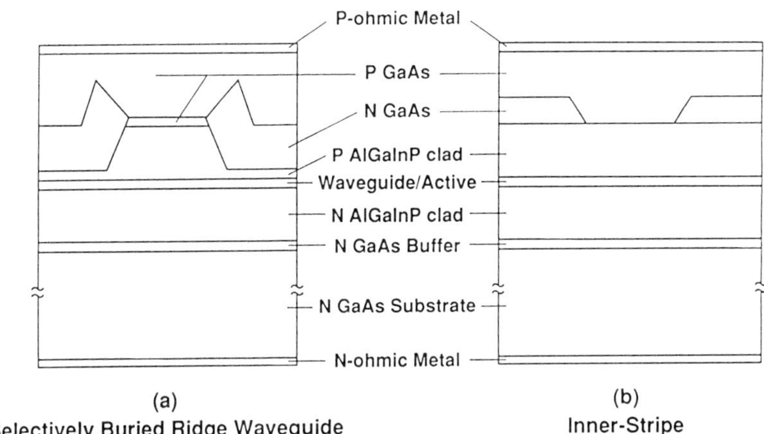


**Fig. 20.** The two most often used $(Al_xGa_{1-x})_{0.5}In_{0.5}P$ laser structures: (a) selectively buried ridge (SBR) waveguide laser; and (b) inner stripe (IS) laser.

lasers (for both CSP and similar structures) [155], it is not feasible in the AlGaInP material system. In particular, the simultaneous growth of lattice-matched films on all the exposed crystal planes creates serious difficulties, as yet unsolved. In consequence, AlGaInP lasers are grown on planar substrates.

The inner stripe (IS, Fig. 20b) laser is gain guided, and so its output beam is somewhat astigmatic. In the IS structure, the N–GaAs serves as a current blocking layer, defining a stripe of width $w \sim 7\ \mu m$. Current spreading is not a very great problem in the p-cladding layer, because of the poor hole-transport properties of AlGaInP [81, 82, 156]. Despite its lack of built-in mode stability compared with the SBR laser, the IS laser is far simpler to fabricate, requiring only one, nonselective regrowth. This structure results in adequate beam quality at low powers, and so it was the first AlGaInP laser available commercially, and it has also been the subject of most realibility experiments.

Conversely, the SBR structure offers superior beam quality, by virtue of its real index guiding and strong mode discrimination [157]. Functionally, it is equivalent to an (inverted) channeled-substrate-planar (CSP) laser, commonly used for AlGaAs lasers [155]. Fabrication involves three growth steps. First, the layers through the N-cladding, active region, P-cladding, and a P–GaAs cap are grown. Alternatively, terminating the first growth with a GaInP "barrier-reduction layer" has been shown to improve the diode's forward current–voltage characteristic [86, 158]. Following the first growth, a narrow stripe (typically 5 $\mu$m) of $SiO_2$ is patterned in the $\langle 01\bar{1} \rangle$ (V-groove) direction, and the ridge waveguide is chemically etched. The $SiO_2$ serves as a mask for both the etching and for the selective regrowth of N–GaAs outside the ridge. Finally, the $SiO_2$ mask is removed and P–GaAs is grown over the entire wafer. The ridge etching step is critical, because the etch depth should be controlled to within ~2000 Å of the waveguide layer. To facilitate this process, structures with etch-stop layers have been developed. Effective etch stops are thin GaInP layers (thin enough that the absorption edge is quantum shifted above the lasing photon energy) [151]; thin (Al)GaAs layers [159]; or $(Al_xGa_{1-x})_{0.5}In_{0.5}P$ layers with low $x$ $(0.1 \leq x \leq 0.2)$ [38, 127].

The N–GaAs burying layer serves several purposes. In addition to acting as a current-blocking layer and as a thermal conductor, it also has a high refractive index relative to the cladding layer. Thus, the field radiates power into the GaAs, and the wavefront of the transverse mode outside the ridge has an upward tilt. This amounts to a lower transverse effective index and provides real lateral index guiding, analogous to that obtained in AlGaAs CSP lasers [155]. Consequently, the beam is far less astigmatic compared with gain guided devices [109, 157]. The N–GaAs is also optically absorbing at the lasing wavelength, similarly favoring oscillation in the fundamental lateral mode, by discriminating againt all the lossy, higher order modes. Strong mode discrimination results in stable, single spatial mode operation, even at high powers. Using SBR lasers with untreated facets, stable, single spatial mode operation has been obtained at powers exceeding 25 mW cw and 100 mW pulsed [35, 109, 151, 158]. For extremely high power, and where a single mode is no longer required, outputs exceeding 300 mW are possible from broad area (40 $\mu$m wide) SBR lasers [90, 107, 158, 160].

Although most demonstrations have used double heterostructure active regions, the performance of SBR lasers containing a quantum well active region is roughly comparable [127, 147]. In addition, the SBR structure exhibits self-pulsations in its output [156]. Such a characteristic is desirable in many optical systems, because low intensity noise can be obtained in spite of any optical feedbacks introduced by the system elements.

The power output of AlGaInP SBR lasers is often limited by catastrophic optical damage (COD), resulting from thermal runaway caused by light absorption at the facet. For untreated facets, the COD power density limit may be as high as 5 $MW/cm^2$, but is typically 1–2 $MW/cm^2$. Several facet treatment techniques have been developed to increase this limit, including sulfur treatment [115], and selective disordering of the GaInP ordered alloy near the facets [68, 69]. The disordering is accomplished by zinc-diffusing the region near the facets, forming a window structure and leaving the as-grown, ordered alloy elsewhere. Since the random alloy has a higher bandgap than the ordered alloy, its absorption edge is shifted above the laser photon energy, so the facet becomes nonabsorbing. Consequently, a fivefold increase in the maximum power of 680-nm SBR lasers has been reported for nonabsorbing facets. Operation at 75 mW cw and 80 mW pulsed has been obtained from a 5 $\mu$m wide SBR, corresponding to a COD limit greater than 10 MW/cm$^2$ [68, 69]. For shorter wavelength diodes, where the active region is already grown as a random alloy, another technique is required for realizing a window structure. By selectively growing GaAs : N current blocking layers within 20–30 $\mu$m of the facets, the COD limit has been increased by a factor of at least five for a 635-nm DH SBR laser [110].

## REFERENCES


1. Minagawa, S., and Kakibayashi, H. (1985). Observations of OMVPE-Grown GaInP/GaAs cross-sections by transmission electron microscopy, *Japan. J. Appl. Phys.* **24**, 1569.
2. Bour, D. P., and Shealy, J. R. (1988). Organometallic vapor phase epitaxial growth of AlGaInP and its heterostructures, *IEEE J. Quantum Electron.* **QE-24**, 1856.
3. Maycock, P. D. (1967). *Solid State Electron.* **10**, 161.
4. Adachi, S. (1983). *J. Appl. Phys.* **54**, 1844.
5. Nozaki, C., Ohba, Y., Sugawara, H., Yasuami, S., and Nakanisi, T. (1988). *J. Crystal Growth* **93**, 406.
6. Onton, A., and Chicotka, R. J. (1970). *J. Appl. Phys.* **41**, 4205.
7. Onton, A., Lorentz, M. R., and Reuter, W. (1971). *J. Appl. Phys.* **42**, 3420.
8. Gomyo, A., Suzuki, T., Kobayashi, K., Kawata, S., Hino, I., and Yuasa, T. (1987). *Appl. Phys. Lett.* **50**, 673.
9. Bellon, P., Chevallier, J. P., Martin, G. P., Dupont-Niven, E., Thiebaut, C., and Andre, J. P. (1988). *Appl. Phys. Lett.* **52**, 567.
10. McKernan, S., Carter, C. B., Bour, D. P., and Shealy, J. R. (1988). *J. Mater. Res.* **3**, 406.
11. Yasuami, S., Nozaki, C., and Ohba, Y. (1988). *Appl. Phys. Lett.* **52**, 2031.
12. Lawaetz, P. (1971). *Phys. Rev. B* **4**, 3460.
13. Agrawal, G. P., and Dutta, N. K. (1986). *Long wavelength semiconductor lasers.* Van Nostrand Reinhold Co., New York.
14. Tanaka, H., Kawamura, Y., and Asahi, H. (1986). Room-temperature operation of MBE-grown InGaP/InGaAlP MQW visible laser diodes, *Electron. Lett.* **22**, 707.

15. Tanaka, H., Kawamura, Y., and Asahi, H. (1987). Observations of room-temperature excitons in InGaP/InGaAlP MQW structures, *Electron. Lett.* **23**, 166.
16. Tanaka, H., Kawamura, Y., Nojima, S., Wakita, K., and Asahi, H. (1987). *J. Appl. Phys.* **61**, 1713.
17. Watanabe, M., and Ohba, Y. (1987). *Appl. Phys. Lett.* **50**, 906.
18. Tanaka, H., Kawamura, Y., and Asahi, H. (1986). Refractive indices of InGaAlP lattice matched to GaAs, *J. Appl. Phys.* **59**, 985.
19. Honda, M., Ikeda, M., Mori, Y., Kaneko, K., and Watanabe, N. (1985). The energy levels of Zn and Se in AlGaInP, *Japan. J. Appl. Phys* **24**, L187.
20. Springthorpe, A. J., King, F. D., and Becke, A. (1975). *J. Electron. Mater.* **4**, 101.
21. Watanabe, M., and Ohba, Y. (1986). *J. Appl. Phys.* **60**, 1032.
22. Ohba, Y., Ishikawa, M., Sugawara, H., Yamamoto, M., and Nakanisi, T. (1986). *J. Crystal Growth* **77**, 374.
23. Ohba, Y., Nishikawa, Y., Nozaki, C., Sugawara, H., and Nakanisi, T. (1988). *J. Crystal Growth* **93**, 613.
24. Ishikawa, M., Ohba, Y., Watanabe, Y., Nagasaka, H., Sugawara, H., Yamamoto, M., and Hatakoshi, G. (1986). *Ext. Abs. 18th Conf. Solid State Devices and Materials*, 153.
25. Nishikawa, Y., Tsuburai, Y., Nozaki, C., Ohba, Y., Kokobun, Y., and Kinoshita, H. (1988). *Appl. Phys. Lett.* **53**, 2182.
26. Nishikawa, Y., Ishikawa, M., Tsuburai, Y., and Kokobun, Y. (1990). *J. Crystal Growth* **100**, 63.
27. Nishikawa, Y., Sugawara, H., Iishikawa, M., and Kokobun, Y. (1991). *J. Crystal Growth* **108**, 728.
28. Ikeda, M., and Kaneko, K. (1989). *J. Appl. Phys.* **66**, 5285.
29. Kobayashi, K., Hino, I., Gomyo, A., Kawata, S., and Suzuki, T. (1987). *IEEE J. Quantum Electron.* **QE-23**, 704.
30. Kawata, S., Kobayashi, K., Gomyo, A., Hino, I., and Suzuki, T. (1986). *Electron. Lett.* **22**, 1265.
31. Asahi, H., Kawamura, Y., and Nagai, H. (1983). *J. Appl. Phys.* **54**, 6958.
32. Hayakawa, T., Takahashi, K., Hosoda, M., Yamamoto, S., and Hijikata, T. (1988). *Japan. J. Appl. Phys.* **27**, L1553.
33. Kikuchi, A., Kishino, K., and Kaneko, Y. (1989). *J. Appl. Phys.* **66**, 4557.
34. Yokotsuka, T., Takamori, A., and Nakajima, M. (1991). *Appl. Phys. Lett.* **58**, 1521.
35. Ishikawa, M., Itaya, K., Watanabe, Y., Hatakoshi, G., Sugawara, H., Ohba, Y., and Uematsu, Y. (1987). *Ext. Abs. 19th Conf. Solid State Devices and Materials*, 115.
36. Hino, I., Gomyo, A., Kawata, S., Kobayashi, K., and Suzuki, T. (1985). *Inst. Phys. Conf. Ser.* **79**, 151.
37. Hino, I., Kawata, S., Gomyo, A., Kobayashi, K., and Suzuki, T. (1986). *Appl. Phys. Lett.* **48**, 557.

38. Kawata, S., Fujii, H., Kobayashi, K., Gomyo, A., Hino, I., and Suzuki, T. (1987). *Electron. Lett.* **23**, 1327.
39. Roentgen, P., Heuberger, W., Bona, G. L., and Unger, P. (1991). *J. Crystal Growth* **107**, 724.
40. Nishikawa, Y., Sugawara, H., Ishikawa, M., and Kokobun, Y. (1990). *Ext. Abs. 22nd Int. Conf. Solid State Devices and Materials*, 509.
41. Wu, M. C., Su, Y. K., Cheng, K. Y., and Chang, C. Y. (1986). *Japan. J. Appl. Phys.* **25**, L90.
42. Kazamura, M., Ohta, I., and Teramoto, I. (1983). *Japan. J. Appl. Phys.* **22**, 654.
43. Kodama, K., Hoshino, M., Kitahara, K., and Ozeki, M. (1986). *Japan. J. Appl. Phys.* **25**, L551.
44. Hayakawa, T., Takahashi, K., Sasaki, K., Hosada, M., Yamamoto, S., and Hijikata, T. (1988). *Japan. J. Appl. Phys.* **27**, L968.
45. Wicks, G. W., Koch, M. W., Varriano, J. A., Johnson, F. G., Wie, C. R., Kim, H. M., and Colombo, P. (1991). *Appl. Phys. Lett.* **59**, 342.
46. Takamori, A., Yokotsuka, T., Uchiyama, K., and Nakajima, M. (1991). *J. Crystal Growth* **108**, 99.
47. Wood, C. E. C. (1982). In *GaInAsP alloy semiconductors*, p. 87. John Wiley and Sons Ltd.
48. Bour, D. P. (1988). Ph.D. Thesis, Cornell University, Ithaca, New York.
49. Gomyo, A., Kobayashi, K., Kawata, S., Hino, I., and Suzuki, T. (1987). *Electron. Lett.* **23**, 85.
50. Kondow, M., Kakibayashi, H., Minagawa, S., Inoue, Y., Nishino, T., and Hamakawa, Y. (1988). *Appl. Phys. Lett.* **53**, 2053.
51. Kondow, M., Kakibayashi, H., Minagawa, S., Inoue, Y., Nishino, T., and Hamakawa, Y. (1988). *J. Crystal Growth* **93**, 412.
52. Morita, E., Ikeda, M., Kumagai, O., and Kaneko, K. (1988). *Appl. Phys. Lett.* **53**, 2164.
53. Kondow, M., and Minagawa, S. (1988). *J. Appl. Phys.* **64**, 793.
54. Gomyo, A., Kobayashi, K., Kawata, S., Hino, I., Suzuki, T., and Yuasa, T. (1986). *J. Crystal Growth* **77**, 367.
55. Nishino, T., Inoue, Y., Hamakawa, Y., Kondow, M., and Minagawa, S. (1988). *Appl. Phys. Lett.* **53**, 583.
56. Kondow, M., Kakibayashi, H., and Minagawa, S. (1989). *Phys. Rev. B* **40**, 1159.
57. Suzuki, T., Gomyo, A., Iijima, S., Kobayashi, K., Kawata, S., Hino, I., and Yuasa, T. (1988). *Japan. J. Appl. Phys.* **27**, 2098.
58. Minagawa, S., and Kondow, M. (1989). *Electron. Lett.* **25**, 758.
59. Gomyo, A., Kawata, S., Suzuki, T., Iijima, S., and Hino, I. (1989). *Japan. J. Appl. Phys.* **28**, L1728.
60. Gomyo, A., Suzuki, T., Iijima, S., Hotta, H., Fujii, H., Kawata, S., Kobayashi, K., Ueno, Y., and Hino, I. (1988). *Japan. J. Appl. Phys.* **27**, L2370.
61. Ueda, O., Takechi, M., and Komeno, J. (1989). *Appl. Phys. Lett.* **54**, 2312.
62. Valster, A., Liedenbaum, C. T. H. F., Finke, M. N., Severens, A. L. G.,

BOERMANS, M. J. B., VANDENHOUDT, D. E. W., and BULLE-LIEUWMA, C. W. T. (1991). *J. Crystal Growth* **107**, 403.

63. SUZUKI, T., GOMYO, A., HINO, I., KOBAYASHI, K., KAWATA, S., and IIJIMA, S. (1988). *Japan. J. Appl. Phys.* **27**, L1549.
64. NISHIKAWA, Y., IISHIKAWA, M., TSUBURAI, Y., and KOKOBUN, Y. (1989). *Japan. J. Appl. Phys.* **28**, L2092.
65. GAVRILOVIC, P., DABKOWSKI, F. P., MEEHAN, K., WILLIAMS, J. E., STUTIUS, W., HSIEH, K. C., HOLONYAK, N., Jr., SHAHID, M. A., and MAHAJAN, S. (1988). *J. Crystal Growth* **93**, 426.
66. MEEHAN, K., DABKOWSKI, F. P., GAVRILOVIC, P., WILLIAMS, J. E., STUTIUS, W., HSIEH, K. C., and HOLONYAK, N., Jr. (1989). *Appl. Phys. Lett.* **54**, 2136.
67. O'BRIEN, S. S, BOUR, D. P., and SHEALY, J. R. (1988). *Appl. Phys. Lett.* **53**, 1859.
68. UENO, Y., FUJII, H., KOBAYASHI, K., ENDO, K., GOMYO, A., HARA, K., KAWATA, S., YUASA, T., and SUZUKI, T. (1990). *Japan. J. Appl. Phys.* **29**, L1666.
69. UENO, Y., ENDO, K., FUJII, H., KOBAYASHI, K., HARA, K., and YUASA, T. (1990). *Electron. Lett.* **26**, 1727.
70. KUO, C. P., FLETCHER, R. M., OSENTOWSKI, T. D., LARDIZABAL, M. C., CRAFORD, M. G., ROBBINS, V. M. (1990). *Appl. Phys. Lett.* **57**, 2937.
71. SUGAWARA, H., ISHIKAWA, M., and HATAKOSHI, G. (1990). *Ext. Abs. 22nd Int. Conf. Solid State Devices and Materials*, 1175.
72. SUGAWARA, H., ISHIKAWA, M., and HATAKOSHI, G. (1991). *Appl. Phys. Lett.* **58**, 1010.
73. HINO, I., GOMYO, A., KOBAYASHI, K., SUZUKI, T., and NISHIDA, K. (1983). *Appl. Phys. Lett.* **43**, 987.
74. KAWAMURA, Y., ASAHI, H., NAGAI, H., and IKEGAMI, T. (1983). *Electron. Lett.* **19**, 163.
75. IKEDA, M., MORI, Y., TAKIGUCHI, M., KANEKO, K., and WATANABE, N. (1984). *Appl. Phys. Lett.* **45**, 661.
76. KOBAYASHI, K., KAWATA, S., GOMYO, A., HINO, I., and SUZUKI, T. (1985). *Electron. Lett.* **21**, 931.
77. KOBAYASHI, K., KAWATA, S., GOMYO, A., HINO, I., and SUZUKI,T. (1985). *Electron. Lett.* **21**, 1162.
78. IKEDA, M., MORI, Y., SATO, H., KANEKO, K., and WATANABE, N. (1985). *Appl. Phys. Lett.* **47**, 1027.
79. IKEDA, M., NAKANO, K., MORI, Y., KANEKO, K., and WATANABE, N. (1986). *Appl. Phys. Lett.* **48**, 89.
80. ISHIKAWA, M., OHBA, Y., SUGAWARA, H., YAMAMOTO, M., and NAKANISI, T. (1986). *Appl. Phys. Lett.* **48**, 207.
81. SHIOZAWA, H., OKUDA, H., ISHIKAWA, M., HATAKOSHI, G.-I., and UEMATSU, Y. (1988). *Electron. Lett.* **24**, 877.
82. OKUDA, H., ISHIKAWA, M., SHIOZAWA, H., WATANABE, Y., ITAYA, K.,, NITTA, K., HATAKOSHI, G., KOKOBUN, Y., and UEMATSU, Y. (1989). Highly reliable InGaP/InGaAlP visible light emitting inner stripe lasers with 667 nm lasing wavelength, *IEEE J. Quantum Electron.* **QE-25**, 1477.

83. Tanaka, T., Ooishi, A., Kajimura, T., and Minagawa, S. (1990). *Ext. Abs. 22nd Int. Conf. Solid State Devices and Materials*, 1177.
84. Ikeda, M., Toda, A., Nakano, K., Mori, Y., and Watanabe, N. (1987). *Appl. Phys. Lett.* **50**, 1033.
85. Bour, D. P., and Shealy, J. R. (1987). High power (1.4W) AlGaInP graded-index separate confinement heterostructure visible ($\lambda \sim 658$ nm) laser, *Appl. Phys. Lett.* **51**, 1658.
86. Itaya, K., Ishikawa, M., Watanabe, Y., Nitta, K., Hatakoshi, G., and Uematsu, Y. (1988). *Japan. J. Appl. Phys.* **27**, L2414.
87. Aoyagi, T., Kimura, T., Yoshida, N., Kadowaki, T., Murakami, T., Kaneno, N., Seiwa, Y., Mizuguchi, K., and Susaki, W. (1990). *Proc. SPIE* **1219**, 8.
88. Ishikawa, M., Shiozawa, H., Itaya, K., Hatakoshi, G., and Uematsu, Y. (1991). *IEEE J. Quantum Elect.* **QE-27**, 23.
89. Casey, H. C., Jr., and Panish, M. B. (1975). *Heterostructure lasers.* Academic Press, Orlando, Florida.
90. Itaya, K., Ishikawa, M., Shiozawa, H., Nishikawa, Y., Suzuki, M., Sugawara, H., and Hatakoshi, G. (1990). *Ext. Abs. 22nd Int. Conf. Solid State Devices and Materials*, 565.
91. Kressel, H., and Butler, J. K. (1977). *Semiconductor lasers and heterojunction LEDs.* Academic Press, New York.
92. Zory, P. S., Reisinger, A. R., Mawst, L. J., Costrini, G., Zmudzinski, C. A., Emanuel, M. A., Givens, M. E., and Coleman, J. J. (1986). *Electron. Lett.* **22**, 475.
93. Zory, P. S., Reisinger, A. R., Waters, R. G., Mawst, L. J., Zmudzinski, C. A., Emanuel, M. A., Givens, M. E., and Coleman, J. J. (1986). *Appl. Phys. Lett.* **49**, 16.
94. Chinn, S. R., Zory, P. S., and Reisinger, A. R. (1988). *IEEE J. Quantum Electr.* **QE-24**, 2191.
95. Ishikawa, M., Ohba, Y., Sugawara, H., Yamamoto, M., and Nakanisi, T. (1985). *Electron. Lett.* **21**, 1084.
96. Takagi, T., Koyama, F., and Iga, K. (1990). *Japan. J. Appl. Phys.* **29**, L1977.
97. Takagi, T., Koyama, F., and Iga, K. (1991). *Electron. Lett.* **27**, 1081.
98. Kishino, K., Kikuchi, A., Kaneko, Y., and Nomura, I. (1991). *Appl. Phys. Lett.* **58**, 1822.
99. Kaneko, Y., Kikuchi, A., Nomura, I., and Kishino, K. (1990). *Tech. Digest, Conf. on Lasers and Electro-Optics (CLEO 1990)*, 12.
100. Kikuchi, A., Kaneko, Y., Nomura, I., and Kishino, K. (1990). *Electron. Lett.* **26**, 1669.
101. Minagawa, S., Tanaka, T., and Kondow, M. (1989). *Electron. Lett.* **25**, 925.
102. Tanaka, T., Minagawa, S., Kawano, T., and Kajimura, T. (1989). *Electron. Lett.* **25**, 905.
103. Ikeda, M., Morita, E., Toda, A., Yamamoto, T., and Kaneko, K. (1988). *Electron. Lett.* **24**, 1094.
104. Ikeda, M., Honda, M., Mori, Y., Kaneko, K., and Watanabe, N. (1984). *Appl. Phys. Lett.* **45**, 964.

105. Kobayashi, K., Hino, I., and Suzuki, T. (1985). *Appl. Phys. Lett.* **46**, 7.
106. Ishikawa, M., Shiozawa, H., Tsuburai, Y., and Uematsu, Y. (1990). *Electron. Lett.* **26**, 211.
107. Itaya, K., Ishikawa, M., and Uematsu, Y. (1990). *Electron. Lett.* **26**, 839.
108. Kobayashi, K., Ueno, Y., Hotta, H., Gomyo, A., Tada, K., Hara, K., and Yuasaa, T. (1990). *Japan. J. Appl. Phys.* **29**, L1669.
109. Uematsu, Y., Hatakoshi, G., Ishikawa, M., and Okajima, M. (1990). *Proc. SPIE* **1219**, 2.
110. Hamada, H., Shono, M., Honda, S., Hiroyama, R., Matsukawa, K., Yodoshi, K., and Yamaguchi, T. (1991). *Electron. Lett.* **27**, 661.
111. Nitta, K., Itaya, K., Nishikawa, Y., Ishikawa, M., Okajima, M., and Hatakoshi, G. (1991). *Appl. Phys. Lett.* **59**, 149.
112. Boermans, M. J. B., Hagen, S. H., Valster, A., Finke, M. N., and van der Heyden, J. M. M. (1990). *Electron. Lett.* **26**, 1438.
113. Ishikawa, M., Okuda, H., Itaya, K., Shiozawa, H., and Uematsu, Y. (1989). *Japan. J. Appl. Phys.* **28**, 1615.
114. Itaya, K., Ishikawa, M., Okuda, H., Watanabe, Y., Nitta, K., Shiozawa, H., and Uematsu, Y. (1988). *Appl. Phys. Lett.* **53**, 1363.
115. Kamiyama, S., Mori, Y., Takahashi, Y., and Ohnaka, K. (1991). *Appl. Phys. Lett.* **58**, 2595.
116. Kawamura, Y., and Asahi, H. (1984). *Appl. Phys. Lett.* **45**, 152.
117. Wang, T. Y., Kimball, A. W., Chen, G. S., Birkedal, D., and Stringfellow, G. B. (1990). *J. Appl. Phys.* **68**, 3356.
118. Hafich, M. J., Lee, H. Y., Robinson, G. Y., Li, D., and Otsuka, N. (1991). *J. Appl. Phys.* **69**, 752.
119. Kondo, M., Domen, K., Anayama, C., Tanahashi, T., and Nakajima, K. (1991). *J. Crystal Growth* **107**, 578.
120. Chinn, S. R. (1984). *Appl. Optics* **23**, 3508.
121. Nagle, J., Hersee, S., Krakowski, M., Weil, T., and Weisbuch, C. (1986). *Appl. Phys. Lett.* **49**, 1325.
122. Shealy, J. R., and Bour, D. P. (1988). *Tech. Digest, Conf. on Lasers and Electro-Optics (CLEO 1988)*, 274.
123. Katsuyama, T., Yoshida, I., Shinkai, J., Hashimoto, J., and Hayashi, H. (1991). *Tech. Digest, Conf. on Lasers and Electro-Optics (CLEO 1991)*, 96.
124. Derry, P. L., Yariv, A., Lau, K. Y., Bar-Chaim, N., Lee, K., and Rosenberg, J. (1987). *Appl. Phys. Lett.* **50**, 1773.
125. Bour, D. P., Carlson, N. W., and Evans, G. A. (1989). *Electron. Lett.* **25**, 1243.
126. Arakawa, Y., and Yariv, A. (1985). *IEEE J. Quantum Electron.* **QE-21**, 1666.
127. Kawata, S., Kobayashi, K., Fujii, H., Hino, I., Gomyo, A., Hotta, H., and Suzuki, T. (1988). *Electron. Lett.* **24**, 1489.
128. Dallesasse, J. M., Nam, D. W., Deppe, D. G., Holonyak, N., Jr., Fletcher, R. M., Kuo, C. P., Osentowski, T. D., and Craford, M. G. (1988). *Appl. Phys. Lett.* **53**, 1826.

129. Valster, A., van der Heijden, J. M. M., Boermans, M. J. B., Hagen, S. H., Finke, M. N., and Acket, G. A. (1990). *Tech. Digest, Conf. on Lasers and Electro-Optics (CLEO 1990)*, 12.
130. Valster, A., Liedenbaum, C. T. H. F., Heijden, J. M. M., Finke, M. N., Severens, A. L. G., and Boermans, M. J. B. (1990). *Tech. Digest, 12th IEEE Int. Conf. on Semiconductor Lasers*, paper **C-1**.
131. Chang-Hasnain, C. J., Bhat, R., and Koza, M. (1991). *Tech. Digest, Conf. on Lasers and Electro-Optics (CLEO 1991)*, 94.
132. Welch, D. F., Wang, T., and Scifres, D. R. (1991). *Electron. Lett.* **27**, 694.
133. Kuo, C. P., Fletcher, R. M., Osentowski, T. D., Craford, M. G., Nam, D. W., Holonyak, N., Jr., Hsieh, K. C., and Fouquet, J. E. (1988). *J. Crystal Growth* **93**, 389.
134. Nam, D. W., Deppe, D. G., Holonyak, N., Jr., Fletcher, R. M., Kuo, C. P., Osentowski, T. D., and Craford, M. G. (1988). *Appl. Phys. Lett.* **52**, 1329.
135. Dallasasse, J. M., Plano, W. E., Nam, D. W., Hsieh, K. C., Baker, J. E., Holonyak, N., Jr., Kuo, C. P., Fletcher, R. M., Osentowski, T. D., and Craford, M. G. (1989). *J. Appl. Phys.* **66**, 482.
136. Wilcox, J. Z., Peterson, G. L., Ou, S. S., Yang, J. J., Jansen, M., and Schechter, D. (1988). *Appl. Phys. Lett.* **53**, 2272.
137. Reisinger, A. R., Zory, P. S., and Waters, R. G. (1987). *IEEE J. Quantum Electron.* **QE-23**, 993.
138. McIlroy, P. W. A., Kurobe, A., and Uematsu, Y. (1985). *IEEE J. Quantum Electron.* **QE-21**, 1958.
139. Haug, A. (1988). *J. Phys. Chem. Solids* **49**, 599.
140. Deppe, D. G., Nam, D. W., Holonyak, N., Jr., Hsieh, K. C., Baker, J. E., Kuo, C. P., Fletcher, R. M., Osentowski, T. D., and Craford, M. G. (1988). *Appl. Phys. Lett.* **52**, 1413.
141. Fujiwara, K., Nakamura, A., Tokuda, Y., Nakayama, T., and Hirai, M. (1986). *Appl. Phys. Lett.* **49**, 1193.
142. Scheps, R., Myers, J. F., Serreze, H. B., Rosenberg, A., Morris, R. C., and Long, M. (1991). *Optics Lett.* **16**, 820.
143. Adams, A. R. (1983). *Electron Lett.* **22**, 249.
144. Yablonovitch, E., and Kane, E. O. (1986). *IEEE J. Lightwave Tech.* **LT-4**, 504.
145. Yablonovitch, E., and Kane, E. O. (1988). *IEEE J. Lightwave Tech.* **LT-6**, 1292.
146. Welch, D. F., Streifer, W., Schaus, C. F., Sun, S., and Gourley, P. L. (1990). *Appl. Phys. Lett.* **56**, 10.
147. Katsuyama, T., Yoshida, I., Shinkai, J., Hashimoto, J., and Hayashi, H. (1990). *Electron. Lett.* **26**, 1375.
148. Hashimoto, J., Katsuyama, T., Shinkai, J., Yoshida, I., and Hayashi, H. (1991). *Appl. Phys. Lett.* **58**, 879.
149. Serreze, H. B., Chen, Y. C., and Waters, R. G. (1991). *Appl. Phys. Lett.* **58**, 2464.
150. Serreze, H. B., Chen, Y. C., Waters, R. G., and Harding, C. M. (1991). *Tech. Digest, Conf. on Lasers and Electro-Optics (CLEO 1991)*, PD571.

151. Fujii, H., Kobayashi, K., Kawata, S., Gomyo, A., Hino, I., Hotta, H., and Suzuki, T. (1987). *Electron. Lett.* **23**, 938.

152. Kamiyama, S., Mori, Y., Mannoh, M., and Ohnaka, K. (1991). *Tech. Digest, Conf. on Lasers and Electro-Optics (CLEO 1991)*, 92.

153. Valster, A., Andre, J. P., Dupont-Nivet, E., and Martin, G. M. (1988). *Electron. Lett.* **24**, 326.

154. van der Poel, C. J., Opschoor, J., Valster, A., Drenten, R. R., and Andre, J. P. (1990). *J. Appl. Phys.* **68**, 868.

155. Evans, G. A., Butler, J. K., and Masin, V. J. (1988). *IEEE J. Quantum Electron.* **QE-24**, 737.

156. Hatakoshi, G., Ishikawa, M., Watanabe, Y., and Uematsu, Y. (1989). *Electron. Lett.* **25**, 125.

157. Nitta, K., Itaya, K., Ishikawa, M., Watanabe, Y., Hatakoshi, G., and Uematsu, Y. (1989). *Japan. J. Appl. Phys.* **28**, L2089.

158. Itaya, K., Watanabe, Y., Ishikawa, M., Hatakoshi, G., and Uematsu, Y. (1990). *Appl. Phys. Lett.* **56**, 1718.

159. Tanaka, T., Minagawa, S., and Kajimura, T. (1989). *Appl. Phys. Lett.* **54**, 1391.

160. Itaya, K., Hatakoshi, G., Watanabe, Y., Ishikawa, M., and Uematsu, Y. (1990). *Electron. Lett.* **26**, 214.

161. Hagen, S. H., Valster, A., Boermans, M. J. B., and van der Heyden, J. (1990). *Appl. Phys. Lett.* **57**, 2291.

# *Chapter 10*

# QUANTUM WIRE SEMICONDUCTOR LASERS

**Eli Kapon**

*Bellcore, Red Bank*
*New Jersey*



## 1. INTRODUCTION

One dimensional (1D) quantum confinement of charge carriers in quantum well lasers (QWL) gives rise to modified band structure and density of states



ISBN 0-12-781890-1

(DOS) distribution, which can significantly improve laser performance [1, 2]. The step-like DOS profile can lead to narrower energy distribution of carriers, resulting in narrower luminescence spectra and higher differential gain. The higher gain, reduced modal absorption (due to low optical confinement factor), and small active region volumes have led to an order of magnitude reduction in threshold currents of QWL lasers compared with bulk devices; GaAs/AlGaAs QWL lasers with threshold current densities as low as 52 $A/cm^2$ [3] and threshold currents as low as 0.35 mA [4] at room temperature have been reported. Furthermore, QWL lasers can exhibit reduced temperature sensitivity of threshold current while retaining superior performance characteristics [5]. The higher differential gain in QWL lasers can lead to higher relaxation oscillation frequency and hence potentially higher modulation bandwidth; relaxation oscillation frequencies as high as 30 GHz have been measured in modulation doped GaAs/AlGaAs QWL lasers [6]. The higher differential gain can also result in a lower linewidth enhancement factor $\alpha$ [2, 7], which can reduce the laser spectral linewidth and chirp [8].

Further improvement in semiconductor laser performance is expected with the introduction of multi-dimensional quantum confinement [9]. The resulting *quantum wire* (QWR) and *quantum dot* (QD) (also called *quantum box*, QB) laser structures employ quasi-1D and 0D carriers, which are quantum confined to wire-like and box-like potential wells of sufficiently small lateral dimensions (less than a few hundred angströms in GaAs/AlGaAs heterostructures). QWR and QD semiconductor lasers were predicted to exhibit extremely low threshold currents, in the microampere range [10–12], as well as higher modulation bandwidth, narrower spectral linewidth [13], and reduced temperature sensitivity [9] compared with their QWL counterparts. Enhancements in electrorefraction, electroabsorption [14], and nonlinear optical properties [15] of QWR and QD heterostructures make them attractive for improving the performance of other types of optoelectronic devices as well, e.g. optical switches and modulators [16]. Low dimensional semiconductor heterostructures may thus form the basis for future optoelectronic devices exhibiting improved features brought about by multi-dimensional quantum size effects. In particular, low dimensional diode lasers are attractive for applications involving densely packed laser arrays and monolithic integration of lasers with low power electronics, including computer optical interconnects, optoelectronic signal processing, and optical computing.

Realization of quantum-confined diode lasers requires development of techniques for fabricating semiconductor potential wells with dimensions in the 10 nm range. These potential wells should be uniform in size and shape to avoid adverse effects of inhomogeneous broadening due to well size fluctuations [17, 18]. Most importantly, virtually defect- and contamination-free potential well interfaces are required in order to ensure sufficiently long carrier lifetimes and high quantum efficiency to allow achievement of high optical gain. In the case of 1D quantum confinement, these requirements have been met by modern epitaxial growth techniques, particularly molecular beam epitaxy (MBE), organometallic chemical vapor deposition (OMCVD), and their variations, which enable growth of thin QWL layers with mon-atomic layer uniformity and accuracy. It is worth emphasizing that the very high interface quality in such QWL heterostructures is a direct result of the fact that all interfaces in these structures are formed *in situ* during the crystal growth step.

Fabrication of QWR and QD semiconductor heterostructures of high optical quality, however, is still a major challenge for current nanofabrication and crystal growth technologies. Elimination of nonradiative recombination centers at the potential well interfaces is even more important in this case because of the increase in surface to volume ratio with increasing dimensionality of quantum confinement. Since the volume of the QWR (and even more so the QD) potential well is extremely small, a practical low dimensional semiconductor laser should incorporate high density arrays of QWRs or QDs, and/or employ low loss, tight optical confinement waveguides and low loss optical cavities in order to compensate for the low optical (modal) gain. In view of these challenges, much of the effort in this field has been concentrated so far in developing effective techniques for producing QWR and QD structures of acceptable quality.

Earlier attempts to simulate QWR and QD laser structures involved operation of double heterostructure (DH) or QWL lasers in high magnetic fields [9, 19–22]. In this case, the Lorentz force yields effective 2D confinement of carriers in the plane normal to the field axis, which can lead to quantization of the energy spectrum in Landau levels. Increase in relaxation oscillation frequency and reduction in spectral linewidth in such simulated QWR and QD lasers was attributed to 2D and 3D quantum confinement effects [20–22]. The first attempts to fabricate "rigid" QWR and QD semiconductor lasers consisted of etching the wires or dots from QWL heterostructures [23–27]. However, damage to the wire and dot interfaces caused by the etching and regrowth steps result in very high threshold currents and laser operation at low temperatures only. Techniques for fabricating QWRs and QDs that avoid post-growth processing of the potential well interfaces, hence having the potential for producing damage-

and contamination-free wire and dot interfaces, have been developed more recently [28]. First demonstration of single- and multiple-QWR lasers, operating with low threshold currents at room temperature and exhibiting 2D quantum confinement effects, was achieved with structures grown by OMCVD on nonplanar substrates [29–33]. QWR laser structures in which post-growth processing is avoided were fabricated also by MBE on vicinal substrates [34, 35].

This chapter reviews recent progress in the evolution of the concept and fabrication technologies of semiconductor QWR lasers. We begin in Section 2 by discussing the effects of multi-dimensional quantum confinement on the lasing properties of QWR lasers, particularly the effect on gain, threshold current, modulation bandwidth, and spectral linewidth. Section 3 describes the different approaches for realizing QWR lasers, including simulation of multi-dimensional confinement using high magnetic fields, etching-and-regrowth of QWL structures, growth on vicinal substrates, and growth on nonplanar, patterned substrates. In Section 4, lasing characteristics of the QWR laser structures reported thus far are reviewed, and possible observation of 2D quantum confinement effects in these devices is discussed. Finally, conclusions and future trends in this field are brought forward in Section 5.

## 2. EFFECT OF MULTI-DIMENSIONAL QUANTUM CONFINEMENT

### 2.1. Density of States

Quantum confinement of charge carriers in semiconductors takes place when the carriers are trapped within potential wells of sufficiently small dimensions. This quantum confinement can give rise to significant modification of the energy band structure and the DOS distribution in these materials. In the discussion that follows, we treat the quantum-confined structure in a simple, single-band model. While this picture is adequate for the conduction band case, more elaborate, multiband models including the effect of band mixing have been developed for the valence band (see, e.g., [36] and Chapter 1 in this book). Nevertheless, our simple model does provide insight into the important consequences of multi-dimensional quantum confinement on laser performance.

In the effective mass approximation, the energy spectrum $E$ of the carriers is obtained by solving Schroedinger's equation

$$\left[-\frac{\hbar^2}{2m^*}\nabla^2 + V(x, y, z)\right]\psi(x, y, z) = E\psi(x, y, z), \tag{1}$$

where $\psi$ is the carrier envelope wavefunction, $m^*$ is the carrier effective mass, and $V(x, y, z)$ is the potential distribution. For potential wells of rectangular shape (see insets in Fig. 1) 1D, 2D, and 3D quantum confinement can be achieved in film-, wire-, and box-like geometries by successively reducing the well dimensions $t_x$, $t_y$, and $t_z$. (In the case of QWLs, the thickness is more commonly denoted by $L_z$.) For infinitely deep potential wells, the energy of the confined carriers (with respect to the band edge) is given by

$$E_l = \frac{\hbar^2\pi^2 l^2}{2m^* t_x^2} + \frac{\hbar^2(k_y^2 + k_z^2)}{2m^*}, \qquad \text{1D confinement;} \tag{2a}$$

$$E_{l,m} = \frac{\hbar^2\pi^2}{2m^*}\left(\frac{l^2}{t_x^2} + \frac{m^2}{t_y^2}\right) + \frac{\hbar^2 k_z^2}{2m^*} \qquad \text{2D confinement;} \tag{2b}$$

$$E_{l,m,n} = \frac{\hbar^2\pi^2}{2m^*}\left(\frac{l^2}{t_x^2} + \frac{m^2}{t_y^2} + \frac{n^2}{t_z^2}\right), \qquad \text{3D confinement;} \tag{2c}$$

where $l$, $m$, $n = 1, 2, \ldots$ are the level quantum numbers and $k_y$, $k_z$ are the wavevector components along the unconfined dimensions. Quantum confinement in such QWL, QWR, or QD structures thus results in charge carriers of quasi-2D, 1D, or 0D nature.

The DOS (per unit volume) functions, including spin degeneracy, are given by

$$\rho_{3D} = \frac{(2m^*/\hbar^2)^{3/2}}{2\pi^2}\sqrt{E}, \tag{3a}$$

$$\rho_{2D} = \frac{m^*}{\pi\hbar^2 t_x}\sum_l \theta(E - E_l), \tag{3b}$$

$$\rho_{1D} = \frac{(2m^*)^{1/2}}{\pi\hbar t_x t_y}\sum_{l,m}(E - E_{l,m})^{-1/2}, \tag{3c}$$

$$\rho_{0D} = \frac{2}{t_x t_y t_z}\sum_{l,m,n}\delta(E - E_{l,m,n}), \tag{3d}$$

for 3D, quasi-2D, 1D, and 0D carriers, respectively. (In (3b), $\theta(x)$ is the Heaviside function; $\theta = 0$ for $x < 0$, $\theta = 1$ for $x > 0$.) The DOS distributions acquire sharper features as the carrier dimensionality is reduced, particularly

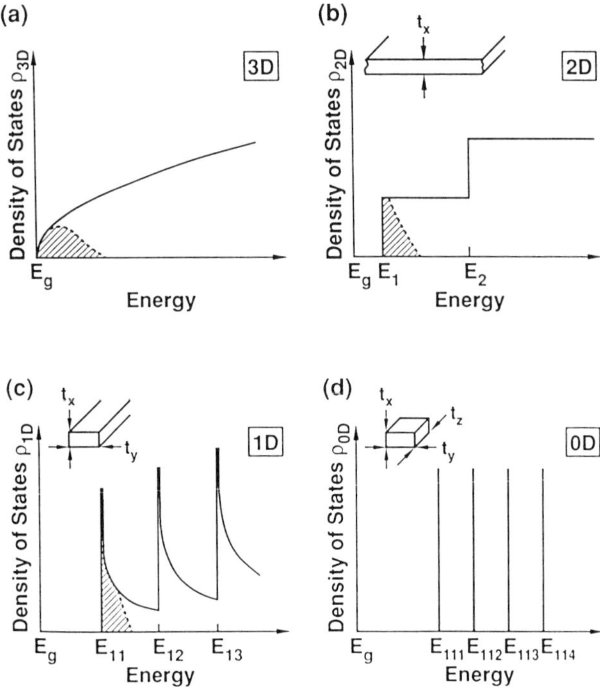


**Fig. 1.** Schematic description of the density of states versus dimensionality. The insets illustrate rectangular potential well configurations of the corresponding quantum-confined structure. (a) Bulk (3D); (b) quantum well (2D); (c) quantum wire (1D); (d) quantum dot (0D). The crossed areas indicate occupied states for similar carrier density. (after [87], © 1992 IEEE.)

in the case of 1D and 0D structures (see Fig. 1). It should be noted, however, that these sharp features can be significantly smoothed out by well size fluctuations, leading to inhomogeneous broadening of the energy spectrum [17, 18]. The broadening effects, in turn, can significantly reduce the impact of the modified DOS on the laser performance that is to be discussed.

## 2.2. Optical Gain

The narrowing of the DOS distribution with reducing dimensionality results in confinement of the carrier energy distribution (at finite temperature) to narrower spectral regions (see Fig. 1). Hence, higher optical gain at given carrier density can be achieved with quantum-confined lasers. This effect can be quantified by evaluating the optical gain function

$$g(N,\omega) = \omega\sqrt{\frac{\mu}{\varepsilon}}\int dE'\,|M|^2\rho_{\rm red}(f_c - f_v)\frac{\hbar/\tau_{\rm in}}{(E'-\hbar\omega)^2 + (\hbar/\tau_{\rm in})^2}, \tag{4}$$

where $E'$ is the transition energy, $\mu$ and $\varepsilon$ are respectively the permeability and dielectric constants, $M$ is the dipole matrix element for the transition, $\rho_{\rm red}$ is the reduced density of states for electrons and holes, $f_c$, $f_v$ are the Fermi functions for the conduction and valence bands, respectively, and $\tau_{\rm in}$ is the intraband relaxation time [37]. [Note the different definition of the matrix element $M$ in Chapter 1. The validity of the Lorentzian lineshape function used in (4) is discussed in Chapter 2.] Figure 2 shows calculated gain spectra

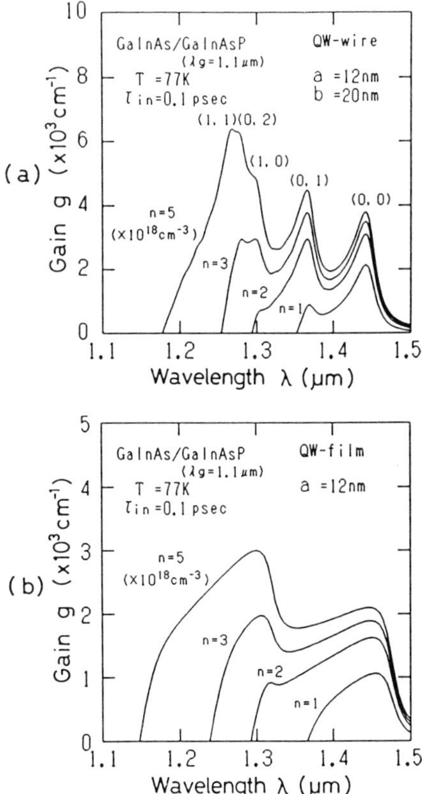


**Fig. 2.** Calculated gain spectra at 77 K for GaInAs/GaInAsP lasers for various values of the carrier density $n$. (a) QWR structure with a rectangular wire, 12 nm wide and 20 nm high; $(l, m)$, with $l, m = 0, 1, \ldots$, indicate QWR subbands. (b) QWL structure, 12 nm thick well (after [23]).

for GaInAs/GaInAsP QWL and QWR lasers at 77 K, at various levels of carrier density $N$ [23]. The features unique to the 1D and 2D DOS distributions of Fig. 1 are reflected in these gain spectra. In the QWR structure, the fundamental QWL level is split into three subbands due to additional, lateral quantum confinement in the 20 nm wide wire. The higher DOS at the QWR subbands results in peak gain values larger by about a factor of two compared with the QWL structure at the same carrier density, with a corresponding increase in differential gain $g' \equiv dg/dN$ [2]. Although the same intraband relaxation time (0.1 ps) is assumed for both structures, the width of the gain spectrum for each subband is smaller for the QWR laser due to the narrower DOS profile. However, smaller increase in $g'$ in GaAs/AlGaAs QWR lasers due to increase in conduction band effective mass has been predicted by tight-binding analysis [38, 39]. On the other hand, possible increase in $\tau_{in}$ due to reduced phase space for scattering in 1D structures [40] may lead to higher gains in QWR lasers [41]. Size fluctuations, particularly of the QWR cross section, can significantly broaden the gain spectra and reduce the attainable peak gain [18]. In addition, the anisotropy of the dipole moment in QWR structures can give rise to higher gain for light polarized along the wire (i.e., light propagation perpendicular to the wire); gain anisotropy ratio of $\sim 2$ was predicted for rectangular wires of equal width and height [41].

## 2.3. Threshold Current

The higher optical gain and the reduced volume of active region in quantum-confined lasers can lead to very low threshold currents in properly designed devices [10–12, 41, 42]. The (material) threshold gain $g_{th}$ in a semiconductor laser is given by

$$\Gamma g_{th} = \alpha_i + \frac{1}{2L} \ln\left(\frac{1}{R_1 R_2}\right) \equiv \alpha_{cav}, \tag{5}$$

where $\Gamma$ is the optical confinement factor (i.e., the fraction of optical mode power in the active region), $\alpha_i = \Gamma\alpha_{fc} + \alpha_{sc}$ is the optical waveguide loss coefficient with $\alpha_{fc}$ and $\alpha_{sc}$ the free carrier and scattering coefficients, $L$ is the cavity length, and $R_1$, $R_2$ are the mirror reflectivities. We assume a linear gain–carrier density relationship in the form $g = g'(N - N_{tr})$, where $N_{tr}$ is the transparency carrier density. (Gain saturation at high carrier densities leads to a deviation from this expression, the consequences of which are discussed subsequently.) The current is given by $I = eV_a N/\eta_i \tau_c$, with $e$ the electron charge, $V_a$ the volume of the active region, $\eta_i$ the quantum efficiency, and $\tau_c$

the carrier lifetime. We consider a collinear laser configuration, with an active region of length $L$ and cross-sectional area $A_{act}$, so that $\Gamma \sim A_{act}/A_{mode}$, $A_{mode}$ being the cross-sectional area of the optical mode. This yields the threshold current

$$I_{th} = I_{tr} + I_{cav}, \tag{6a}$$

$$I_{tr} \equiv \frac{eV_a N_{tr}}{\eta_i \tau_c}, \tag{6b}$$

$$I_{cav} \equiv \frac{eV_a \alpha_{fc}}{\eta_i \tau_c g'} + \frac{eA_{mode}}{\eta_i \tau_c g'} [\alpha_{sc} L + \ln(1/\sqrt{R_1 R_2})]. \tag{6c}$$

Examination of the various terms in (6) gives valuable information about the ultimate limit in threshold current as well as the design rules for achieving very low threshold currents. The transparency term $I_{tr}$ is proportional to the volume of the active region and hence can be extremely small for QWR or QD structures. As an example, consider a GaAs/AlGaAs single-QWR structure of $10 \times 10\ nm^2$ cross section and $L = 100\ \mu m$. Typical transparency carrier density $N_{tr} \approx 1.10^{18}\ cm^{-3}$ [11], $\eta_i = 1$, and $\tau_c = 3$ ns give $I_{tr} \approx 0.8\ \mu A$. $I_{tr}$ is the lowest achievable threshold current, obtained in the limit of vanishing cavity losses. The transparency current can be further reduced simply by reducing the active volume $V_a$. Reduced heavy-hole mass in QWR structures due to mixing of heavy-hole and light-hole bands [39] can decrease $N_{tr}$ (and hence $I_{tr}$), since it yields larger separation of the quasi-Fermi levels at a given carrier density.

In the more realistic case of finite cavity losses, the current $I_{cav}$ required to offset these losses should also be accounted for. For sufficiently low threshold carrier densities $N_{th}$, the free carrier term in (6c) is negligible compared with $I_{tr}$. The dominant term of $I_{cav}$ can be minimized by employing tight optical confinement (to decrease $A_{mode}$), increasing $g'$, maintaining high $\eta_i$ and long $\tau_c$, and reducing the cavity losses. This illustrates the significance of the higher differential gain in quantum-confined lasers, and the importance of fabricating potential wells with near ideal interfaces for maintaining long $\tau_c$ and high $\eta_i$. As an example, consider a single-QWR GaAs/AlGaAs laser with $A_{mode} \sim 1\ \mu m^2$, $\tau_c = 3$ ns, $g' \sim 10^{15}\ cm^2$ [2], and $\eta_i = 1$; then, $eA_{mode}/\eta_i \tau_c g' = 0.53$ mA. For $L = 100\ \mu m$ and readily achievable cavity loss parameters $\alpha_{sc} = 10\ cm^{-1}$ and $R = 0.9$, we obtain $I_{cav} = 109\ \mu A$; for a lower loss cavity with $\alpha_{sc} = 1\ cm^{-1}$ and $R = 0.99$, $I_{cav} = 11\ \mu A$. Clearly, low loss, tight optical confinement cavities are essential for achieving threshold currents in the microampere range with QWR structures. Such optical cavities can be realized by applying high reflection (HR) mirror coatings [43] and employing low loss, separate-confinement optical waveguides [12, 42]. Low internal losses $\alpha_i$ are particularly important in HR coated lasers for maintaining useful differential efficiencies.

Whereas the preceding discussion serves to illustrate the roles of reduced active volume and enhanced differential gain in quantum-confined lasers for threshold current reduction, the omission of gain saturation effects at high carrier densities can lead to underestimation of $I_{th}$. The decrease in $g'$ with increasing $N$ implies that there is an optimum number of QWRs (or QDs) for minimizing the threshold current [11, 12, 42]. This optimum number is a result of a tradeoff between the reduced $I_{tr}$ for a small number of wires and the higher $\Gamma$ (hence higher modal gain) for a large number of wires. The microampere range threshold currents expected in well designed QWR lasers should be compared with the lowest threshold currents demonstrated to date for QWL edge emitting and vertical cavity lasers, which are 0.35 [4] and 0.7 mA [44], respectively, at room temperature.

The spectral confinement of charge carriers in quantum-confined lasers limits the spread in carrier distribution to higher energies at elevated active region temperatures. This should result in reduced temperature sensitivity of the threshold current as the carrier dimensionality is reduced. The effect is illustrated in Fig. 3, which compares the calculated dependence of the threshold current density of GaAs/AlGaAs bulk, QWL, QWR, and QD lasers on temperature $T$ [9]. The threshold current density $J$ exhibits exponential dependence on temperature in the form $J_{th} = J_0 \exp(T/T_0)$. For QWR lasers, the calculation of Fig. 3 predicts $T_0 = 481$ K, substantially higher than the characteristic temperature for typical bulk or QWL lasers. In the case of ideal QD lasers with subband separation $\gg k_B T$, the $\delta$ function distribution of the DOS does not allow for significant spread in carrier distribution at elevated temperatures, and hence no variation in $J$ is expected. Broadening effects that result in wider spectral gain distributions, however, will reduce the $T_0$ values shown in Fig. 3 and will lead to finite $T_0$ for QD lasers. Semiconductor lasers with reduced temperature sensitivity are important because of reliability considerations, especially in the case of high power operation.

### 2.4. Direct Modulation Bandwidth

The relaxation oscillation frequency, which sets the limit on the direct modulation bandwidth of semiconductor lasers, is given by

$$f_r = \frac{1}{2\pi}\left(\frac{g' P_0}{\tau_{ph}}\right)^{1/2}, \tag{7}$$

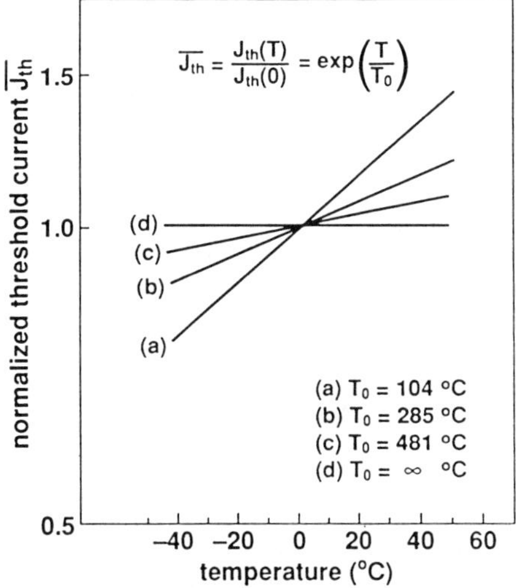


**Fig. 3.** Calculated temperature dependence of the threshold current density of low dimensional GaAs/AlGaAs semiconductor lasers: (a) bulk (double heterostructure); (b) QWL; (c) QWR; (d) QD (after [9]).

where $P_0$ is the photon density in the cavity and $\tau_{ph} = n_g/(c\alpha_{cav})$ is the photon cavity lifetime ($n_g$ is the group index) [45]. The larger differential gain resulting from the narrower DOS distribution can thus increase the modulation bandwidth in quantum-confined lasers [13]. An example of the calculated relaxation frequency $f_r$ for QWL and QWR GaAs/AlGaAs lasers versus the width of the well or wire is shown in Fig. 4 [46]. Since $g'$ attains

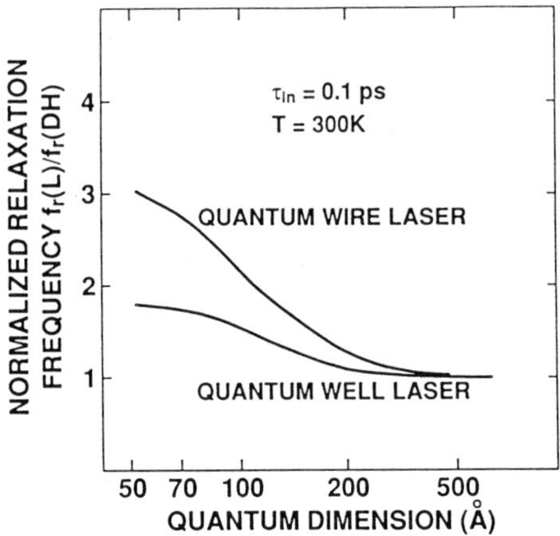


**Fig. 4.** Calculated relaxation frequency versus width of potential well for GaAs/AlGaAs QWL and QWR lasers. The number of wells or wires is optimized at each well width (after [46]).

its maximum for a specific carrier density, highest modulation bandwidth is achieved by selecting an optimal number of QWRs and QWLs. The curves in Fig. 4 were generated by such an optimization procedure. Bandwidth enhancement by a factor of $\sim 3$ compared with a bulk structure is predicted for a properly designed QWR laser. The reduction in $g'$ at high carrier densities would usually require multi-QWR laser structures for achieving enhanced modulation speeds. The need to operate QWR lasers with reduced cavity losses (i.e., long $\tau_{ph}$) may set a limit to the increase in $f_r$ in these structures. Enhancement in $f_r$ for QWL lasers has been demonstrated; relaxation frequencies as high as 30 Ghz have been observed in highly p-doped multi-QWL GaAs/AlGaAs devices [8].

## 2.5. Spectral Linewidth

The field spectrum linewidth of a single mode semiconductor laser is given by [47]

$$\Delta \nu = \frac{R_{sp}}{4\pi s}(1 + \alpha^2), \tag{8}$$

where $R_{sp}$ is the rate of spontaneous emission coupling into the lasing mode, and $s$ is the number of photons in the lasing mode. The linewidth enhancement factor $\alpha = \delta n'/\delta n''$, where $\delta n'$, $\delta n''$ are the changes in real and imaginary part of the refractive index at the active region caused by coupling of spontaneous emission photons into the lasing mode. The deviation of the spectral linewidth (8) from the conventional Schawlow–Townes expression (by the factor $1 + \alpha^2$) reflects the strong coupling between phase and intensity fluctuations in semiconductor lasers, and it results from the detuning of the spectrum of $\delta n''$ from that of $\delta n'$. Typical values of $\alpha$ in bulk lasers are 4–7, resulting in output power–linewidth products of 50–100 Mhz mW [48]. Large values of $\alpha$ also cause enhanced chirping of the laser wavelength during direct high speed modulation [49]. The increase in differential gain $g'$ in quantum-confined lasers increases $\delta n''$ and thus can reduce $\alpha$. The effect is illustrated in Fig. 5, which shows the calculated $\alpha$ for bulk, QWL, and QWR GaAs/AlGaAs lasers as a function of the conduction band quasi-Fermi level [46]. The $\alpha$ parameter is expected to reduce to $\sim 1$ in QWR lasers, thus resulting in linewidths more than an order of magnitude narrower than in bulk lasers. The trend of reduction in $\alpha$ and linewidth is expected to continue by resorting to QD structures. In the case of ideal QDs, the DOS distribution approaches that of gas or other solid state lasers, and $\alpha$ approaches zero value. It should be noted, however, that the higher nonlinear gain coefficient

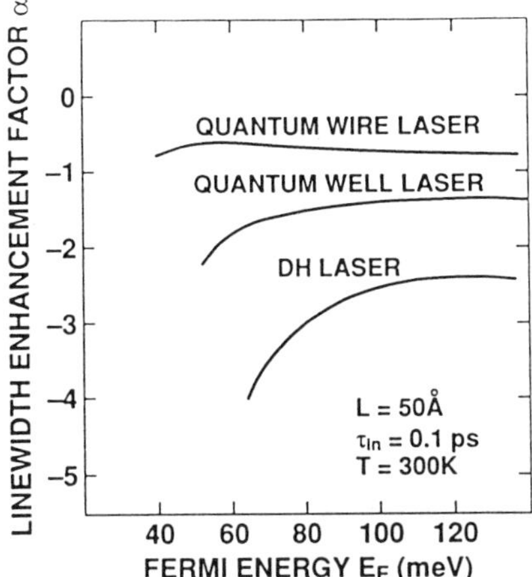


**Fig. 5.** Calculated linewidth enhancement factor $\alpha$ for GaAs/AlGaAs bulk, 50-Å QWL, and 50 × 50 Å QWR lasers as a function of the conduction band Fermi energy (after [46]).

expected in quantum-confined lasers [39] may significantly degrade the improvement in modulation bandwidth and spectral linewidth discussed previously.

## 3. FABRICATION TECHNIQUES

### 3.1. Fabrication of QWR Structures for Optoelectronics

The first attempts to study the effect of 2D and 3D quantum confinement on the performance of diode lasers employed high magnetic fields to produce quasi-1D and 0D carrier states (see Section 3.1 for details) [9, 19–22, 46]. Although such simulation can provide useful insight into the nature of multi-dimensional quantum confinement, it is highly desirable to realize structure incorporating "rigid" QWR and QD potential wells. These structures offer greater flexibility in the design of the quantum structures and are more suitable for device applications.

Perhaps the most direct approach for fabricating rigid QWRs and QDs has been by etching the wire or dot patterns through etch masks placed on QWL wafers [50–52]. This approach has been quite successful in studies of electrical transport in low dimensional systems [53]. In the case of structures used for optical studies, on the other hand, damage incorporated during the etching process onto the lateral wire and dot surfaces can severely degrade the luminescence efficiency due to formation of nonradiative electron–hole recombination centers. Reduction in luminescence efficiency by several orders of magnitude can occur in structures with lateral dimensions below 100 nm, where multi-dimensional quantum size effects are expected to emerge [54]. It should be noted that the effect of interface damage is considerably less severe in the case of electric transport involving a single carrier-type. Low temperature photoluminescence excitation (PLE) spectra of relatively wide (70 nm) GaAs/AlGaAs QWR structures, reactive ion etched (RIE) in $SiCl_4$ plasma, exhibited heavy-hole related QWR subbands separated by 2.5 meV, ~15% increase in excitonic binding energy, and strong polarization effects [52]. Subbands associated with 2D quantum confinement have been identified also in low temperature PLE spectra of 30 nm wide, etched InGaAsP/InP QWR structures, in which surface recombination velocity is considerably lower than in the GaAs/AlGaAs system [51]. However, effective application of this approach to quantum-confined semiconductor lasers requires developing techniques for low-damage etching and/or damage removal. In addition, epitaxial regrowth for surface passivation is desirable for eliminating effects of depletion layers and for providing surface protection. Decrease in surface recombination velocity from $1.3 \times 10^5$ cm/s for etched surfaces to $1.6 \times 10^4$ cm/s for regrown interfaces was observed in GaAs/AlGaAs QWR structures fabricated by $SiCl_4$ RIE and OMCVD [55]. *In situ* processing and regrowth methods seem particularly useful in this context, since they can avoid surface oxidation and contamination between etching and growth steps [56].

Quantum wire and dot structures suitable for optical devices have also been fabricated by patterned QWL disordering [57, 58]. In this technique, interdiffusion of group III elements is enhanced via implantation or diffusion of various species through a mask, leading to intermixing of the well and barrier material and thus forming the lateral potential barriers. GaAs/AlGaAs QWR and QD structures made by Ga implantation-induced QWL disordering showed features attributed to 2D quantum confinement in cathodoluminescence (CL) spectra [57]. Although this method avoids interface damage caused by etching steps, it often involves the introduction of impurities to the wire and dot interfaces [58], which is undesirable for some applications. In addition, the diffusion process involved in defining the interfaces of the quantum structures may set a limit to the spatial resolution of this method.

More recently, considerable effort has been directed at developing QWR and QD fabrication techniques that avoid any post-growth processing steps for defining the multi-density potential wells. These include strain-induced quantum confinement [59], growth on patterned, masked substrates [60], side growth on cleaved QWL heterostructure [61], growth on vicinal substrates [62–64], and growth on patterned nonplanar substrates [29, 65–67]. The two latter approaches are particularly attractive, since they allow formation of all interfaces of the potential wells entirely during the growth step and make possible the fabrication of quantum structures embedded in semiconductor heterostructure host material.

Growth of QWL heterostructures on vicinal substrates has been proposed as an approach for fabricating QWR arrays of sub–10 nm lateral dimensions without the use of lithography [62]. In this technique, fractions of monolayers of, e.g., GaAs or AlGaAs are alternately grown on the stepped surface of a slightly (a few degrees) misaligned (100) GaAs substrate. Under certain growth conditions [63, 64], nucleation at the edges of the monatomic steps results in the formation of lateral arrays of low bandgap wires separated by higher bandgap barriers. Typical substrate misalignments of 2° yield lateral periodicity of 8 nm. Fabrication of such lateral superlattices has been demonstrated with both MBE [63] and OMCVD [64] techniques. Polarization anisotropy of PLE spectra has been interpreted as evidence for 2D quantum confinement in these structures [36, 68].

This approach is attractive, since it does not involve post-growth processing steps for defining the wire potential well, and thus it can potentially yield QWR structures of high interface quality. Furthermore, the lateral dimensions of the wires achievable with this technique are beyond the capabilities of most lithography methods currently in use. However, its application requires preparation of vicinal surfaces with regular steps and control of growth rate with extremely high accuracy (on the order of 1% of a monolayer). In addition, the resulting lateral bandgap modulation can be limited by diffusion effects, which might prevent complete segregation of group III atoms [69].

In the patterned, nonplanar substrate approach, lateral thickness variations of QWL layers grown on nonplanar surfaces are employed to form lateral potential wells [29, 65]. Formation of a sufficiently narrow, thicker QWL section laterally bounded by thinner QWL layers can give rise to QWR and QD potential wells due to the strong dependence of confinement energy on QWL thickness [65]. As in the case of growth on vicinal substrates, all interfaces of the potential wells are formed entirely during the growth step, and they are thus potentially free of damage and contamination. The use of lithography in defining the nonplanar patterns, however, provides more flexibility in the design of the quantum structure. Evidence for lateral bandgap vaiations in patterned QWLs grown on nonplanar substrates was provided by TEM and low temperature CL studies of structures grown by MBE [66] and by OMCVD [67]. Lateral potential wells produced in this way have been employed in low threshold QWL semiconductor lasers [4, 70, 71] and laser arrays [72]. This technique is particularly useful for producing extremely narrow potential wells ($\sim$10 nm wide) when it is coupled with formation of stable crystalline facets that evolve during the growth of the QWL barriers [29, 73–75a]. In this case, the wire size and shape can be independent of the initial, lithographically defined surface profile, which reduces well size fluctuations. Stable facets in the $\{111\}$ or (100) planes have been employed for fabricating GaAs/AlGaAs QWR structures by OMCVD on (100) or 6°-off (100) GaAs substrates grooved along the $[01\bar{1}]$ direction [73, 74]. The smallest widths of wires fabricated with this method might be limited to $\sim$10 nm by the finite curvature of the lower barrier layer. Arrays of GaAs/AlGaAs QWRs 50–100 nm wide and with 250-nm pitch have been produced by MBE [75b, 76] and OMCVD [77] on periodically corrugated substrates.

## 3.2. Realization of QWR Laser Structures

### *3.2.1. Simulation by Strong Magnetic Fields*

The carrier energy in a magnetic field $B$ oriented in the $z$ direction is quantized in Landau levels

$$E_l = (l + 1/2)\hbar\omega_c + \hbar^2 k_z^2/2m^*, \tag{9}$$

where $\omega_c = eB/m^*$ is the cyclotron frequency, and $k_z$ is the carrier wavevector along the unconfined direction. The DOS distribution in this case is given by

$$\rho_B(E) = \hbar\omega_c\left(\frac{2m^*}{\hbar^2}\right)^{3/2} \sum_l \frac{1}{\sqrt{E - (l + 1/2)\hbar\omega_c}} \tag{10}$$

and is similar to that in a QWR structure (Fig. 1c), with the QWR subbands represented by the Landau levels. The Lorentz force can thus give rise to effective 2D quantum confinement provided level broadening due to carrier scattering and higher level thermal population are negligible, i.e., $\hbar\omega_c > k_B T$ and $\omega_c\tau_{sc} > 1$, where $\tau_{sc}$ is the carrier scattering time. Similarly, applying high magnetic fields in QWL structures could lead to 0D carrier states, simulating the situation in QD material; carrier confinement in the third dimension is then provided by the QWL potential distribution.

### 3.2.2. *Etching and Regrowth*

QWR and QD diode laser structures based on the InP/InGaAsP material system were fabricated using etching and epitaxial regrowth [23–27]. A schematic diagram of a QWR laser structure formed in this way is shown in Fig. 6 [23]. The structure was fabricated by first growing a conventional QWL laser structure (less its upper cladding layers) by low pressure OMCVD. Next, a wire pattern was formed by wet-chemically etching the QWL active region through an electron-beam written grating mask. The upper part of the laser structure was then formed in a second OMCVD growth step. Scanning electron microscope (SEM) cross sections of the structure at the last two fabrication steps are shown in Fig. 7. The cavity mirrors were cleaved so that light propagates normal to the grating grooves to utilize the higher gain for light polarized along the wires. The wires in this structure were 30 nm wide and 10 nm thick, and the grating pitch was

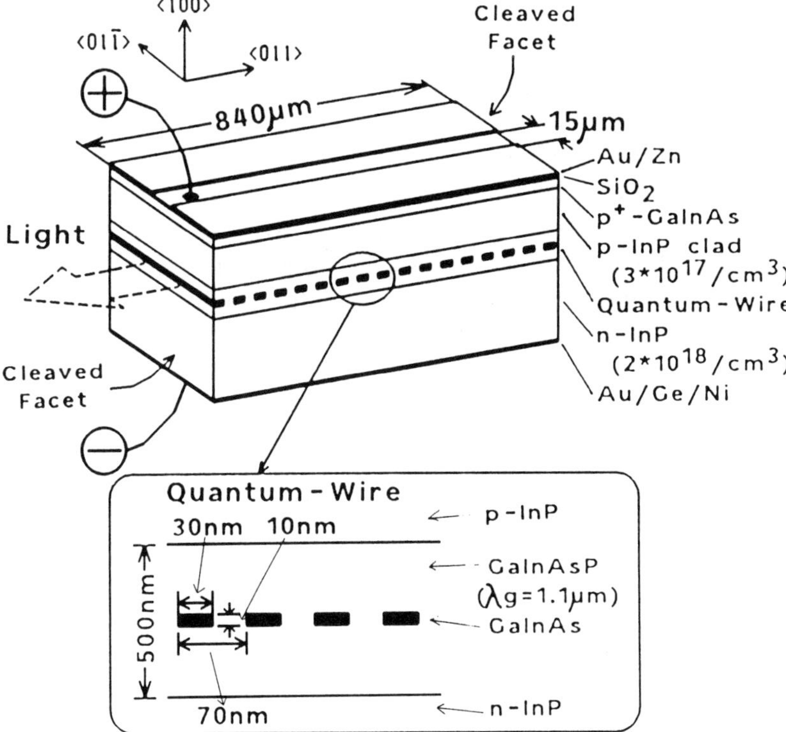


**Fig. 6.** Schematic view of GaInAs/GaInAsP QWR laser fabricated by etching and regrowth (after [23]).

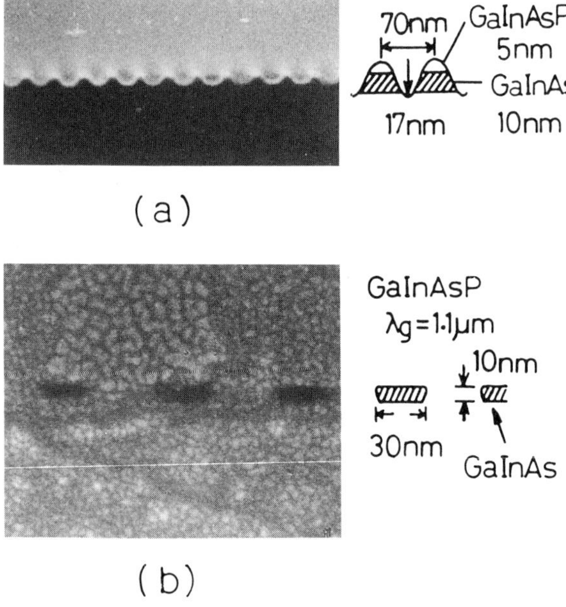


**Fig. 7.** SEM cross sections of GaInAs/GaInAsP QWR lasers: (a) corrugated quantum well structure after wet chemical etching; (b) embedded quantum wires after OMCVD regrowth (after [23]).

70 nm. Wet chemical etching was used in this process to minimize etching damage to the lateral interfaces of the wire. In addition, low temperature regrowth (560°C) was attempted in order to reduce thermal damage to the active region [23, 27]. A similar process was used to make QD laser structures with lateral dot dimensions of $<100$ nm [24, 25].

### *3.2.3. Growth on Vicinal Substrates*

Figure 8 shows a schematic view of a GaAs/AlGaAs QWR laser structure grown by MBE on 2° off-(100) GaAs substrate; the inset shows a transmission electron microscope (TEM) cross section of the laser core [34, 35]. The laser active region consisted of a 5 nm thick GaAs layer and a 5 nm thick lateral superlattice grown by alternating (nominally) half monolayer depositions of GaAs and $Al_{0.25}Ga_{0.75}As$. This forms a GaAs QWL active region laterally modulated by the lateral superlattice with 8-nm pitch. The weakly modulated structure is expected to be less sensitive to lateral size fluctuations compared with an active layer consisting entirely of the lateral superlattice

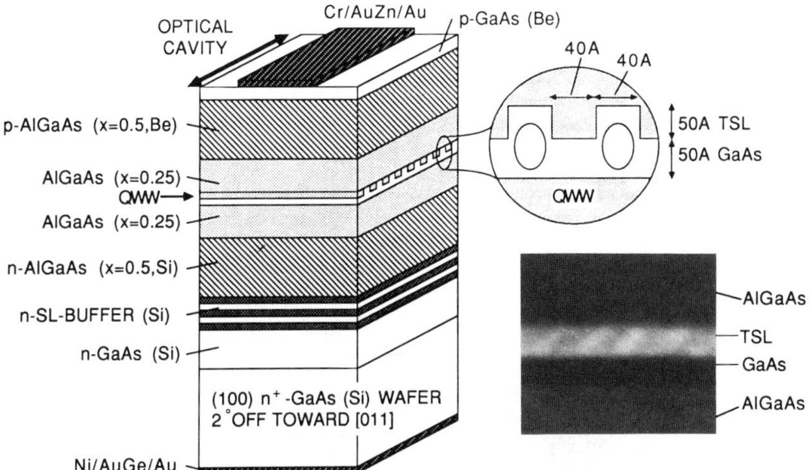


**Fig. 8.** Schematic illustration of GaAs/AlGaAs QWR laser fabricated by MBE growth on a vicinal substrate. The QWRs are perpendicular to the contact stripe. The inset is a transmission electron micrograph of the QWRs; brighter regions correspond to compositions with higher Al content (after [35]).

[35]. The active region was cladded by optical waveguide layers, and the contact stripe was oriented perpendicular to the wires.

### *3.2.4. Growth on Patterned Nonplanar Substrates*

Cross sections of a single-QWR GaAs/AlGaAs semiconductor laser grown by OMCVD on a V-grooved (100) GaAs substrate are shown in Fig. 9 [32]. Single QWL, graded index, separate confinement heterostructure (GRIN-SCH) laser was grown in the $[01\bar{1}]$ oriented V-shaped channel. A very sharp corner develops between two $\{111\}$A planes during the growth of the AlGaAs cladding layers, as can be seen in the SEM cross section of Fig. 9b. The GaAs active region, on the other hand, grows thicker at the bottom of the groove as a result of diffusion of Ga species to the concave groove corner, forming a crescent-shaped QWR $\sim$100 nm wide and $\sim$10 nm thick at its center (see TEM cross section in Fig. 9c). The tapering in the QWL crescent, however, yields a lateral potential well of considerably smaller effective width, as discussed in what follows. The optical waveguide layers are laterally tapered as well as a result of preferential migration of Ga species during the growth of the graded index waveguide layer, giving rise to a 2D optical waveguide surrounding the QWR (see Fig. 9b). This unique GRIN-SCH

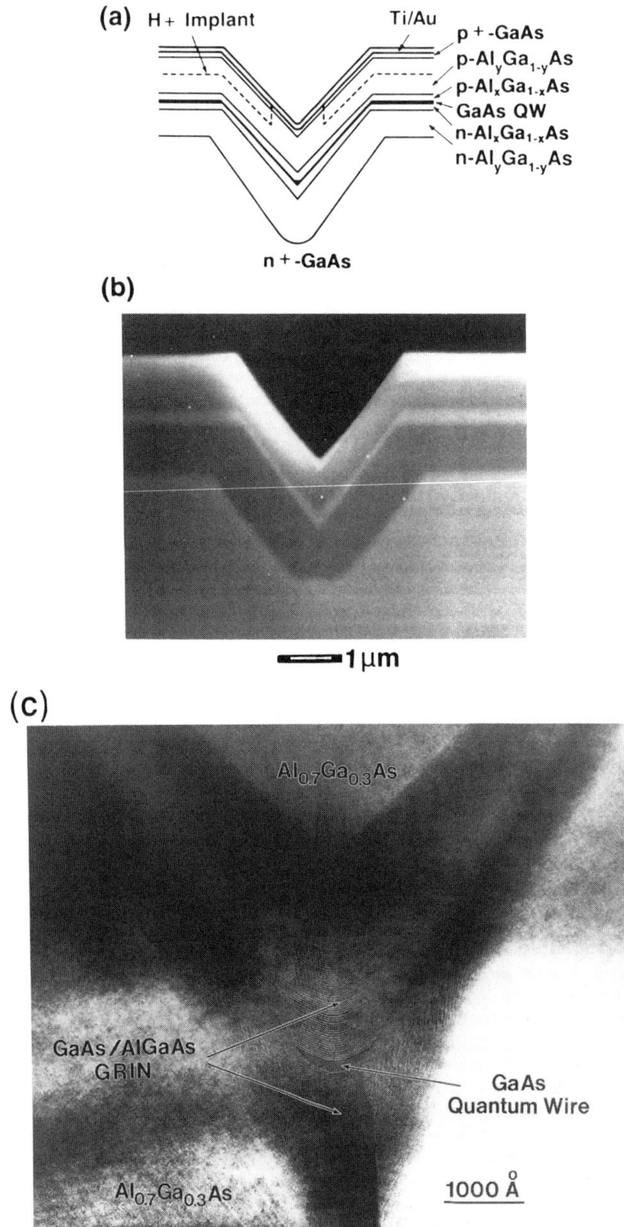


**Fig. 9.** Cross sections of GaAs/AlGaAs single-QWR laser grown by OMCVD on V-grooved substrate: (a) schematic cross section; (b) SEM cross section showing the two dimensional optical waveguide; (c) TEM cross section showing the QWR crescent at the center of the optical waveguide (after [32]).

configuration, the 2D equivalent of 1D GRIN-SCH QWL structures [78], is important for achieving low threshold QWR laser structures, since it significantly increases the optical confinement factor Γ. It may also play a role in improving the efficiency of carrier capture by the QWR crescent. Proton implantation was employed for confining the injected current to the vicinity of the QWR active region (Fig. 9a).

In the QWR crescent structures, the lateral bandgap variations due to the QWL tapering are much more gradual than the ones resulting from the transverse, abrupt variations in the Al-mole-fraction differences. This allows the evaluation of their 2D subband structure using a perturbation method [31]. In this approach, the wavefunctions and eigenvalues for each value of QWL thickness are first obtained by solving a set of 1D Schroedinger equations for each lateral position at the crescent. The eigen-energies thus calculated are then employed to construct a 1D lateral potential well, which yields the QWR subbands and wavefunctions. Figure 10a displays the

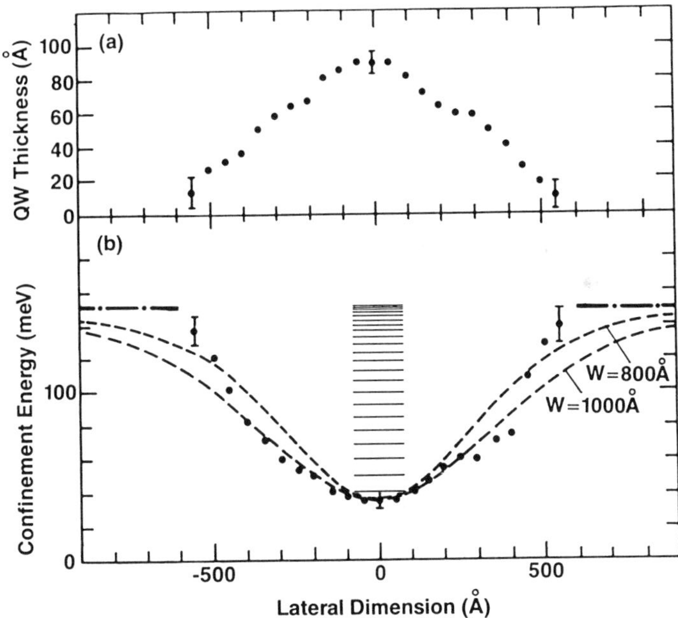


**Fig. 10.** Quantum wire subbands of a GaAs/AlGaAs crescent shaped QWR structure. (a) Measured lateral thickness distribution in the QWR crescent; (b) lateral distribution of electron confinement energies calculated from (a). The horizontal lines are the QWR subbands calculated from the resulting lateral potential well. The dashed lines are model potential wells (after [31]).

measured lateral variation of QWL thickness at the GaAs QWR crescent of a single-QWR laser structure similar to that of Fig. 9c [31]. The corresponding lateral distribution of confinement energy for electrons, forming the QWR conduction band potential well, is shown in Fig. 10b. In this case, the lateral potential profile can be conveniently approximated by $\varepsilon_{el}(0) + \Delta\varepsilon_{el}\tanh^2(y/2W)$, where $y$ is the lateral coordinate, $\varepsilon_{el}(y)$ is the confinement energy, $\Delta\varepsilon_{el} \equiv \varepsilon_{el}(\infty) - \varepsilon_{el}(0)$, and $W$ is a measure of the width of the potential well. The QWR energy levels are then given by [31]

$$E_{el,l} = \varepsilon_{el}(0) + \Delta\varepsilon_{el} - \frac{\hbar^2}{2m^*W^2}\left[-(1+2l) + \left(1 + \frac{2m^*W^2\,\Delta\varepsilon_{el}}{\hbar^2}\right)^{1/2}\right]^2, \quad l = 0, 1, 2, \ldots. \tag{11}$$

A similar procedure yields the subbands for the heavy and light holes. Figure 10b shows the QWR subbands for electrons in the V-groove QWR laser structure. This model predicts ~10 meV separation between adjacent transition energies for 2D confined electrons and holes in the crescent-shaped wire [31].

Resharpening of the V-groove during the growth of the AlGaAs cladding layers above the GaAs QWR allows vertical stacking of QWRs having similar cross sections, which is useful for increasing the confinement factor Γ. Vertically stacked QWR lasers incorporating two and three wires have been fabricated [33, 79–81]. TEM cross section of the core of a 3-QWR laser structure is shown in Fig. 11 [80]. The three crescent-shaped wires are each ~10 nm thick and <100 nm wide, and they are placed at the center of the 2D GRIN optical waveguide.

## 4. LASING CHARACTERISTICS

### 4.1. Semiconductor Lasers in High Magnetic Fields

Reduction in temperature sensitivity of threshold current in GaAs/AlGaAs double heterostructure (DH) diode lasers immersed in high magnetic fields has been demonstrated; increase of $T_0$ from 155 K for $B = 0$ to 313 K for $B = 24$ T was obtained for active region temperatures between 230 and 300 K [9]. Increase in $T_0$ from 200 K for $B = 0$ to 260 K for $B = 20$ T (active region at room temperature) was measured for GaAs/AlGaAs QWL lasers, in which three dimensional quantum confinement is expected [19]. It is interesting to note that the absolute value of the threshold current of semiconductor lasers in high magnetic fields was found to *increase* with increasing field $B$, in agreement with a theoretical model [19].

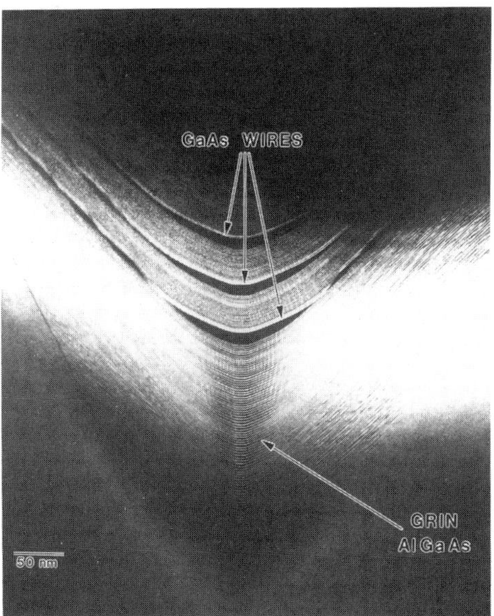


**Fig. 11.** Cross-sectional transmission electron micrograph showing the core of a 3-QWR GaAs/AlGaAs laser. The three QWR crescents are at the center of a 2D graded index (GRIN) optical waveguide (after [80]).

The dynamic properties of DH and QWL lasers placed in high magnetic fields have been investigated [20–22]. Figure 12 shows the measured relaxation oscillation frequency $f_r$ versus the square root of the output power for a GaAs/AlGaAs DH laser at $B = 0$ and 20 T. A 1.4 $\times$ increase in relaxation oscillation frequency was obtained for $B = 20$ T at room temperature as compared with $B = 0$ [20]. The measured twofold increase in the differential gain at $B = 20$ T is consistent with the expected increase due to 2D-confinement effects. In addition, these lasers exhibited blue shift in lasing wavelength corresponding to $\sim$9 meV in photon energy at 20 T, compared with an expected $(1/2)\hbar\omega_c$ zero point energy of 17 meV.

The variation of the measured spectral linewidth with inverse output power for a GaAs/AlGaAs DH for several values of $B$ is shown in Fig. 13. Spectral linewidth reduction by a factor of 0.6 was observed for $B = 19$ T at 190 K [21]. Quantum well GaAs/AlGaAs lasers exhibited 2 $\times$ reduction in linewidth for $B = 19$ T at 165 K [22].

### 4.2. QWR and QD Lasers Fabricated by Etching and Regrowth

The GaInAs/GaInAsP QWR laser structure shown in Fig. 6 was operated under pulsed conditions at 77 K [23]. Lasing from the optical waveguide

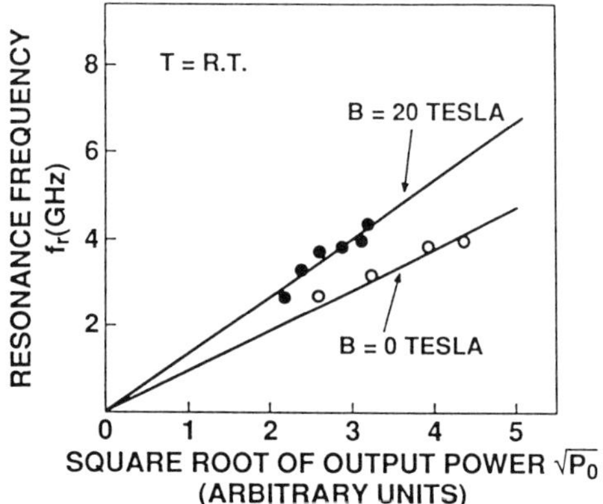


**Fig. 12.** Measured relaxation oscillation frequency with and without magnetic fields versus the square root of the output power for a GaAs/AlGaAs double heterostructure laser at room temperature (after [20]).

region (at 1.01-$\mu$m wavelength) was observed with $I_{th} = 250$ mA (threshold current density $J_{th} = 2$ kA/cm$^2$), and lasing from the wires (at 1.226-$\mu$m wavelength) occurred with $I_{th} = 5$ A ($J_{th} = 39.7$ kA/cm$^2$) for an 840 $\mu$m long optical cavity. Two nonlasing peaks were also observed at 1.35- and 1.42-$\mu$m wavelengths. Lasing from the QWRs was attributed to one of the higher subband transitions. In a similar structure, for which the wires were only partly etched, the threshold current was greatly reduced, to 320 mA ($J_{th} =$ 3.8 kA/cm$^2$), and lasing occurred from the wires only. The spectra of these QWR lasers are shown in Fig. 14, where the assignments of QWR subbands

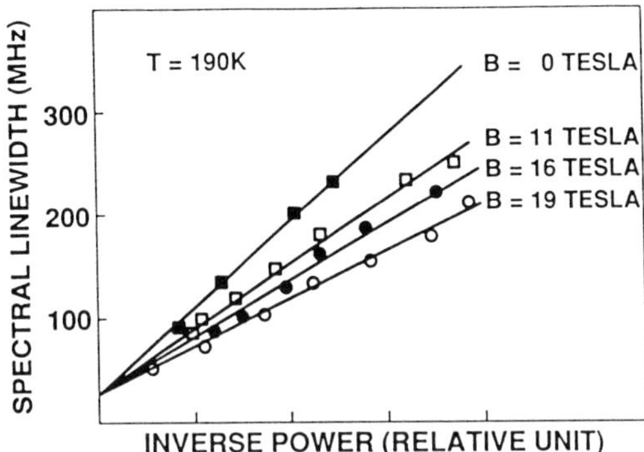


**Fig. 13.** Measured spectral linewidth versus reciprocal output power at various magnetic field values $B$ for a double heterostructure GaAs/AlGaAs laser at 190 K (after [21]).

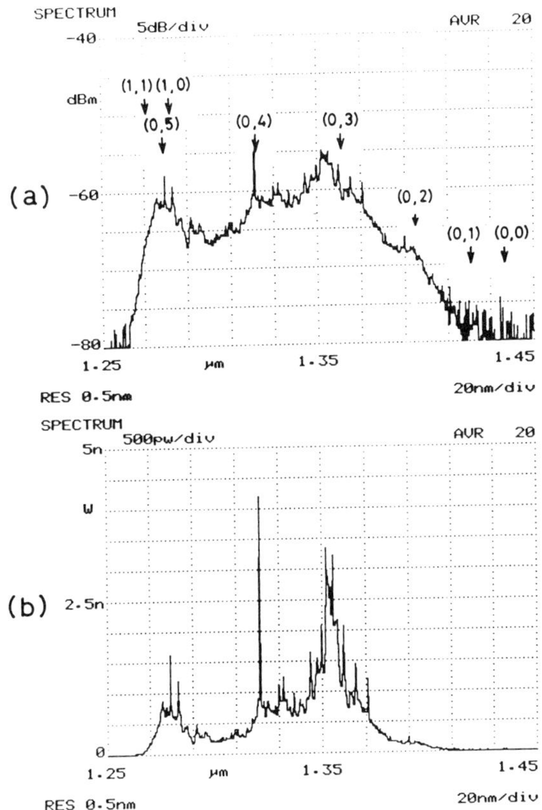


**Fig. 14.** Spectra of GaInAs/GaInAsP QWR lasers fabricated by etching and regrowth; T = 77 K. (a) Logarithmic scale; (b) linear scale. The assigned QWR subbands are indicated by $(l, m)$, with (0, 0) being the fundamental energy level (after [23]).

determined by a model are also indicated [23]. Lasing started at 1.36-$\mu$m wavelength, which corresponds to QWR subband (0, 3) (here, (0, 0) indicates the fundamental subband). At higher injection currents, lasing occurred at higher subbands, at 1.32- and 1.28-$\mu$m wavelengths.

QWR lasers fabricated in a similar way but with wires of larger cross sections (120 nm wide, 30 nm thick) operated cw at 77 K at 1.375-$\mu$m wavelength and with $I_{th} = 63$ mA ($J_{th} = 810$ A/cm$^2$) for $L = 390$ $\mu$m cavity length. However, no evidence for 2D quantum-confined subbands was observed [27]. Electroluminescence (and possibly amplified spontaneous emission) from GaInAsP QD structures fabricated by etching and regrowth was observed at 77 K under pulsed conditions at 1.37- and 1.42-$\mu$m wavelengths, but lasing was not achieved [25].

### 4.3. QWR Lasers Grown on Vicinal Substrates

GaAs/AlGaAs QWR lasers grown on vicinal GaAs substrates were operated at room temperature under pulsed conditions [34, 35]. Broad area (50 $\mu$m stripe width) lasers exhibited threshold currents as low as 130 mA for 400 $\mu$m cavity lenth and threshold current densities as low as 470 A/cm$^2$ for 990 $\mu$m cavity length, and differential efficiency was 22.5% per facet. Characteristic temperature $T_0$ was 80–130 K near room temperature. The lasing wavelength was 8277 Å; lasing spectra did not show features different from those of conventional QWL lasers. Low temperature photoluminescence excitation (PLE) spectra showed anisotropy in the heavy-hole versus the light-hole transitions, attributed to 2D quantum confinement of carriers in the active region (see Fig. 15). The lasing wavelength, extrapolated to the low temperature range using the temperature dependence of the GaAs band edge (also shown in Fig. 15), corresponds to transition between the lowest energy QWR

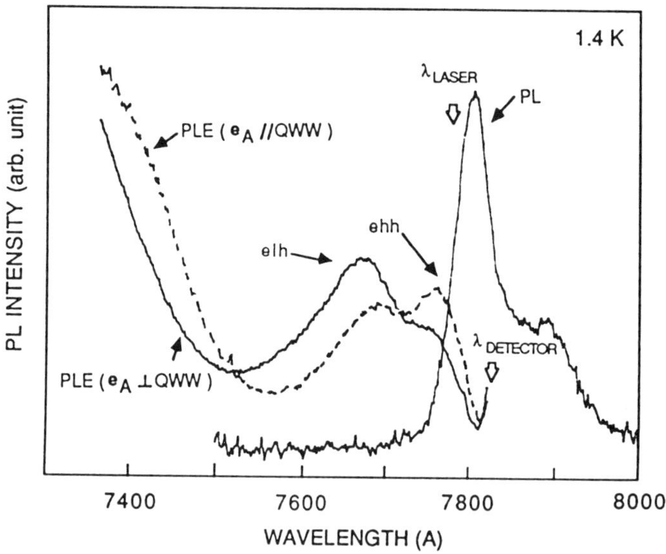


**Fig. 15.** Low temperature photoluminescence (PL) and photoluminescence excitation (PLE) spectra of the GaAs/AlGaAs QWR laser structure of Fig. 8, grown on a vicinal substrate; $\mathbf{e}_A$ indicates the polarization vector. The lasing wavelength, extrapolated from the value measured at room temperature, is also indicated (after [35]).

subbands [35]; this was interpreted as evidence for lasing from the fundamental QWR subband in this structure. However, this procedure did not account for bandgap shrinkage effects, which would blue shift the lasing wavelength with respect to the QWR subbands.

### 4.4. QWR Lasers Grown on Patterned Nonplanar Substrates

Amplified spontaneous emission (ASE) spectra of QWR lasers grown by OMCVD in a V-groove are shown in Fig. 16 for various cavity lengths; ASE spectrum of a QWL laser grown simultaneously on a planar substrate is also shown [31]. Whereas the QWL laser spectra exhibited widely spaced features associated with the QWL subband structure, the QWR laser spectra showed more closely spaced peaks attributed to transitions between QWR subbands. The spacing between these peaks, ~10 meV, is in agreement with the calculated separation for the QWR subband transitions (see Fig. 10) [31].

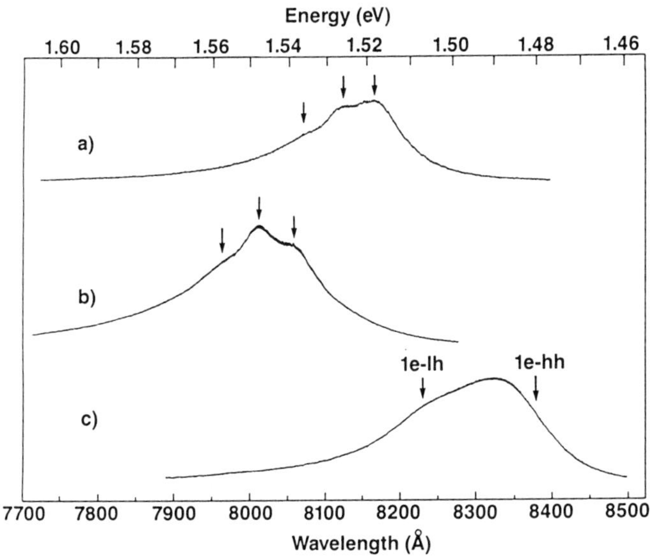


**Fig. 16.** Amplified spontaneous emission spectra of QWR and QWL lasers. (a) QWR laser, $L = 1.68$ mm, $I = 34$ mA ($I_{th} = 35$ mA); (b) QWR laser, $L = 0.54$ mA, $I = 17$ mA ($I_{th} = 18$ mA). Arrows indicate peaks due to transitions between the QWR subbands; (c) QWL laser. Arrows indicate calculated ground state electron–heavy-hole and electron–light-hole transitions (room temperature, pulsed operation). The QWR conduction band structure is shown in Fig. 10; the QWL structure was grown simultaneously on a planar substrate (after [31]).

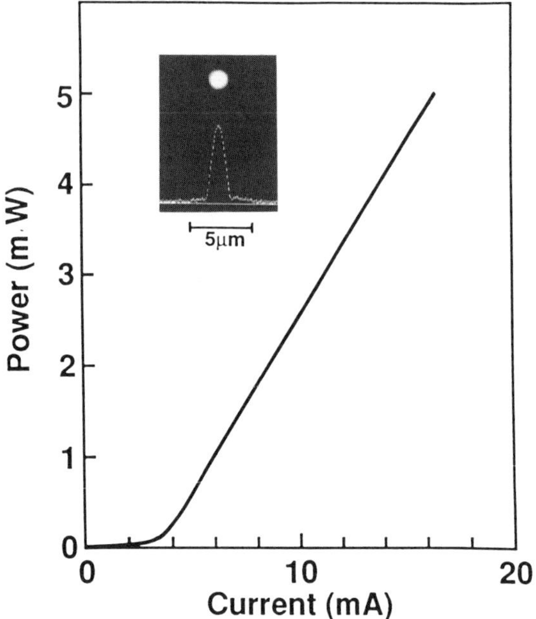


**Fig. 17.** Light versus current characteristics of the single-QWR GaAs/AlGaAs laser structure shown in Fig. 9; room temperature, pulsed operation, cavity length $L = 350\ \mu\text{m}$, uncoated facets. The inset shows a photograph and the lateral distribution of the near field pattern (after [32]).

The peaks are a result of the higher DOS, and hence higher optical gain, at the QWR subband energies. QWR lasers with shorter cavities exhibited higher order subband transitions, since their higher cavity losses (per unit lenth) result in higher carrier density at threshold and hence blue-shifted Fermi levels (see Fig. 16). Above threshold, the QWR lasers oscillated at one of the peaks observed in their ASE spectra, with shorter lasers lasing at higher energy peaks due to band filling.

The extremely small cross section of the crescent-shaped QWR results in small confinement factor $\Gamma$ and hence reduced modal gain. The QWR lasers whose ASE spectra are shown in Fig. 16 had threshold currents as low as 18 mA at room temperature. Higher $\Gamma$ was obtained by increasing the Al mole fraction in the waveguide cladding layers from 0.5 to 0.7 and decreasing the thickness of each GRIN layer from 0.2 to 0.15 $\mu$m (see Fig. 9) [32]. The light versus current characteristic of such an improved single-QWR laser is shown in Fig. 17. Threshold currents as low as 3.5 mA and total differential efficiency of 45% were achieved under pulsed operation at room temperature for uncoated mirrors; cw operation was also achieved with threshold currents 10–15% higher due to thermal effects. The near field intensity distribution was circular, with $\sim 1\ \mu$m full width at half maximum (FWHM), and the lasing mode was polarized parallel to the (100) plane (TE-like polarization), indicating lasing at the fundamental waveguide mode (see inset of Fig. 17). The fact that lasing occurred from the wire was established by comparing the measured lasing photon energy with the calculated bandgap of the QWL regions as determined by the TEM data [31, 32, 82].

Further evidence for larger DOS and higher optical gain at the QWR subbands was provided by examining the evolution of the QWR laser spectra above threshold, displayed in Fig. 18 [32]. Near threshold ($I = 4.7$ mA),

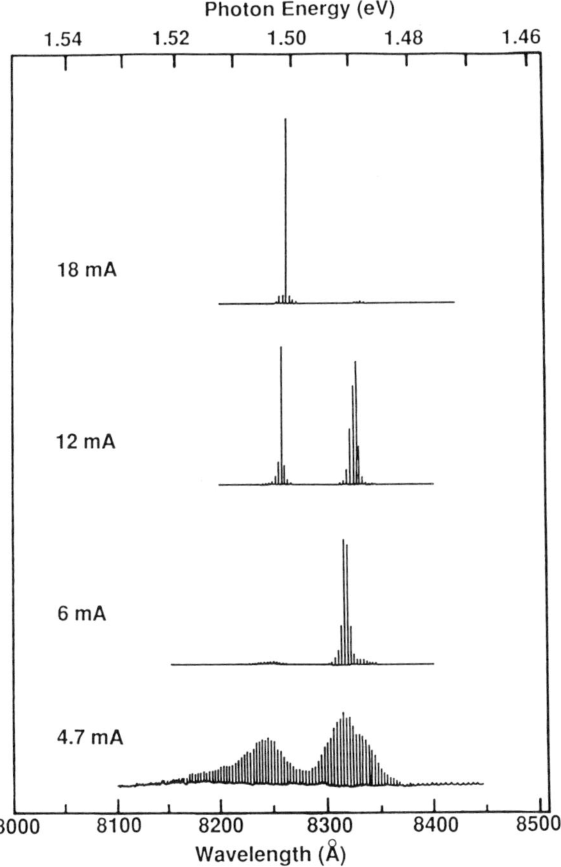


**Fig. 18.** Evolution of emission spectra above threshold of the single-QWR laser structure shown in Fig. 9 exhibiting lasing at two adjacent QWR subbands; $L = 270\ \mu$m, $I_{th} = 4.3$ mA (after [32]).

many longitudinal modes were observed. The envelope of these modes reflects the spectral distribution of optical gain, which is enhanced at the two QWR subbands, separated by ~13 meV. The lasing shifted to the low energy peak slightly above threshold. At higher currents, the lasing mode switched to the higher energy subband due to band filling. Note that no lasing takes place in between the subbands because of the lower DOS there. These effects are similar to mode hopping between adjacent QWL subbands, which were induced in QWL lasers by varying the cavity loss [83] or active region temperature [5].

The low modal gain in single-QWR laser structures requires extremely low cavity losses for achieving low threshold currents and lasing from the lowest QWR subbands. Reduction of the mirror losses can be achieved by HR facet coatings. Spectra of single-QWR GaAs/AlGaAs lasers grown on V-grooved substrates are shown in Fig. 19 for various mirror reflectivities $R$ and cavity lengths $L$ [80]. The ASE spectrum of a long (3.48 mm), uncoated laser is also shown for comparison. The subband structure in this ASE spectrum is more pronounced than in the spectra of Fig. 16. This might be due to the tighter optical confinement, which results in lower carrier density

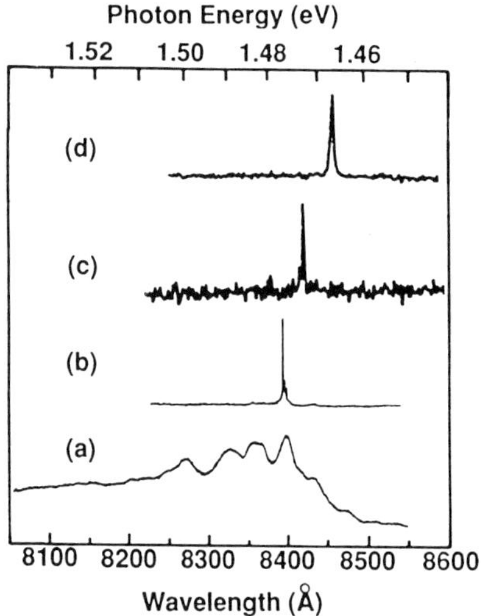


**Fig. 19.** Spectra of single-QWR lasers grown by OMCVD on V-grooved substrate (see Fig. 9) for different values of the mirror loss parameter $\alpha_M = (1/L)\ln(1/R)$; pulsed operation, room temperature. (a) $L = 3.48$ mm, uncoated facets ($\alpha_M = 3.5\ \text{cm}^{-1}$); $I = 22$ mA, $I_{th} = 23$ mA; (b) same as (a), except $I = 25$ mA; (c) $L = 180\ \mu$m, HR coated facets ($\alpha_M \sim 2\ \text{cm}^{-1}$); $I = 1.3$ mA, $I_{th} = 1.05$ mA; (d) $L = 350\ \mu$m, HR coated facets ($\alpha_M \sim 1\ \text{cm}^{-1}$); $I = 2$ mA, $I_{th} = 1.7$ mA (after [80]).

at threshold and hence can reduce broadening effects associated with carrier–carrier scattering. It can be seen that successive reduction of mirror loss coefficient $\alpha_M = (1/L)\ln(1/R)$ leads to monotonic red shift in the lasing wavelength and lasing from lower QWR subbands. A similar effect is observed in evolution of lasing spectra as functions of cavity length for constant mirror reflectivity. Threshold currents as low as 1.05 mA were obtained for HR coated single-QWR lasers ($L = 180\ \mu$m) at room temperature.

The higher modal gain in vertically stacked multi-QWR lasers (see Fig. 11) was employed to further reduce the threshold current and carrier density (per wire). Threshold current as a function of cavity length is displayed in Fig. 20 for single- and triple-QWR laser structures [33]. A minimum threshold current of 2.5 mA (uncoated facets) was obtained for the 3-QWR devices. Furthermore, this minimum value is obtained at a shorter cavity length, $L \sim 100\ \mu$m, which results in higher differential efficiency (55%) [33, 79, 81]. Although TEM cross sections of the 3-QWR laser core show that the wires are not identical in size and shape, the variations in subband structure from wire to wire were small enough to allow observation of 2D quantum size effects in the laser spectra (Fig. 21). The near field patterns of the 3-QWR lasers were virtually identical to those of the single-QWR devices ($\sim 1\ \mu$m FWHM, see inset in Fig. 21), implying the expected increase in the total power filling factor. The reduced carrier density per wire at threshold resulted in further red shift in the lasing wavelength of

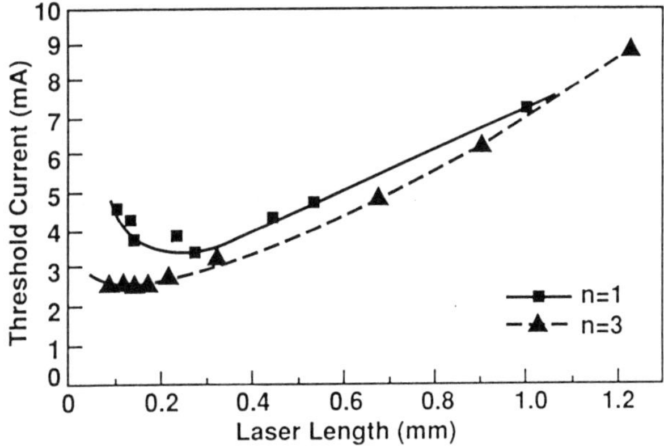


**Fig. 20.** Threshold current versus cavity length for the single-QWR and 3-QWR laser structures of Figs. 9 and 11; uncoated facets, pulsed operation, room temperature (after [33]).

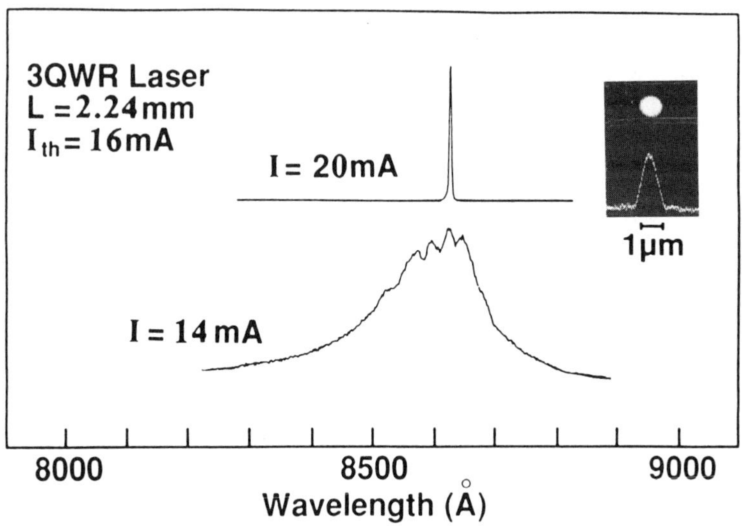


**Fig. 21.** Spectra of the 3-QWR laser of Fig. 11 below and above threshold; room temperature, pulsed operation, uncoated facets; $L = 2.24$ mm, $I_{th} = 16$ mA. Inset shows the near field pattern above threshold (after [33]).

the 3-QWR devices compared with the single-QWR ones. Submilliampere threshold currents were achieved with short cavity (135 $\mu$m) HR coated 3-QWR lasers [33]. Threshold currents as low as 0.6 mA were obtained at room temperature (see Fig. 22); lasing occurred at 859- and 863-nm wavelengths, compared with 840 nm for longer cavity ($L = 3.48$ mm) single-QWR lasers with uncoated facets.

## 5. CONCLUSIONS AND FUTURE TRENDS

Recent progess in the development of QWR lasers has been driven by expectations of achieving improved semiconductor laser performance as well as the desire to gain better understanding of quasi-1D systems. However, realization of these novel laser structures has required major advances in semiconductor epitaxial growth and processing technologies. As a result, much of the effort in this field has been aimed at developing methods for fabricating QWR heterostructures of acceptable optical quality. Although no single fabrication method of choice has emerged yet, it has been recognized that avoiding post-growth processing steps for defining the QWR potential well is important for achieving high QWR interface quality. A

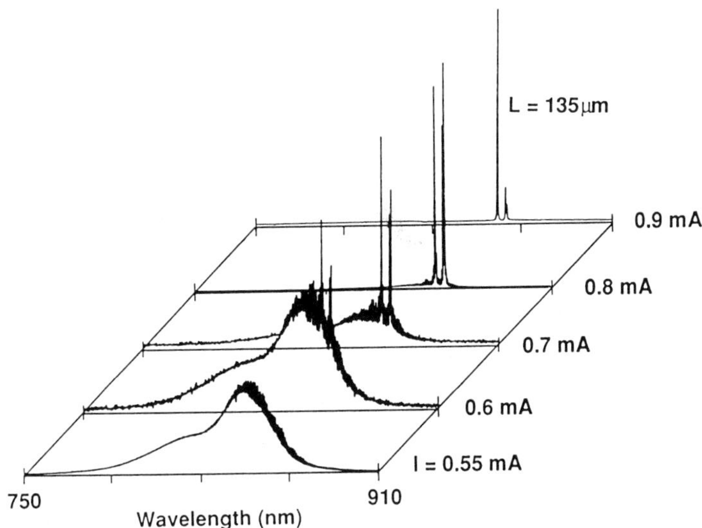


**Fig. 22.** Spectra of HR coated 3-QWR laser whose cross section is shown in Fig. 11, for various currents $I$; $L = 135\ \mu m$, $I_{th} = 0.6$ mA (room temperature, pulsed operation) (after [33]).

number of techniques involving formation of QWRs *during* epitaxial growth have thus been developed, including growth on vicinal substrates [62–64] and growth on patterned, nonplanar substrates [65–67, 75–77]. In addition, modified etching-and-regrowth techniques incorporating steps of damage removal and regrowth for surface passivation are being pursued [56].

Attempts have been made to fabricate single- or multi-QWR laser structures by etching-and-regrowth [23–27], growth on vicinal substrates [34, 35], and growth on nonplanar patterned substrates [30–33, 79–82]. The first approach is the most direct one, and its advantage lies in providing simplicity and flexibility in the design of quantum structures. While indications for 2D quantum confinement effects have been sugested by the measured spectra of these lasers, etching-and-regrowth related damage to the wire interfaces results in operation at low temperatures only and with very high threshold currents. Further progress, especially in techniques for damage-layer removal and improved regrowth processes (possibly *in situ*) could lead to improved QWR laser structures produced with this method.

QWR laser structures fabricated by growth on vicinal substrates exhibit reasonably low threshold current densities at room temperature. However, no clear evidence for 2D quantum confinement effects in the lasing features (or any difference from comparable QWL laser structures) has been found.

Further work is required to develop QWR laser structures with uniform wires and well-confined carriers with this approach. Its main advantage is in the ability to produce extremely dense and narrow QWR arrays without the need to rely on lithography techniques. On the other hand, the fact that the technique avoids employing lithography may limit its flexibility in designing the quantum structures.

QWR laser structures grown on nonplanar patterned substrates exhibit the lowest threshold currents for QWR lasers (as low as 0.6 mA), comparable with the lowest threshold QWL lasers reported to date, and show spectral features suggesting 2D quantum confinement effects. Although the threshold currents are still some two orders of magnitude higher than the theoretical predictions for ideal QWR lasers, optimization of the structure is expected to further reduce the current values. The structures demonstrated thus far incorporate tight optical waveguides in which light propagates along the wire axis. This collinear configuration may result in lower (material) gain due to gain anisotropy in the QWR active region. On the other hand, the tight optical waveguide, whose formation relies on the peculiar OMCVD growth in a $[01\bar{1}]$ oriented groove as well, leads to relatively high confinement factor Γ. This makes the collinear configuration attractive for achieving very low threshold currents. Development of densely packed, smaller size wires is required in order to achieve lasing from the fundamental wire subband, so that the advantage of 2D quantum confinement effects can be fully utilized. This will also allow construction of transverse waveguide-wire configurations. Recent reports demonstrated the feasibility of producing such QWR arrays by MBE [75b, 76] and OMCVD [77] on submicron period corrugations.

Clearly, further progress is required to realize QWR laser structures exhibiting better performance as compared with the current generation of QWL lasers. Smaller size, better interface quality, higher density, and better uniformity of the QWRs is necessary for achieving higher optical gain. Furthermore, lower loss optical cavities that will be compatible with the inherently low modal gain in these devices are essential for minimizing threshold currents and maximizing differential efficiency. Such optical cavities could be realized by developing low loss laser waveguides or resorting to vertical cavity configurations.

One particular future direction in the development of QWR lasers is further reduction in threshold current. A perspective on this issue is shown in Fig. 23, which displays progress in reduction of threshold current of the semiconductor laser since its first demonstration three decades ago. It is interesting to note that a reduction by about one order of magnitude per

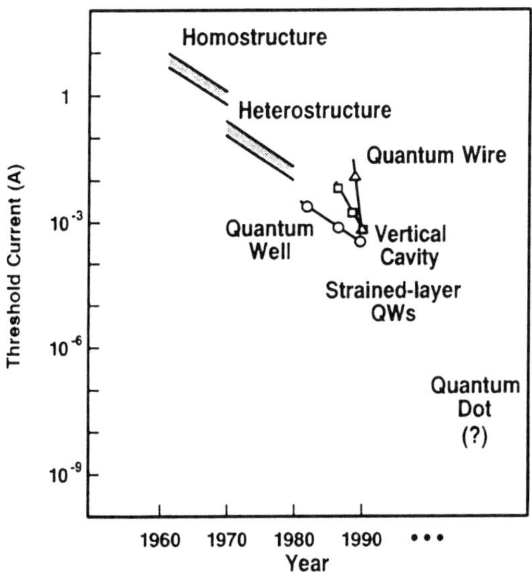


**Fig. 23.** Threshold current as function of year, showing evolution of low threshold semiconductor laser structures. Circles: QWL lasers; squares: vertical cavity lasers; triangles: QWR lasers.

decade has occurred, and that this constant reduction has been maintained owing to the introduction of new concepts in device materials and configurations. The two first major advances were achieved with the advent of heterostructures, followed by the introduction of quantum well materials. Currently, there are a number of new concepts being explored for further threshold current reduction, namely, strained layer quantum wells, vertical cavity surface emitting lasers (VCSELs), and QWR (and QD) lasers. Strained QWLs should exhibit lower internal absorption, higher material gain, and lower transparency currents due to modified band structure and reduced heavy-hole effective mass [84, 85]. The small active volume and low cavity losses in VCSELs are beneficial for extremely low threshold currents as well [44]. All three types of lasers are rapidly approaching the performance of conventional QWL lasers in terms of threshold current. A combination of all three concepts might lead to the "ultimate" low threshold semiconductor laser. The resulting structure, consisting of strained QWRs or QDs embedded in a high finesse vertical cavity, will have an extremely small active volume and reduced heavy-hole mass for minimizing transparency current. The reduced heavy-hole mass will also be effective for obtaining multi-dimensionally quantum-confined holes with wire or dot dimensions comparable with those required for electron quantum confinement. Higher differential gain due to strain and quantum confinement effects and low cavity losses would allow operation with the low modal gain achievable with a small number of wires or dots. In properly designed microcavities, control of spontaneous emission into the lasing mode might also play an important role in threshold current reduction [86].

The rapid progress in the field of multi-dimensional quantum-confined lasers is expected to continue in the near future. Further improvement in the fabrication techniques should lead to reduction in threshold currents and will allow the investigation of the other potential advantages of QWR lasers predicted by theory. The development of these advanced structures might thus lead to semiconductor lasers with better performance and to better understanding of optical phenomena in low dimensional semiconductors.

## ACKNOWLEDGMENT

It is a pleasure to thank my collaborators R. Bhat, D. M. Hwang, S. Simhony, E. Colas, and N. G. Stoffel for their contributions to the work performed at Bellcore described in this review.

## REFERENCES

1. Holonyak, N., Jr., Kolbas, R. M., Dupuis, R. D., and Dapkus, P. D. (1980). *IEEE J. Quantum Electron.* **QE-16**, 170.
2. Arakawa, Y., and Yariv, A. (1986). *IEEE J. Quantum Electron.* **QE-22**, 1887.
3. Alferov, Zh. I., Vasil'ev, A. M., Ivanov, S. V., Kop'ev, P. S., Ledentsov, N. N., Lutsenko, M. E., Mel'tser, B. Ya., and Ustinov, V. M. (1988). *Sov. Tech. Phys. Lett.* **14**, 782.
4. Kapon, E., Simhony, S., Harbison, J. P., Florez, L. T., and Worland, P. (1990). *Appl. Phys. Lett.* **56**, 1825.
5. Zory, P. S., Reisinger, A. R., Waters, R. G., Mawst, L. J., Zmudzinski, C. A., Emanuel, M. A., Givens, M. E., and Coleman, J. J. (1986). *Appl. Phys. Lett.* **49**, 16.
6. Uomi, K., Mishima, M., and Chinone, N., (1987). *Appl. Phys. Lett.* **51**, 78.
7. Ogasawara, N., Itoh, R., and Morita, R. (1985). *Japan. J. Appl. Phys.* **24**, L519.
8. Derry, P. L., Chen, T. R., Zhuang, Y. H., Paslaski, J., Mittelstein, M., Vahala, K., and Yariv, A. (1988). *Appl. Phys. Lett.* **53**, 271.
9. Arakawa, Y., and Sakaki, H. (1982). *Appl. Phys. Lett.* **40**, 939.
10. Asada, M., Miyamoto, Y., and Suematsu, Y. (1986). *IEEE J. Quantum Electron.* **QE-22**, 1915.
11. Yariv, A. (1988). *Appl. Phys. Lett.* **53**, 1033.

12. Miyamoto, Y., Miyake, Y., Asada, M., and Suematsu, Y. (1989). *IEEE J. Quantum Electron.* **QE-25**, 2001.
13. Arakawa, Y., Vahala, K., and Yariv, A. (1984). *Appl. Phys. Lett.* **45**, 950.
14. Miller, D. A. B., Chemla, D. S., and Schmitt-Rink, S. (1988). *Appl. Phys. Lett.* **52**, 2154.
15. Schmitt-Rink, S., Chemla, D. S., and Miller, D. A. B. (1989). *Adv. Phys.* **38**, 89.
16. Matsubara, K., Ravikumar, K. G., Asada, M., and Suematsu, Y. (1989). *Trans. IEICE* **E72**, 1179.
17. Vahala, K. J. (1988). *IEEE J. Quantum Electron.* **QE-24**, 523.
18. Zarem, H., Vahala, K., and Yariv, A. (1989). *IEEE J. Quantum Electron.* **QE-25**, 705.
19. Berendschot, T. T. J. M., Reinen, H. A. J. M., Bluyssen, H. J. A., Harder, C., and Meier, H. P. (1989). *Appl. Phys. Lett.* **54**, 1827.
20. Arakawa, Y., Vahala, K., Yariv, A., and Lau, K. (1985). *Appl. Phys. Lett.* **47**, 1142.
21. Arakawa, Y., Vahala, K., Yariv, A., and Lau, K. (1986). *Appl. Phys. Lett.* **48**, 384.
22. Vahala, K., Arakawa, Y., and Yariv, A. (1987). *Appl. Phys. Lett.* **50**, 365.
23. Cao, M., Miyake, Y., Tamura, S., Hirayama, H., Arai, S., Suematsu, Y., and Miyamoto, Y. (1990). *Trans. IEICE* **E73**, 63.
24. Miyamoto, Y., Cao, M., Shingai, Y., Furuya, K., Suematsu, Y., Ravikumar, K. G., and Arai, S. (1987). *13th European conference on optical communication*, Helsinki, Finland, September 13–17, 1987. *ECOC '87 Digest*, 35–38.
25. Miyamoto, Y., Cao, M., Shingai, Y., Furuya, K., Suematsu, Y., Ravikumar, K. G., and Arai, S. (1987). *Japan. J. Appl. Phys.* **26**, L225.
26. Cao, M., Miyake, Y., Tamura, S., Hirayama, H., Arai, S., Suematsu, Y., and Miyamoto, Y. (1989). *Proceedings of the seventh international conference on integrated optics and optical fiber communication (IOOC '89)*, Kobe, Japan, July 18–21, 1989, paper **20PDB-10**, Vol. 5, 56–57.
27. Cao, M., Daste, P., Miyamoto, M., Miyake, Y., Nogiwa, S., Arai, S., Furuya, K., and Suematsu, Y. (1988). *Electron. Lett.* **24**, 824.
28. For a recent review, see Kash, K. (1990). *J. Luminescence* **46**, 69.
29. Kapon, E., Yun, C. P., Hwang, D. M., Tamargo, M. C., Harbison, J. P., and Bhat, R. (1988). *Proc. SPIE* **944**, 80.
30. Kapon, E., Simhony, S., Hwang, D. M., Kash, K., Bhat, R., and Colas, E. (1989). *Conference on quantum electronics and laser science (QELS '89)*, Baltimore, Maryland, April 24–28, 1989, paper **PD-15**, 264–265.
31. Kapon, E., Hwang, D. M., and Bhat, R. (1989). *Phys. Rev. Lett.* **63**, 430.
32. Kapon, E., Simhony, S., Bhat, R., and Hwang, D. M. (1989). *Appl. Phys. Lett.* **55**, 2715.
33. Simhony, S., Kapon, E., Colas, E., Hwang, D. M., Stoffel, N. G., and Worland, P. *Appl. Phys. Lett.* (submitted).
34. Tsuchiya, M., Petroff, P. M., and Coldren, L. A. (1989). *47th annual research*

*conference*, Massachusetts Institute of Technology, Cambridge, Massachusetts, June 19–21, 1989, paper **IVA-1**.

35. Tsuchiya, M., Coldren, L. A., and Petroff, P. M. (1989). *Proceedings of the seventh international conference on integrated optics and optical fiber communication (IOOC '89)*, Kobe, Japan, July 18–21, 1989, paper **19C1-1**, Vol. 2, 104–105.
36. Sercel, P. C., and Vahala, K. J. (1990). *Appl. Phys. Lett.* **57**, 545.
37. Asada, M., and Suematsu, Y. (1985). *IEEE J. Quantum Electron.* **QE-21**, 434.
38. Yamauchi, T., Arakawa, Y., and Schulman, J. N. (1990). *Appl. Phys. Lett.* **57**, 1224.
39. Arakawa, Y. (1990). *Extended abstracts of the 22nd (1990 international) conference on solid state devices and materials (SSDM '90)*. Sendai, Japan, August 22–24, 1990, 745–748.
40. Sakaki, K. (1980). *Japan. J. Appl. Phys.* **19**, L735.
41. Asada, M., Miyamoto, Y., and Suematsu, Y. (1985). *Japan. J. Appl. Phys.* **24**, L95.
42. Kapon, E. (1990). *Optics Lett.* **15**, 801.
43. Derry, P. L., Yariv, A., Lau, K., Bar-Chaim, N., Lee, K., and Rosenberg, J. (1987). *Appl. Phys. Lett.* **50**, 1773.
44. Geels, R. S., and Coldren, L. A. (1990). *Appl. Phys. Lett.* **57**, 1605.
45. Lau, K. Y., Bar-Chaim, N., Ury, I., Harder, C., and Yariv, A. (1983). *Appl. Phys. Lett.* **43**, 1.
46. Arakawa, Y., Vahala, K., and Yariv, A. (1986). *Surf. Sci.* **174**, 155.
47. Henry, C. H. (1982). *IEEE J. Quantum Electron.* **QE-18**, 259.
48. Fleming, M. W., and Mooradian, A. (1981). *Appl. Phys. Lett.* **38**, 511.
49. Koch, T. L., and Linke, R. A. (1986). *Appl. Phys. Lett.* **48**, 613.
50. Kash, K., Scherer, A., Worlock, J. M., Craighead, H. G., and Tamargo, M. C. (1986). *Appl. Phys. Lett.* **49**, 1043.
51. Gershoni, D., Temkin, H., Dolan, C. J., Dunsmuir, J., Chu, S. N. G., and Panish, M. B. (1988). *Appl. Phys. Lett.* **53**, 995.
52. Kohl, M., Heitmann, D., Grambow, P., and Ploog, K. (1989). *Phys. Rev. Lett.* **63**, 2124.
53. See, e.g., Roukes, M. L., Scherer, A., Allen, S. J., Jr., Craighead, H. G., Ruthen, B. M., Beebe, E. D., and Harbison, J. P. (1987). *Phys. Rev. Lett.* **59**, 3011; and references therein.
54. Clausen, E. M., Jr., Craighead, H. G., Worlock, J. M., Harbison, J. P., Schiavone, L. M., Florez, L., and van der Gaag, B. (1989). *Appl. Phys. Lett.* **55**, 1427.
55. Izrael, A., Sermage, B., Marzin, J. Y., Ougazzaden, A., Azoulay, R., Etrillard, J., Thierry-Mieg, V., and Henry, L. (1990). *Appl. Phys. Lett.* **56**, 830.
56. Clausen, E. M., Jr., Harbison, J. P., Florez, L. T., and van der Gaag, B. P. (1990). *J. Vac. Sci. Technol.* **B8**, 1960.
57. Cibert, J., Petroff, P. M., Dolan, G. J., Pearton, S. J., Gossard, A. C., and English, J. H. (1986). *Appl. Phys. Lett.* **49**, 1275.

58. Zarem, H. A., Sercel, P. C., Hoenk, M. E., Lebens, J. A., and Vahala, K. J. (1989). *Appl. Phys. Lett.* **54**, 2692.
59. Kash, K., Van der Gaag, B. P., Worlock, J. M., Gozdz, A. S., Mahoney, D. D., Harbison, J. P., and Florez, L. T. (1990). In *Localization and confinement of electrons in semiconductors* (Kuchar, F., Heinrich, H., Bauer, G., eds.), Springer Series in Solid State Sciences. Springer-Verlag, Berlin, Vol. 97, 39–50.
60. Fukui, T., Ando, S., Tokura, Y., and Toriyama, T. *Extended abstracts of the 22nd (1990 international) conference on solid state devices and materials (SSDM '90)*, Sendai, Japan. August 22–24, 1990, 99–102.
61. Gershoni, D., Weiner, J. S., Chu, S. N. G., Baraff, G. A., Vandenberg, J. M., Pfeiffer, L. N., West, K., Logan, R. A., and Tanbun-Ek, T. (1990). *Phys. Rev. Lett.* **65**, 1631.
62. Petroff, P. M., Gossard, A. C., and Wiegmann, W. (1984). *Appl. Phys. Lett.* **45**, 621.
63. Gaines, J. M., Petroff, P. M., Kroemer, H., Simes, R. J., Geels, R. S., and English, J. H. (1988). *J. Vac. Sci. Technol.* **B6**, 1378.
64. Fukui, T., and Saito, H. (1990). *Japan. J. Appl. Phys.* **29**, L731.
65. Kapon, E., Tamargo, M. C., and Hwang, D. M. (1987). *Appl. Phys. Lett.* **50**, 347.
66. Clausen, E. M., Jr., Kapon, E., Tamargo, M. C., and Hwang, D. M. (1990). *Appl. Phys. Lett.* **56**, 776.
67. Colas, E., Clausen, E. M., Jr., Kapon, E., Hwang, D. M., and Simhony, S. (1990). *Appl. Phys. Lett.* **57**, 2472.
68. Tsuchiya, M., Gaines, J. M., Yan, R. H., Simes, R. J., Holtz, P. O., Coldren, L. A., and Petroff, P. M. (1989). *Phys. Rev. Lett.* **62**, 466.
69. Fukui, T., Saito, H., and Tokura, Y. (1989). *Appl. Phys. Lett.* **55**, 1958.
70. Kapon, E., Harbison, J. P., Yun, C. P., and Stoffel, N. G. (1988). *Appl. Phys. Lett.* **52**, 607.
71. Kapon, E., Yun, C. P., Harbison, J. P., Florez, L. T., and Stoffel, N. G. (1988). *Electron. Lett.* **24**, 985.
72. Kapon, E., Harbison, J. P., Yun, C. P., and Florez, L. T. (1989). *Appl. Phys. Lett.* **54**, 304.
73. Bhat, R., Kapon, E., Hwang, D. M., Koza, M. A., and Yun, C. P. (1988). *J. Crystal Growth* **93**, 850.
74. Colas, E., Kapon, E., Simhony, S., Cox, H. M., Bhat, R., Kash, K., and Lin, P. S. D. (1989). *Appl. Phys. Lett.* **55**, 867.
75a. Cox, H. M., Lin, P. S., Yi-Yan, A., Kash, K., Seto, M., and Bastos, P. (1989). *Appl. Phys. Lett.* **55**, 472.
75b. Kojima, K., Mitsunaga, K., and Kyuma, K. (1990). *Appl. Phys. Lett.* **56**, 154.
76. Turco, F. S., Simhony, S., Kash, K., Hwang, D. M., Ravi, T. S., Kapon, E., and Tamargo, M. C. (1990). *J. Crystal Growth* **104**, 766.
77. Colas, E., Simhony, S., Kapon, E., Bhat, R., Hwang, D. M., and Lin, P. S. D. (1990). *Appl. Phys. Lett.* **57**, 914.

78. Tsang, W. T. (1981). *Appl. Phys. Lett.* **39**, 134.
79. Simhony, S., Kapon, E., Colas, E., Bhat, R., Stoffel, N. G., and Hwang, D. M. (1990). *Photon. Tech. Lett.* **2**, 305.
80. Kapon, E., Simhony, S., Bhat, R., Colas, E., and Hwang, D. M. (1990). *Extended abstracts of the 22nd (1990 international) conference on solid state devices and materials (SSDM '90)*, Sendai, Japan, August 22–24, 749–752.
81. Kapon, E., Simhony, S., Hwang, D. M., Colas, E., and Stoffel, N. G. (1990). *Digest of 12th IEEE international semiconductor laser conference*, Davos, Switzerland, September 9–14, 1990, paper **F-2**, 80–81.
82. Kapon, E., Harbison, J. P., Bhat, R., and Hwang, D. M. (1989). In *Optical switching in low dimensional systems* (Haug, H., and Banyai, L., eds.), pp. 49–59. Plenum, New York.
83. Mittelstein, M., Arakawa, Y., Larsson, A., and Yariv, A. (1986). *Appl. Phys. Lett.* **49**, 1689.
84. Adams, A. R., (1986). *Electron. Lett.* **22**, 250.
85. Yablonovitch, E., and Kane, E. O. (1986). *J. Lightwave Technol.* **LT-4**, 504.
86. Yokoyama, H., and Brorson, S. D. (1989). *J. Appl. Phys.* **66**, 4801.
87. Kapon, E. (1992). *Proc. IEEE* **80**, 398.

# Index

# Quantum Electronics—Principles and Applications



N. S. Kapany and J. J. Burke, *Optical Waveguides*
Dietrich Marcuse, *Theory of Dielectric Optical Waveguides*
Benjamin Chu, *Laser Light Scattering*
Bruno Crosignani, Paolo Di Porto, and Mario Bertolotti, *Statistical Properties of Scattered Light*
John D. Anderson, Jr., *Gasdynamic Lasers: An Introduction*
W. W. Duly, *$CO_2$ Lasers: Effects and Applications*
Henry Kressel and J. K. Butler, *Semiconductor Lasers and Heterojunction LEDs*
H. C. Casey and M. B. Panish, *Heterostructure Lasers: Part A, Fundamental Principles; Part B, Materials and Operating Characteristics*
Robert K. Erf, editor, *Speckle Metrology*
Marc D. Levenson, *Introduction to Nonlinear Laser Spectroscopy*
David S. Kliger, editor, *Ultrasensitive Laser Spectroscopy*
Robert A. Fisher, editor, *Optical Phase Conjugation*
John F. Reintjes, *Nonlinear Optical Parametric Processes in Liquids and Gases*
S. H. Lin, Y Fujimura, H. J. Neusser, and E. W. Schlag, *Multiphoton Spectroscopy of Molecules*
Hyatt M. Gibbs, *Optical Bistability: Controlling Light with Light*
D. S. Chemla and J. Zyss, editors, *Nonlinear Optical Properties of Organic Molecules and Crystals, Volume 1, Volume 2*
Marc D. Levenson and Saturo Kano, *Introduction to Nonlinear Laser Spectroscopy, Revised Edition*
Govind P. Agrawal, *Nonlinear Fiber Optics*
F. J. Duarte and Lloyd W. Hillman, editors, *Dye Laser Principles: With Applications*
Dietrich Marcuse, *Theory of Dielectric Optical Waveguides, Second Edition*
Govind P. Agrawal and Robert W. Boyd, editors, *Contemporary Nonlinear Optics*
Peter S. Zory, Jr., editor, *Quantum Well Lasers*